31/6/1998

Photoperiodism in Plants

Photoperiodism in Plants

SECOND EDITION

BRIAN THOMAS
Horticulture Research International, Wellesbourne, Warwick, UK

and
DAPHNE VINCE-PRUE

ACADEMIC PRESS
San Diego   London   Boston
New York   Sydney   Tokyo   Toronto

Academic Press, Inc.
525 B Street, Suite 1900, San Diego, California 92101-4495, USA
http://www.apnet.com

Academic Press Limited
24–28 Oval Road, London NW1 7DX, UK
http://www.hbuk.co.uk/ap/

ISBN 0-12-688490-0

A catalogue record for this book is available from the British Library

Typeset by J&L Composition Ltd, Filey, North Yorkshire
Printed in Great Britain by Hartnolls Ltd, Bodmin, Cornwall

96 97 98 99 00 01 EB 9 8 7 6 5 4 3 2 1

Contents

Abbreviations

ABA	abscisic acid
B	blue light, 400–500 nm
BA	benzyladenine
CK	cytokinin
CT	circadian time
CDL	critical daylength
CNL	critical nightlength
D	dark
d	days
DNP	day-neutral plant
FR	far red light, 700–750 nm
GA, GA_1 GA_2 etc.	gibberellin, gibberellin A_1, A_2 etc.
h	hours
HIR	high irradiance response
IAA	β-indolyl acetic acid
iP	iso-pentenyladenine
iPA	isopentyladenosine
L	light
LD	long-day
LDP	long-day plant
LSDP	long-short-day plant
min	minutes
NB	night-break
NBmax	time of maximum night-break sensitivity
P	total phytochrome
Pr	red-absorbing form of phytochrome
Pfr	far red-absorbing form of phytochrome
PA	polyamine
PGR	plant growth regulator
R	red light, 600–700 nm
s	seconds
SD	short-day
SDP	short-day plant

SLDP	short-long-day plant
TF	tungsten filament (incandescent lamp)
Z	zeatin
ZR	zeatin riboside

Illuminance values given in the original papers as lux or foot candles have been converted throughout to W m^{-2} using the following conversion factors (Cockshull, 1984).

Light source	Factor: W m^{-2} per lux
Cool white fluorescent lamp	2.9×10^{-3}
Daylight fluorescent lamp	3.9×10^{-3}
Tungsten-filament lamp	4.2×10^{-3}
Natural daylight	4.0×10^{-3}
1 foot candle = 10.76 lux	

Introduction

Photoperiodism is one of the most significant and complex aspects of the interaction between plants and their environment. The word itself is derived from the Greek roots for 'light' and 'duration of time', and can be defined as responses to the length of the day that enable living organisms to adapt to seasonal changes in their environment. Such a response could clearly confer selective advantages to the organism. It can be used as a means of anticipating, and consequently preventing, the adverse effects of a particular seasonal environment. For example, at high latitudes autumnal short days precede winter low temperatures. The shortening autumn days act as a signal to induce bud dormancy and cold hardiness, responses which enable the plant to survive the unfavourable winter environment. Similarly, in some desert species dormancy is induced by the long-day conditions which accompany the unfavourable environment of water stress (Schwabe and Nachmony-Bascombe, 1963). A photoperiodic response can enable a plant to occupy an ecological niche in space and time. For example, a response to short days can enable a woodland plant to flower and seed before the dense leaf canopy is formed. Even in tropical latitudes where the seasonal daylength changes are small, many plants are photoperiodic, using daylength to synchronise reproductive or other activities with seasonal events such as dry or rainy periods. Synchronisation through photoperiodic sensitivity can confer advantages independently of whether flowering is matched with a particular favourable environment. In particular, coincident flowering in individuals of a species increases the chances of outbreeding and thus genetic recombination.

The broad definition of photoperiodism given above, while adequate for ecophysiological questions, requires refining when one considers the mechanisms in the plant which give rise to photoperiodic behaviour. Perhaps the most useful proposal is that of Hillman (1969a), who defined photoperiodism as a response to the *timing* of light and darkness. Implicit in this definition is that total light energy, above a threshold level, is relatively unimportant, as is the *relative* lengths of the light and dark period. What is important is the timing of light and dark periods or, to think of it another way, the times at which the transitions between light and darkness take place. Implicit in this definition is that as long as light is above a particular threshold, so that it is perceived as light by the plant, the actual level is relatively immaterial. Because plants respond directly to light through photosynthesis and photomorphogenesis, it is to be expected that varying combinations of light and dark periods might directly affect plant development. Where this happens, the response is likely to be quantitatively related to the durations of the light and dark periods over a range of values. This type of response is distinct from photoperiodism where development or metabolism proceeds in one of two alternative states depending on the relationship between the daylength received and a threshold or 'critical' daylength. It is not enough, therefore, to show that changing the daylength has an effect on a particular response for that response to be considered to be photoperiodically regulated. This book is concerned with true as opposed to quasi-photoperiodic processes and we have tried where possible to limit description and discussion to the former. However, experimental design frequently fails to discriminate between the two, particularly with respect to some of the more marginal processes discussed in Chapter 13.

DISCOVERY OF PHOTOPERIODISM

Henfrey suggested as long ago as 1852 that the natural distribution of plants was due, at least partly, to latitudinal variations in summer daylength (Henfrey, 1852). However, the first experiments in which the daily duration of light was controlled were carried out by Kjellman in the Arctic Circle (see Naylor, 1961). In these experiments plant development was faster in longer light periods but a clear distinction was not made between photoperiodic and photosynthetic effects. Experiments to extend the daily light period were made possible by the invention of the incandescent electric light in the latter part of the nineteenth century (Bailey, 1893; Rane, 1894). Many of these attempts at *electrohorticulture* were successful in causing the acceleration of flowering in a number of summer annuals. Although the illumination levels used in these experiments were relatively low, photosynthetic effects were not rigidly excluded and the importance of the photoperiod was not recognised.

That the duration, rather than the quantity of light, in the daily cycle is a major factor in plant development was proposed independently by Julien Tournois and Hans Klebs at the beginning of the twentieth century (Tournois, 1912, 1914; Klebs, 1913). Tournois, in his studies with the SDP *Humulus* and *Cannabis*, found that the plants flowered precociously in winter under glass. He eliminated temperature, humidity, and seed origin as causal factors and began his critical experiments on daylength in 1912. He established that plants given only 6 h of light each day flowered most rapidly, even though they grew more slowly. Initially, Tournois believed that the determining factor was the reduction in light quantity, but in his last paper on sexuality in *Humulus*, he showed that this had only a small effect. He concluded that the precocious flowering was due to the short periods of daily illumination and furthermore concluded that it was 'not so much caused by the shortening of the days but by the lengthening of the nights'. Although Tournois planned further experiments he was killed in action as a soldier shortly after the publication of his last paper.

At about the same time Klebs was carrying out carefully controlled experiments on flowering in *Sempervivum funkii*, an LDP. He succeeded in inducing the rosettes to flower in the middle of winter by giving a few days of continuous illumination from incandescent lamps. Non-irradiated rosettes always remained vegetative. He concluded that 'in nature, flowering is probably determined by the fact that from the equinox (21 March) the length of the day increases . . . when it reaches a certain length flowering is initiated. Light probably acts as a catalytic rather than a nutritive factor'. Klebs thus recognised that flowering could be accelerated by long days. It was Garner and Allard (1920, 1923), however, who first saw clearly that flowering and many other responses in plants could be accelerated either by long days (LD) or short days (SD), depending on the plant. They introduced the terms *photoperiod* and *photoperiodism* and classified plants into the photoperiodic groups we know today.

They were led to their discoveries by observations on two species of plants being used in breeding programmes at the time. In certain varieties of *Glycine max*, particularly the late maturing strain Biloxi, flowering tended to occur at the same time, independent of planting date (Table I.1). Secondly, the Maryland Mammoth variety of *Nicotiana tabacum* grew to a prodigious size out of doors in summer in Washington DC but failed to flower. However, plants growing in pots under glass flowered while still quite small in winter and early spring. The tobacco result, in particular, suggested a seasonal factor and after eliminating temperature and light intensity as causal factors, Garner and Allard concluded (apparently with extreme diffidence) that the only remaining seasonal phenomenon was the relative length of the day and night. To investigate this they transferred *Glycine max* cv Peking and *Nicotiana tabacum* cv Maryland Mammoth to a darkened, ventilated hut for part of the daily light period, in order to limit the exposure to light to 7 h, during the summer and compared the response with the plants grown in the open. Plants which received the shortened day flowered promptly, while those exposed to natural summer day-lengths remained vegetative. They proposed that the answer to the tobacco and soyabean problems was the same: the varieties in question would only flower if the duration of the daily light period was sufficiently short.

After their initial experiments on flowering in *Nicotiana* and *Glycine*, Garner and Allard extended their observations to a wide range of species and responses. They established that daylength influences many aspects of plant behaviour, including flowering, tuberisation and dormancy and considered its possible effects on plant distribution and crop yield. Further, their experiments with plants suggested to Garner and Allard that the time of bird migration might also be controlled by daylength, which was later confirmed. The ability of organisms to measure time, often with a high

TABLE I.1 Time of flowering of *Glycine max* cv Biloxi planted at different times and grown under natural conditions in Washington DC.

Date of germination	Date of anthesis (first flower)	Number of days to flowering
2 May	4 September	125
2 June	4 September	94
16 June	11 September	92
30 June	15 September	77
15 July	22 September	69
2 August	29 September	58
16 August	16 October	61

Data of Garner and Allard (1920), from Hillman (1969a).

degree of accuracy, is now fully accepted and photoperiodism is established as a major factor in the seasonal regulation of plant and animal behaviour, although it is only one of many examples of biological time measurement.

IMPACT OF PHOTOPERIODISM

It is difficult to quantify the benefits of basic studies on photoperiodism, but there is little doubt that modern agricultural and horticultural practice has been profoundly influenced by the appreciation of its biological importance. The initial studies of Garner and Allard over a 10-year period were estimated to have cost approximately $10 000, but brought benefits to farmers, horticulturists and plant breeders of billions of dollars (Sage, 1992). Since the time of Garner and Allard, testing plants for their photoperiodic requirements has become standard practice in breeding programmes, and a major reason for crop failures can now be avoided. Plant breeders can manipulate daylength to obtain multiple generations of plants per year and obtain seed from plants that fail to reproduce out of doors because the days are too long, by growing them in shortened days in the greenhouse. Also, by using daylength to synchronise flowering in plants that normally flower at different times, new crosses between varieties are possible.

Photoperiodism has had a particular impact on horticulture. Daylength manipulation by the use of blackouts or supplementary lighting to promote or enhance flowering has been used for a wide range of ornamental species. In the case of the major horticultural crop, chrysanthemum, daylength sensitive cultivars of this SDP have been selected so that plants can be maintained in a vegetative state or brought to flowering as required by the grower through varying daylength in combination with other environmental conditions. Production of poinsettia, the most valuable ornamental crop in the USA, is also dependent on daylength management. Timing of production for a specific market, e.g. Christmas, Easter or Mother's Day, is very important in floriculture as there is a price premium at these times and the price is dramatically lower for crops produced too early or late. Daylength regulation in combination with appropriate temperature regimes make this possible for a range of crops such as begonias, Christmas cactus, chrysanthemums and poinsettias.

Almost all of the considerable benefits flowing from photoperiodism research have so far been to do with cultural practices and breeding. These have been achieved by using an understanding of the basic physiological processes and genetic variation in the response to daylength. As understanding proceeds from the whole plant to the molecular and genetic level the initial benefits will be related to the ability to use molecular markers for daylength sensitivity in breeding programmes and the direct manipulation of dosage of key genes by breeding and genetic manipulation. In the longer term, the introduction or removal of daylength sensitivity to particular species or cultivars will be feasible. One can predict that it will be possible to couple new processes, e.g. synthesis of high-value secondary products, to particular daylength regimes or to alter the response type of a particular species e.g., convert LDP to SDP or vice versa. It is against this background of new potential opportunities that this book has been written.

AIMS OF THIS BOOK

Photoperiodism is a complex and pervasive phenomenon, which has been the subject of scientific investigation for more than three quarters of a century. There is a huge volume of literature pertaining to all aspects of the subject. Experiments have been carried out on a range of processes in a range of species using a range of techniques and approaches. An account of photoperiodism research up until 1975 was published in the first edition of this book. This revised edition aims to update the work in the light of the 20 years of research that have subsequently been performed. As the new powerful technologies of molecular genetics are brought to bear on photoperiodism, it becomes particularly important to place new work in the context of the considerable amount of physiological information which already exists on the subject. Given the volume of literature that has been published on photoperiodism, it is no longer possible for a book of this sort to be encyclopaedic. However, while we cannot include everything that has been published, we have tried to ensure that a self-sufficient account of all the important subjects is included and the key literature references are there for those who want to follow topics up in detail. One thing that became increasingly clear to the authors as the book progressed is that despite the apparently diverse phenomena which are under photoperiodic control, a number of clear themes based on common underlying mechanisms emerge.

- Photoperiodism involves the coupling of the capacity for daylength perception to a number of target processes. The most important and best studied target is flowering, but vegetative reproduction and dormancy are also major processes which can be under photoperiodic control.
- It is increasingly clear that photoperiodic timekeeping is based on a circadian oscillator or oscillators, and these are coupled to rhythms in light sensitivity to form daylength detection mechanisms.
- From studies of a range of phenomena such as flowering, tuberisation, stem extension and dormancy we consistently find evidence for two distinct and apparently rather different daylength perception mechanisms. The first of these is a *dark dominant* mechanism, where the response is determined primarily by the length of the dark period and conditions in the light period are relatively unimportant. The second is a *light dominant* mechanism, where the primary response is to light given in the photoperiod and there is a requirement for particular spectral properties, specifically a strong effect of FR and a sufficiently high irradiance, to elicit a positive response. In some cases, dark dominant and light dominant mechanisms may coexist within the same species.
- Phytochrome, the main photoreceptor for daylength perception in higher plants, is now known to consist of a family of molecules, with different physiological roles. There is now strong evidence that one of these, phytochrome A, plays a central role in light-dominant responses, being responsible for the FR requirement. In contrast, dark dominant responses are not enhanced by FR and phytochrome A appears not to be involved.

We point out these themes to those who may wish to limit their reading to specific chapters and also hope that these recurring themes suggest themselves to readers of the whole book.

PART I
PHOTOPERIODIC CONTROL OF FLOWER INITIATION

1 Some General Principles

PHOTOPERIODIC RESPONSE GROUPS

The classification of plants according to their photoperiodic responses has usually been made on the basis of flowering (Appendix 1). Similar categories might, of course, be set up for other processes affected by daylength such as the onset of tuberisation or dormancy. With respect to flower initiation, the responses to daylength seem to be of three basic types, with some modifications. The three main categories are:

- **short-day plants** (SDP), which only flower, or flower most rapidly, with fewer than a certain number of hours of light in each 24 h period;
- **long-day plants** (LDP), which only flower, or flower most rapidly, with more than a certain number of hours of light in each 24 h period;
- **daylength indifferent** or **day-neutral plants** (DNP), which flower at the same time irrespective of the photoperiodic conditions.

Plants that are responsive to daylength may be further subdivided into obligate (or qualitative) types, where a particular daylength is essential for flowering, and facultative (or quantitative) types, where a particular daylength accelerates but is not essential for flowering. It is not always easy to discriminate between these two subcategories because a photoperiodic requirement may be more pronounced under some conditions than others. For example, a plant can have an obligate photoperiodic requirement at one temperature but only a facultative response at another. It is probably better to consider these two response types not as two distinct groups but as a continuum from a slight acceleration of flowering by favourable daylengths at one end to an indefinite delay in flowering by unfavourable daylengths at the other.

Some plants have dual daylength requirements and, for flowering, must be exposed to both short days (SD) and long days (LD). The sequence in which these daylengths are experienced is important and both LD–SD (e.g. *Bryophyllum daigremontianum*, Zeevaart, 1969b) and SD–LD plants (e.g. *Dactylis glomerata*, Heide, 1987) are known. In these and several other dual-daylength plants, floral initiation at the apex does not take place during exposure to the first or **primary** induction but only after

3

transfer to the **secondary** induction treatment. In contrast, floral primordia are formed during SD in some northern grasses (e.g. in northern ecotypes of *Poa pratensis*), LD being required only for further development of the inflorescence (Heide, 1994). Whether this is strictly a dual-daylength requirement is a matter of interpretation since other species are known for which the daylength requirement for initiation of floral primordia differs from that for their further development to anthesis. An example is the cultivated strawberry where floral primordia are initiated in response to autumnal SD but anthesis is promoted by LD. Such plants are usually classified simply as short-day plants (SDP), although with respect to final flowering they could be considered as short–long-day plants (SLDP).

A few plants appear to have rather specialised daylength requirements which have not yet been investigated in any detail (Vince-Prue, 1975). Some flower only when the day is neither too long nor too short. Such **intermediate-day plants** may only be a variation of a short-day type of response since SDP also need some light each day in order to flower. The daylength limitations for most SDP are, however, fairly wide. (*Xanthium strumarium* will flower with between 2–3 and 15.5 h light in a 24 h period) whereas those for true intermediate-day plants are quite narrow (*Mikania scandens* will flower only with between 12.5 and 16 h light in every 24 h). Another possibility is that intermediate-day plants really have a dual daylength requirement but that the LD and SD ranges overlap; a constant intermediate daylength would then satisfy both the LD and SD requirement. The range for the LD and SD parts of photoperiodic induction are known to overlap in some dual daylength plants (Lang, 1965; Heide, 1994). These different hypotheses have not been explored for a range of plants but, based on experiments with *Salsola komarovii* (which behaves as an intermediate-day plant with respect to the time of anthesis), it was concluded that this particular intermediate-day response is a variation of a short-day type of response with a requirement for a long daily light period (Takeno *et al.*, 1995). Another special category is found in plants in which flowering occurs rapidly in either SD or LD but is delayed at intermediate daylengths, This type of behaviour has been called **ambiphotoperiodism**, and is shown by *Madia elegans* which flowers in 8 and 18–24 h days, but not in 12–14 h days. The basis of this response category does not appear to have been investigated.

ASSESSMENT OF FLOWERING RESPONSE

Studying the effects on flowering of different treatments requires some kind of quantitative measurement of the response. Several approaches have been used. Many are open to criticism, although they can all be useful if handled in the right way. Possibly the simplest method is to express the results as the percentage of plants in any given treatment which have flowered within some arbitrary time. This requires the use of large numbers of replicate plants in order to determine the statistical validity of small differences and is often not possible where space is restricted. The extreme photoperiodic sensitivity of some ecotypes of Chenopodium rubrum allows flowering in a petri dish and, in this case, large numbers can readily be accommodated in a small space (Cumming, 1959). In *Lemna*, flowering is usually scored as the percentage of all fronds with any recognisably floral stage; flowering usually takes place in about a week from treatment and large numbers of plants can be grown and scored (Hillman, 1959). It is also necessary to define flowering. The presence of macroscopically visible flower buds is often taken as the criterion that the plant has flowered, but critical physiological studies require dissections of all apices in order to determine whether or not floral initiation has taken place, as post initiation development may also be affected by the conditions being imposed. A second commonly used method is to present data as the number of days taken to reach some arbitrary stage of floral development, such as days to the appearance of a macroscopically visible flower bud or to an open flower (**anthesis**). This is a useful technique for the study of floral development but can also be used to study floral initiation if repeated dissections are carried out in order to determine when the apex becomes recognisably floral. Again large numbers of plants are needed.

Another method is to examine plants at an arbitrary time and assign a stage of floral development to them (e.g. *Xanthium*; Salisbury, 1969). Such stages are usually established from a series of dissections made at various times after giving a treatment known to cause flowering. The determination of the stage of development reached at an arbitrary time will be a measure of the rate of floral development when all plants initiate floral primordia at the same time (Fig. 1.1), but differences in the stage of development can also arise from changes in the time of floral initiation. A single determination of floral stage may, therefore, confound effects on the rate of floral development and on the time of initiation, unless either one is known to be unaffected by the treatment. Consequently such determinations should be treated with care when evaluating treatment effects. In particular species other methods have been found to be useful measures of the flowering response, such as the number of flowering nodes

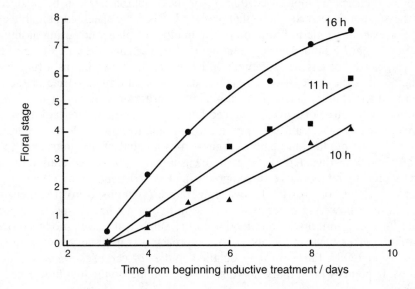

FIG. 1.1. Rate of development of the terminal male inflorescence of *Xanthium strumarium* as affected by the length of the inductive dark period. The time of floral initiation was not affected by the treatment and the floral stage at an arbitrary time would be a measure of the effect of the treatment on the rate of early development of the inflorescence (Salisbury, 1963b).

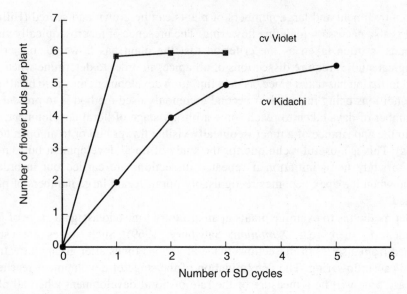

FIG. 1.2. Effect of the number of inductive short-day cycles on the number of flower buds per plant in *Pharbitis nil*. All plants were flowering after one short day in both cultivars. In the cultivar Violet, which is extensively used for photoperiodic studies, the number of buds saturates at six or seven because the terminal bud flowers; the number can vary below this value according to the effectiveness of the inductive treatment (Imamura *et al.*, 1966).

(*Glycine*) or the number of flower buds (*Pharbitis*; Fig. 1.2). In several *Gramineae*, apex (spike) length may be used (Fig. 1.3).

Several workers have emphasised that all of these approaches can be misleading because they do not necessarily show that observed effects are specific to the flowering process. For example, time to initiation or the number of plants flowering at any one sample time might be influenced by treatments which change the rate of vegetative growth. This is true of any measurement which expresses flowering as a function of time, whether it is the stage of development reached at an arbitrary time or the time taken to reach an arbitrary stage of development. A criterion specific for flowering is to determine the number of leaves produced before flower initials are laid down. If a determinate plant grows at all, the apex must continue to produce leaf primordia or change over to the production of floral organs; thus a count of the number of leaves produced on the flowering stem is a specific measurement of the acceleration or delay of the transition to reproductive development and is independent of effects of the treatment on the rate of growth. A low temperature (**vernalisation**) treatment may delay the onset of flowering in terms of time to initiation but have a specific effect to accelerate flowering in terms of a reduction in the number of leaves produced before floral initiation takes place at the apex. This approach was used very successfully by F.G. Gregory and O.N. Purvis in their detailed studies of vernalisation in winter rye (Chapter 6). In plants with axillary flowers, the leaf number to the first flowering node can be used.

Many different methods have been used in the experiments described in this book. It is important when evaluating the results presented to remember that, where differences in the rate of vegetative growth could have influenced the assessment, the observed

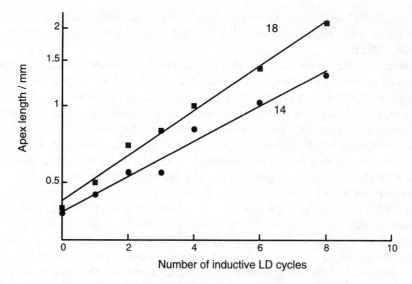

FIG. 1.3. Effect of the number of inductive long-day cycles on the length of the apex in *Lolium temulentum*. Plants were given various numbers of long-day cycles, as indicated, before being returned to short-day conditions. The age of the plants (in days) at the beginning of the LD treatment is shown on the figure. The rate of development increases in proportion to the logarithm of the number of LD cycles given and is most simply presented as the inflorescence length (on a logarithmic scale) at a given time from the beginning of the inductive treatment (2 weeks in this case). There is no fixed relationship between apex length and the stage of floral differentiation because the vegetative apex continues to elongate slowly in non-inductive conditions (Evans, 1960).

effects may not be completely specific to the flowering process. Moreover, when considering the timekeeping aspects of photoperiodism (Chapter 2) special problems occur which require methods of assessing the timing of events during induction, rather than considering only the magnitude of the flowering response to the inductive treatment.

WHERE IS DAYLENGTH PERCEIVED?

It was recognised quite early that the main site for the perception of daylength is the leaf, even though the observed responses usually take place elsewhere in the plant. This was first demonstrated by Knott (1934) for the LDP spinach, where exposing only the leaves to long photoperiods resulted in the initiation of floral primordia at the shoot apex, while plants remained vegetative when the apical bud alone was treated. Some of the earliest work on the site of perception was concerned with vegetative responses. Tuberisation on underground stolons occurred in Jerusalem artichoke (*Helianthus tuberosus*) when the stem tips were covered to give SD, provided that leaves longer than 2 inches were included (Hamner and Long, 1939). Other plants where the site of perception of daylength has been shown to be in the leaves include the LDP *Anethum graveolens*, *Brassica crenata*, *Hordeum vulgare* and *Lolium temulentum*; the SDP *Xanthium strumarium*, *Glycine max*, *Kalanchoë blossfeldiana*, *Dendranthema grandiflora*, *Pharbitis nil* (cotyledons) and *Perilla* (Lang, 1965; Vince-Prue, 1975; Halevy, 1985); and the LSDP *Bryophyllum* (Zeevaart, 1969b). In the LSDP *Cestrum nocturnum*, the LD and SD treatments must be given to the same leaf; floral induction does not occur when some leaves receive LD and others on the same branch receive SD (Sachs, 1985).

Depending on the plant, not all the leaves need to be exposed to favourable photoperiods in order to achieve a response. Treatment of a single leaf of *Xanthium strumarium* with SD was shown to cause the development of macroscopically visible flower buds, even though the remainder of the plant was growing in LD. In *Xanthium*, it is not necessary to defoliate the remainder of the plant but, in many cases, treating a single leaf will only result in flowering when other leaves are removed; it is particularly important to remove leaves between the treated leaf and the apex as these often interfere with the flowering response to the photoperiodically induced leaf.

Leaves may be removed from induced plants and grafted to vegetative receptors. Zeevaart (1958) showed that such grafted donor leaves of the SDP *Perilla* (red-leaved) and *Xanthium* will cause a flowering response in receptors maintained in LD. (The nomenclature of *Perilla* is confusing, and different workers have used a variety of names for the red-leaved and green-leaved forms (Zeevaart, 1969c; 1985b). Both are SDP with somewhat different response characteristics and, for the sake of simplicity, they are referred to in the present text by leaf colour.) In *Perilla*, the leaves remain in good condition for a long time and it was possible to remove and re-graft them to other receptor plants; they continued to cause receptors to flower even when more than 3 months had elapsed since the last SD cycle. That the leaf is independently capable of perceiving daylength is shown by experiments in which leaves of *Perilla* were detached from a vegetative parent plant, exposed in isolation to the appropriate daylength for flowering and then grafted to vegetative receptor plants maintained in unfavourable daylengths. Such leaves in *Perilla* readily caused flowering (Zeevaart, 1958). In *Begonia*, the exposure of isolated leaves to SD resulted in flowering in subsequently developing adventitious buds (Rünger, 1957). Even in *Xanthium*, where leaves do not remain in good condition when they are excised, a small number of detached leaves exposed to SD cycles caused flowering when grafted to receptors in LD (Zeevaart, 1958). Failure to induce excised leaves has been reported in some cases (e.g. *Glycine max* Biloxi; Heinze *et al.*, 1942), but this may have been due to rapid senescence or giving an inadequate number of inductive cycles (Zeevaart, 1958). Induction of excised leaves does not appear to have been reported for LDP. In red *Perilla*, it has been demonstrated that leaves induced on shoots without roots can cause flowering when grafted to receptor shoots also without roots (Table 1.1); since roots were continually removed (daily) from the receptors, it is clear that they are not necessary for the actual formation of floral primordia. Thus the perception of daylength is possible in isolated, detached leaves and such leaves are also able to cause the initiation of floral primordia in the absence of roots on either the donor or receptor shoots.

It is evident from a number of different physiological approaches that the machinery for photoperiodic perception and signal generation is located in the leaves, although the precise location within the leaves is not known and has hardly been investigated. Bünning and Moser (1966) reported that the upper leaf surface in *Kalanchoë* was more sensitive than the lower surface to inhibition by LD, using a low-intensity (approximately 2.3×10^{-2} W m^{-2}) day-extension treatment. Phytochrome, the photoreceptor for the perception of daylength signals (see Chapter 3) is known to be distributed throughout the leaf and such differential sensitivity could be explained by the screen-

TABLE 1.1. Induction of red *Perilla* leaves in the absence of roots and buds.

Donor	Receptor (in LD)	Days to appearance of flower bud
SD −roots	−roots	19.7
SD −roots	+roots	19
SD +roots	−roots	18
SD +roots	+roots	14
LD +/−roots	+/−roots	Vegetative

Single leaves on long debudded stems with or without roots were exposed to 29 SD; the leaves were then grafted to intact plants (+roots) or to stems with two receptor shoots without roots (−roots). In all treatments without roots, a 1 cm piece was removed from the base of the stem each day to prevent root formation.
After Zeevaart and Boyer, 1987.

ing effects of chlorophyll, especially at such low irradiances. Schwabe (1968) suggested that the epidermis may, at least partly, be the locus of perception but, as this was based on the damaging technique of stripping the epidermis and observing a reduction in response, the conclusion must be in some doubt.

An important corollary of the work with individual and isolated leaves is that they are not only capable of perceiving daylength but must also export a stimulus which is capable of evoking the observed photoperiod response at the receptive site. It was, indeed, the work with leaves which led to the concept of a photoperiodic stimulus being exported from them; in particular the idea of a specific floral hormone, or **florigen**, emerged from such studies (see Chapter 6). We speak of leaves that are capable of causing flowering when grafted to a receptor plant as being induced. This is not restricted to photoperiodic induction; leaves of the DNP *Glycine max* cv Agate are able to cause flowering in the SD cultivar Biloxi (Heinze *et al.*, 1942). Following the proposal of Evans (1969b), the terms **induced leaf** and **induction** will be used in this sense and the term **evocation** will be used for the events that occur at the receptive shoot meristem following the arrival of the floral stimulus and lead to the formation of floral primordia.

Although the leaves appear to be the main site for photoperiodic perception, other parts of the plant have been shown to be capable of perceiving daylength under certain conditions. A striking example is *Chenopodium amaranticolor* where defoliated plants can be induced to flower by SD (Lona, 1949); either stems or the very young leaves of the apical bud must be capable of daylength perception. Work with cultured excised tissue bears on this question. For example, Nitsch and Nitsch (1967) showed that internode segments of *Plumbago indica* produced flower buds if maintained in SD for 4 weeks before returning them to non-inductive LD; at this time adventitious buds were starting to become visible so that the perception of photoperiod must have occurred in either stem tissue or, more likely, in the regenerating buds. Certainly no expanded leaves were present during induction. Another interesting example is the cotyledons of pea cv Greenfast which become competent to perceive photoperiod after about 5 days from soaking; the epigeal cotyledons of several plants are also sensitive to daylength (e.g. *Pharbitis nil*) but, in pea, the hypogeal cotyledons remain buried in the soil and presumably do not normally contribute to photoperiodic induction (Paton, 1971). Penetration of light below the soil surface may, however, be important in some cases. The leafless flowering desert geophyte *Colchicum tunicatum* is a quantitative SDP which flowers in autumn and perceives the SD signal when the dry bulb lies well below the soil surface (Gutterman and Boeken, 1988). The dry tubular cataphyll which reaches to the soil surface may be sensitive to the daylength signal or may possibly allow light penetration to the bulb, although the leaves are still very small and more than 10 cm deep in the soil at this time.

The apex itself has also been shown to be capable of daylength perception in some cases. Using optical fibres and red light (660 nm), flowering in *Pharbitis* was inhibited to 39% (SD controls = 100%) by irradiating the apex alone (Gressel *et al.*, 1980). Some apical sensitivity was also indicated for the LDP *Silene coeli-rosa*, where the excised apical dome and youngest pair of leaf primordia showed an increase in G2 on days 0–3 of LD induction as found in intact plants; this increase did not occur in SD. However, the characteristic increase in G2 proportion on days 7 and 8 which also occurs in intact plants was not observed in the cultured apices and they remained vegetative (Francis, 1987). Similarly, in the LDP *Sinapis alba*, the great majority of the ultrastructural changes associated with flowering did not occur when only the shoot tip and leaves less than 0.5 cm were exposed to LD and it was concluded that most of these changes were under the control of a leaf-generated signal (Havelange and Bernier, 1991). In *Sinapis*, the only apical response to direct irradiation was an increase in mitochondrial number.

These experiments confirm that the leaves normally play the dominant role in photoperiodic perception and generate one or more signals which evoke flowering at the shoot apical meristems. However, the possibility that light reaching the apex itself may contribute to the overall response in the intact plant should be taken into account when interpreting results, even though it has been demonstrated that photoperiodic induction and stimulus generation can occur in the leaf alone and do not require a contribution from inductive cycles perceived directly by the apex.

WHAT IS PERCEIVED?

The factor which governs classification as a SDP or LDP is whether flowering occurs (or is accelerated) only when the daylength exceeds (LDP) or only when it is less than (SDP) a certain critical duration in each 24 h cycle (Fig. 1.4). Thus plants are able to measure the duration of light and/or darkness in each 24 h period. In order to be able to do this, photoperiodically sensitive plants must possess both a clock (to measure time) and a photoreceptor (to discriminate between light and darkness). The identity and functioning of these components of the photoperiodic mechanism form the subjects of Chapters 2, 3, 4 and 5 but some general points are first discussed here.

The value of the critical daylength (CDL) which marks the transition between vegetative growth and flowering in obligate photoperiodic plants of both SD and LD types varies considerably between species and cultivars (Table 1.2). For example, the CDL for the induction of flowering in the LDP *Calendula* is only 6.5 h whereas, for the prevention of dormancy in a high latitude population of *Betula*, the critical daylength is longer than 20 h. The CDL may also alter with environmental conditions and plant age. In the SDP, *Pharbitis*, adult plants will flower on longer days than seedlings and, in the LDP, *Hyoscyamus*, the CDL is decreased when the night

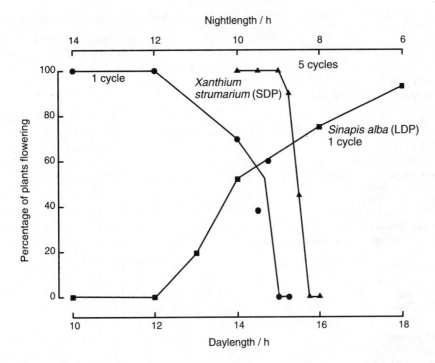

FIG. 1.4. Flowering responses of short-day and long-day plants to different photoperiods. Data for *Sinapis* from Bernier (1969) and for *Xanthium* from Salisbury (1963b).

temperature is lowered (Table 1.2). Similarly, in a range of nine soyabean genotypes (SDP), the CDL decreased from 10–12 h at a mean temperature of 24 °C to 9–10.5 h at 20°C (Hadley *et al.*, 1984). In some plants, the change between flowering and non-flowering is very abrupt. In *Xanthium*, all plants remained vegetative when grown in photoperiods of 15.75 h and all flowered when this was shortened to 15 h (see Fig. 1.4), while a difference of only 15 min can determine whether or not flowering occurs in some tropical SDP. However, the transition may be much more gradual, as in the LDP *Sinapis*, where all plants remained vegetative in 12 h photoperiods but required 18 h light each day to achieve flowering in most of the plants (see Fig. 1.4). The range of photoperiods over which plants change from vegetative growth to maximum flowering can also vary with the number of inductive cycles given, with the range often being greater when the number of inductive cycles is decreased (see Fig. 1.4). Thus increasing the number of inductive cycles (as would occur naturally) increases the precision of time measurement; this is important since an error of 20 min in time measurement might influence flowering time within a population by 2–3 weeks. In nature, flowering in *Xanthium* is more uniform than this (Salisbury, 1963b).

Facultative LDP and SDP will eventually flower in unfavourable photoperiods and so cannot be said to have a CDL, as defined above. It has been suggested that, for such plants, the CDL might be considered as that photoperiod above or below which the time to flower is minimal. Photoperiods longer than this value would delay flowering in facultative SDP, whereas those shorter than this value would delay flowering in facultative LDP. This concept of CDL could also be applied to obligate types and has

1. SOME GENERAL PRINCIPLES

TABLE 1.2. Critical daylengths for floral initiation in some short-day and long-day plants.

	Critical daylength[a] in 24 h cycles (h)
Short-day plants	
Dendranthema grandiflora (15.7°C)	
cv White Wonder	16
cv Encore	14.5
cv Snow	11
Coleus frederici	13–14
Euphorbia pulcherrima	12.5
Fragaria fresca semperflorens	10.5–11.5
Fragaria × ananassa (16–17°C)	11–16
Glycine max cv Biloxi	13.5
Kalanchoë blossfeldiana	12
Lemna paucicostata 6746	14
Nicotiana tabacum cv Maryland Mammoth	14
Perilla (red)	14
Perilla (green)	16
Pharbitis nil cv Violet	15–16
Stevia rebaudiana	14
Xanthium strumarium	15.5
Long-day plants	
Anagallis arvensis	12
Anethum graveolens	10–14
Arabidopsis thaliana early summer races	4–5
Brassica carinata	18
Calendula officinalis	6.5
Dactylis glomerata	12
L. perenne	
early strain	9
late strain	13
L. temulentum	
cv Ceres: 1 cycle	14–16
cv Ceres: multiple cycles	9
Ba 6137-21	14
Hyoscyamus niger	
(28.5°C)	11.5
(15.5°C)	8.5
Rudbeckia bicolor	10
Sedum spectabile	13
Silene armeria	
strain N	12–13
strain E	11
strain L	13.5
Sinapis alba	14
Spinacea oleracea	13
Trifolium pratense	12

[a] Short-day plants will flower in photoperiods shorter than the critical daylength; long-days plants will flower in photoperiods longer than the critical daylength. As indicated, the critical daylength may vary with environmental conditions and also with other cultivars or strains.

been proposed as an alternative definition (Roberts and Summerfield, 1987). The CDL has also been defined as the daylength at which 50% of the plants flower. It is evident that different authors have taken different approaches to the concept of CDL and it is, therefore, important to define clearly what is meant when using the term. Since this book primarily addresses the basic mechanism(s) of photoperiodism, the term CDL is here defined as that photoperiod which marks the transition between flowering and non-flowering during the period of the experiment; in this sense it can be applied to both obligate and facultative photoperiod response types.

Under natural conditions, daylength and nightlength are absolutely related to form a 24 h cycle of light and darkness. Thus plants could detect the critical daylength by measuring either the duration of light, or of darkness, or even their relative durations. Julien Tournois already suspected in 1914 that the precocious flowering which he observed in young plants of hemp and hop was caused not so much by shortening the days as by lengthening the nights. To establish the relative significance of the two, however, it was necessary to vary them independently. Using this approach, Karl Hamner and James Bonner (1938) were able to show that photoperiodic timekeeping in the SDP *Xanthium* is essentially a question of measuring the absolute duration of darkness. Flowering occurred only when this exceeded 8.5 h, irrespective of the relative durations of light and darkness in the experimental cycle. With sufficiently long nights, flowering occurred even when these were coupled with LD; in contrast, SD did not cause flowering when they were coupled with short nights (Fig. 1.5).

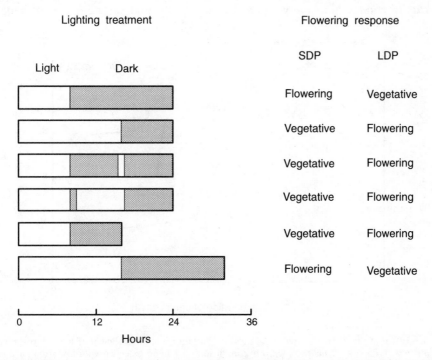

FIG. 1.5. Effect of the duration of the dark period on flowering. Short-day plants flower with long dark periods; long-day plants flower with short dark periods. A short light interruption given in the middle of a long dark period nullifies its effect. In contrast, a short dark break during the photoperiod does not modify the flowering response.

One possibility that has received relatively little attention is that plants might respond to a changing rather than to an absolute duration of darkness. The perception of lengthening versus shortening days (or nights) would enable plants to discriminate between spring and autumn when absolute daylength would give an ambiguous signal. In sugar cane, for example, flowering occurs in autumn in both northern and southern hemispheres even when crops are separated by only a few hundred miles and with similar temperatures. In growth rooms, a comparison between nights lengthening from 10 h 40 min to 12 h 04 min or shortening over the same range resulted in good flowering with lengthening nights but little or none with shortening nights (Clements, 1968). However, these results do not prove that sugar cane is able to perceive directly whether nights are lengthening or shortening, since it has been shown that successive stages of floral development have longer critical night lengths (Moore, 1985); this would result in earlier flowering when the nights were lengthening. A similar situation occurs in *Dendranthema*, where initiation of the capitulum occurs in shorter nights than those required for the initiation and development of florets (Fig. 1.6). Other strategies are known which confer the ability to discriminate between autumn and spring without involving the perception of lengthening versus shortening days. In the SDP strawberry, for example, floral initiation does not take place in spring because the previous exposure to winter low temperatures results in insensitivity to photoperiod (Guttridge, 1969).

A further component of the natural photoperiod that varies with season and latitude is the **rate of change** in its daily duration (see Fig. 2.1) and, during the last decade or

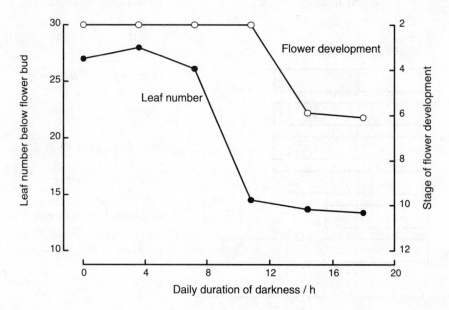

FIG. 1.6. Effect of the daily duration of darkness on initiation and development of the inflorescence in *Dendranthema* cv Polaris. Plants received 8 h in high-intensity light, extended with low-irradiance light from TF lamps. Initiation of the terminal inflorescence is indicated by the number of leaves formed below the flower bud; with 10.5 h darkness, the acceleration of initiation is almost saturated but there is essentially no further development of the flower bud beyond stage 2 (initiation of capitulum) unless the duration of darkness is increased further. Data of K.E. Cockshull, taken from Vince-Prue (1983c).

so, it has been suggested that this may also influence developmental processes. For example, it has been claimed that the mean rate of leaf appearance in wheat is better correlated with the rate of change than with the absolute photoperiod (Baker et al., 1981). However, the hypothesis that rate of change of daylength influences photoperiodic responses largely rests on time of sowing experiments in which it is difficult to separate this factor from others, such as absolute daylength and temperature. When different rates of change (2.5, 9.8 and 13.1 min d^{-1}) were imposed experimentally in two cultivars of wheat, there was no effect of the rate of change of daylength on development (from emergence to terminal spikelet initiation) that was independent of its average duration (Slafer et al., 1994a). Similarly there was no effect attributable to the rate of change of photoperiod on the final leaf number in these wheat cultivars (Slafer et al., 1994b). Different rates of daylength shortening have also been examined with respect to their effect on frost hardening in *Pinus* (Greer et al., 1989); the results were somewhat ambiguous, however, as hardiness also developed under constant photoperiods at a rate that was influenced by the absolute daylength, making it difficult to separate effects specifically due to rate of change from those dependent on average duration. Thus, the ability to respond to a changing daylength *per se* remains unproven for plants, although some animals are known to do this.

The importance of the absolute duration of darkness is well illustrated in the SDP *Pharbitis nil* cv Violet. In this plant, seedlings otherwise grown throughout in continuous light will flower in response to a single dark period, provided this exceeds about 9 h, and the response is essentially independent of the duration of the preceding period of continuous light (Vince-Prue and Gressel, 1985). Thus, in SDP at least, it appears that the absolute duration of darkness is the primary determinant of the photoperiodic response, whereas the effects of light can usually be interpreted as interactions with this essential component. Under natural cycles of light and darkness, the result is an apparent critical daylength which, if exceeded, prevents flowering in SDP, although it is in fact a critical nightlength which is important (see Fig. 1.4).

Originally it was thought that the measurement of nightlength was also the critical factor in determining the response of LDP. This was based on the observation that some LDP would flower in SD coupled with short nights but not in LD coupled with long nights (see Fig. 1.5). However, more recent investigations have shown that the situation in LDP is complex and that nightlength may not be the overriding factor in determining the direction of the response (see Chapter 5).

A further feature which underlines the importance of the dark period is that it can be rendered ineffective by a short light interruption or night-break (Lang, 1965). Only a few minutes of light will prevent flowering completely in many SDP and 30 min is usually more than adequate. Examples are *Perilla* (red), *Glycine* cv Biloxi, *Kalanchoë*, *Pharbitis* and *Xanthium*.

The commercially important SDP chrysanthemum (*Dendranthema grandiflora*) is an exception and requires several hours of light from tungsten filament (TF) lamps; however, a single exposure to high-intensity red light for only 1 min completely inhibited flowering (Cathey, 1969). Many LDP are also responsive to a relatively brief night-break under some conditions and will flower with long nights when these are interrupted with light of 30 min duration or less. Examples are *Hordeum vulgare* cv Wintex, *Hyoscyamus niger* and *Anagallis arvensis*, Paris strain. Night-break

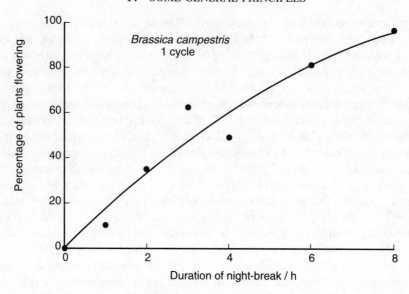

FIG. 1.7. Flowering of *Brassica camprestris* in response to various durations of night-break The night-break treatments were given with sunlight at approximately 34 W m^{-2} (Friend, 1969).

sensitivity has also been recorded for the photoperiodic responses of some species of algae (Cunningham and Guiry, 1989).

Several LDP, however, do not flower with night-breaks of this kind, even though they may be closely related to the ones already mentioned. The GO strain of *Anagallis arvensis* remained vegetative with a 1 h night-break repeated for several cycles, although a single LD induced flowering (Ballard, 1969). In both *Brassica campestris* (Fig. 1.7) and carnation (Harris, 1968) a 30 min night-break had little or no effect and the flowering response increased as the light interruption was lengthened over several hours. The LD promotion of erect gametophyte axes in the marine red alga *Nemalion* was also found to be insensitive to night-breaks (Cunningham and Guiry, 1989). There are, in addition, a few SDP for which night-breaks are relatively ineffective. Thus, in some photoperiodic responses (especially in LDP) it seems that the light interrupting a long dark period may act in a manner different from that in the majority of SDP. We shall return to this point again in Chapter 5.

A night-break was found to be most effective when given near the middle of a 12–16 h dark period (Fig. 1.8) and, consequently, it was thought that the action of light was to divide the long night into two periods of darkness, each of which would be shorter than the critical. This explanation proved inadequate, however, when it was discovered that the time of maximum sensitivity to a night-break, NBmax, was not altered by prolonging the dark period. For example, in experiments with the SDP *Pharbitis* and *Xanthium* in non-24 h cycles, the time of NBmax still occurred about 8–9 h after the beginning of a 40–48 h night, despite the fact that the remaining 30–40 h period of unbroken darkness was far longer than the critical nightlength (Fig. 1.9). LDP such as *Hyoscyamus niger* (Hsu and Hamner, 1967) and *Lolium temulentum* (Vince-Prue, 1975) were found to behave in a similar way. Such experiments have

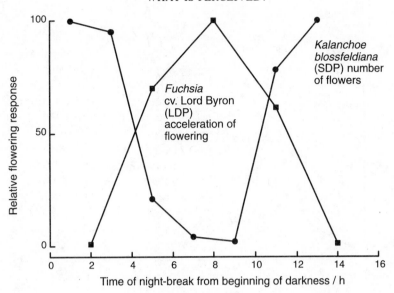

FIG. 1.8. The flowering responses of long-day and short-day plants to the time of giving a night-break. *Fuchsia* received a 1 h exposure to light in a 16 h dark period; *Kalanchoë* received a 1 min exposure in a 15 h dark period (Vince-Prue, 1975).

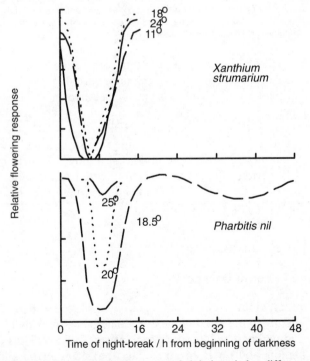

FIG. 1.9. Time of night-break sensitivity in an extended dark period at different temperatures. In both *Pharbitis* and *Xanthium*, the time of maximum sensitivity to a night-break occurred at about 8 h from the beginning of darkness, even though the subsequent unbroken dark period of 40 h was much longer than the critical nightlength. The time of maximum night-break sensitivity was not affected by temperature (Salisbury and Ross, 1991).

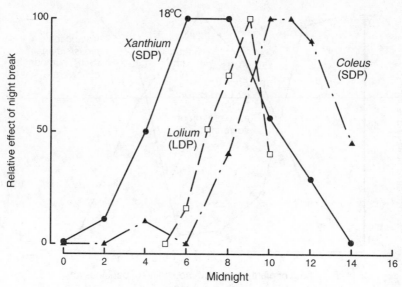

FIG. 1.10. The effect of giving a night-break at different times during a 16 h dark period. *Xanthium strumarium* was given 5 min and *Coleus frederici* and *Lolium temulentum* Ba3081 were given 1 h night-breaks. Flowering was inhibited by the night-break in the SDP and promoted in the LDP; the figure shows the relative effect to promote or inhibit flowering at the times indicated. *Xanthium* data from Salisbury (1963b); *Coleus* from Halaban (1968); *Lolium* from Vince (1965).

established that the night-break response is a transient period of sensitivity to light, which is related in time to the beginning of the dark period.

The way in which the time of NBmax is established and the relationship between night-break timing and the critical nightlength concern the biological timing mechanism(s) and are discussed in detail in Chapters 2 (SDP) and 5 (LDP). Here it need only be said that, in a natural 24 h cycle, a light interruption is usually most effective somewhere near but not necessarily at the middle of a long dark period. The time of NBmax in a 16 h dark period can vary considerably with species (Fig. 1.10) and it is unwise to assume, as is often done for the commercial regulation of flowering, that the 'best' time to give a night-break is at midnight.

THE ROLES OF LIGHT

Counteracting the Effect of Darkness

As discussed in the previous section, photoperiodic induction in SDP depends on exposure to a sufficiently long period of darkness. There is also evidence that a longer than critical period of darkness is inhibitory to flowering in LDP. Thus one important role for light in photoperiodism is to counteract the effect of darkness. We have already noted that quite a short exposure to light can nullify the flowering response when this night-break is given at a particular time of night, designated NBmax. At other times, the effect of light is much less or there is no effect. Not only is the magnitude of

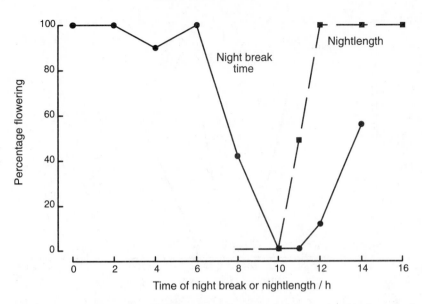

FIG. 1.11. The relationship between nightlength and the time of giving a night-break on the flowering response of *Coleus frederici*. The 1 h night-break treatments were given at different times during a 16 h dark period following an 8 h photoperiod. The values have been plotted from the beginning of darkness (Halaban, 1968).

the response to light different at different times but the sensitivity to light also varies. For example, the dose for half maximum response to a red-light pulse in *Pharbitis* varies by a factor of 20–40 during the course of the night (Thomas, 1991).

One action of light is thus the night-break, which nullifies the effect of darkness. Under natural conditions, however, plants do not experience a night-break and flowering is determined by whether or not light (dawn) occurs before, or after the critical nightlength for the plant is question. Is a night-break, therefore, simply equivalent to the dawn signal which ends the night? The answer to this question is that they appear to be two distinct components of the photoperiodic mechanism although, operationally, they may be difficult to distinguish. For example, in *Pharbitis*, with a long photoperiod and/or repeated cycles, NBmax occurs 8–9 h after the onset of darkness and the duration of the critical nightlength is also 8–9 h. A similar situation is found in *Coleus* (Fig. 1.11). However, under other conditions the critical nightlength in *Pharbitis* is several hours longer than the time to NBmax (Fig. 1.12). In this case, a night-break and dawn are clearly not the same signals as far as the photoperiodic mechanism is concerned. We will return to this point again in Chapter 2.

The variations in sensitivity make it virtually impossible to establish a threshold value for the intensity of light needed to counteract darkness, since this depends on when and for how long light is given. When light from TF lamps is given continuously throughout the night, threshold values have been shown to vary considerably between species (Table 1.3). A question often asked is whether moonlight can influence flowering. The answer is obviously complicated by the changes in sensitivity to light during the course of the night but, even for the most sensitive species, it has been calculated that the threshold level of photoperiodically effective red light is higher

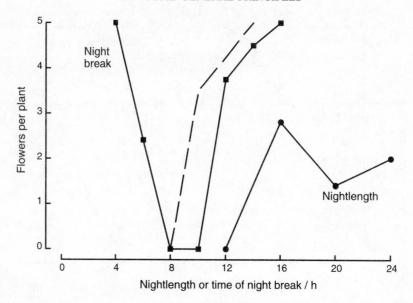

FIG. 1.12. The relationship between nightlength and the time of sensitivity to a night-break in *Pharbitis nil*. A single photoperiod of 12 h in white light was followed by a 72 h inductive dark period interrupted by a 10 min night-break (■) or by various durations of darkness (●). The critical nightlength occurred 3–4 h after the time of maximum sensitivity to a night-break (NBmax). However, when the photoperiod duration increased to 24 h (dashed line), the critical nightlength decreased to 8 h; the time of NBmax was unchanged and the critical nightlength and NBmax coincided (Lumsden, 1984).

TABLE 1.3. Approximate threshold light values for the suppression or induction of floral initiation in some long-day and short-day plants.

Short-day plants	Threshold light value[a] for the inhibition of flowering (mW m^{-2})
Dendranthema grandiflora	90.3
Kalanchoë blossfeldiana	90.3
Euphorbia pulcherrima	21
Pharbitis nil	4.2–42.0
Cannabis sativa	1.3
Glycine max	0.4
Xanthium strumarium	0.4
Long-day plants	Threshold light value[a] for the promotion of flowering (mW m^{-2})
Brassica campestris	4,515
Lolium temulentum	44.1
Silene armeria	31.5–90.3
Hordeum vulgare	10.5–21.0
Callistephus chinensis	4.2–12.6

[a] Using tungsten filament lamps continuously throughout the night or for the greater part of the night. From Vince-Prue (1975).

TABLE 1.4. Effect of moonlight on flowering in the short-day plant *Pharbitis nil*.

	16 h dark		13–14 h dark	
	% Flowering	Flower number	% Flowering	Flower number
Experiment 1				
Moonlight	100	1.5	73	0.7
Shaded	100	2.1 (NS)	90	1.5 (NS)
Experiment 2				
Moonlight	100	1.7	83	1.0
Shaded	100	3.4	100	1.5

Plants were exposed to or shaded from moonlight during a single dark period of the duration indicated.
Data of Kadman-Zahavi and Peiper (1987).
NS, not significant.

than the red light content of moonlight (Salisbury, 1969). Moreover, mutual shading effects and nocturnal leaf movements (Bünning and Moser, 1969) may reduce the amount of moonlight reaching the leaves, in some cases to only 5–10% of direct moonlight. It seems unlikely, therefore, that light from the full moon influences flowering, even in very sensitive plants. Experimental evidence on the subject is scanty. In some early work, very slight (2–3 days) differences in flowering time were recorded between covered plants and those exposed to the sky, but factors other than moonlight (such as temperature) may have been involved. More recently, the question has been re-examined using the highly sensitive SD species, *Pharbitis nil* (Kadman-Zahavi and Peiper, 1987). With a single 16 h dark period, exposure to moonlight had almost no effect on the percentage of plants flowering but slightly decreased the number of flower buds; with a 13–14 h dark period the inhibitory effect was slightly greater (Table 1.4). It is evident that, although moonlight may be perceived (and again the shading treatment could have affected other environmental factors), the effect is very slight even in this extremely sensitive SDP. Since the moon is full for only a few days each month, its effect must be of negligible importance under natural conditions. However, street lamps which remain on for long periods may influence flowering or other photoperiodic responses such as leaf fall and dormancy in nearby trees. Such effects have been recorded.

A Requirement for Light

Although flowering in SDP requires exposure to a sufficient duration of darkness, the light period is not without effect. Perhaps the most important attribute of the photoperiod is that a preceding exposure to light is necessary for a subsequent dark period to be inductive. The requirement for light before an inductive dark period was first demonstrated by Karl Hamner (1940). After several cycles of 3 min light/3 h dark, *Xanthium* plants completely failed to flower in response to a single 12 h dark period unless it was preceded by several hours of light. The effect of a preceding period of light cannot be replaced by extending the dark period. In *Begonia*, for example, exposure to 4 cycles of 8 h light/16 h dark resulted in flowering, whereas plants receiving a single cycle of 8 h light/96 h dark remained vegetative (Krebs and Zimmer, 1983). Initially, it was considered that a relatively high intensity of light was necessary for this effect and it was called the **high-intensity light reaction** of photoperiodism (Lang, 1965). Later, it was found that less than 0.02 W m^{-2} from a TF lamp was sufficient to produce a strong promoting effect on flowering in *Xanthium* and there was no obvious effect of irradiance when examined in the range above about 3 W m^{-2} from fluorescent lamps (Salisbury, 1965). Dark-grown seedlings of *Pharbitis* have been shown to flower with only 2 × 1 min exposures to red light separated by 24 h and followed by an inductive dark period (Friend, 1975), or even a single 5 min exposure provided that seedlings were simultaneously sprayed with a cytokinin (Ogawa and King, 1979a). Controls without light did not flower. In at least one multicycle SDP, the requirement for light has also been shown to be satisfied by quite small amounts; *Kalanchoë* flowered with as little as 1 s of light each day over several cycles, whereas darkness for the same duration was without effect (Schwabe, 1969). Thus, in several SDP plants, the so-called high-intensity light reaction appears to be a **low-intensity** reaction.

Nevertheless, there are many examples where the magnitude of flowering is strongly influenced by light quantity during the photoperiod and it has been suggested on several occasions that this indicates a requirement for photosynthesis or photosynthetic products. For example, in *Glycine* cv Biloxi, 4.5 W m^{-2} was below the threshold for flowering in 5 h days and more than 2.2 W m^{-2} was required with 10 h days (Hamner, 1940). Carbon dioxide must be present during induction in several, but not all SDP. Among these are *Kalanchoë* and *Xanthium*, where removing the carbon dioxide supply to the induced leaf depressed flowering, even with normal carbon dioxide to all other leaves (Ireland and Schwabe, 1982). However, it is not clear that the carbon dioxide effect is photosynthetic since removing carbon dioxide during the photoinductive dark period also depressed flowering in both plants. In some cases the immediate products of a period of light appear to be more important than stored assimilates. This has been shown in the glasshouse chrysanthemum where the intensity of the daily light period strongly influences the response to inductive long nights and flowering occurs earlier and at a lower leaf number as the irradiance during the short day is increased (Cockshull, 1984). The considerable delay in flowering which occurs in poor light conditions makes a winter crop quite difficult in more northerly latitudes. However, the general carbohydrate status of the plant apparently has little effect for plants previously grown at low irradiances flowered with no delay provided a high irradiance was given during the period of SD induction. This has been used as a basis for supplementary lighting techniques to improve crop quality in commercial practice since the treatment needs to be given for only a relatively short period; if additional light was given during the first two weeks of SD induction, flowering was not delayed even when plants were subsequently transferred to low intensity conditions (Cockshull and Hughes, 1972).

As photosynthesis must ultimately supply the substrate for all plant processes, a relationship between photosynthesis and photoperiodism is to be expected. It seems unlikely, however, that photosynthesis has any role in the photoperiodic mechanism itself for, as already discussed, there are several SDP where the photoperiodic

induction has been shown to operate with amounts of light which are photosynthetically insignificant (e.g. *Kalanchoë, Xanthium, Pharbitis*). Photoperiodic competence
and normal timekeeping (both in terms of NBmax time and CNL) have also been
demonstrated in photobleached cotyledons of *Pharbitis* (King *et al.*, 1978); green
chloroplasts were not present, although storage carbohydrate would presumably have
been available. Thus, although the dark period processes which lead to flowering do
not occur without prior exposure to a period of light, from these detailed studies with a
few SDP it seems reasonable to conclude that the light requirement for photoperiodic
induction in SDP is not primarily photosynthetic. At least two components of this light
requirement have now been identified. Firstly, light acts as a signal to set the phase of
the circadian rhythm which underlies photoperiodic timekeeping and, secondly, light
acts in a reaction which has been called the **Pfr-requiring reaction**. These two actions
of light are discussed in detail in Chapters 2 and 4.

In SDP where a single inductive cycle is sufficient to induce flowering, there seems
to be no upper limit to the duration of the photoperiod provided that the critical dark
period is exceeded. This is not so, however, when several inductive cycles are
required. When *Glycine* was given seven cycles composed of different light and
dark durations, there was a sharply defined critical nightlength of 10 h, which was
independent of photoperiod duration (Fig. 1.13). However, with an inductive 16 h
dark period, the longest photoperiod which allowed flowering was between 18 and
20 h. It is not clear from this experiment, however, if 18–20 h represents an absolute
upper limit or whether the cycle lengths examined between 36 and 46 h (20–30 h light
plus 16 h dark) were unfavourable. Later experiments (see Fig. 2.8) tend to support the

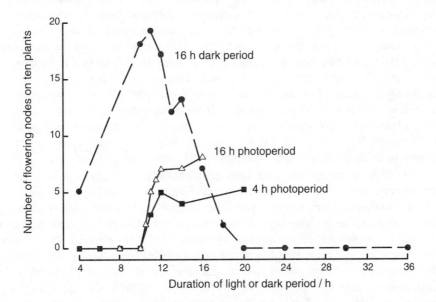

FIG. 1.13. The effect of varying the duration of the light and dark periods on flowering in soya bean
(*Glycine max* cv Biloxi). Plants received 7 cycles in which the duration of the light period accompanying a
16 h dark period (l), or the dark period accompanying a 4 hour (n) or 16 hour (△) light period was varied.
From Hamner (1940).

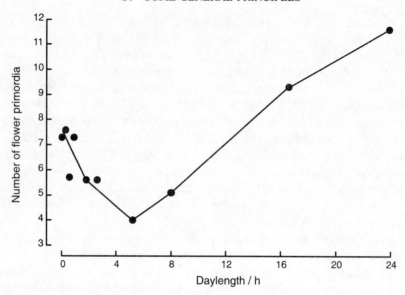

FIG. 1.14. Daylength and flowering in the long-day plant *Hyoscyamus niger*. The daylength treatments (in 24 h cycles) were given for 4 days in fluorescent light (60 W m^{-2}), followed by 13 long days in mixed fluorescent and tungsten filament light (Joustra, 1969).

latter conclusion, so that the apparent upper limit to daylength in multicycle plants is probably a property of the rhythmic nature of the time-measuring system.

There is some evidence that inhibitory processes which occur in darkness in LDP also depend on previous exposure to a certain amount of light. In *Hyoscyamus*, the maximum inhibition of flowering occurred with a daylength of 5 h and flowering was actually accelerated when shorter daylengths were used (Fig. 1.14). In these experiments the 6 days of treatment were followed by inductive LD so that the action of the experimental daylengths could be assessed. Flowering in *Hyoscyamus* was also delayed more when 2 LD were alternated with 2 SD than when they were alternated with 48 h of darkness. There is evidence from other types of physiological experiments such as defoliation, that part, at least, of the control of flowering in *Hyoscyamus* is inhibition by SD cycles. We would, therefore, expect to find a similar mechanism operating in these inhibitory SD cycles as in the promotive SD cycles for SDP.

Many LDP have requirements for long daily exposures to light in order to promote flowering. This type of response has been called **light-dominant** to distinguish it from the **dark-dominant** response in which the control of flowering appears to reside primarily in exposure to long periods of darkness (see Chapter 5). The requirement for long daily exposures is not necessarily associated with photosynthesis since, in some cases, it can be satisfied at low irradiances or by exposures to intermittent pulses of light (Fig. 1.15). It is evident that the latter effect is not entirely photosynthetic since the most effective pulses (FR–R) are photosynthetically equivalent to less effective ones (R–FR). Photoperiodic competence has also been demonstrated in hypogeal cotyledons of the LDP *Pisum*, which have no green chloroplasts (Paton, 1971). In barley, chlorophyll-deficient mutants and plants bleached with the herbicide Norflurazon were also found to have normal photoperiodic responses (Friend, 1984).

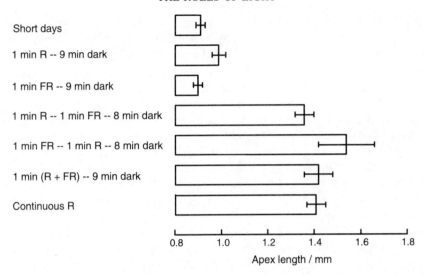

FIG. 1.15. Flowering response of *Lolium temulentum* to a day-extension with pulses of red and far-red light. Plants received a single LD cycle consisting of an 8 h day in sunlight followed by a 16 h experimental light treatment as indicated. Pulses with both red (R) and far red (FR) light were more effective than R or FR alone. Photosynthetically equivalent treatments (R–FR and FR–R) did not have the same effect on flowering (Wall and Vince-Prue, unpublished results).

On the other hand, plants requiring long daily exposures to light often flower more rapidly if light is given at high irradiances for at least a part of the day. In *Sinapis alba*, for example, high-intensity light promotes flowering during the first 8 h of a long photoperiod and carbon dioxide is required for this effect suggesting the involvement of photosynthesis (Bodson *et al.*, 1977). However, neither high-intensity light nor carbon dioxide is necessary during the day-extension which lengthens the photoperiod beyond the CDL, indicating that they are not essential components of the photoperiodic timekeeping mechanism. Plants remained vegetative when high intensity light was given during a single SD, although certain changes normally observed during floral initiation at the apex did occur (Havelange and Bernier, 1983). There was, for example, an increase in the number of mitochondria, a response which also occurred when only the apex was exposed to LD.

The situation in LDP thus appears to be complex. Part of the effect of light is similar to that in SDP; i.e. it is in some way associated with the effect of the subsequent dark period which, in LDP, is inhibitory. Part seems to be a promotive effect of long daily exposures to light which is not photosynthetic, and part may be an effect of photosynthetically active light. In different LDP, the relative importance of these various actions of light is likely to be different. The possible actions of light in LDP are considered in more detail in Chapter 5.

Light After an Inductive Dark Period

In some early experiments, there were indications that light conditions following an inductive dark period were important for flowering in SDP. Thus the existence of a **postinductive light reaction** has sometimes been postulated as part of the photoperiodic mechanism. If such a reaction occurs, it can only easily be demonstrated in single-cycle plants since in multicycle plants any postinductive effect of light is confounded with its effect in the subsequent cycle (except for the last cycle in the series).

A typical experiment with *Xanthium* showed that plants flowered more when returned to daylight after the end of a single inductive dark period than when plants were given 10 min light followed by a further period of darkness. Various interpretations have been given for these results, including an effect on translocation of the floral stimulus or on stabilising it in some way. Some support for the latter suggestion comes from evidence that the products of an inductive dark period in *Xanthium* can be destroyed by high temperature. At 30°C, flowering was decreased as the night-length was increased beyond 11 h and almost completely inhibited when plants remained in darkness for a further 5 h (Salisbury, 1963b). Thus the stimulus synthesised after the 8.5 h critical night appeared to be subject to destruction or inactivation at high temperatures if the dark period was prolonged. Such an effect could account for the apparent postinductive light requirement and explain some of the variations in results. For example, Searle (1961), who was unable to observe any effect of postinductive light conditions, maintained his plants in darkness at 24°C; at this temperature Salisbury did not observe any apparent destruction or inactivation when the dark period was extended (see Fig. 2.18).

The transport of a floral stimulus out of an induced leaf could be influenced by a period of high intensity light if, as has been suggested (Chapter 6) the 'floral stimulus' and sucrose are co-transported. However, as Salisbury (1969) has pointed out, it is unlikely that translocation effects alone could account for the inhibitory effect of giving a period of high temperature in darkness before returning plants to light; inhibition of translocation during the second dark period would be expected only to delay flowering, not to inhibit it entirely. Consequently, some effect on stabilisation or destruction of a postinduction product is indicated. However, in many cases (in *Perilla*, for example), there is little or no effect of postinduction conditions on the flowering response and so a specific postinductive high-intensity light process as part of the basic photoperiodic mechanism seems unlikely.

CONCLUSIONS

From this discussion, it is evident that light has a number of effects in photoperiodism and, when considering how the process operates, it is important not to confound the different roles of light. From the results of many experiments it appears that there are two essential components of the photoperiodic process in SDP. Time is measured in darkness and the critical factor for floral induction is a sufficiently long dark period or, in the majority of SDP, a succession of such dark periods. Exposure to a nightlength longer than the critical does not, however, result in flowering unless it is preceded by a photoperiod whose duration does not appear to be critical. Although it has been postulated that this light requirement is a high-intensity light process, the weight of evidence suggests that only small amounts of light are required and that photosynthesis is not an essential component of the photoperiodic mechanism. Nevertheless, in some SDP, flowering is increased or accelerated when the SD are given at higher irradiances suggesting that, as might be expected, photosynthesis may contribute to the magnitude of the overall flowering response.

The situation in LDP is less clear. Where long nights have been shown to be inhibitory to flowering, the overall mechanism appears to be similar to that in SDP. However, in many LDP long exposures to light are necessary for flowering and this light requirement does not appear to be simply to counteract the effect of darkness. In some cases, flowering is accelerated when at least part of the long exposure is given at high intensities and photosynthesis may be involved.

It is difficult to determine the contribution of photosynthesis in plants that require an induction period of many weeks as, for example, in the absolute LDP *Marjorana syriaca* which requires at least 28 LD for floral initiation (Dudai *et al..*, 1989). Here photosynthesis must obviously be required just to maintain growth over this period. It is also evident that photosynthesis influences the number of flowers that can develop and be supported by the plant; this may complicate the interpretation of the effects of light in photoperiodic induction, especially where 'flower-number' is used as a criterion of the response. The basic question is whether the process of photosynthesis is part of the photoperiodic induction mechanism or whether it is peripheral, being required in some way for the **expression** of induction. It is not yet possible to determine that induction has occurred in the leaf until its effects are realised in some recognisable event(s) at the apex. These could be blocked at several stages between induction in the leaf and the initiation of floral primordia (e.g. loading, transport and evocational events) and photosynthesis, or its products could be required at any stage. A further complication in some cases is that high intensity light may accelerate floral induction under suboptimal or non-inductive conditions. In rice, low light intensity during the photoperiod resulted in a lower floral stage in suboptimal cycles but only 7% sunlight was fully effective under optimal photoinductive conditions (Ikeda, 1974) while, in the SDP *Dendranthema*, the transition to reproduction in continous white light occurred more rapidly and with a lower leaf number as the irradiance increased from 7.5 to 120 W m^{-2} (Cockshull, 1979). It is possible to induce near maximal flowering in *Pharbitis* in continuous light by giving 3 days of high-intensity light at 23°C followed by 7 days at 13/14°C at normal intensities (Shinozaki *et al.*, 1982); DCMU inhibited flowering when given during the period of high-intensity light indicating the involvement of photosynthesis. However, it is evident that photosynthesis is not part of the **photoperiodic** induction mechanism in *Pharbitis* since DCMU had no effect when given before or after a 16 h inductive dark period.

It is evident from a number of approaches (e.g. the use of very small quantities of light or chlorophyll-deficient plants) that the photoperiodic timing mechanism in the leaf can operate in the absence of concurrent photosynthesis in both SDP and LDP. However, the realisation of the flowering response does not occur unless assimilates are available; these can be derived from stored assimilate (e.g. cotyledons of *Pisum*, *Pharbitis*) or from sucrose feeding (e.g. *Hordeum*; and see Fig. 5.4). It seems likely, therefore, that photosynthesis plays no direct part in the photoperiodic mechanism in the leaf but is primarily an energy source for the realisation of the response. Among the suggestions that have been made are that photosynthetic products are necessary for export of the floral stimulus (Evans and King, 1985) or for the completion of the inductive process in the leaf (Ireland and Schwabe, 1982). The possibility that sugars constitute part of the signal that is involved in triggering the transition to flowering in receptive meristems is considered in Chapter 7. Even if such a role is indicated, it does not follow that photosynthesis is directly involved in the photoperiodic induction mechanism in the leaf since soluble carbohydrates could be derived from stored assimilate.

Other effects of light that may influence flowering are clearly not part of the process of photoperiodic induction in the leaf. For example, a direct effect of light reaching the apical bud has been shown to promote flower development in the DNP *Rosa* by increasing sink strength (Halevy, 1984); this has been shown to be a phytochrome-mediated response. It has been suggested that effects of light on flowering in day-neutral plants could give information about the roles of light that are **not** part of the photoperiodic induction mechanism. However, this approach must be viewed with some caution since, even in day-neutral species, there are considerable variations in the responses to light and the direct effect of light to increase sink-strength in the apical bud is not found in all plants. In the SDP *Dendranthema*, for example, maintaining the apical bud (including leaves <1 cm) in darkness did not reduce their sink-strength nor cause the developing inflorescence to atrophy: even when buds were covered from the beginning of SD induction, there was little or no lag in their development (Steffen *et al.*, 1988).

Under natural conditions, photoperiodically sensitive plants respond to an alternation of periods of light and darkness. When these are varied independently, the results suggest that the transition to flowering in both SDP and LDP is determined primarily by the duration of the dark period and a more or less sharply defined critical night-length has been found to be a characteristic feature of photoperiodically sensitive species. Several lines of approach, however, have shown that photoperiodism cannot be analysed in terms only of dark reactions that are counteracted by light. A positive requirement for light has been found for flowering in SDP and, while this may partly be for photosynthesis, light also has a direct positive role in the photoperiodic mechanism. In many LDP, flowering does not appear to be governed solely by inhibitory events which take place in darkness and are prevented by light. The nature of these various light reactions and the identity and functioning of the photoreceptors form the subjects of Chapters 3, 4 and 5 but, first, it is necessary to consider the nature of the timing mechanism which underlies photoperiodism and consider how light interacts with this.

2 Photoperiodic Timekeeping

In order to locate the time of year accurately, the timekeeping mechanism in photoperiodism must operate with a considerable degree of precision. It must also be relatively insensitive to random variations in the environment. This applies particularly to changes in temperature, or to light quality where this is due to factors such as sunflecks passing over a woodland floor. As the rate of change of daylength is not constant but is fastest in spring and autumn (Fig. 2.1), the absolute precision will vary with the time of year. The rate of change is lower in the tropics during much of the year and timekeeping needs to be more precise than at higher latitudes in order to locate a seasonal event with the same degree of accuracy.

For many years, it was rather generally assumed that time is measured by a kind of hourglass consisting of a series of catenary steps which must proceed to completion in order to measure the durations of light and darkness, the latter being by far the most important factor. As this critical nightlength is often largely independent of the

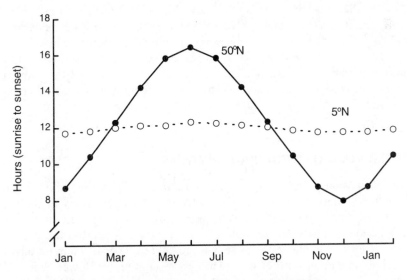

FIG. 2.1.   Daylength as a function of the time of year for two different latitudes.

duration of the associated photoperiod, transfer to darkness was thought to initiate a non-cyclic process (or sequence of processes) functioning as an hourglass. In such a system, time-measurement stops after a critical nightlength has been measured and a minimum duration of light is needed to re-start dark timekeeping (to 'turn the hourglass over'). The most specific proposal for the nature of the hourglass was made in 1960 by Sterling Hendricks (Hendricks, 1960). Based on many experiments with dark-grown seedlings, it was known that one of the properties of the photoreceptor pigment, phytochrome (see Chapter 3), was to change in darkness from the active form (Pfr) to a form which is biologically inactive (Pr). Consequently, it was proposed that the time-measuring hourglass might be the time taken for Pfr to fall to a critical threshold below which it no longer inhibited flowering in SDP, nor promoted it in LDP. In this way a critical nightlength could be measured. The nature of phytochrome changes in darkness and their possible function in photoperiodism is discussed in detail in Chapters 3, 4 and 5. Here, it is necessary only to say that the idea that, in plant photoperiodism, phytochrome is directly involved in both timekeeping and photoperception has been shown to be incorrect. One of the main arguments against loss of Pfr being the main photoperiodic timer comes from experiments in which [Pfr] is rapidly reduced in a photochemical reaction before plants are transferred to darkness. This is done by exposing plants briefly to far red light at the end of the photoperiod. It has been found that such a rapid photochemical reduction in [Pfr] has rather little effect on either the critical nightlength or the time of NBmax (see Chapter 4). Other suggestions for an 'hourglass' timer have been much less precise.

Although the possibility of an hourglass component of the critical nightlength has not been entirely excluded, most attention is now focused on the hypothesis that photoperiodic timekeeping depends on a circadian oscillator of the kind that underlies many endogenous rhythmic phenomena (Bünning, 1936). The difference between an 'hourglass' clock and a 'circadian' clock is that a circadian clock restarts nightlength measurement spontaneously after completing each cycle, whereas an hourglass is incapable of restarting nightlength measurement in prolonged darkness. It is evident, however, that 'hourglass' behaviour can also be generated by a circadian clock under some conditions (Vaz Nunes *et al.*, 1991) and there is now a considerable body of experimental evidence for the involvement of a circadian oscillator in photoperiodic timekeeping. In order to consider how this clock may operate to control flowering under natural day/night cycles, it is first necessary to understand the properties of endogenous circadian systems and how they interact with light and other environmental variables, such as temperature.

CHARACTERISTICS OF CIRCADIAN RHYTHMS

Organisms are normally subjected to daily alternations of light and darkness and often exhibit rhythmic behaviour in association with these changes. Many of the rhythmic responses to day and night continue even in a constant light or dark environment, at least for a period of time. An example of such rhythms is seen in the daily 'sleep' movements of leaves. In many plants, leaves are held more or less horizontally during the day and assume an upward or downward position at night. However, in 1729, the astronomer De Mairam observed that the movements persisted (at least for a time)

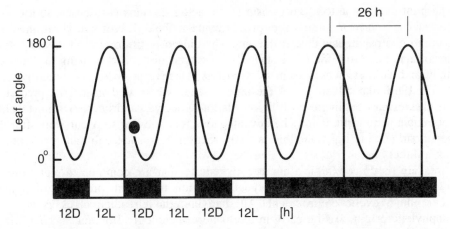

FIG. 2.2. A typical circadian rhythm of leaf movement. The rhythm is shown entrained to 24 h (12 h light/12 h dark) cycles, followed by reversion to the free-running period (26 h in this example) following transfer to continuous darkness. (•) Phase-point.

when plants were transferred from daily light/dark cycles to continuous darkness, even though the changes between day and night no longer occurred (Fig. 2.2); he interpreted this as evidence of some kind of biological timing device. It is now generally accepted that the rhythm in leaf movement is based on an unseen, endogenous pacemaker which is self-sustaining and continues to oscillate under constant conditions. Thus, like the hands of a clock, the observed rhythm is coupled to an underlying oscillatory process, called a **circadian oscillator** (Fig. 2.3). At the present time the operation of the underlying oscillator can only be deduced from studies of the behaviour of the observed, coupled rhythm. In Fig. 2.3, two rhythms are shown coupled to the same oscillator but, in some cases (e.g. the leaf movement and photoperiodic rhythms in *Pharbitis*) studies of the rhythms indicate that they are coupled to different circadian oscillators. It seems likely, therefore, that more than one clock is involved in driving circadian rhythms.

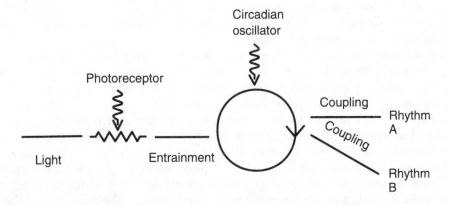

FIG. 2.3. The action of light on a circadian oscillator and coupling to circadian rhythms. In this example, two different rhythms are shown coupled to the same circadian oscillator.

The presumed value to the organism of circadian rhythms is to ensure that events occur at the 'correct' time of day. Horizontally held leaves are positioned for maximum photosynthesis during the day. The closed position at night could reduce water or heat loss from the foliar surfaces. However, direct and immediate responses to light and darkness would seem to be just as effective as an endogenous rhythmic one. The particular advantage of the latter is seen when we consider its predictive value; leaves can begin to close before darkness arrives and be beginning to open before dawn. Similarly, it may be advantageous for an event to occur well after the onset of darkness, or light, and this can only occur if the response to light, or darkness, is not a direct one.

Circadian rhythms exhibit a number of features that are common to many organisms. This might be due to convergent evolution under the pressures of natural selection in a cyclic environment. Alternatively, the circadian clock may be of monophyletic origin, arising early in the history of biological organisms; if so, then it is interesting to speculate for which cell function the time of day was originally important. In either case, the daily light/dark cycle was probably the primary agent of selection and, under natural conditions, synchronization of circadian rhythms is usually to the daily 24 h cycle of light and darkness.

Rhythms are derived from cyclic phenomena and the time required for an oscillation to make a complete cycle and return to the original starting position is known as the **period** (Fig. 2.4a), while the magnitude of the oscillation is known as the **amplitude**. Although the rhythms are innate, they may need a transition between light and darkness (or sometimes a change in temperature) to start them; thereafter, they often continue for several cycles in darkness. Under these conditions, the rhythm is said to be **free-running** and its periodicity is then close to, but not exactly, 24 h. Hence the term **circadian** (from the Latin for 'about one day') is generally used. The period of the free-running rhythm has a Q_{10} close to 1, with observed values ranging from 0.8 to 1.3. A change in temperature may alter the periodicity transiently but the normal free-running period (FRP) is usually established after a few cycles. The clock mechanism is, therefore, temperature compensated in some way, so that timekeeping can continue accurately under different temperature conditions: compensation is usually better within the 15−30°C range. Temperature may, however, affect the amplitude of the oscillation. The **timing mechanism** in plant photoperiodism is also fairly insensitive to changes in temperature, although non-timing components may be affected and lead to modifications of the overall response.

Because the period of the free-running rhythm is not exactly 24 h, the rhythm drifts in relation to solar time when organisms are maintained in a constant environment. In the leaf movement example shown in Fig. 2.2, the FRP is longer than 24 h and so the peak time is later each day. Other rhythms have a period shorter than 24 h and so would gain time. The rhythm, therefore, has to be **entrained** (synchronised) to a 24 h cycle by a periodic environmental signal, usually light, called a **zeitgeber** (the German word for 'time-giver'). This action of light is considered to be on the circadian oscillator to which the observed rhythm is coupled (see Fig. 2.3) and, under natural conditions, the main environmental zeitgebers are the daily light/dark transitions at dawn and dusk. A change in temperature may, in some cases, also act as a zeitgeber, although the **period** of the rhythm is temperature compensated, as already noted.

The term **phase** is used for any point or stage in the cycle (see Fig. 2.4) and different

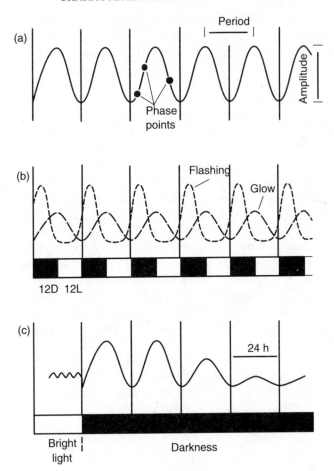

FIG. 2.4. Some characteristics of circadian rhythms. A free-running rhythm (a) has a characteristic period length and magnitude (amplitude); any specific point in the rhythm is known as a phase point. When entrained to light/dark cycles (b), rhythms show different phase relationships with the underlying circadian oscillator and, therefore, peak at different times of day. Rhythms frequently damp out in continuous bright light (c) but are restarted with the characteristic free-running period following transfer to darkness.

rhythms have different **phase relationships** with the entraining light signals, resulting in peaks of activity at different times of day.

For example, the single-celled alga *Gonyaulax polyedra* has two rhythms of bioluminescence, a constant 'glow' and a rhythm of discreet 'flashes'. These two rhythms have different phase relationships with the entraining 12 h light/12 h dark cycles in such a way that maximum flashing luminescence occurs near the middle of the night and maximum glowing occurs at the end (see Fig. 2.4b). Thus, daily timekeeping is a consequence of the phase relationships between the rhythm and the entraining environmental zeitgebers. Under natural conditions, the daily durations of light and darkness are continually changing with the seasons and so the entraining light signals do not always occur at the same solar time. Organisms must, therefore, entrain adaptively under different photoperiods. As an example, take a particular phase point which occurs 1 h after the onset of darkness in a 12 h light/dark cycle (see Fig. 2.2). If the daylength is extended by 1 h, this phase point will appear to have drifted forwards in relation to the dusk (light-off) zeitgeber and the appropriate response is a correcting backwards shift (or phase delay) such that the phase point still occurs 1 h after the end of the light period. Similarly, the appropriate response to light perceived at a phase point which would normally occur 1 h before dawn would be a phase advance. All circadian rhythms in plants and animals appear to operate in this way and exhibit characteristic **phase response curves** (PRC) to light. This phasing action of light is considered to be on the circadian oscillator (see Fig. 2.3) and the PRC describes the time-course of the oscillator's sensitivity. The phase of the rhythm itself is a function of the phase relationship between the observed rhythm and the oscillator (known as the **phase angle** of the rhythm).

The experimental protocol normally used to elucidate the oscillator's response to dawn and dusk signals is to place the organism in continuous darkness, give a relatively short exposure to light at different phases of the free-running rhythm, and observe the direction and magnitude of the resulting phase shift. Although there are considerable differences in the shape of the PRC, all agree in having delay responses when light is given at the beginning of the night phase of the rhythm (the **subjective night**) with a change to advance responses near the middle of the subjective night, and a prolonged dead zone of near insensitivity to light during the **subjective day** (Fig. 2.5). It is evident that the phase shifting response to light is complex, with both the direction and magnitude of the resulting phase shift depending on when the signal is given (i.e. on the phase of the oscillator at that time). The sensitivity to light also varies with the phase of the oscillator. Experimentally determined phase response

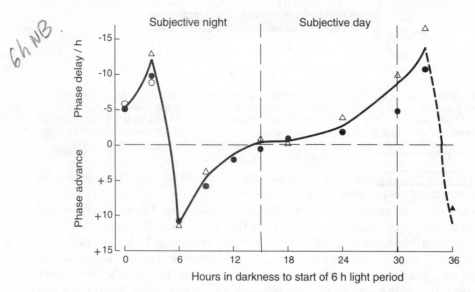

FIG. 2.5. A phase response curve for photoperiodic timekeeping in *Chenopodium rubrum*. Plants received a 6 h light interruption at different times in a 30 h free-running rhythm of flowering in darkness. When compared with the uninterrupted control rhythms, the response curve shows a phase delay early in the subjective night and a sudden change to a phase advance later in the subjective light. After King and Cumming, 1972a.

free-running period

curves of this kind allow a fairly accurate picture of the relationships between the plant or animal's circadian timekeeping and local time. Organisms with a day-active strategy have a FRP which is longer than 24 h and a large advance/delay ratio in the phase response curve (Pittendrigh, 1979). Since the FRP is longer than 24 h, the pacemaker must undergo a net advance each day to entrain to a 24 h cycle and this is ensured by the greater advance than delay shifts. The combination of a long FRP with a large advance/delay ratio in the PRC allows organisms to conserve an ecologically adaptive phase relationship between the clock (after reset has occurred) and dawn of the next cycle, irrespective of the duration of the photoperiod. This allows them correctly to anticipate dawn at all seasons (Johnson and Hastings, 1989). The phase relationship can also be maintained between the clock and dusk by an appropriate PRC shape and FRP (Pittendrigh, 1979).

Endogenous biological rhythms are, by definition, self-sustaining, but the oscillations frequently damp out fairly rapidly when the environmental zeitgebers are withdrawn. The rhythm of emergence of the fruit fly *Drosophila* from pupal cases appears to stop after 12 h in continuous bright light (see Fig. 2.4c) so that, when the organism is transferred to darkness at any time after 12 h in the light, the oscillation is immediately set in motion at the same phase point of the cycle irrespective of the duration of the previous light period (Pittendrigh, 1966). Several plant rhythms are known to behave similarly as, for example, carbon dioxide output in *Bryophyllum*, where continuous white light (LL) inhibits the rhythm which then restarts at the light/dark transition (Wilkins, 1959). As we shall see later, this observation appears to be highly relevant to photoperiodic timekeeping in SDP. However, many rhythms persist in continuous light, although period length is usually altered compared with the FRP in darkness. Such rhythms may also be reset by transfer to darkness; for example the acid-stimulated bioluminescence rhythm of *Gonyaulax* continues for many days in continuous bright light and is always reset to **circadian time** (CT) 12–14 at the light/dark transition (Sweeney, 1979). The period of the rhythm and the shape of the PRC may also be influenced by light quality and/or quantity (Johnson and Hastings, 1989; Lumsden, 1991; Nongkynrih and Sharma, 1992) and this should be taken into account when attempting to interpret complex effects of light on photoperiodic behaviour.

Oscillations may continue for only a few cycles when organisms are maintained in continuous darkness; in this case the oscillation is usually restarted by transfer to light. In some plants, this cessation of oscillation in darkness may be a consequence of carbohydrate depletion as the application of sucrose often enhances and sustains an endogenous rhythm in continuous darkness. An example is the petal movement rhythm in isolated flowers of *Kalanchoë*; this ceases rapidly when flowers are maintained in water but continues for some time if sugars are added. Sucrose may, however, be having an effect other than as a simple nutrient.

Although the mechanism of entrainment to particular light/dark cycles is not understood, it is obvious that a photoreceptor is needed for perception of the entraining light signals. Circadian rhythms occur widely in both plants and animals and a variety of different photoreceptor pigments appear to have been utilised in different organisms (Ninneman, 1979). For at least some rhythms in plants (e.g. carbon dioxide evolution in *Bryophyllum* and *Lemna* and leaf movements in *Phaseolus*, *Robinia* and *Samanea*), phytochrome (see Chapter 3) has been shown to be the photoreceptor, but it is clearly not involved in the photocontrol of the circadian system in other cases (e.g.

the bioluminescence rhythms in *Gonyaulax*). Many organisms (e.g. *Neurospora*, *Drosophila*) respond to blue light. It is evident from some studies that more than one photoreceptor may input to the oscillator (Lumsden, 1991). For example, blue light and the R/FR reversible reaction of phytochrome are both effective as entraining signals in the leaf movement rhythm in *Samanea* (Satter *et al.*, 1981).

In some plants, the phase of rhythms is shifted only after quite long exposures (2 h in *Phaseolus*) indicating a low sensitivity of the oscillator to light. This might be expected because, under natural conditions, the full photoperiod is associated with entrainment and so a highly sensitive photoreceptor system is unnecessary. It might even be deleterious since, if the system is too sensitive, the clock could respond to spurious signals such as moonlight. However, there are examples of plant rhythms which have a high sensitivity to light; the phase of the *Samanea* leaf movement rhythm requires only a single 5 min exposure to R in order to effect a phase shift (Simon *et al.*, 1976). The photoperiodic rhythm in *Pharbitis* is also very sensitive to light at certain times (see Fig. 2.15). An important general point is that there is a marked variation in the sensitivity to light, depending on the phase of the oscillator when the phase-shifting signal is given. This is evident both in photoperiodic responses (Lumsden and Furuya, 1986; Thomas, 1991) and in overt circadian rhythms.

The basic mechanism of the circadian clock is not yet understood, although a number of proposals have been made, including the involvement of membranes, biochemical feedback oscillators of various kinds and DNA transcription (Nongkynrih and Sharma, 1992; and see Chapter 8). Evidence is emerging that both the periodicity of the oscillator as well as its phase control may be regulated by phyochrome. In the LDP *Arabidopsis*, the periodicity of a rhythm in bioluminescence was longer in darkness than in red light; a mutant deficient in phytochrome had lengthened periods in continuous red, while the period length in darkness was decreased in mutants in which the photoreceptor pathways were active in the dark (Millar *et al.*, 1995b). Control of the oscillator by phytochrome may be achieved *via* changes in the intracellular calcium content since a 2 h pulse of calcium chloride or a calcium ionophore resulted in the same type of PRC for the leaflet rhythm in *Robinia* as was obtained with a 15 min exposure to R (Gomez and Simon, 1995).

Although the underlying clock mechanism(s) awaits elucidation, it is evident that the observed circadian rhythms have a number of formal characteristics in common. These can be summarised as follows.

- The rhythms oscillate with a period length close to 24 h, which is largely independent of the ambient temperature.
- They are synchronised (entrained) to 24 h cycles by environmental zeitgebers, most commonly the light/dark transitions at dawn and dusk but also, in some cases, by a change in temperature.
- The rhythm is coupled with a particular phase relationship (phase angle) to an underlying pacemaker, the circadian oscillator, and the phasing action of light is on the latter.

However, rhythms also differ from one another in a number of ways.

- The patterns of the PRCs vary, as does the sensitivity to light.
- Some rhythms are rephased rapidly (within one cycle in *Gonyaulax*), while others

may show transients before rephasing is complete (Johnson and Hastings, 1989). The rhythm is said to be reset when the new phase has been established.

- Some rhythms damp quickly, while others persist for several cycles in a constant environment.
- When rhythms persist in continuous light, the characteristics of the rhythm (period and PRC) may vary with the colour and/or intensity of the background illumination.
- In several cases, the rhythm appears to damp, or be suspended, at a particular circadian time in continuous bright light; it then restarts at this circadian time following transfer to darkness, or dim light. Rhythms which continue in bright light may also be reset on transfer to darkness.
- In some systems, a light treatment given at the appropriate phase of the rhythm apparently abolishes its subsequent expression (Lumsden and Furuya, 1986).

All of these points should be borne in mind when considering how circadian rhythms may operate in photoperiodic timekeeping.

CIRCADIAN RHYTHMS AND PHOTOPERIODIC RESPONSES

Night-Break Experiments

It is evident from the results of many experiments that photoperiodism involves a rhythmic change of sensitivity to light. The most direct approach has been to combine a short photoperiod with a very long dark period and to scan the latter with a night-break of the kind known to modify the flowering response (see Chapter 1). Under these conditions, maximum inhibition (SDP) or promotion (LDP) of flowering has been observed to occur at circadian (i.e. approximately 24 h) intervals. Results with *Lolium, Sinapis* and *Glycine* are shown in Fig. 2.6 and it is evident that the peaks of promotion of flowering in the two LDP approximately coincide with the peaks of inhibition of flowering in the SDP *Glycine*. There is clearly a difference between LDP and SDP in the timing of their responses to light. Similar circadian rhythms in the response to a night-break have been demonstrated in several other plants including the SDP *Perilla* (Carr, 1952), *Kalanchoë blossfeldiana* (Engelmann, 1960), *Chenopodium rubrum* (see Fig. 4.4), and the alga *Acrochaetium* (Abdel-Rahman, 1982); and the LDP *Hyoscyamus niger* (Hsu and Hamner, 1967) and *Lolium temulentum* Ceres (Périlleux *et al.*, 1994). Experiments have been carried out both with bidiurnal (48 h) and tridiurnal (72 h) cycles. Results with the latter are more convincing evidence for a circadian rhythm because the response seen in the second circadian cycle is clearly independent of the preceding and subsequent photoperiods. The observation that, in tridiurnal cycles, the middle NBmax is often somewhat less pronounced than the first and third (see Fig. 2.6) suggests there is some interaction between the effects of the night-breaks and the preceding and succeeding photoperiods. For example, in *Sinapis alba*, the second NBmax was only 50% effective compared with the other two in a single 72 h cycle (Kinet *et al.*, 1973). In this example, it is the magnitude of the response to light which was reduced in the middle cycle. The timing of the night-break response may also be modified. In *Lolium temulentum*, the second maximum occurred some 10 h earlier in a tridiurnal cycle (Périlleux *et al.*, 1994) than in a bidiurnal cycle (see Fig. 2.6), where interaction with the subsequent photoperiod might be expected.

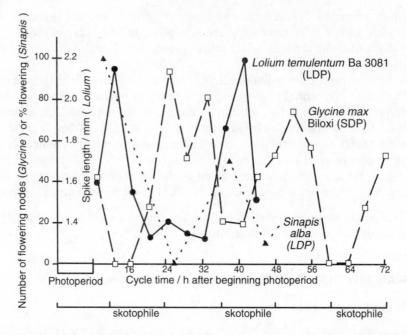

FIG. 2.6.   Rhythmic flowering responses of long-day and short-day plants to light-breaks given at different times during a long dark period. The light breaks were low intensity red light for *Lolium temulentum* (4 h at 0.45 W m^{-2}) and high intensity white fluorescent light for *Glycine max* (4 h at 50 W m^{-2}) and *Sinapis alba* (8 h at 25 W m^{-2}). A photoperiod of 8 hours was followed by a dark period of 40 hours (bi-diurnal cycle; *Lolium*) or 66 hours (tri-diurnal cycle; *Glycine, Sinapis*). The points have been plotted to correspond with the beginning of the light-break. For *Sinapis*, only the maximum and minimum points for the first two cycles are shown. Results for *Lolium* from Vince-Prue 1975, *Glycine* from Coulter and Hamner, 1964, and *Sinapis* from Kinet *et al.*, 1973.

Night-break experiments have not yielded clear-cut rhythmic responses in all species, including SDP and LDP. In *Xanthium* (Moore *et al.*, 1967) and *Pharbitis* (Fig. 2.7), a night-break was only inhibitory 8–9 h after transfer to darkness (from a 24 h photoperiod or LL), with no further periods of inhibition. Although, at first sight, this type of response seems to be an hourglass, there is evidence that a circadian rhythm is the underlying timer in both cases. Dropping the temperature from 20 to 18.5°C revealed a second slight inhibition in *Pharbitis* (Takimoto and Hamner, 1964). The presence of an underlying rhythm in *Pharbitis* is more convincingly demonstrated in dark-grown seedlings. If the 24 h photoperiod is replaced by 10 min R, the response to a night-break becomes rhythmic with minimum flowering occurring at circadian intervals (see Fig. 2.7). As the timing mechanism is almost certainly the same in these treatments, the apparent hourglass is thought to be caused by a damping of the rhythmic response following the long photoperiod. In *Pharbitis*, this damping may be due to the fact that, when the photoperiod is sufficiently long, the induction response in this cultivar, Violet, is already saturated in the first photoperiodic cycle and so no second inhibition point is possible. This conclusion is supported by the temperature experiment, where the rhythm is revealed at low temperatures which would be expected to decrease the induction response. In *Xanthium*, the photoperiodic rhythm always appears to be highly damped but features such as phase shifting (see

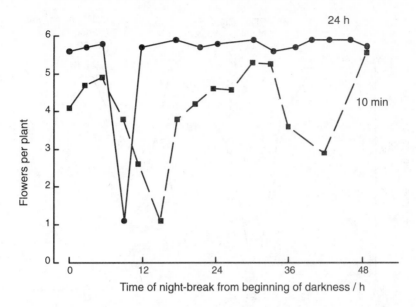

FIG. 2.7. The effect of duration of the photoperiod on the pattern of flowering response to a night-break in *Pharbitis nil*. Dark-grown plants were given a photoperiod of 10 minutes (■) or 24 hours (●), followed by an inductive dark period which was interrupted with a single night-break with red light at the time shown on the abscissa. After Vince-Prue and Lumsden, 1987.

Fig. 2.10), which are characteristic of circadian rhythms, indicate the underlying rhythmicity of night-break timekeeping in this plant.

The rhythmicity of the night-break response is frequently less clear cut in LDP, especially in tridiurnal cycles. For example, *Coleus blumei* (Kribben, 1955) showed only one NBmax in the middle of the 70 h night whereas, in *Anagallis arvensis*, the middle peak was missing (Hussey, 1954). As discussed later (Chapter 5), the photoperiodic timekeeping mechanism is less well understood in LDP, although it has been established that there is a circadian rhythm in the flowering response to a night-break in several species.

Cycle-Length Experiments

Another approach which has revealed a circadian rhythm in photoperiodic time-keeping is to examine the flowering response under different cycle lengths: these are sometimes called **resonance experiments**. It has been shown that the flowering response may change rhythmically with increasing duration of the dark period which follows a short photoperiod. For example, when a constant short photoperiod was combined with dark periods of various durations up to 64 h, flowering in the SDP *Glycine* and *Impatiens* was inhibited in cycle lengths of 36 and 60 h, while optimum flowering occurred in cycle lengths of 24, 48 and 72 h (Fig. 2.8 and Nanda *et al.*, 1969b). With a single dark period interrupting continuous light, the SDP *Chenopodium rubrum* also showed a circadian rhythm (with a periodicity of about 30 h) in the flowering response to the duration of darkness. This rhythm was shown to originate in the leaf (the site of photoperiodic perception) and required only one leaf to be exposed

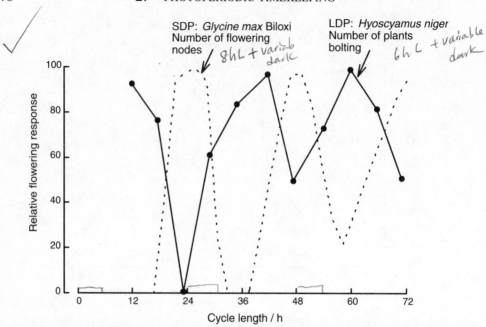

FIG. 2.8.   Effect of cycle length on the flowering response of short-day and long-day plants. Photoperiods of 8 hours (*Glycine*) or 6 hours (*Hyoscyamus*) were combined with dark periods of various durations to achieve the cycle lengths indicated on the abscissa (e.g. 24 h cycle = 8 or 6 h light + 16 or 18 h dark). *Glycine* plants received 7 cycles and *Hyoscyamus* plants 42 cycles. Data from Hamner and Takimoto, 1964 (*Glycine*) and Hsu and Hamner, 1967 (*Hyoscyamus*).

(King, 1975). A similar rhythm in the response to the duration of the dark period has been observed in LDP. In *Hyoscyamus*, when a 6 h photoperiod was combined with dark periods of various durations, a high level of flowering occurred in cycle lengths of 12, 36 and 60 h, with minimum flowering in 24, 48 and 72 h cycles (see Fig. 2.8). Thus, a circadian rhythm in the flowering response to cycle length has been observed in both LDP and SDP (including the alga *Acrochaetium*) and, as with the night-break experiments, there is a clear difference between the two photoperiodic categories in the timing of their responses to light (see Fig. 2.8). It is evident that, if flowering is to occur in these plants, the external light/dark pattern must be synchronised in some way with an internal circadian oscillation. The results illustrate particularly well that, contrary to the conclusions drawn from early experiments in 24 h cycles (see Fig. 1.5), the absolute length of the dark period is not always the controlling factor in photoperiodic induction. The LDP *Hyoscyamus* failed to flower when a 6 h photo- period was combined with 18 h of darkness (a 24 h cycle) but flowering was restored if the duration of the dark period was either increased or decreased (see Fig. 2.8). It is apparent that, in *Hyoscyamus*, a SD inhibits flowering in a 24 h cycle not because the night is too long, but because the light/dark cycle is unfavourable. However, although SDP also show rhythmic responses to cycle length, they do not flower with less than a critical duration of darkness.

Not all photoperiodically sensitive plants show clear-cut rhythms in cycle length. *Xanthium* appears to be totally non-rhythmic under these conditions (Fig. 2.9),

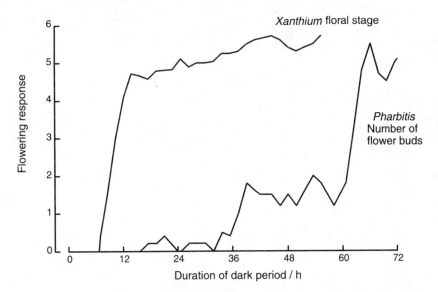

FIG. 2.9. Flowering responses of two short-day plants, *Xanthium strumarium* and *Pharbitis nil*, to a single dark period of various durations. *Xanthium* plants were grown in long days prior to the experiment. *Pharbitis* plants were grown in continuous light and then received 8 hours of darkness followed by 12 hours of light before transfer to the experimental dark period; plants that were transferred from continuous light to the various durations of darkness behaved like *Xanthium*. After Vince-Prue, 1975 (Data for *Xanthium* from Moore *et al.*, 1967 and for *Pharbitis* from Takimoto and Hamner, 1964).

although as already mentioned, it exhibits other features associated with circadian rhythms. *Pharbitis* shows yet another type of response. When a single inductive dark period followed a long period in continuous light, the response was non-rhythmic and resembled *Xanthium*; if, however, an inductive dark period was preceded by a non-inductive dark–light cycle, the response to increasing duration of darkness showed a stepwise increase with a periodicity of about 24 h (see Fig. 2.9).

Phase Shifting Experiments

A characteristic feature of overt circadian rhythms such as the sleep movements of leaves, is that the phase of the rhythm can be advanced or delayed by exposure to light, the precise response depending on when the light is presented. Similar phase shifts have been demonstrated in photoperiodism. Examples include the rhythm of flowering response to the duration of darkness in *Chenopodium rubrum*, for which a complete PRC has been constructed showing the classical features of phase delay by light given early in the subjective night and phase advance by light given late in the subjective night (see Fig. 2.5). For other SDP, the information is less complete but similar phase shifts have been demonstrated in *Xanthium* (Fig. 2.10) and *Pharbitis* (Fig. 2.15). In the latter, the phase of the rhythm has been shown to undergo delays or advances depending on whether the phase shifting signal is given earlier or later in the subjective night (Lumsden and Furuya, 1986).

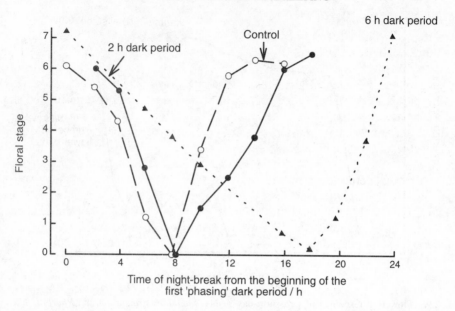

FIG. 2.10. Change in time of maximum sensitivity to a night-break (phase shift) in *Xanthium*. Control plants received a long inductive dark period interrupted at various times by a brief night-break. Other plants received a brief exposure to light after the first 2 or 6 hours in darkness, followed by a night-break at various times. After 6 hours of darkness, a brief exposure to light resulted in a marked phase shift in the night-break rhythm, when compared with the control plants. A much smaller phase-delay is also indicated when light was given after 2 hours in darkness. After Papenfuss and Salisbury 1967.

GENERAL MODELS FOR CIRCADIAN TIMEKEEPING IN PHOTOPERIODISM

Although admittedly based on results from rather few species, evidence of the kind outlined has shown that a circadian oscillator is involved in the photoperiodic processes of plants. Erwin Bünning (1936) was the first to suggest that the measurement of time in photoperiodism is dependent on an endogenous circadian oscillation. His original proposal was that photoperiodic timekeeping involves a regular oscillation of phases (i.e. portions of the rhythm) with different sensitivities to light. He suggested that transfer to light set in motion a 12 h **photophile** (light-requiring) phase, which is followed by a 12 h **skotophile** (dark-requiring) phase. Light was required during the photophile phase but flowering was inhibited by light given during the skotophile phase, thus giving rise to the inhibitory night-break effect in SDP and the need for both a critical duration of darkness and for light during the photoperiod, as had previously been demonstrated (see Chapter 1). In order to account for the behaviour of LDP, it was necessary to assume that the beginning of their photophile phase was delayed by about 12 h after transfer to light; light would then fall in the photophile phase only in long photoperiods. This would suggest that the photophile and skotophile phases in LDP are displaced by about 12 h compared with SDP, which seems to accord with some of the experimental data (Figs 2.6 and 2.8). The hypothesis also received support from results with the LDP dill, which was shown to flower better when a period of darkness was given during the postulated skotophile phase. This rather general model has become known as **Bünning's hypothesis**. It has, however, since been substantially modified from its original form, by both Bünning and others, and more explicit models based on known properties of the circadian system have been developed.

One problem with the original hypothesis was that many LDP did not appear to show a promotion of flowering by a period of darkness during the postulated skotophile phase (i.e. the first 12 h of the photoperiod in LDP). Consequently, the scheme for LDP was modified to propose that the difference between LDP and SDP is that light during the skotophile phase promotes flowering in the former, but inhibits flowering in SDP. Thus coincidence or non-coincidence of light with the skotophile phase was considered to be the determining factor in the induction of flowering in photoperiodically sensitive plants, with light having opposite effects in the two response groups (Bünning, 1960). Much of the later work followed the lead set by Bünning in supposing that there is a light-sensitive phase in the photoperiodic rhythm (Bünning's skotophile phase) and that the response is determined by coincidence or non-coincidence of light with this phase (Andrade and La Motte, 1984; Bollig *et al.*, 1976; Fukshansky, 1981; Thomas and Vince-Prue, 1987; Vince-Prue and Lumsden, 1987). This type of scheme is known as **external coincidence**. It assumes that there is a single photoperiodic rhythm and that light has direct effect to prevent the induction of flowering in SDP (or induce it in LDP) when it is coincident with a particular light-sensitive phase of this rhythm. Bünning's term skotophile relates to a full half-cycle of the rhythm (180°). However, it has been observed that photoperiodic induction is contingent on the illumination of a much smaller fraction of the cycle and that light does not affect induction during many parts of the presumed skotophile phase. Consequently, Pittendrigh (1966) proposed the introduction of the term **inducible phase** (φ_i), as it is more restrictive in time and explicitly distinguishes between the inducing and entraining actions of light in the photoperiodic rhythm. He postulated that 'photoperiodic induction (in the SDP *Lemna paucicostata*) is contingent on the (non-)coincidence of light and a specific inducible phase, φ_i, in the oscillation'. He also emphasised that, in such a model, light has two different kinds of effect, firstly as an entraining or rephasing agent and secondly as a photoperiodic inducer/inhibitor. As discussed for overt circadian rhythms, the entire subjective night (or skotophile half-cycle) is sensitive to light as it affects the phase control of the oscillation but the photoperiodic induction process is sensitive to light only at a particular time.

The crucial factor of any external coincidence mechanism for photoperiodic time-keeping is that induction, or non-induction, depends on the coincidence of an external signal (light) with an internal light-sensitive phase of a circadian rhythm. Much of the evidence obtained with SDP, especially those which can be induced with a single short-day/long-night cycle, is consistent with an external coincidence model of the type originally proposed by Pittendrigh. In this model, it is argued that the oscillation assumes a definite phase relationship with the light cycle so that the light-sensitive inducible phase (φ_i) is, or is not illuminated as the light/dark regime changes. Thus, according to this theory, the annual variations in daylength illuminate φ_i at some seasons but not others. The way in which the oscillation may be controlled by the light signals to achieve this result has been the subject of many experiments and is discussed in the next section with reference to SDP. A detailed discussion of timekeeping in LDP is deferred to Chapter 5.

An alternative type of scheme, called **internal coincidence**, ascribes photoperiodic responses to the interaction of two rhythms, with induction occurring only when critical phase points in the two rhythms coincide. An example would be a rhythm of enzyme activity together with a rhythm of substrate availability. In such a mechanism, light would not interact directly with a circadian rhythm. The inhibition of flowering in unfavourable cycles would then be due to the rephasing of one oscillation so that it is no longer in phase with another. In this way the critical phase points would only coincide under particular photoperiods to give rise to short- or long-day responses. This approach has not yet been explored in any detail in plants.

CIRCADIAN TIMEKEEPING IN SHORT-DAY PLANTS

Any model for circadian timekeeping in SDP must be capable of explaining how an oscillation with a 24 h periodicity can measure the critical duration of darkness which determines their photoperiodic response. A number of explicit models have been developed, based mainly on results with single-cycle SDP such as *Pharbitis*, *Xanthium* and *Chenopodium*. *Pharbitis* is a particularly useful subject for studying the behaviour of the photoperiodic rhythm because dark-grown seedlings will respond to a single light/dark cycle. This means that the relationship between photoperiod duration and dark-timekeeping can be studied without the complicating problem of entrainment to non-inductive light/dark cycles during growth of the plant to experimental size. However, this problem can be partly overcome in other plants (e.g. *Xanthium*; Papenfuss and Salisbury, 1967) by exposing them to a 'neutral', non-inductive dark period before giving the single, experimental cycle.

Timing the Night-Break

One approach to understanding how the circadian rhythm operates to control photoperiodic timekeeping has been to examine the time of maximum sensitivity to a night-break (NBmax) following a single photoperiod of varying duration. Dark-grown seedlings of *Pharbitis* will flower in response to a 5 min pulse of R followed by an inductive dark period before transfer to LL, provided that the seedlings are sprayed with benzyladenine at the time of giving the brief photoperiod (Ogawa and King, 1979a). Under these conditions, a circadian rhythm of responsiveness to a night-break occurred with the first NBmax at approximately 15 h after the initial exposure to light (King *et al.*, 1982; Lumsden *et al.*, 1982; Lee *et al.*, 1987). The time of the first NBmax varied between 10 and 18 h (in different experiments and seed lots) but the second NBmax always occurred 24 h after the first. Rhythm phasing was established by the initial exposure to R, irrespective of the age of the dark-grown seedlings at that time. When the duration of the photoperiod was varied, it was found that the first NBmax occurred at a constant time (approximately 15 h) from the beginning of the photoperiod when this was less than about 6 h long. When the photoperiod was longer than 6 h, the time of sensitivity to light was delayed and NBmax always occurred at a constant time after the end of the photoperiod (Fig. 2.11). With longer photoperiods the rhythm was gradually damped and, after 24 h, there was essentially only a single NBmax response at 25°C (Fig. 2.7).

These results with *Pharbitis* have been interpreted in the following way (Fig. 2.12).

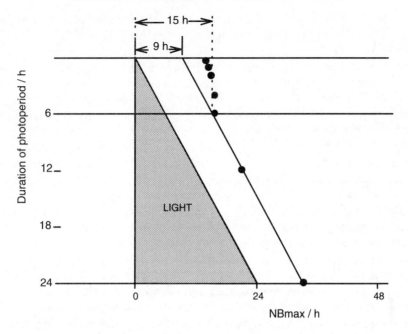

FIG. 2.11. The effect of photoperiod duration on the time of night-break sensitivity in dark-grown seedlings of *Pharbitis*. Dark-grown seedlings received a photoperiod of various durations as indicated on the figure, followed by an inductive dark period interrupted at various times by a 10 minute night-break. Plants were returned to continuous white light 72 h after the beginning of the photoperiod. The time of the maximum sensitivity to a night-break (NBmax) was constant (15 h) from light-on with photoperiods of < 6 h duration; with photoperiods of > 6 h, NBmax was constant (9 h) from light-off. After Lumsden *et al.*, 1982.

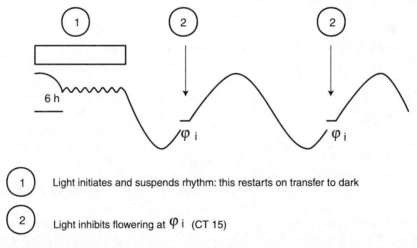

① Light initiates and suspends rhythm: this restarts on transfer to dark

② Light inhibits flowering at φ_i (CT 15)

FIG. 2.12. Scheme showing the actions of light in the photoperiodic control of flowering in *Pharbitis nil*. φ_i represents a specific light-sensitive phase of the circadian photoperiodic rhythm when light acts directly to inhibit flowering. The action of light to initiate and suspend the rhythm is considered to be on the underlying circadian oscillator to which the rhythm in light sensitivity is coupled.

A single photoperiodic rhythm of sensitivity to light is initiated by transfer to light at dawn (i.e. CT = 0 at light-on) and the light-sensitive phase (φ_i) of this rhythm occurs about 15 h after the light-on signal (i.e. at CT 15). The rhythm initially continues to run in continuous light so that, in real time, NBmax always occurs 15 h after light-on. A light-on signal is thus sufficient to induce flowering under these conditions unless a second pulse of light is given at φ_i. After about 6 h in continuous light (i.e. at CT 6), the rhythm appears to become 'suspended' and remains at CT 6 for as long as the plant stays in continuous light. It is then released by a light/dark transition. Since the rhythm is suspended at CT 6, the time of NBmax (in real time) is always about 9 h after transfer to darkness (i.e. at CT 15), as would be predicted from the original light-on rhythm. At temperatures sufficiently high to maintain growth, the daylength under natural conditions would always be longer than 6 h so that the flowering rhythm would go into suspension during the photoperiod and be released by the light/dark transition at the end of the day; this would result in NBmax always occurring at a constant time from dusk. Induction would then depend on whether or not φ_i is reached before the dawn signal is experienced.

This model for *Pharbitis* is based on results obtained with dark-grown seedlings where the first exposure to light initiates the circadian rhythm of light sensitivity. Under natural conditions, plants are repeatedly exposed to photoperiods of several hours duration and, according to this model, would have a dominant rhythm phased from the dusk signal. One can question, therefore, whether the mechanism operates under these conditions and also whether it operates in other SDP. In *Pharbitis*, although the results were slightly different from those obtained in dark-grown plants, there is evidence that the model is applicable to light-grown seedlings. Seedlings given a preliminary photoperiod of 24 h, followed by 8 h dark, showed a phase shift (advance) of the light-off rhythm following a second photoperiod from 10 min to 2 h in duration, while light for more than between 2 and 6 h resulted in the first NBmax at a constant time from the new light-off (Fig. 2.13). Earlier experiments with seedlings grown for 3–4 weeks in LL or long days produced similar results: following a 6 h exposure to R, NBmax always occurred at 9–9.5 h from light-off, whereas a 2 h exposure resulted in a phase shift (Bollig, 1977).

In *Pharbitis*, there is clear evidence of a rhythmic sensitivity to night-break light, at least when the photoperiods are short (Fig. 2.7). In *Xanthium*, in contrast, attempts to demonstrate the existence of a rhythm by giving light perturbations at different times during a long dark period have been wholly unsuccessful. Nevertheless, it seems likely that photoperiodic timekeeping in *Xanthium* operates in a way wholly comparable to that proposed for *Pharbitis*. Following a neutral, non-inductive dark period of 7.5 h, NBmax occurred at a constant time (14.0 h) from the beginning of the photoperiod when this was less than 5 h long. After longer photoperiods, NBmax occurred at a constant time (about 8.5 h) after the end of the photoperiod (Fig. 2.14). These results are consistent with the concept that a rhythm is initiated or rephased (at CT 0) by the light-on transition at dawn with a light-sensitive phase, φ_i, occurring 14.0 h later (at approximately CT 14). After 5 h in the light, this rhythm goes into suspension and the time of NBmax occurs about 8.5 h after transfer to darkness in real time (i.e. close to CT 14) as predicted from the original light-on rhythm).

The results with *Xanthium* (Fig. 2.14) were obtained when a photoperiod of varying duration was given after a dark period of 7.5 h. In other experiments, it was found that

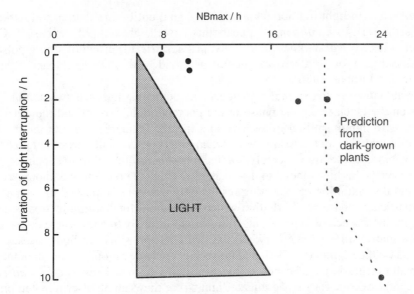

FIG. 2.13. Effects of light to phase shift the photoperiodic rhythm in light-grown plants of *Pharbitis*. Seedlings were grown in darkness and exposed to a 24 hour photoperiod before transfer to a 72 hour dark period. A second photoperiod in red light of various durations was given beginning at the 6th hour of darkness. The time of maximum sensitivity to a subsequent 10 minute night-break (NBmax) is shown. Photoperiods of 10 minutes and 40 minutes caused a phase shift in the position of NBmax, and a photoperiod of 6 hours or longer was sufficient to give a new light-off signal with NBmax occurring 8 h later. After Lumsden and Furuya, 1986.

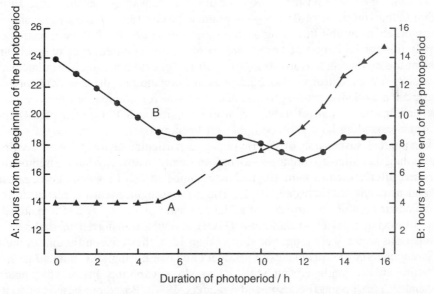

FIG. 2.14. Time of maximum sensitivity to a night-break in *Xanthium* as affected by the duration of the preceding photoperiod. Plants were given a 'phasing' dark period of 7.5 hours followed by a photoperiod of various durations. After this an inductive dark period of 14 hours was interrupted by a short night-break at various times. The figure shows the times at which this night-break was most effective (NBmax) in terms of the number of hours from (A) the beginning or (B) the end of the photoperiod. The time of NBmax was constantly related in time to light-on with photoperiods of ≤ 5 hours duration and was a constant (8.5 h) from light-off after photoperiods of ≥ 6 hours. After Papenfuss and Salisbury, 1967.

a short exposure to light did not act as a dawn signal until more than 6 h of darkness had elapsed. Thus, in *Xanthium*, it appears that dawn (light-on) will rephase to CT 0 only after at least 6 h of darkness. Similarly, in *Pharbitis*, the effect of a light pulse on the rhythm phase of light-grown seedlings varied with the time from light-off (Lumsden and Furuya, 1986).

A crucial feature of an external coincidence model is that light has two actions: one action is on the clock to set the phase of the photoperiodic rhythm and the other is a direct action to inhibit (in SDP) flowering at a specific inducible phase of the rhythm. However, in interpreting the results with *Xanthium*, it was initially proposed that light may only have one action, namely on the phasing of the clock (Papenfuss and Salisbury, 1967). In this model a night-break always acts to rephase the photoperiodic rhythm and the inhibition of flowering results only when it is rephased in such a way that the particular phase of the rhythm which is essential for floral induction is never reached. In the absence of experiments designed specifically to address the question, it is not possible to differentiate between the two concepts since, in both cases, dawn would rephase the rhythm (to CT 0) and flowering would occur only when an inducible phase for flowering is reached before the next dawn occurs. However, in *Pharbitis*, there is good evidence for two actions of light, since they can be discriminated on the basis of their dose–response characteristics (Lumsden and Furuya, 1986). For example, at the 8th hour of an inductive dark period following transfer from 24 h LL (when light is strongly inhibitory to flowering), a marked reduction of flowering was obtained with an exposure which had no effect on the phase of the night-break rhythm (Fig. 2.15). At other times, for example at the 6th hour, the rhythm could be phase shifted with no effect on flowering. Using a different protocol with a brief photoperiod, a considerably greater exposure of R (312 μmol m^{-2}) was needed to phase shift the rhythm at the 16th hour of darkness, than the exposure (25–30 μmol m^{-2}) shown by other workers to inhibit flowering at that time (Lee *et al.*, 1987). These results argue strongly that the inhibition of flowering by light is not a consequence of a shift in the phase of the rhythm and afford strong support for an external coincidence mechanism for photoperiodic control in *Pharbitis*. It is unfortunate that there is not more direct experimental evidence relating to this question, which has important implications for understanding the actions of light to control flowering in SDP (see Chapter 4) as well as for understanding how photoperiodic timekeeping operates.

The apparent suspension of the rhythm at a particular phase point in continuous light is not peculiar to photoperiodism. In many overt rhythms, rhythmicity is abolished after about 12 h in LL and is resumed at CT 12 when the organism is returned to darkness (Lumsden, 1991). The most straightforward explanation is that the circadian oscillator is arrested at a characteristic phase in LL and is restarted (or reset) following transfer to darkness. However, when transferred to darkness after photoperiods which were longer or shorter than 12 h, the eclosion rhythm in the flesh fly *Sarcophaga agyrostoma* showed residual circadian fluctuations, indicating that the oscillation in LL continued but with only a slight variation around the apparently 'suspended' phase point (Peterson and Saunders, 1980). Based on these results, it was suggested that the apparent suspension in continuous light may result from a change in the dynamics of the pacemaker such that it oscillates within a much reduced cycle (**a light-limit cycle**), effectively occupying a small time domain. Following transfer to darkness, the circadian oscillator moves to a **dark-limit cycle** taking essentially the

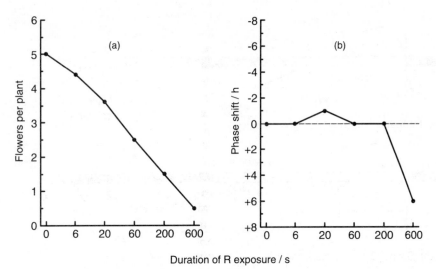

FIG. 2.15. Dose–response curves for the action of light to inhibit flowering and to phase shift the night-break rhythm in *Pharbitis*. Dark-grown seedlings were given a 24 hour photoperiod followed by a 72 hour dark period. Various durations of light as shown on the abscissa were given 8 hours after the beginning of the dark period. (a) shows the direct inhibition of flowering and (b) the magnitude of the phase shift in NBmax obtained by scanning the subsequent dark period with a second light pulse. After 8 hours in darkness a strong inhibition of flowering occurred with light exposures of 200 s, which had no effect on the phase of the rhythm. After Lumsden and Furuya, 1986.

same time to reach it regardless of the starting point in the light-limit cycle. The rhythm would then appear to resume from a more or less constant phase point irrespective of the duration of the preceding photoperiod.

The fact that, in *Xanthium*, the time to NBmax is not precisely constant but varies between 7 and 8.5 h (Fig. 2.14) can be interpreted in terms of the light-limit cycle concept, with a small oscillation such that the time taken to reach the dark-limit cycle varies slightly according to the position in the light-limit cycle when the plant is transferred to darkness. However, the results have been interpreted differently by Salisbury (Fig. 2.16), who proposed that the clock begins to oscillate into the 'night-mode' after about 9 h in continuous light, going as far as it can after about 12 h. If plants are placed in darkness at this time, the clock would continue to oscillate and NBmax would occur earlier (after about 7 h). However, if the plants are left in the light, the clock is forced back into the suspended 'day mode', with dusk required to restart the oscillation and NBmax occurring 8.5 h later. It is not clear, however, why more than 12 h of light should be required to suspend the clock; it appears already to be suspended after 5 h in the light since NBmax is a constant time (8.5 h) from light-off after a 5 h photoperiod (Fig. 2.14). Overall, the results with *Xanthium* seem to be more consistent with the model presented for *Pharbitis*, with a light-on oscillation which requires about 5 h of light to move into a light-limit cycle; the time taken to move into a dark-limit cycle will vary slightly according to the position in the light-limit cycle that has been reached when plants are transferred to darkness, leading to a small residual oscillation in the time to NBmax, as observed in Fig. 2.14. In *Pharbitis*, the time to NBmax was essentially constant after a 6 h photoperiod with no clear evidence

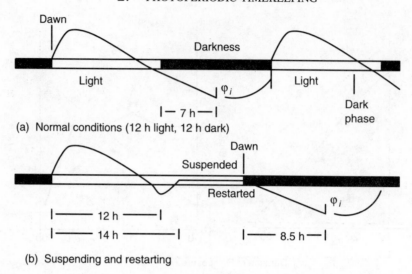

FIG. 2.16. A model for photoperiodic timekeeping in *Xanthium*. It is asssumed that the rhythm is started by transfer to light and begins to oscillate into the 'dark' phase after about 9 hours. If plants are transferred to darkness after 12 hours in the light (a), the clock continues to oscillate. Consequently the phase point which allows flowering (φ_i) is reached after about 7 h (based on the data of Fig. 2.17 for the time of NBmax). After 12–14 hours in the light, the clock is forced into a state of suspension (b); the oscillation is then restarted by transfer to darkness and φ_i is reached 8.5 hours later. Based on Salisbury, 1990, and data of Fig. 2.14.

for an oscillation during the photoperiod. However, a small oscillation in flower number was revealed when seedlings were transferred from continuous light at different times after sowing and tested with a dark period of constant duration, which was only slightly longer than the CNL (Spector and Paraska, 1973): it was thought that the initial zeitgeber in this case was the time of seedling emergence into LL. If the NBmax and CNL are timed in the same way (see below), these results could also be interpreted in terms of a light-limit cycle such that the CNL is influenced by the position in the light-limit cycle at the time of transfer to darkness.

Timing the Critical Nightlength

The model presented above proposes that the critical nightlength is measured by the time taken to reach a light-sensitive, inducible phase of a photoperiodic rhythm which is itself controlled by light. Since this has been based on studies of the time of NBmax, it can be criticised on the grounds that a night-break is an artificial treatment which is never encountered in nature where, as we have seen, a critical duration of darkness is the primary determinant of flowering in SDP. Consequently, direct studies of critical nightlength are essential in order to determine if this behaves in the same way as NBmax. In dark-grown seedlings of *Pharbitis*, this question has been addressed by varying the duration of both the photoperiod and the (uninterrupted) dark period. When the photoperiod was no more than 8 h, the flowering response (number of flower buds) showed a sharp increase at approximately 20 h after the beginning of the photoperiod with a second sharp increase about 24 h later (Saji *et al.*, 1984). These results confirmed earlier experiments with light-grown seedlings (Takimoto and

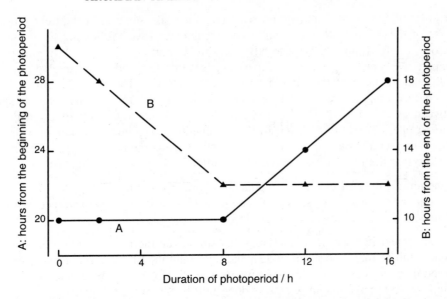

FIG. 2.17. The effect of photoperiod duration on the critical nightlength in dark-grown seedlings of *Pharbitis nil*. Dark-grown seedlings received a photoperiod of various durations as indicated on the figure, followed by a dark period also of various durations before transfer to continuous light. The figure shows the critical nightlength in terms of the number of hours from (A) the beginning or (B) the end of the preceding photoperiod. The critical nightlength was constantly related in time to light-on with photoperiods of ≤ 8 hours duration and was a constant (12 h) from light-off after photoperiods of ≥ 8 hours. After Saji *et al.*, 1984.

Hamner, 1964; Fig. 2.9). After longer photoperiods, the rhythmic response to dark duration largely disappeared as was observed with the NBmax rhythm. When the photoperiod was no more than 8 h long, the sum of the light period and the critical nightlength (as measured by the first stepwise increase in flower number) was constant at about 20 h (Fig. 2.17) suggesting that, as with the night-break sensitivity response, the rhythm began with a light-on signal and continued to run in the light. After photoperiods of 8 h or more, the CNL was a constant (12 h) from the end of the light period. Similar results were obtained independently by Lumsden, who also observed that the rhythm in response to an increasing duration of darkness disappeared after a photoperiod of several hours (Vince-Prue and Lumsden, 1987). However, in this case, the rhythm began from light-on with photoperiods up to 6 h but, between 6 and 18 h, the CNL showed no constant relationship with either light-on or light-off. However, after photoperiods of 18 h, the CNL was constant (about 9 h) from the end of the light period and coincided with the time of NBmax. Therefore, the CNL in *Pharbitis* also appears to be controlled by the rhythm of light sensitivity that gives rise to the night-break response, with timekeeping running from light-on or light-off depending on the duration of the photoperiod. The fact that both rhythms are damped with increasing duration of the photoperiod also suggests that the same rhythm controls both NBmax and CNL.

Although NBmax and CNL appear to be manifestations of the same circadian rhythm, they do not always coincide. Following short photoperiods, the CNL in *Pharbitis* was somewhat longer than the time of NBmax (Fig. 1.12) indicating that, for floral induction, additional dark reactions must take place after the light-sensitive phase of the rhythm (NBmax) has been reached. However, following longer photoperiods, the CNL and the time to NBmax were essentially the same. Experiments on temperature sensitivity during darkness suggest that the CNL may involve reactions additional to those of the rhythm which times NBmax. The time of NBmax is not affected by temperature in *Xanthium* and this also applies to the CNL at temperatures between 15 and 30°C (Fig. 2.18). However, there appears to be a two-phase response; the first phase is relatively insensitive to temperature leading to a common value for CNL, whereas the second phase is much more temperature dependent. The results indicate that the response to increasing nightlength has two components:

• the temperature-insensitive component is concerned with timing
• a second, temperature-sensitive, component appears to be concerned with the magnitude of the flowering response

In this context it is worth emphasising that studies on the magnitude of the flowering response alone may lead to misunderstandings about the timekeeping mechanism; it is essential, always, to include experiments specifically to examine the timekeeping component(s) of the response as was done in this experiment.

In *Pharbitis*, the time of NBmax is unaffected by temperature but, in contrast, the CNL is markedly temperature dependent (Fig. 2.18). From the *Xanthium* results, it is possible that the temperature-dependent component of the CNL in *Pharbitis* merely represents the time required to achieve a measurable flowering response after a common temperature-insensitive timing point has been reached. The correspondence

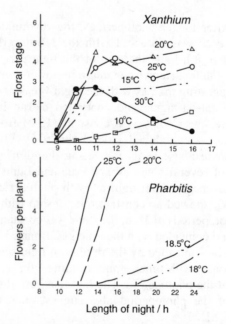

FIG. 2.18. Effects of temperature and the duration of the dark period on flowering in *Xanthium strumarium* and *Pharbitis nil*. Plants were exposed to the indicated temperatures during a single dark period of different durations as shown on the abscissae. After Thomas and Vince-Prue, 1984 (Adapted from Salisbury and Ross, 1969).

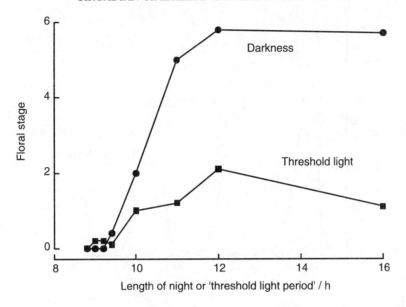

FIG. 2.19. The flowering response of *Xanthium strumarium* to the duration of darkness of low intensity light. The 'night' of various durations in a 24 h cycle was given either in darkness or at 0.1 W m^{-2} from tungsten-filament lamps (threshold light). After Salisbury, 1963a.

between the CNL and NBmax under other conditions, together with the rhythmic nature of both processes and their apparently similar relationships with photoperiod duration, make it likely that the timekeeping component of both is the same circadian rhythm which, under natural conditions, is released at the transition to darkness. However, the nature of the additional light-sensitive reactions which appear to be involved in the induction of flowering await elucidation. That these reactions can be distinguished from the measurement of the critical night length is evident from the fact that the magnitude of the flowering response in *Xanthium* was strongly inhibited by light at intensities below the threshold which allowed normal dark timekeeping to proceed (Fig. 2.19).

In contrast to *Pharbitis* and *Xanthium*, *Chenopodium rubrum* shows a marked rhythmic response to the duration of a single dark period inserted into LL, which largely parallels a rhythm in the response to a brief night-break (Fig. 2.20). Only the nightlength rhythm has been studied with respect to its phase control (King and Cumming, 1972a). Following transfer to darkness from LL, the light-off rhythm can be rephased by a relatively short exposure to light and is reset to a new light-off signal by exposures of 12 h or more. With 6 h of light a typical PRC is obtained, with advance or delay of the rhythm depending on when the light is given (Fig. 2.5). Rhythm rephasing is rather insensitive to light requiring an exposure of more than 2.5 h whereas, in *Pharbitis*, only a few seconds may be sufficient. With an exposure of more than 12 h, the *Chenopodium* rhythm was reset to give a new light-off signal and φ_i (as determined by maximum flowering response) occurred after 13 h darkness. The second circadian peak in the flowering response was also displaced to the same extent. Thus, as the duration of the photoperiod increases, there is a change in effectiveness of

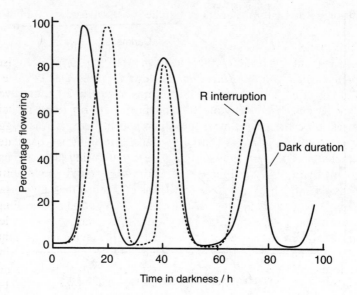

FIG. 2.20.   Rhythmic flowering response of *Chenopodium rubrum* to variations in dark duration and time of giving a night-break. Plants were given a dark period varying from 3 to 96 h or 72 h dark period interrrupted at different times by 4 min of red light. Plants were maintained in continuous light from tungsten-filament lamps (47 W m^{-2}) before and after the experimental dark period. After Cumming *et al.*, 1965.

dawn and dusk signals for the control of rhythm phase, with the rhythm being apparently suspended at a fixed phase with long photoperiods and subsequently being phased from the dusk signal. It is also evident that with photoperiods of between 15 and 30 h, there is a residual fluctuation in the first peak of the flowering rhythm (King and Cumming, 1972a), which varies between 12 and 15 h from light-off, as would be expected if the rhythm enters a light-limit cycle during long photoperiods (Peterson and Saunders, 1980). Thus photoperiodic timing in the SDP *Chenopodium* strongly resembles that in *Pharbitis* and *Xanthium*.

Is There a Common Mechanism for Timekeeping in Short-Day Plants?

Based largely on results obtained with *Pharbitis,* a model is proposed which assumes a single photoperiodic rhythm, the essential characteristics of which are that it runs from a light-on signal, is 'suspended' after several hours in continuous light and is then subsequently released, or rephased, at the transition to darkness. The CNL is then measured by the time from dusk to the inducible phase, φ_i, of the rhythm. A light-on signal alone (i.e. an exposure of 1 h or less) is sufficient to induce flowering under some conditions, provided that a second pulse of light is not given at φ_i (King *et al.*, 1982; Oota, 1983a,b). However, once the duration of the daily light period exceeds a certain value (from 5–12 h depending on the plant) a certain fixed phase is achieved and timing is then coupled to the light-off signal at dusk. Thus the characteristics of the system lead to the measurement of a critical nightlength under normal conditions of growth, although the rhythm is actually initiated (or rephased) by a light-on (dawn) signal.

Although this model for timekeeping in SDP is consistent with results obtained with other single-cycle plants such as *Xanthium* and *Chenopodium*, problems arise for some multicycle plants. For example, *Kalanchoë* (Schwabe, 1969) and *Glycine* (Hamner, 1969) failed to flower with photoperiods longer than about 16–20 h, however long the associated dark period. Thus a rhythm running from light-off to an inducible phase does not appear to be the only timekeeping component here. It was suggested, for *Glycine*, that a circadian rhythm is started by transfer to light and oscillates through two inducible phases (Hamner, 1969). Light is required during the first phase, which reaches a maximum after 8–10 h light, while maximum sensitivity to inhibition during the second phase occurs 16 h after light-on. It was concluded that, under natural 24 h cycles, long days prevent flowering because the photoperiod is too long and not because the extension of the photoperiod shortens the dark period to less than a critical value. However, this goes against much of the evidence presented above which strongly indicates that, following a sufficiently long photoperiod, timekeeping in SDP runs from the dusk signal. It is, therefore, important to re-examine more carefully how timing is controlled in these two plants before the light-on model proposed for *Glycine* (which is very close to Bünning's original hypothesis) is accepted. The failure to flower with more than 16–20 h light could be explained if the rhythm entered a refractory period after this time so that it was not restarted by the dusk signal.

Few other SDP have been examined with respect to the effect of photoperiod duration on timekeeping. Departures from the model proposed for *Pharbitis* have been reported, but the reasons for this have not always been clear. In *Lemna pauci-costata*, timekeeping both in terms of NBmax and nightlength was coupled to the light-off signal when plants were transferred from continuous light (LL) to darkness, with NBmax occurring after 7 h (strain 441) or 8 h (strain 6746) and a CNL of 7 h (441) or 12 h (6746). Both responses exhibited a circadian rhythm (Hayashi *et al.*, 1992). However, the timing of the NBmax response was modified when a non-inductive dark period preceded the photoperiod (similar to the 'phasing' dark-period used in the *Xanthium* experiment shown in Fig. 2.14). Under these conditions, NBmax did not occur at a constant time from light-off if photoperiods of 3 or 4 h are compared with photoperiods of 9 or 12 h (Fig. 2.21). Based on a whole series of experiments with strain 6746, it was concluded that with short photoperiods NBmax occurs at a constant time (about 14 h) from light-on (Oota, 1983a); this is not inconsistent with the results for strain 441, where NBmax occurred about 16 h after light-on following a 4 h photoperiod (Fig. 2.21). Since timekeeping clearly runs from light-off when transferred from LL, it seems that *Lemna* may also respond in a manner similar to *Pharbitis* with a rhythm initiated by a light-on signal and an inducible phase after about 14 h. Such a rhythm would have to be 'suspended' after 6–7 h in LL (i.e. at CT 6–7) in order for NBmax to occur 7–8 h from light-off, but the evidence for this is lacking (Oota, 1985). However, in earlier experiments, the maximum inhibition of flowering occurred at a constant time (9 h) from light-off, following either a 7 h or 10 h photoperiod (Purves, 1961) suggesting that, as predicted, rhythm 'suspension' may already have occurred after 7 h in continuous light. Hillman (1969b) also concluded that only 4–6 h light was needed in order to give a new light-off signal in strain 6746.

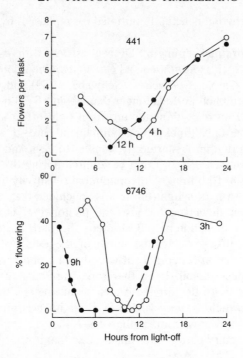

FIG. 2.21. Effect of photoperiod duration on the time of responsivity to a night-break in *Lemna paucicostata*. In strain 441 (upper figure) plants from stock cultures in continuous light were exposed to 8 hours dark followed by a photoperiod of 4 or 12 hours as indicated; they were then given a single 48 hour inductive dark period interrupted by a single 15 minute night-break at various times. In strain 6746 (lower figure) plants from long-days (15 h light/9 h dark) were given 5 cycles of 3 h light/21 h dark or 9 h light/15 h dark, with a 15 minute night-break during the dark period (a 1 or 2 hour night-break was given at certain times). In both cases, the flowering response is plotted from the beginning of the inductive dark period. Data of Hayashi *et al.*, 1992 (strain 441) and Oota, 1983a (strain 6746).

Thus many of the results with *Lemna* are not inconsistent with the model developed for *Pharbitis*, although other interpretations have been proposed. For example, it has been suggested that different clocks function under different conditions of entrainment (Oota, 1985); it seems more likely, however, that the entrainment conditions may modify the expression of the coupled rhythm.

Timing from light-on with 8 h and 11 h photoperiods has been claimed for the commercially important florists' chrysanthemum (*Dendranthema*) and in *Begonia* (Rünger and Patzer, 1986). However, other experiments with chrysanthemum suggest that it may also respond to a photoperiod duration in a manner similar to *Pharbitis*. When a 60 s night-break was given, induction occurred if the CNL was achieved before the night-break but not when a longer than critical dark period followed the night-break. In contrast, a dark period longer than the critical value was inductive both before and after a 4 h night-break indicating that a 4 h photoperiod may have been sufficient to reset the clock and lead to normal light-off timing (Horridge and Cockshull, 1989).

More than one NB$_{max}$ in a single cycle has been observed in some SDP. In *Begonia boweri*, this was attributed to two separate rhythms phased by dawn and dusk (Krebs and Zimmer, 1984). However, it is also possible that there are two actions of light on a single rhythm. The first NBmax early in the dark period may result from phase shifting the rhythm (delay) so that φ_i no longer occurs before dawn, while the second NBmax may result from the action of light at φ_i (Saunders, 1981). A similar conclusion was reached for the SDP *Lemna paucicostata* 6746, where the phasing of the assumed oscillator was deduced by extrapolation to determine whether particular test cycles (which preceded several inductive SD) were inductive or non-inductive (Oota, 1984). The existence of two NBmax may, therefore, be considered as evidence in favour both of a single rhythm and of an external coincidence model for timekeeping.

Bünning's original proposal for photoperiodic timekeeping proposed that light in the photophile phase promoted flowering, whereas more recent external coincidence models for SDP have assumed that the only direct effect of light is to inhibit flowering at φ_i. There are certainly many observations that light given during the subjective day phase can promote flowering compared with control plants maintained in darkness, but this does not necessarily imply that the rhythm goes through two inducible phases, with light either promoting or inhibiting flowering, as suggested for *Glycine* (Hamner, 1969). The promoting effect of light is much stronger when the exposures are relatively long; in *Glycine*, for example, there was no promotion above the control value with light exposures of 30 min duration, whereas exposures of 2 and 4 h increased flowering considerably (Fig. 2.22). It is likely that the long light period

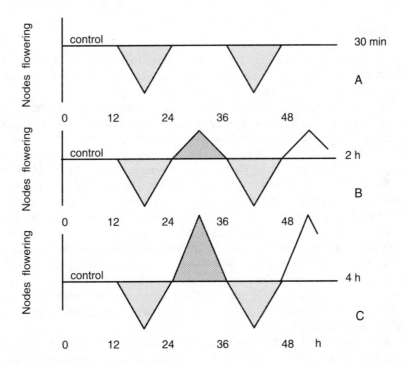

FIG. 2.22. Schematic representation of the rhythmic flowering responses of *Glycine max* Biloxi to light exposures of different durations. Light exposures of 30 minutes (A) inhibited flowering at circadian intervals but did not increase flowering above the level of the uninterrupted control treatment. Exposures of 2 h (B) or 4 h (C) also inhibited flowering at the same times as in (A); when given at other times these longer exposures to light promoted flowering above the control value (represented by the horizontal line). After Carpenter and Hamner, 1964.

acts as a second photoperiod to generate a new light-off rhythm which is in phase with the original light-off rhythm and so reinforces it. A photosynthetic component to increase the magnitude of flowering could also be involved. This cannot be the explanation in other cases, however, where flowering is promoted above the control value by exposures to light of only a few minutes duration (see Fig. 4.4). A possible explanation for this promotion response lies in the demonstrated requirement for a Pfr-requiring reaction in the induction of flowering in SDP. This reaction (which is discussed in detail in Chapter 4) has been shown to be non-rhythmic and is apparently not involved in timekeeping. When the reaction has not been completed, a short exposure to R during a long dark period would be expected to promote flowering except at the inducible, light-sensitive phase of the photoperiod, as has been found in *Pharbitis* (Saji *et al.*, 1983). The observed effect would be an apparent rhythm in promotion by R, which actually results from the circadian rhythm in the inducible phase during which R inhibits flowering. The broad windows of R promotion over a 60 h period in *Chenopodium* (Fig. 4.4) are consistent with this interpretation and it is clear that the night-break (inducible phase) rhythm is much more precisely defined. Thus, the single inducible phase model for photoperiodic timekeeping in SDP is not inconsistent with the observed promotion of flowering under some conditions by light given during part of the daily cycle.

SEMIDIAN RHYTHMS

Not all endogenous rhythms are circadian. Higher frequency **ultradian** rhythms have been demonstrated in several organisms, with frequencies which range from a few seconds to several hours. Of these, the only ones that have been claimed to affect photoperiodic timekeeping have a periodicity of approximately 12 h – a half cycle of a circadian rhythm. Such **semidian** rhythms have been observed in some properties of the photoperiodic photoreceptor, phytochrome (King *et al.*, 1982). A series of experiments on *Pharbitis* has also revealed a semidian periodicity in the flowering response to a near CNL when the preceding photoperiod was interrupted with either 90 min FR or 5 min FR followed by 85 min dark (Heide *et al.*, 1986b); the effect of these two treatments was substantially the same and they are referred to here as FRD. When treatments began 2–4 h before the end of the photoperiod, FRD strongly promoted flowering when the dark period was close to the critical; this promotion was shown to result from an acceleration of dark timing, such that NBmax occurred earlier, i.e. φ_i occurred before dawn (Heide *et al.*, 1988). Increasingly earlier treatment with FRD revealed a 12 h periodicity in the flowering response (Fig. 2.23) and the rhythm was shown to be phased from the beginning of the FRD treatment. From these and other similar experiments, it was concluded that photoperiodic timekeeping in *Pharbitis* was controlled by this semidian rhythm.

It was suggested that the circadian rhythmicity in NBmax evident in other experiments (e.g. King *et al.*, 1982) was due to the fact that the first NBmax is controlled by the semidian rhythm and is altered by FRD given during the photoperiod, while the second and subsequent inhibition points result from a circadian, i.e. different, rhythm (Heide *et al.*, 1988). If this is so, then giving FRD during the photoperiod would be expected to influence only the first NBmax whereas, if the NB response is a function of a single oscillator, FRD would affect both the first and second cycles. It was shown in

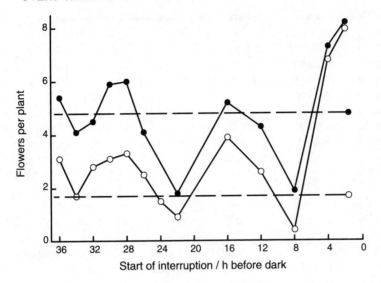

FIG. 2.23. Flowering responses of *Pharbitis nil* to an interruption with far red light during the photoperiod. Dark-grown seedlings were given 90 minutes far red light at various times during a 72 hour photoperiod in white fluorescent light. They were then given an inductive dark period of 11.3 (○) or 12.3 (●) hours. The abscissa shows the number of hours between the beginning of the far red interruption and the end of the photoperiod. The dashed lines show the uninterrupted control values. After Heide *et al.*, 1986b.

later experiments that a FRD treatment beginning 4 h before the end of a 24 h photoperiod rephased both the first and second NBmax to the same degree (Fig. 2.24). It seems likely, therefore, that the rhythm of NBmax in *Pharbitis* is controlled by a single circadian oscillator as originally proposed. The large and consistent promotion of flowering observed by Heide and co-workers with FRD beginning about 4 h or less before the end of the photoperiod (Fig. 2.23) can be explained by an advance in the phase of this circadian rhythm (Lumsden *et al.*, 1995). Although they do not appear to have a role in dark timekeeping, the effects of the earlier FRD treatments are clearly interesting and may be important in relation to the mechanisms involved in induction (oscillator and/or phytochrome) which are operating during the photoperiod.

OVERT CIRCADIAN RHYTHMS AND THE PHOTOPERIODIC CLOCK

The difficulties inherent in studying the operation of the photoperiodic clock are fairly obvious. It is necessary to give complex treatments over one or several cycles and observe effects on flowering often several weeks later. The induction mechanism operates in the leaves but translation into a flowering response requires the information to be relayed to the apical meristems. Thus several events must take place after the photoperiodic clock has led to a switch between induction and non-induction in the leaf and any one of these may be influenced by postinductive conditions. Because of these difficulties, several attempts have been made for both LDP and SDP to identify

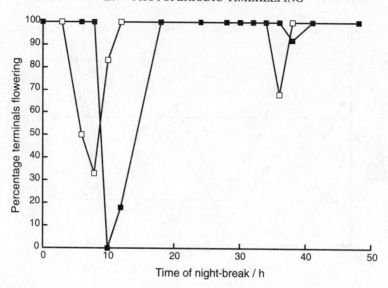

FIG. 2.24.   The effect of an interruption with far red light during the photoperiod on the phase of the night-break rhythm in *Pharbitis nil*. Dark-grown seedlings were given a photoperiod of 24 hours followed by an inductive dark period of 48 hours, scanned by a 10 minute red night-break. When a 90 minute exposure to far red light was given beginning 2 hours before the end of the photoperiod (■), both maxima of the night-break inhibition response (NBmax) were displaced to the same amount compared with the uninterrupted controls (□). After Lumsden *et al.*, 1994.

an overt circadian rhythm that is coupled to the photoperiodic clock and whose entrainment behaviour can be studied directly.

In *Lemna paucicostata* 6746, the phasing of a rhythm in carbon dioxide output was similar to that of NBmax, in relation to light-on and light-off signals (Hillman, 1976). However, this was true only when grown on ammonia or nitrate; on aspartate, there were differences between the two rhythms. Moreover, in other strains of *Lemna*, the phasing patterns of the carbon dioxide rhythm did not reflect photoperiodic timing on any nitrogen source. Coupled rhythms of leaf movement and flowering response have been proposed for several species, including *Coleus* (Halaban, 1968) and *Glycine* (Brest *et al.*, 1971). However, in *Coleus frederici*, the leaf movement rhythm appeared to be coupled to a light-on signal, whereas, after photoperiods of 4–12 h, NBmax for the inhibition of flowering showed a more constant relationship with dusk (9–12 h) than with dawn (14–21 h). The fixed temporal relationship between the minimum leaf position and the inductive phase for flowering was based on the time at which light began to inhibit flowering, rather than on the time of NBmax. The broader zone of light inhibition which occurred after some photoperiods accounts for the discrepancy. In the most studied SDP, *Pharbitis*, *Chenopodium* and *Xanthium*, it is clear that the leaf movement rhythm and photoperiodic responses are not entrained in the same way (Salisbury, 1985).

To date, no overt rhythm has been unequivocally shown to be coupled to the photoperiodic clock under a wide variety of conditions and so their use in studying the behaviour of the photoperiodic rhythm is limited. If, as discussed below, each rhythm is coupled through its own 'slave' oscillator to a master clock, or pacemaker, comparative studies of other rhythms are unlikely to assist in understanding the specific behaviour of the oscillator which controls photoperiodic responses. However, where the two rhythms appear to be closely coupled, a study of the behaviour of the overt rhythm can be a useful preliminary approach to the photoperiodic rhythm. Nevertheless, conclusions about the behaviour of the latter must be based, finally, on direct studies.

CONCLUSIONS

The precise mechanism for photoperiodic timekeeping is still not entirely understood, although it is now widely accepted for both plants and animals that it is based on a circadian oscillator. Apparent hourglass timing would result if the oscillator is very strongly damped and ceases to oscillate in darkness after a single cycle; thus hourglass clocks may only be very strongly damped circadian oscillators (Vaz Nunes *et al.*, 1991). Both hourglass and circadian responses could thus be generated by varying the degree of damping of the oscillator clock. The involvement of a semidian (12 h) rhythm has also been proposed for plants, although it is difficult to see how this might operate under natural conditions since expression of the rhythm requires exposure to 90 min FR light, or FR/dark, during the photoperiod, a situation which would not be expected to occur under natural conditions.

A model for circadian timekeeping in *Pharbitis* proposes that the underlying time-keeping process in SDP is the circadian interval between a light-on signal (CT 0) and an inducible light-sensitive phase of the rhythm, φ_i, which (in *Pharbitis*) occurs at CT 15. A similar basis for photoperiodic time measurement has also been proposed for the SDP *Lemna paucicostata* where, as in *Pharbitis*, flowering is prevented when a light-on signal which starts the rhythm is followed after 14 h by a second light pulse (Oota, 1983b). The model developed for *Pharbitis* further assumes that following photoperiods of several hours duration, the clock is 'suspended'. The rhythm is then set at a constant phase point at the transition to darkness and the photoperiodic response of the plant (flowering or non-flowering) depends on whether φ_i (which now occurs at a fixed number of hours after the onset of darkness) is reached before the next exposure to light (dawn) occurs. In this way the mechanism measures a critical nightlength which, if exceeded, results in floral induction. Timekeeping would thus appear to run from either a dawn or a dusk signal depending on conditions, although only a single rhythm is involved. The light-on signal at dawn then rephases the rhythm for the next cycle. This model explains the function of the apparently 'unnatural' night-break response in SDP and shows how the circadian timekeeping mechanism could operate under natural day/night cycles.

In several other SDP (e.g. *Xanthium* and *Chenopodium*), photoperiodic timekeeping appears to operate in a similar manner to that proposed for *Pharbitis*. However, there are others (such as *Glycine* and *Kalanchoë*, where flowering does not occur with very long photoperiods) which do not conform to this pattern.

The model for timekeeping in SDP proposed in this chapter is based on the concept that a circadian rhythm of sensitivity to light is coupled to a a single circadian oscillator (or pacemaker) which is entrainable by light. Such a single-oscillator model is able to simulate several known features of circadian rhythms such as, for example, damping in continuous bright light (Vaz Nunes *et al.*, 1991). It may, however, be an oversimplified representation. It has been suggested that each circadian rhythm may be directly controlled by a 'slave' oscillator, which, in turn, is coupled to a master clock, or pacemaker (Johnson and Hastings, 1986). Such a pacemaker/slave system would explain why rhythms may apparently be correlated in some experimental conditions but not in others since, if the master clock is reset, one slave oscillator may take longer to adjust than another. A model based on the pacemaker/slave concept has been developed for the photoperiodic clock in insects and mites (Vaz Nunes *et al.*, 1991). It assumes that the measurement of nightlength is performed by the slave oscillator and, by varying values such as the coupling strength between the two oscillators and their light sensitivity, model-generated response curves were able to simulate several of the photoperiodic features that have been observed in insects and mites. A number of these features are also found in plants, including variations in the number of night-break maxima and cycle-length (resonance) effects. However, it remains to be determined whether the clock in plant photoperiodism is a single oscillator as proposed here, or some form of multioscillator system such as the pacemaker/slave.

3 Photoperiodic Photoreceptors

Although critical studies of the photoperiodic pigment awaited the discovery that flowering could be controlled in certain plants by a brief exposure to light given during a long night, earlier experiments had already revealed similarities in the responses to light of LDP and SDP and indicated the importance of the red part of the visible spectrum. One of the earliest investigators was Rasumov (1933) who concluded that LD responses occurred when the red and yellow parts of the spectrum were used to extend short natural photoperiods, whereas green, blue or violet light were ineffective. Withrow and Benedict (1936) found that earliest flowering occurred in the LDP *Viola tricolor*, *Matthiola incana* and *Callistephus chinensis* when natural SD were extended with orange or red light; blue and green light had little effect, except on *Callistephus*, which is now known to be particularly sensitive to light and was found to be responsive to all wavebands at the irradiances used. When these investigations were extended to SDP, the red waveband which was most effective in promoting flowering in LDP was also most effective in preventing flowering in SDP (Withrow and Biebel, 1936; Withrow and Withrow, 1940).

DISCOVERY OF PHYTOCHROME AND ITS FUNCTION IN PHOTOPERIODISM

The history of phytochrome research, particularly in its early years, was intimately intertwined with the history of photoperiodism. Inevitably this meant that concepts of phytochrome function in photoperiodism were established against an incomplete understanding of the complexity of the phytochrome system. Many of the early ideas still form the core of explanations of photoperiodic mechanisms as presented in teaching and research texts. For these reasons it is useful to consider the properties of phytochrome and its function in photoperiodism in an historical context.

The first true action spectra for photoperiodism were determined at the US Department of Agriculture laboratories in Beltsville, Maryland between 1946 and 1950. The spectrograph used in these experiments was designed to produce a highly dispersed spectrum of high intensity light; single leaves, parts of leaves, or leaflets were irradiated and usually subtended a waveband of not more than 15 nm. The action spectra for flowering obtained using this spectrograph (Parker *et al.*, 1946, 1950;

63

Borthwick *et al.*, 1948; Nakayama *et al.*, 1960; Kasperbauer *et al.*, 1963b) remained the most detailed available until the construction in the early 1980s of the large spectrograph facility at Okazaki (Watanabe *et al.*, 1982) which has subsequently been used on several occasions to obtain action spectra for particular photoperiodic responses (Lumsden *et al.*, 1987).

The first action spectra were constructed for night-break treatments given near the middle of a long dark period with various durations and intensities of light. The durations were kept as short as possible (usually not more than 25 min) and it was established that reciprocity held over the range used, i.e. the response was proportional to the total photon exposure. The effect on the initiation of floral primordia was determined by dissecting plants a few days after the experimental treatments had been given. The action spectra for the prevention of flowering in the SDP *Glycine max* cv Biloxi and *Xanthium strumarium* (Parker *et al.*, 1946) and for the promotion of flowering in the LDP *Hordeum vulgare* (Borthwick *et al.*, 1948) and *Hyoscyamus niger* (Parker *et al.*, 1950) were remarkably similar, particularly in the red wavelengths (Fig. 3.1). The cutoff at wavelengths longer than 720 nm, the position of maximum effect in the red between 600 and 660 nm, the wavelength of rapid change in sensitivity between 500 and 560 nm and the position of minimum effect near 480 nm coincided for all four plants, indicating that the same pigment was responsible. Furthermore, the absolute amount of energy needed to effect a response in the region of maximum sensitivity was nearly the same in all cases. The relative effectiveness of blue light, however, varied considerably; if red is set at 1.0, the relative amount of energy needed in blue to elicit the same response was 20 for *Glycine*, 150 for *Xanthium* and 250 for *Hordeum*. Action spectra for other light-dependent growth responses were determined by the same team using the Beltsville spectrograph. The action spectra for the promotion of leaf growth in etiolated pea seedlings (Parker *et al.*,

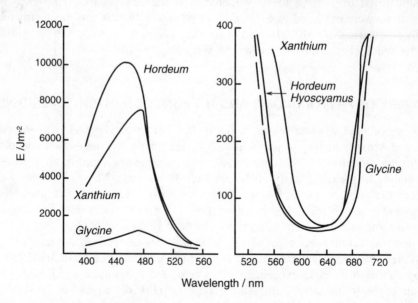

FIG. 3.1. Action spectra for night-break effects in LDP and SDP. After Vince-Prue, 1975 (Data from Borthwick *et al.*, 1948, and Parker *et al.*, 1946, 1950).

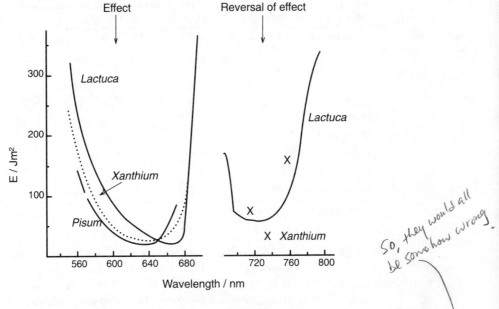

FIG. 3.2. Action spectra for phytochrome-mediated responses. Effects measured: *Xanthium*, 50% floral initiation; *Lactuca*, 50% germination; *Pisum*, promotion of leaf growth by 45%. After Hendricks and Siegelman, 1967.

1949) for the inhibition of stem growth in dark-grown seedlings of *Hordeum* (Borthwick *et al.*, 1951) and for the promotion of germination of light-sensitive lettuce seeds cv Grand Rapids (Borthwick *et al.*, 1952b) were found to be essentially the same as those for the photoperiodic night-break response. The similarity of the action spectra (Fig. 3.2) indicated that the same pigment was the primary photo-receptor in each case.

Earlier experiments of Flint and McAlister (1935) had shown that the germination of lettuce seeds was suppressed by light of wavelengths longer than about 700 nm. In the early papers, radiation in the active region between 700 and 800 nm was called **infrared**; later the term **far red** (FR) was coined for this part of the spectrum. The demonstration that the response to a few minutes of red light (R) could be prevented if it was immediately followed by a similar duration of light of longer wavelengths (see Fig. 3.2) then followed. At the time, it was thought that the photosynthetic pigments might be involved, but the classic paper which established the characteristic photo-reversibility of these responses led to a different conclusion. The important new feature was to show that, after a series of alternating exposures to R and FR, the physiological response depended on the last exposure in the sequence. When the sequence ended with R, most of the seeds subsequently germinated in darkness; when the series ended in FR, many of the seeds remained dormant (Table 3.1). The authors concluded that the response must be mediated by a single photoreceptor which could exist in reversibly interconvertible R- and FR-absorbing forms.

Clearly, the next step was to test for R/FR reversibility in flowering. The night-break reaction controlling floral initiation was indeed found to have the same rever-sibility characteristics as did germination (Borthwick *et al.*, 1952a, and see Table 4.1)

TABLE 3.1. Germination of lettuce seeds cv Grand Rapids following consecutive exposures to R (1 min) and FR (4 min).

Treatment	% Germination
Dark control	8.5
R	98.0
R–FR	54.0
R–FR–R	100.0
R–FR–R–FR	43.0
R–FR–R–FR–R	99.0
R–FR–R–FR–R–FR	54.0
R–FR–R–FR–R–FR–R	98.0

After Borthwick *et al.* (1952b).

substantiating the conclusion from the action spectra that the same pigment controlled both responses. The action spectrum for the FR-reversal of the R night-break response was determined for *Xanthium*, by irradiating near the middle of an inductive night first with unfiltered radiation from a tungsten filament (TF) lamp adequate to prevent floral initiation (as shown by the controls) and then in the spectrograph with radiation in the 720–745 nm region. The action spectrum for the re-promotion of flowering in *Xanthium* is shown in Fig. 3.2 together with that for the inhibition of flowering in the same plant. As far as the authors are aware, an action spectrum has not been determined for reversal of the R-potentiated promotion of flowering in LDP, but FR of similar wavelengths to those effective in *Xanthium* was shown to reverse the induction of flowering by R in the LDP *Hordeum* and *Hyoscyamus* (Downs, 1956) and there is no reason to suppose that it would differ markedly from Fig. 3.2. The similarities in action spectra and absolute quantum requirements, together with the discovery that the reaction was reversible in each case, were the steps which led to the realisation that the photoperiodic control of flowering by a night-break, light-stimulated germination and de-etiolation processes were under the control of a single pigment. The pigment was named **phytochrome**. Using R/FR reversibility as a criterion, many plant responses to light have since been shown to be under the control of the phytochrome system (Smith, 1975; Schopfer, 1977). Action spectra have not been constructed for many of these but the effective wavelengths for the induction of the response lie between 600 nm and 700 nm, and for suppression between 700 nm and 760 nm. The red 'peak' is usually near 660 nm but the flowering responses tend to have a broader effective waveband with a maximum nearer 640 nm than 660 nm. This shift appears to be due to screening by chlorophyll in the leaves, which are the photoperceptive organs in photoperiodism. A comparison of action spectra for *Lemna* (see Fig. 4.15) clearly demonstrates this shift from a sharp peak at 660 nm in etiolated plants to a broader band at 620–640 nm in green ones.

The photoreversibility of responses by R and FR not only indicated that phytochrome represented a new class of photoreceptor but also established a distinctive property of the pigment which could be used in all subsequent work to isolate and identify it. Because the physiological response depends on the light quality of the final exposure before transferring plants to darkness, the early workers deduced that the controlling, light-absorbing pigment must be photochromic, i.e. that it must exist in two forms which are converted from one to the other by light in the following reaction:

```
                    ------red light------>
   Pr                                                      Pfr
                 <------far-red light------
```

The red-absorbing form of phytochrome, called Pr, absorbs light and is consequently converted to Pfr, the far red absorbing form. In the reverse reaction Pfr is converted back to Pr when it absorbs light. It was further deduced that Pfr, formed by red light, was the active form resulting in the observed physiological response and that phytochrome must be synthesised in darkness as Pr which was the biologically inactive form:

```
                    ------red light------>
synthesis in darkness ---> Pr                       Pfr ---> observed response
                 <------far-red light------
```

It was also predicted from the action spectra for responses and their reversal that the absorbance characteristics of Pr and Pfr must be significantly different, Pr absorbing maximally in the red part of the spectrum and Pfr in the far red part. Photoconversion from Pr to Pfr and back again should therefore produce photoreversible changes in the relative absorbance at these wavelengths in any tissues or extracts which contained phytochrome. Using an instrument called a 'ratiospec', a photometer which measured absorbance at two different wavelengths, Butler and his colleagues were able to show the predicted photoreversible absorbance changes in etiolated seedling tissues following alternating treatments with R and FR light (Butler *et al.*, 1959). Within hours of the first detection of phytochrome *in vivo* by this method they were able to demonstrate similar changes in an aqueous extract from the same tissues, thus confirming the biochemical existence of phytochrome. Phytochrome was first purified from dark-grown oat seedlings seven years later and shown to be a chromoprotein (Mumford and Jenner, 1966).

THE PHYTOCHROME FAMILY OF MOLECULES

In addition to proving the existence of a molecule with the predicted photoreversible absorbance properties, the ratiospec method formed the basis for a dual-wavelength spectrophotometry assay for total phytochrome and the proportion of phytochrome present as Pfr *in vivo* and *in vitro*. In this technique the difference in absorbance (ΔA) at two preselected wavelengths in the red and far red parts of the spectrum, usually 660/720 nm or 730/815 nm, are measured before and after irradiation of the sample with alternating R and FR actinic light of sufficient intensity to saturate the photoconversion reaction. The relative amount of phytochrome present can then be calculated from the reversible change in (ΔA) i.e. ($\Delta(\Delta A)$) which is a function of the average concentration of phytochrome in the tissue. The absolute signal depends upon the sample thickness, the light-scattering factor and the molar absorption coefficient as well as the phytochrome content. This innovative method was highly effective for dark-grown tissues but had limited usefulness for green tissues where chlorophyll, which strongly absorbs red light, precludes the assay. Alternative methods were needed to study phytochrome in light-grown tissues which are for the most part the relevant ones for photoperiodism. To this end phytochrome proved to be amenable to study by immunochemical methods, being a large protein which is very effective in eliciting antibody production. However, when antibodies were raised against phyto-chrome isolated from dark-grown seedlings and then tested against phytochrome in extracts of light-grown plants, little or no reaction was obtained (Shimazaki *et al.*, 1983; Shimazaki and Pratt, 1985; Thomas *et al.*, 1984; Tokuhisa *et al.*, 1985). This led to the conclusion that the phytochrome protein in light-grown plants was not the same as that in dark-grown seedlings, even though spectrophotometrically (i.e. in their R/FR reversibility) they were very similar. This discovery, made independently by several groups, triggered an explosion of research into the question of the differences between the phytochrome in light-grown plants and that in dark-grown seedlings (Jordan *et al.*, 1986; Furuya, 1989). This was much aided by the availability of monoclonal antibodies against phytochrome. Monoclonal antibodies are clones of single antibodies which, therefore, recognise a specific site (epitope) on the protein. Specific monoclonals which recognized different epitopes provided molecular probes for comparing proteins from different sources for similarities and differences. From such studies, it became evident that antigenic differences between phytochromes in dark- and light-grown tissues were found across the entire molecule. It also became evident there were at least three phytochrome proteins and probably more (Furuya, 1989; Wang *et al.*, 1991; Pratt, 1995).

Conclusive confirmation of the diversity of phytochromes came from phytochrome gene studies. With *Arabidopsis*, Sharrock and Quail (1989) identified five different phytochrome-related sequences in Southern blot analyses and sequenced cDNAs representing three of the genes which they designated *phyA*, *phyB* and *phyC*. Subsequently, two other genes, *phyD* and *phyE*, were also isolated and sequenced (Quail, 1994; Clack *et al.*, 1994). Sequences for *phyA*, *phyB* and *phyC* show about 50% similarity at the amino acid level. Phytochrome genes have now been isolated and sequenced from at least 15 plant species, including monocotyledons, dicotyledons, ferns, mosses and algae. The homologues of *phyA* and *phyB* have been fully or partially sequenced from rice, potato and tobacco, and immunological evidence points to at least three phytochrome gene products in *Avena* (Wang *et al.*, 1991). The sequences of *phyA* or *phyB* genes are more similar between species, even between monocotyledon and dicotyledon species, than are the sequences of *phyA* and *phyB* within a single species (Fig. 3.3). This indicates that the divergence of the gene families pre-dated the evolutionary dichotomy between monocotyledons and dicotyledons and that the organisation of the phytochrome gene family is an evolutionarily conserved feature of higher plants. *Arabidopsis* has one of the smallest and least complex plant genomes and the identification of five distinct genes in this species suggests that all angiosperms will have a family of phytochrome genes which is at least as big. This immediately raises the question of whether these molecules have different properties and functions which, in turn, markedly influences our approach to physiological questions. We have to consider the possibility that a specific phyto-

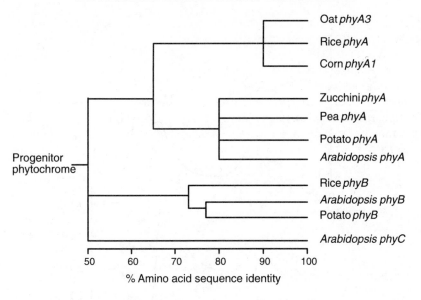

Oat *phyA3*
Rice *phyA*
Corn *phyA1*

Zucchini *phyA*
Pea *phyA*
Potato *phyA*
Arabidopsis *phyA*

Rice *phyB*
Arabidopsis *phyB*
Potato *phyB*

Arabidopsis *phyC*

% Amino acid sequence identity

FIG. 3.3. Phylogeny of phytochrome polypeptides as deduced from sequence similarity. After Quail, 1994.

chrome may function in photoperiodism and that it may have properties different from those of other phytochromes.

PHOTOCHEMICAL PROPERTIES OF PHYTOCHROMES

The diagnostic property of a phytochrome is that it exists in two photo-interconvertible isomeric forms called Pr and Pfr which have different absorbance properties. Pr has a major absorbance band at about 665 nm and a secondary maximum at about 380 nm, as compared with 730 nm for the major peak and 400 nm for the secondary maximum in Pfr (Fig. 3.4). Both forms absorb weakly in the blue part of the spectrum. Subtraction of the Pr absorbance from the Pfr absorbance spectrum gives a difference spectrum which is highly characteristic for phytochrome (see Fig. 3.4). When light energy of any wavelength is absorbed by either Pr or Pfr, it drives the phototransformation of the molecule to the alternative form. The effectiveness of light at any wavelength in driving phototransformation is given by the photoconversion cross-section, defined as the product of the absorbance of Pr or Pfr at that wavelength multiplied by the quantum efficiency, which is 0.15–0.17 for Pr → Pfr and 0.7–1.0 for Pfr to Pr (Fig. 3.5). Because, for the most part, the absorbance spectra of Pr and Pfr overlap, photoconversion takes place simultaneously in both directions resulting in a dynamic photoequilibrium in which the rate of conversion from Pr to Pfr is balanced by the reverse reaction. The proportion of Pr and Pfr in the photoequilibrium mixture depends on the relative photoconversion cross-sections and is therefore wavelength dependent (see Fig. 3.8). It should be appreciated that, in comparing light treatments, differences may exist either in the photoequilibrium established by each treatment

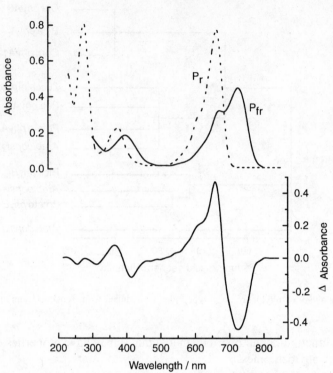

FIG. 3.4. Absorbance and difference spectra of *Avena* phytochrome. The upper spectra are those for phytochrome following a saturating R (Pfr) or FR (Pr) irradiation. The difference spectrum for Pr–Pfr is shown in the lower part of the figure. Data from Vierstra and Quail, 1983.

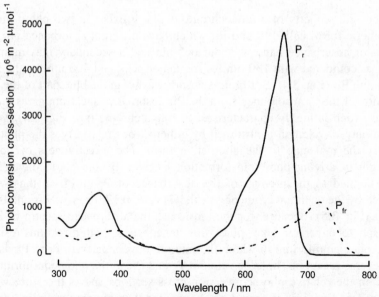

FIG. 3.5. Photoconversion cross-sections for *Avena* Pr and Pfr at different wavelengths. From calculated values of Mancinelli, 1988.

Gene sequence

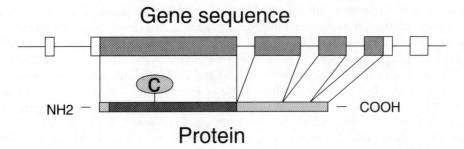

Protein

FIG. 3.6. Relationship between phytochrome gene and protein structure. Shaded rectangles in the gene represent 4 exons; these are joined by 3 introns. The amino terminal domain bearing the chromophore is coded for by a single exon, while the carboxy terminal half of the molecule is derived from 3 exons. Modified from Quail, 1994.

and/or in the rate of interconversion between Pr and Pfr, sometimes called the **cycling rate**.

The photochemical properties of phytochrome have been obtained from studies of highly purified phytochrome *in vitro*. Where comparison with the properties of *in vivo* phytochrome are possible (e.g. difference spectra or spectral change ratio (ΔA Pr/ ΔA Pfr) the *in vivo* and *in vitro* properties are virtually indistinguishable. This suggests that the photochemical properties determined *in vitro* are very similar to those of the functional molecule *in situ*. Although the data from *in vitro* studies can, therefore, be used with some confidence in predicting the effects of a particular radiation treatment on phytochrome *in vivo*, what is less certain is the light environment experienced by phytochrome molecules in green plants under natural or experimental conditions. Light gradients and differential chlorophyll screening within tissues should lead to variation in phytochrome photoequilibria in cells in different locations within a leaf which is receiving natural irradiation. The action of phytochrome appears to be cell autonomous (Neuhaus *et al.*, 1993) and thus cells in different parts of a leaf might be expected to respond differently to a particular light treatment. Nevertheless, these internal factors are not sufficient to prevent plants detecting quite small changes in the spectral distribution of their immediate external light environment, such as that caused by shading or reflection from neighbouring plants. In fact, Holmes and Smith (1977) were able to establish a very robust relationship between the responses of light-grown plants to changes in the ratio of R to FR wavelengths caused by vegetational shading, based on the phytochrome photoequilibria established in dark-grown achlorophyllous tissues by the same light treatments.

When Pr absorbs light and is photoconverted to Pfr or vice versa, it passes through a series of intermediates (Kendrick and Spruit, 1977). Photoconversion can be separated into two phases, an initial photochemical event leading within nanoseconds to rearrangement of the chromophore. This is followed on a millisecond timescale by conformational readjustment of the protein component. Under conditions of high irradiance (e.g. on a sunny day) any Pr or Pfr molecules will be rapidly photoexcited but, because the decay of intermediates is slow compared to the initial photochemical step, photoconversion intermediates will accumulate. This means that, for the same photoequilibrium, the absolute amount of phytochrome as Pfr may differ at different irradiances. It has also been shown that the absorbance of light by intermediates contributes to the photoequilibrium established under continuous illumination. The proportion of the phytochrome molecules present in the form of photoconversion intermediates increases with the level of illumination (Kendrick and Spruit, 1977).

Photochemical differences between different phytochrome types are not known in detail. Gene sequences of phytochromes A, B and C show a high degree of conservation around the chromophore attachment site and the overall domain structure of the phytochromes appear similar. Thus, photochemical differences between different phytochromes are expected to be small. In partially purified phytochrome from light-grown tissues, difference spectra show maxima at about 650 nm in R and at about 720 nm in FR as compared with 665 and 730 nm respectively for phytochrome from etiolated tissues. A similar wavelength shift is seen in the difference spectra of light-grown plants in which chlorophyll accumulation has been prevented by the use of the herbicide Norflurazon. Attachment of phytochromobilin, the tetrapyrrole chromophore precursor, to phytochrome A is autocatalytic, i.e. phytochrome has bilin lyase activity (Lagarias and Lagarias, 1989). Reconstitution experiments with recombinant apoprotein and synthetic chromophores are therefore possible. In such experiments phytochromes A and B behave in a comparable manner (Furuya and Song, 1994), which again suggests that the photochemical characteristics of different members of the phytochrome family are quite similar.

SYNTHESIS AND DESTRUCTION

For phytochrome A in particular, synthesis and destruction are major factors in determining biological effects. In etiolated seedlings, phytochrome is synthesised in darkness as P_r and phytochrome levels increase until synthesis is balanced by destruction. As for all proteins, loss of phytochrome through protein degradation is a function of the pool size and the half-life of the molecule and because the half-life of P_r is relatively long, approximately 100 h, phytochrome can accumulate in darkness to a high level. The apparent concentration may fall during growth but, in the meristematic regions, the levels remain constantly high. Exposure to light and the consequent photoconversion of Pr to Pfr is followed by a rapid loss of spectrophotometrically and immunochemically detectable phytochrome. Phytochrome is thus rapidly proteolytically degraded as Pfr in a series of reactions which involve the ubiquitination pathway (Cherry and Vierstra, 1994). Destruction is a complex process involving biochemical modification and subcellular relocation. In etiolated tissues, phytochrome is dispersed evenly through the cytoplasm (Pratt, 1994). However, following irradiation, phytochrome is rapidly sequestered into discrete areas of the cell. This appears to be an early stage in the destruction process as, under the right conditions, immunostaining also shows enrichment for ubiquitin in the sequestered particles.

Destruction in darkness following a light treatment occurs predominantly from the Pfr form of the molecule, a typical Pfr half-life being 1–2 h. (Jabben and Holmes, 1983). This 'unstable' phytochrome has been designated **type I phytochrome** by several workers (Vince-Prue, 1991). Comparison of microsequences of proteolytically derived fragments of type I phytochrome with the *phyA* gene sequence confirms that *phyA* encodes type I phytochrome. It may be that the two 'stable' **Type II phytochrome** proteins that have been distinguished immunochemically (Wang *et* *al.*, 1991) are the products of the *phyB* and *phyC* genes. The term 'type I phytochrome' will be used here specifically to denote the phytochrome which is the predominant form in etiolated tissues and when isolated and purified is confirmed immunochemically, or by other methods such as microsequencing, to be phytochrome A. Where investigations have identified labile phytochromes or transient effects by physiological or spectrophotometric methods, the term **unstable phytochrome** is used since these methods alone cannot determine which phytochrome gene product is involved.

Destruction of type I phytochrome occurs predominantly from the Pfr form. Smith *et al.* (1988) showed that, at high fluence rates, phytochrome is protected from the destruction process in etiolated seedlings of *Amaranthus caudatus*, *Avena sativa*, *Pisum sativum* and *Phaseolus aureus*. This was attributable to the proportion of phytochrome present as photoconversion intermediates rather than to an effect of light on the destruction mechanism. Photoprotection was observed at light levels as low as 30 μmol m^{-2} s^{-1}. Extrapolation to the situation with green plants, which may not have initially elevated phytochrome levels as found in etiolated seedlings, or where screening by chlorophyll may decrease irradiance effects, is difficult. However, quantitative immunoassay data from light-grown wheat indicate that the level of phytochrome may be at least partly dependent on the irradiance in which the plant is grown (Carr-Smith, 1990).

In light-grown plants, the amount of phytochrome A is usually very low and, in some cases, undetectable. In some species, such as *Arabidopsis*, phytochrome A synthesis continues at about the same rate in light and darkness and levels remain low in light because of the rapid destruction of phytochrome A Pfr. Light treatments which establish a low R/FR photoequilibrium, such as vegetational shade, may allow accumulation of phytochrome A because Pfr levels and hence destruction rates will be lower than in white light. In other species, such as *Avena*, phytochrome A synthesis is inhibited by light and, in these cases, phytochrome A levels in the light should be negligible (Quail, 1994).

In the light, the total amount of photoreversible phytochrome, though small, stays approximately constant. Immunochemical studies have shown that this phytochrome consists predominantly of different proteins made up of a mixture of phytochrome B and other phytochrome gene products, in addition to any phytochrome A which is present. Although presumably a mixture of different proteins, the phytochrome present in light-grown plants (other than phytochrome A) is sometimes referred to as type II phytochrome and, because it is degraded from the Pfr form much more slowly than type I phytochrome, it is also called stable phytochrome. The term 'type II phytochrome' will be used in this book to denote molecular species of phytochrome which can be shown by immunochemical or other means to be other than the *phyA* gene product, and the term 'stable phytochrome' will be used to designate phytochrome in which the Pfr form is not rapidly lost *in vivo*, but where the molecular species has not been identified. This distinction between stable and unstable and types I and II phytochrome is important for a proper evaluation of the molecular species involved in photoperiodism.

REVERSION

Pfr can revert to Pr in darkness through a thermal reaction. This property is thought to be limited to the phytochrome of dicotyledonous plants (although it is not present in the Centrospermae; Kendrick and Hillman, 1970). Reversion has been shown to be absent in dark-grown seedlings of the Graminae and is assumed not to occur in monocotyledons in general. Dark reversion can be quite rapid, with a half-life of 8 min, although this is slow when compared with the rates of photoreactions under natural conditions. The degree of reversion does not relate directly to the amount of Pfr formed but rather shows an optimum, with maximum reversion at Pfr/Ptot ratios of 0.4–0.6 (Schmidt and Schäfer, 1974). The reversion process has been of particular interest in photoperiodism for many years since, in early studies, it was postulated as being part of the timing mechanism and, more recently, as being important for the light-off signal which releases the circadian time-measuring rhythm (see Chapter 4). However, the significance of reversion has been something of a puzzle as many members of the Graminae show photoperiodic responses, even though reversion does not occur (at least in dark-grown seedlings).

Measurement of reversion requires spectrophotometry and is consequently largely limited to dark-grown tissues. Spectrophotometric studies show that, in dicotyledons, the Pfr form of phytochrome undergoes both destruction and reversion (Hillman, 1967b). In contrast, equivalent measurements made on light-grown, non-chlorophyll containing tissues, such as cauliflower florets (Butler, 1964) and the white parts of variegated *Cornus alba* leaves (Spruit, 1970), indicate that the total Pfr pool undergoes reversion to P_r with no detectable destruction. One can speculate that the differential display of reversion in different plant types and at different stages of development may reflect differences in the cellular environment or result from the interaction of phytochrome with another cellular component. To this end it is worth noting that reversion of Pfr *in vitro* is influenced by pH and reducing reagents, and can also be induced by interaction with certain monoclonal antibodies (Lumsden *et al.*, 1985).

Unfortunately, we do not yet know if reversion is characteristic of type I or type II phytochrome, or both. Spectrophotometry gives information about conversion of phytochromes between Pr or Pfr but does not discriminate between different types. On the other hand, immunochemical methods measure the total phytochrome protein and can distinguish between different molecular species but do not generally distinguish between Pr and Pfr. At present, therefore, we can only make assumptions about reversion based on stability data. For example, it seems likely that the Pfr undergoing reversion in cauliflower is type II since it appears to be highly stable in the light.

PHYTOCHROME BIOGENESIS

The availability of cDNA probes has enabled the detailed study of phytochrome mRNA expression by Northern or slot-blot analysis and by run-on transcription. Classic models of the phytochrome system, based on spectrophotometric measurements made mostly on *Sinapis* seedlings, showed Pfr synthesis following zero-order kinetics. It was thus surprising when the down-regulation of phytochrome mRNA by light was found in etiolated *Avena* seedlings (Gottmann and Schäfer, 1983; Colbert *et al.*, 1983, 1985). The inhibition was mediated by phytochrome and was highly sensitive, requiring <1% Pfr in dark-grown *Avena* tissues (Lissemore and Quail, 1988). Down-regulation of phytochrome has since been demonstrated for pea seedlings (Otto *et al.*, 1984; Sato, 1988) and *Cucurbita* (Lissemore *et al.*, 1987) but was not observed in tomato (Sharrock *et al.*, 1988). Down-regulation was also found in germinating embryonic axes of both pea and *Avena* (Sato and Furuya, 1985; Thomas *et al.*, 1989). In both of the latter cases, the decrease in mRNA levels appeared to be insufficient to account for the full decrease in phytochrome apoprotein biosynthesis. An effect of light at a point other than mRNA levels may thus also help to determine the rate of phytochrome synthesis (Colbert, 1988).

Down-regulation of phytochrome A by light is commonly referred to as autoregulation (e.g. Colbert *et al.*, 1985), although there is no direct evidence that this is the case. In germinating embryos of *Avena*, the inhibition of synthesis was shown to be mediated through a stable population of Pfr. Based on this, it was proposed that the synthesis of type I phytochrome may be regulated by type II phytochrome (Hilton and Thomas, 1987; Thomas *et al.*, 1989). A re-evaluation of the earlier work with *Avena* seedlings also indicates that control is exercised through a stable Pfr, assumed to be type II (Colbert *et al.*, 1985). It may be the case that type I and type II phytochromes are functionally equivalent and are both capable of regulating phytochrome A synthesis. Furuya (1989) has suggested that stable subpopulations of type I and unstable subpopulations of type II could exist in plant tissues.

MOLECULAR STRUCTURE

The amino acid sequences for phytochrome from several species, including *Avena*, *Cucurbita*, *Oryza*, *Pisum* and *Arabidopsis*, have been deduced from cDNA sequencing (Hershey *et al.*, 1985; Lissemore *et al.*, 1987; Sato, 1988; Kay *et al.*, 1989a; Sharrock and Quail, 1989). The subunit molecular size of phytochrome is in the range 120–125 kDa. In *Arabidopsis*, sequence A codes for a protein with a molecular size of 124.5 kDa whereas B and C are 129.3 and 123.7 kDa respectively. The chromophore is a single linear tetrapyrrole covalently linked to a cysteine residue *via* a thioether linkage. Spectral differences between Pr and Pfr originate partly from a *cis–trans* isomerisation of the chromophore and partly from protein–chromophore interactions.

Analysis of proteolysis patterns indicates that phytochrome is comprised of three major domains, a 10 kDa fragment cleavable from the N-terminus of the molecule, a 55 kDa domain at the carboxy C-terminus and a protease-resistant residual core of 60 kDa which contains the single chromophore per subunit (Vierstra *et al.*, 1984; Jones *et al.*, 1985). Comparison of genomic sequences reveals a conserved exon–intron organisation (Fig. 3.6). The 70 kDa N-terminal domain of the protein is coded for by a single exon, while the C-terminal domain is coded for by three exons with intron length variable between species. The purified protein is a dimer, with the subunits linked through the C-terminal domain (Jones and Quail, 1986). Based on data obtained using small-angle X-ray scattering and shadowing electron microscopy, Tokutomi *et al.* (1989) proposed a 'four-leaved shape' model for the pea phytochrome dimer, as shown in Fig. 3.7. An immediate question raised by the dimeric nature of phytochrome is whether both elements in the dimer have to be in

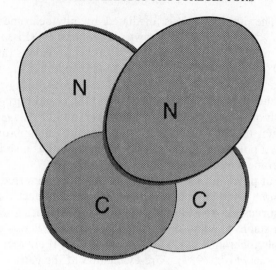

FIG. 3.7. Visual representation of the dimeric phytochrome molecule as seen from 'above' deduced from the small-angle X-ray scattering data of Tokutomi *et al.*, 1989. Dimerisation takes place through the carboxy-terminal domains (C), which are partially sandwiched between the amino terminal domains (N).

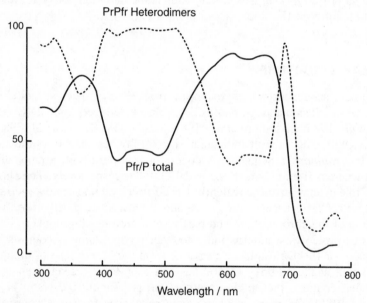

FIG. 3.8. Proportions of total phytochrome molecules as Pfr and of phytochrome dimers as the PrPfr heterodimer established at different wavelengths. The values are expressed relative to the theoretical maximum possible (100% for Pfr and 50% for the heterodimers) set as 100. From calculations of Mancinelli, 1988.

the same conformation, i.e. whether photoconversion is a property of the individual subunits or of the entire structure. By using a monoclonal antibody specific for the Pr form of phytochrome, Holdsworth and Whitelam (1987) were able to show that Pr:Pr and Pfr:Pfr homodimers and Pr:Pfr heterodimers could all coexist *in vivo*. The dimerisation of phytochrome may be significant for its action. Van der Woude (1985) proposed that responses to very low fluences of light are mediated by the heterodimer Pr:Pfr and ideas of how phytochrome might act as a dimer *in vivo* have been developed by Brockmann *et al.* (1987). They suggested that Pr:Pfr heterodimers are more unstable than Pfr:Pfr homodimers and, unlike the homodimers, undergo rapid dark reversion. Maximum numbers of Pr/Pfr heterodimers are present when 50% of the total phytochrome pool is as Pfr. The proportion of phytochrome as heterodimers varies with wavelength but the pattern is different from that of the Pfr/Ptotal ratio (see Fig. 3.8).

PHYTOCHROME-MEDIATED RESPONSES

Phytochrome regulates almost every aspect of plant development from seed germination through to flowering and senescence. Responses can be subdivided into **inductive responses** and **high-irradiance responses**. Inductive responses usually require low levels of red light, are repeatedly R/FR reversible and show reciprocity between actinic intensity and the duration of exposure (Pratt, 1979); they are designated 'inductive' because Pfr action continues in darkness (i.e. while Pfr is present) after the irradiation treatment has ceased. In some cases reversibility is not shown because the amount of Pfr required to saturate the response is very low and sufficient Pfr is formed even in far red treatments. These are called **very low fluence** (VLF) responses to distinguish them from **low fluence responses** (LF) in which the amount of Pfr required is higher and reversibility can be shown (Mandoli and Briggs, 1981; Kaufman *et al.*, 1984). The criteria for inductive responses are essentially the same as for Pfr formation and maintenance, implying strongly that Pfr is the biologically active form of the molecule. When plants are grown for extended periods in light, development is modified by light quality, in particular by changes in the relative amounts of light energy in the R and FR parts of the spectrum brought about by leaf shading or reflectance of the incident sunlight (Smith, 1992). These **shade-avoidance responses** include increased stem extension, enhanced apical dominance and changes in patterns of partitioning of assimilates. Specific responses, such as stem extension, can be directly related to the proportion of total phytochrome (Ptot) calculated to be present as Pfr in the tissues: the **Pfr/Ptot ratio**. As the predominant phytochromes in light grown tissues are type II (i.e. stable as Pfr and relatively constant in amount) shade avoidance responses are consistent with regulation by the amount of Pfr, rather than the Pfr/Ptot ratio as such. Many aspects of the shade avoidance response can be mimicked by giving a brief exposure to FR at the beginning of the daily dark period, to reduce the amount of Pfr present in darkness. This so-called **end-of-day** treatment with FR is reversible by R in a manner typical of inductive responses (Downs *et al.*, 1957).

The characteristics of the HIR include irradiance dependency, lack of reciprocity between irradiance and duration of the light treatment, and action maxima in the blue and FR parts of the spectrum (Mancinelli and Rabino, 1978; Fig. 3.9). The blue action maximum may be due to a specific blue-absorbing photoreceptor (Thomas and Dickinson, 1979) but the FR action maximum appears to derive from an interaction

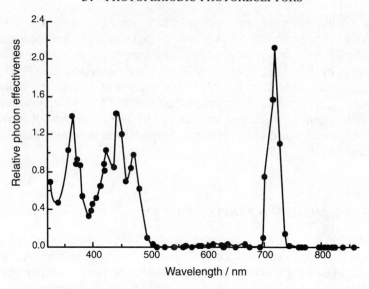

FIG. 3.9. Action spectrum of the 'High Irradiance Response' of phytochrome mediating the inhibition of hypocotyl lengthening in seedlings of *Lactuca sativa*. After Hartmann, 1967.

between the photochemical and dark reactions of the phytochrome system during extended light treatments (Hartmann, 1966; Schäfer, 1975; Wall and Johnson, 1983).

MECHANISM OF PHYTOCHROME ACTION

The primary mechanisms by which phytochrome brings about biological effects have remained surprisingly elusive. Some light-regulated responses such as de-etiolation clearly involve changes in the patterns of gene expression within plant tissues, but these are often relatively slow and could be secondary events. Some responses mediated by phytochrome can be observed very rapidly following light treatments. For example, the ability of plant roots to stick to negatively charged glass was shown to be reversibly affected by exposure to R or FR within seconds (Tanada, 1968), while R-induced depolarisation of oat parenchyma cell membranes occurred within 30 s and was reversible by FR almost as quickly (Newman, 1981). These rapid responses, involving changes in membrane properties, supported the proposal of Hendricks and Borthwick (1967) that the primary action of phytochrome was through the modification of membrane properties. However, analysis of the amino acid sequences of phytochromes as deduced from cDNA sequences indicates that phytochrome is a soluble cytosolic protein without any membrane-spanning domains. This is supported by immunolocalisation studies on dark-grown seedlings which shows phytochrome to be uniformly distributed through the cytoplasm (Pratt, 1994). Although the bulk of phytochrome in extracts of dark-grown seedlings is soluble, a small amount co-pellets with membrane preparations. This has been ascribed to the 'sticky' nature of phyto-chrome (Nagatani et al., 1988) but there is some evidence that the membrane-associated fraction does not change in parallel with the cytoplasmic concentration,

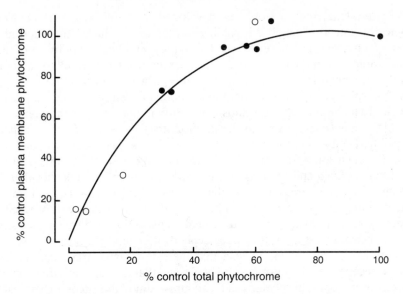

FIG. 3.10. The relationship between phytochrome associated with the plasma membrane and total phytochrome in experiments where phytochrome levels have been depleted through destruction (○) or synthesis inhibited by FR pulses (●). Data are based on quantitative immunoassays with antibodies raised to phytochrome A and are plotted as a percentage of the dark levels. After Terry *et al.*, 1992.

which may indicate an association of part of the phytochrome pool with a membrane-located component (Terry *et al.*, 1992) (Fig. 3.10).

Not all rapid responses involve changes in membrane properties. Phytochrome down-regulation of *phyA* mRNA levels in *Avena* is measurable within 5 min of irradiation (Colbert *et al.*, 1985). The response does not require protein synthesis, indicating that it is a primary target for phytochrome action rather than being mediated by the induction of intermediate regulatory genes. However, immunocytochemical and cell fractionation studies give no sustainable evidence for phytochrome being associated with nuclei; neither is there evidence that Pr or Pfr can directly associate with the *phyA* promoter. Thus down-regulation of *phyA* must involve at least one transduction component between Pfr action in the cytoplasm and interaction with the *phyA* promoter (Quail, 1994).

An alternative, elegant approach to signal transduction in gene regulation has been developed by Neuhaus *et al.* (1993) who used the *aurea* mutant of tomato in which levels of functional phytochromes are depleted giving rise to a phenotype in which anthocyanin formation and chloroplast development are poorly expressed. By micro-injection into epidermal cells, purified phytochrome A from *Avena* was found to stimulate both anthocyanin formation and chloroplast development in a manner that was R/FR reversible and limited to the injected cell. This approach allowed the experimenters to inject with compounds which might replace or block components of the signal transduction chain. Their results suggested that activation of hetero-trimeric G-proteins was a necessary early event in the light-signalling chain for both anthocyanin formation and chloroplast development. However, while activation of the synthesis of several chloroplast proteins involved the calcium–calmodulin system, anthocyanin formation was regulated by a calcium-independent pathway.

An alternative single-cell system was developed by Harter *et al.* (1994a,b) using physiologically intact evacuolated parsley protoplasts, which still showed light regulation of chalcone synthase. When lysed they produce a highly concentrated cytoplasmic preparation containing cytosolic membranes but only minor amounts of proplastids, mitochondria and nuclei. Remarkably, it was still possible to show light-dependent phosphorylation of specific polypeptides in these preparations. Their data suggest that phytochrome regulates phosphorylation in these tissues through a soluble cytosolic kinase/phosphatase system. In addition, they have shown the presence of a pool of specific DNA-binding regulatory proteins (G-box binding activity and factors, GBFs) in this cytosolic fraction. The DNA binding activity of this factor was modulated both *in vivo* and *in vitro* by light and phosphorylation/dephosphorylation activities and GBFs were transported to the nucleus in response to the light treatment. Consequently, they proposed that the subcellular displacement of cytosolic GBFs to the nucleus is an important step in the signal transduction pathway which couples photoreception within the cytosol to light-dependent gene regulation.

Taken together these studies indicate that there is a signal transduction network linking photoreceptor with light-regulated genes and that the pathways may include G-protein activation, phosphorylation and calcium–calmodulin steps. There is some evidence for the modulation of the inositol triphosphate pathway by phytochrome. Guron *et al.* (1992) reported R/FR photoreversible changes in the turnover of PIP2 in leaves of *Zea mays*, although these were not related to any specific physiological response. Indirect evidence for the involvement of IP3 in light-mediated responses comes from the work of Shacklock *et al.* (1992). They showed that the R-inducible and calcium-dependent swelling of wheat protoplasts, which is known to be under phytochrome control (Blakely *et al.*, 1983), could be mimicked by IP3 released from a chemically caged form inside the protoplasts. It is worth noting that the protoplast response does not involve gene activation and it may be that different transduction pathways operate for different classes of response.

Although there is now a body of evidence based on elegant cellular physiological approaches showing that photoregulation can operate through a range of what are becoming the standard signal transduction mechanisms in both plants and animals, the primary link between phytochrome and the transduction chain remains elusive. One possibility is that phytochrome acts as a light-activated ligand which binds to a receptor to initiate the signal transduction chain. Alternatively, phytochrome could be an enzyme which is active in the Pfr configuration. Based upon its ability to show autophosphorylation, it has been proposed that highly purified phytochrome from *Avena* has protein kinase activity. However, this was later shown to be an artefact, attributable to the co-purification of kinase activity (Grimm *et al.*, 1989; Kim *et al.*, 1989). As yet the only enzyme activity which is known to be associated with phytochrome is the bilin lyase activity of the apoprotein which catalyses chromophore attachment (Lagarias and Lagarias, 1989). However, the enzyme hypothesis should not be prematurely relinquished until an alternative is proved. Although it may not be of direct relevance to phytochromes in higher plants, a phytochrome sequence isolated from the moss *Ceratodon* contains a 300 amino acid region in the non-chromophore-bearing domain that bears a striking resemblance to the catalytic domain of eukaryotic protein kinases (Thummler *et al.*, 1992). Although there is no homology with other phytochromes, the results with *Ceratodon* suggest that the N-terminal domain, which confers light-responsivity, may be coupled with C-terminal regions with diverse biochemical properties, thus allowing more than one type of action.

The properties of the phytochrome molecule which might be important for its biological activity have been studied and several differences in the biochemical properties of Pr and Pfr have been observed. Pr and Pfr react differently to protein-modifying reagents and to phosphorylation by protein kinases. Sites of proteolysis, which are presumed to be located at the protein surface, are different for and specific to each form. Monoclonal antibodies to defined epitopes and site-directed antibodies raised to peptides specific for parts of the phytochrome sequence have also identified parts of the molecule preferentially exposed as Pr or Pfr. As a whole the data indicate that differences between Pr and Pfr are not localised to particular regions, which would then be candidates for active sites, but rather consist of conformational changes across the whole of the phytochrome molecule. It is assumed that some of the changes in the structural configuration of the protein that occur when Pr is photoconverted to the Pfr form bring about its biological activation, whereas others are related to protein–chromophore interactions which are necessary for the contrasting light-absorbing properties of Pr and Pfr.

A more promising line of experimentation to establish which parts of the phyto-chrome molecule have functional importance is to examine the biological activity of modified phytochrome sequences in transgenic plants. When *phyA* genes from *Avena* and *Oryza* are transformed into a suitable dicotyledon host such as tobacco, tomato or *Arabidopsis*, the resulting transgenic mutants show a range of exaggerated photomor-phogenetic responses, typified by semi-dwarfing, reduced apical dominance, thicker, dark green leaves and enhanced responsivity to FR (Keller *et al.*, 1989; Boylan and Quail, 1989, 1991; Kay *et al.*, 1989b). This has provided an experimental framework within which the effects of mutated phytochrome molecules can be evaluated. A number of transgenic mutants have been generated with *phyA* genes from which portions have been deleted. In the majority of these cases there is a loss of the mutant phenotype in the resultant transgenic plants. That this was not due in most cases to the inability of the truncated phytochrome to autocatalyse chromophore attachment was demonstrated by Cherry and Vierstra (1994) who reported that only amino acid residues 69–399 were needed for chromophore lyase activity. Deletion experiments indicate that regions at both the N-terminal and the C-terminal are important for biological activity. Deletion of amino acids 7–69 from oat phytochrome resulted in the loss of the light-exaggerated phenotype in transgenic tobacco, as did removal of 35 amino acids from the C-terminal (Cherry *et al.*, 1992, 1993). In the case of the N-terminal deletion, the resulting phytochrome had slightly modified spectral absorbance properties although it still showed R/FR photoreversibility. The properties of the phytochrome produced by the C-terminal deletion were, however, indistinguishable from the wild type molecule. In contrast to the lack of biological activity in most deletion mutants, Stockhaus *et al.* (1992) found that modifying the serine-rich N-terminus of rice phytochrome A, by replacing the first ten serines with alanine, resulted in increased sensitivity to R in tobacco plants that were transformed with this mutated phytochrome. Boylan *et al.* (1994) found that, although certain deletion mutants were themselves unable to mediate photomorphogenetic responses in trans-genic *Arabidopsis*, they interfered with the action of the endogenous phytochrome.

Three deletion mutant phytochrome As (lacking the first 52 amino acids from the N-terminal (ΔN52); lacking the entire C-terminal domain (ΔC617): or lacking amino acids 617–686 (Δ617–686) all interfered with the ability of the endogenous phytochrome A to mediate the inhibition of *Arabidopsis* hypocotyl extension by far red light. The ΔC617 and Δ617–686 mutant phytochromes also interfered with the inhibition of hypocotyl extension by white and red light but, under these conditions, ΔN52 was as active as unmutagenised oat phytochrome in enhancing the response to light. Thus, at least three separate domains on phytochrome A are required for different aspects of light sensing and response. The chromophore-bearing domain between amino acid residues 53 and 616 is required, but is not in itself sufficient, for light-induced initiation of signal transduction. The C-terminal domain between 617 and 1129 is needed for initiating the signal transduction chain under all irradiation conditions. The N-terminal region is required for initiating signal transduction under FR high irradiance conditions, but is not required for the response to R or white light. This is important as it implies that the FR-mediated HIR involves a mechanism of action which is distinct from that of the R/FR reversible response.

BIOLOGICAL ROLES OF DIFFERENT PHYTOCHROMES

When considering the mechanisms of action by which phytochrome brings about its effects, it is necessary to establish whether the different phytochromes in higher plants have different biological functions. This has been greatly helped by the generation of a series of photomorphogenetic mutants of *Arabidopsis*, designated *hy1–hy8* and characterised in the most part by an insensitivity of hypocotyl extension to inhibition by light. Of these, *hy3* is a mutation in the *phyB* gene (Somers *et al.*, 1991) and *hy8* in the *phyA* gene (Parks and Quail, 1993; Nagatani *et al.*, 1993; Whitelam *et al.*, 1993). *hy3* mutants are characterised by insensitivity to R, lack the shade avoidance response to R/FR ratio and do not show the end-of-day R/FR response. The *hy8* mutants, on the other hand, lack the FR-HIR, but show the normal wild type response to continuous R or white light. Thus, although both phytochrome A and phytochrome B can have overlapping functions (they both control hypocotyl extension), they appear to have distinct roles in photoperception, namely that phytochrome A senses FR light, whereas R wavelengths are detected by phytochrome B (Whitelam and Harberd, 1994).

BLUE LIGHT EFFECTS

Early studies using action spectroscopy which implicated phytochrome as the photoreceptor for photoperiodism were based on night-break experiments in which relatively short light treatments were involved. When more extended light treatments were used to test the responses of LDP it was noted that blue was generally of low effectiveness, except in members of the Cruciferae (Wassink *et al.*, 1950; Stolwijk, 1952a,b). Interpretation of the response to blue light is problematic because phytochrome absorbs in this region of the spectrum, albeit weakly, and the low Pfr/Ptot established in blue light would be expected to be favourable for induction in LDP (see Chapter 5). However, the strong sensitivity of the members of the Cruciferae, but not other LDP, to induction by blue light implied the action of a separate blue-absorbing photoreceptor. Support for this idea comes from work with *Arabidopsis*, a typical cruciferous LDP, where dichromatic irradiation with blue and monochromatic light at 589 nm was shown to be inductive, whereas irradiation with 589 nm light alone was ineffective. The Pfr/Ptot ratios and cycling rates between Pr and Pfr established by the two treatments are approximately equal and thus the blue response cannot be due to phytochrome (Mozley and Thomas, 1995). Although phytochromes are probably the most important photoperiodic photoreceptors in higher plants, blue light is frequently the most effective waveband in lower plants.

Speculation about the nature of the blue-absorbing photoreceptor has occupied photobiologists for several decades. Action spectra for responses sensitive to blue light, such as phototropism, are characterised by an action maximum at about 450 nm, with shoulders at about 420 and 490 nm, accompanied by a broad action peak near 370 nm. Candidates for the blue photoreceptor are flavoproteins and carotenoproteins; both flavins and carotenoids are capable of generating an absorbance spectrum to match the action spectrum under appropriate conditions. The blue-absorbing photo-receptor has yet to be isolated from plants, but the gene corresponding to the *HY4* locus of *Arabidopsis* has recently been cloned and sequenced (Ahmad and Cashmore, 1993). *Hy4* mutants show greatly reduced sensitivity to blue light for inhibition of hypocotyl extension and are thus postulated to be deficient in a blue-absorbing photoreceptor. The *HY4* gene codes for a protein which shows significant homology to microbial DNA photolyases, which are flavoprotein enzymes activated by blue light. This offers very strong evidence that one higher plant blue-absorbing photo-receptor has now been identified. The *hy4* mutant shows reduced sensitivity for long-day photoperiodic induction (Mozley and Thomas, 1995), suggesting that the same photoreceptor is responsible for blue-light regulation of hypocotyl growth and photoperiodic effects in *Arabidopsis*.

CONCLUSIONS

The detection of light for photoperiodic responses in Angiosperms is primarily accomplished by members of the phytochrome family of photoreceptors, with the exception of members of the Cruciferae where a blue-light photoreceptor also plays a photoperiodic role. Lower plants, on the other hand, primarily use blue-absorbing photoreceptors for photoperiodic perception. Much of the physiological evidence for phytochrome involvement is based on action spectroscopy of night-break responses, where very similar action maxima are found in the red for the inhibition of flowering in SDP and the promotion of flowering in LDP. Reversal of the effect of R in inhibiting flowering by FR is diagnostic for regulation by phytochrome. However, an extended family of phytochrome genes, encoding a number of distinct regulatory proteins with R/FR properties, has recently been revealed. Mutants lacking phyto-chrome A or phytochrome B have shown that these two phytochromes have distinct, albeit possibly overlapping, functions as far as mediating the inhibition of extension growth by light of different wavelengths is concerned. By extrapolation, we may expect that different members of the phytochrome family will have different physio-logical roles and some of these will include light detection for photoperiodic responses.

Phytochromes mediate a wide range of response to light. Some of these are very rapid, while others are slow or persistent. Many, but not all, involve gene regulation, while in others biophysical changes involving ion transport or other changes in membrane properties are apparently the targets. The phytochrome regulatory system appears to be very flexible and capable of being coupled to almost any key aspect of cellular regulation. From this, it is possible to envisage that the action of phytochrome in photoperiodism could include regulation at one or more levels of cellular organisation. Extrapolation from non-photoperiodic situations could be dangerous and misleading and ultimately the action of phytochrome in photoperiodism will only be discovered by working with model photoperiodic systems. Where a photoreceptor other than phytochrome (i.e. the blue-light photoreceptor) has been implicated in daylength sensing in higher plants, it is interesting to note that regulation by phytochrome is still retained. This implies that there is redundancy in photoreceptor action in daylength sensing. This could, of course, extend to redundancy in phytochrome participation, with more than one member of the phytochrome family co-regulating the same photoperiodic responses. Initially, the increasing complexity of the phytochrome family raises a number of new questions about the identity and mechanism of photoreceptors in photoperiodism, but it is likely that this complexity will itself help explain how phytochromes accomplish the specific and highly specialised functions of daylength sensing, while at the same time mediating the wide range of other developmental responses that plants show to light.

4 Daylength Perception in Short-Day Plants

In Chapter 2, two actions of light were distinguished in the photoperiodic control of flowering in the SDP *Pharbitis nil*. Light acts to control the phase of the photoperiodic rhythm. Light also interacts with a specific phase of that rhythm, φ_i, to inhibit flowering. Although these two actions of light have not clearly been differentiated in the majority of SDP, there is much circumstantial evidence for their existence and they are considered as two separate reactions in the following discussion. Additionally, it is well documented that removal of the Pfr form of phytochrome early in the night may inhibit flowering in many SDP. The photoreceptor thus has multiple actions in the photoperiodic control of flowering (Vince-Prue and Takimoto, 1987). These are shown schematically in Fig. 4.1.

THE NIGHT-BREAK REACTION

This R/FR reversible reaction was one of the first to be shown to be under the control of phytochrome (see Chapter 3). It is interesting that the association of flowering control with phytochrome was first observed in a reaction which, at the time, appeared to be an artificial treatment with little relevance to natural conditions. It is now thought that the night-break reaction represents the action of light at the inducible phase of the photoperiodic rhythm and that the position of this phase in circadian time underlies the timekeeping mechanism (Chapter 2). From studies with *Pharbitis*, it is likely that the night-break effect in SDP is a direct action of light to prevent the flowering response which would otherwise be induced, and that this can be separated from any phase-shifting action of light (Lumsden and Furuya, 1986). Therefore, studies of the night-break are of major importance in understanding how flowering is controlled in SDP.

A brief illumination with R has little effect on flowering during the first few hours of darkness. After several hours, however, a night-break with R inhibits flowering in SDP; the timing of this night-break inhibition is controlled by a circadian rhythm of sensitivity to light and is a property of the photoperiodic clock. In many SDP the inhibition of flowering which would occur with an R night-break is prevented by a

85

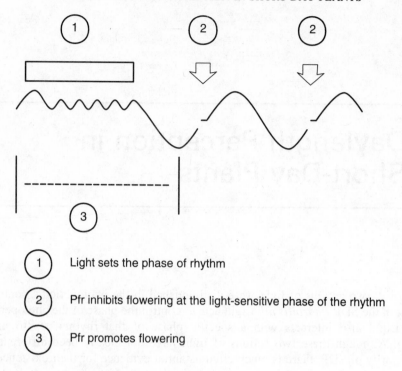

(1) Light sets the phase of rhythm

(2) Pfr inhibits flowering at the light-sensitive phase of the rhythm

(3) Pfr promotes flowering

FIG. 4.1.   Scheme showing the several actions of light in the photoperiodic control of floral induction in the SDP, *Pharbitis nil*.

brief subsequent exposure to FR, showing that this 'night-break' action of light depends on the formation of the Pfr form of phytochrome. First seen in *Xanthium* (Borthwick *et al.*, 1952a), reversibility of the R night-break effect by FR has been demonstrated in several SDP, including *Amaranthus caudatus*, *Chenopodium rubrum*, *Dendranthema grandiflora* (chrysanthemum), *Glycine max* cv Biloxi, *Kalanchoë blossfeldiana*, *Lemna paucicostata* and *Pharbitis nil*. The reaction is repeatedly reversible with the effect depending on whether R or FR light is given as the final exposure, as for other Pfr-mediated responses (Table 4.1).

The time taken to escape from reversibility indicates for how long Pfr must be present in order to complete the inhibitory reaction. Flowering in *Xanthium* was completely inhibited by a R night-break given for 2 min. A 3 min exposure to FR almost completely restored flowering if given immediately afterwards but was less effective when a period of darkness was interposed (Fig. 4.2). FR no longer had any effect after about 40 min darkness at 20°C. At 5°C reversibility was maintained for longer. In *Glycine max* (Downs, 1956), *Chenopodium rubrum* (Kasperbauer *et al.*, 1963b) and chrysanthemum (Cathey and Borthwick, 1957), reversibility by FR was also maintained for some time in darkness (up to about 45, 70 and 90 min respectively). In some species, however, FR was not effective unless given almost immediately after the R night-break. Reversibility was lost within 1 min in *Chenopodium album* (Borthwick, 1964), within 2 min in *Pharbitis* (Fredericq, 1964) and about 6 min in Kalanchoë (Fredericq, 1965). Thus the night-break action to prevent flowering appears

TABLE 4.1.   Red/far red reversibility of the effect of a night break on flowering in SDP

Night-break treatment	*Xanthium strumarium*[a]	*Dendranthema grandiflora*[b] Honeysweet	*Glycine max*[c]
			Number of flowering
	Floral stage	Floral stage	nodes
R	0.0	0.0	0.0
R–FR	5.6	1.3	1.6
R–FR–R	0.0	0.0	0.0
R–FR–R–FR	4.2	1.2	1.0
R–FR–R–FR–R	0.0	0.0	–
R–FR–R–FR–R–FR	2.4	0.7	0.6
Control (no night-break)	6.0	3.0	4.0

[a]Red (R) and far red (FR), each 3 min; Downs (1956).
[b]R and FR, each 2 min; Cathey and Borthwick (1957).
[c]R, 2 min; FR, 8 min; Downs (1956).

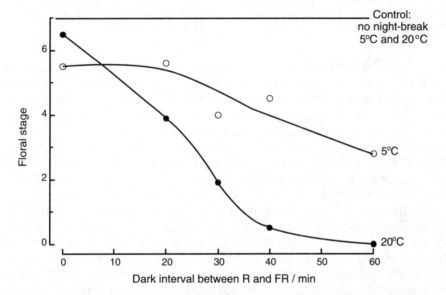

FIG. 4.2.   Loss of FR effectiveness for the re-promotion of flowering in *Xanthium strumarium* at two different temperatures. Plants were maintained at 20°C except during the dark interval between the R and FR exposures in the 5°C treatment. The night-break treatments were given near the middle of a 12-h inductive dark period; R, 2 min; FR, 3 min. After Downs, 1956.

to occur fairly rapidly following the formation of Pfr at the appropriate time, although the final response can only be observed after some considerable time, often several weeks. This has to be born in mind when investigating the biochemistry of the night-break inhibition (see Chapter 8).

   The amount of light necessary to saturate the night-break inhibition of flowering varies with species, conditions and the time of exposure (Lumsden and Furuya, 1986). In etiolated, dark-grown seedlings of *Pharbitis*, flowering was completely inhibited by

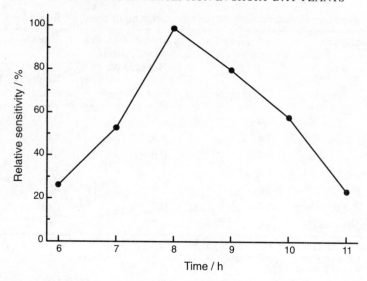

FIG. 4.3. Changes in sensitivity for the inhibition of flowering by night-breaks in *Pharbitis nil*. Relative sensitivity was calculated from the reciprocal of the number of seconds of R required to give 50% inhibition of flowering at different times over the first 12 h of darkness, taking 8 h as 100%. After Thomas, 1991.

25–30 μmol m^{-2} R at NBmax. However, green seedlings that had been de-etiolated for 24 h in the light required more than 6000 μmol m^{-2} at NBmax. Seedlings of *Chenopodium* required about 3000 μmol m^{-2} of broadband R to reduce flowering to zero whereas light-grown plants of *Lemna paucicostata* 441 required only 120 μmol m^{-2}. In chrysanthemum, which appears to be rather opaque to R, a 1 min exposure to white fluorescent light was only completely inhibitory at an irradiance of 18.7 W m^{-2} (Cathey and Borthwick, 1964). One way of expressing the sensitivity of the response is to plot the reciprocal of the exposure required to produce a given level of flowering (Thomas, 1991). When calculated in this way for *Pharbitis* (Fig. 4.3), it is evident that the sensitivity to light changes substantially during the night, increasing between 5 and 8.5 h and then decreasing again. This changing sensitivity to light has not been satisfactorily explained in molecular terms, but is also seen in other circadian rhythms. In dark-grown *Pharbitis nil*, the flowering rhythm is accompanied by changes in the quantum yield of photoconversion between Pfr and Pr (King *et al.*, 1982), although the magnitude of these was too small to account for the sensitivity changes. Changes in the availability of receptors are also possible.

The night-break reaction appears to be a threshold type of response to [Pfr]. Evidence for this comes from experiments with *Chenopodium rubrum* where, with night-breaks of 70 min duration, flowering was prevented by both FR and R, which would establish about 0.03 and 0.8 Pfr/Ptot, respectively. The reaction leading to the inhibition of flowering appeared to be completed in about 70 min and to be independent of the amount of Pfr over the range from about 0.03 to 0.8 Pfr/Ptot, provided that plants were given a final exposure to FR in order to establish a common value of Pfr before returning them to darkness (Table 4.2). This correlates well with the loss of reversibility by FR which also occurred after about 70 min in *Chenopodium*. FR can also be an effective night-break in chrysanthemum if continued for sufficiently long:

TABLE 4.2. Comparison of the effects of red (R) and far red (FR) light given as a night-break on flowering in *Chenopodium rubrum*.

Duration of night-break (min)	Floral stage		
	FR	R until final 4 min, then FR	5 min R, then dark; final 4 min FR
10	8.3	7.9	8.2
20	7.5	7.3	7.2
30	6.8	6.5	6.7
40	5.9	5.4	5.5
50	3.0	2.7	3.1
60	1.0	1.2	1.1
70	0.3	0.3	0.4

R, 6.0 W m^{-2}; FR, 7.5 W m^{-2}.
After Kasperbauer *et al.* (1963b).

here, 80 min FR was found to be equivalent to a few min R (Cathey and Borthwick, 1957). The difference in the response to R and FR when short exposures were used appears to depend on the fact that, following an exposure to FR, [Pfr] rapidly falls below a threshold while, after R, [Pfr] remains above the threshold for a longer time in darkness and continues to act. The results of several physiological experiments with *Chenopodium rubrum* indicated that, when transferred to darkness from FR, [Pfr] fell in <5 min to a level that was ineffective in the night-break reaction. In contrast, it appeared to take at least 80 min to reach an ineffective value after a R night-break (Kasperbauer *et al.*, 1964).

Although these results afford evidence for a threshold type of Pfr action, there are problems in interpreting the inhibitory effect of FR light on flowering. Under certain conditions Pfr is required for flowering in the **Pfr-requiring response** discussed below. Flowering may then be inhibited by FR not because Pfr is maintained above a threshold but because the Pfr concentration is actually lowered. However, in the experiments with chrysanthemum and *Chenopodium*, this explanation for the inhibitory action of relatively long exposures to FR can be eliminated since a short exposure to FR (which would be expected to have the same effect on reducing the Pfr level) does not inhibit flowering and reverses the inhibitory effect of a prior night-break with R.

The effectiveness of intermittent (or cyclic) lighting may also indicate that the inhibition of flowering is not proportional to the amount of Pfr present in the tissue but depends only on maintaining the concentration above some threshold during a particular time. With tungsten filament (TF) lamps (which would establish approximately 0.5 Pfr/Ptot), intermittent lighting over a 4 h period in the middle of an inductive night prevented flowering in chrysanthemum provided that the dark intervals did not exceed about 30 min; when higher (or lower) Pfr levels were established with fluorescent (or ruby-red) lamps, intermittent lighting was effective with longer (or shorter) dark periods as would be expected if a threshold type of response is involved (Table 4.3). However, when low energies of R were used, intermittent light over a 2 h period did not significantly increase the inhibitory effect compared with the same total energy given continuously over 30 min (Kadman-Zahavi and Yahel, 1971).

TABLE 4.3. The effect of intermittent lighting on flowering in *Dendranthema grandiflora* cv White Pink Chief

Cycle length (min)	Floral stage		
	White fluorescent	Tungsten filament	Ruby-red
0.7 L/14.3 D	0.0	0.0	2.0
1.3 L/28.7 D	0.0	0.0	5.0
2.4 L/57.6 D	<1.0	4.0	10.0
3.0 L/77.0 D	1.0	5.0	10.0
4 h tungsten-filament		0.0	
Control (no night-break)		10.0	

Plants were lit for 5% of the time in intermittent schedules of different frequencies as indicated over a total period of 4 h. L, light; D, dark.
The irradiance was 2–3 W m^{-2}.
After Borthwick and Cathey (1962).

Since a subsaturating R exposure would also be expected to produce a relatively low Pfr/Ptot, an increase in effectiveness by giving light intermittently over a long period would be expected if maintenance of [Pfr] above a threshold for inhibition were involved. It is known that the inhibition of floral initiation in chrysanthemum has characteristics associated with the control of flowering in LDP (Vince-Prue, 1982), where a mixture of R and FR light is more effective than R at certain times in the dark period (see Chapter 5). This complicates the analysis of responses to R and FR in this SDP and may account for the apparent inconsistencies. One important consequence of the effectiveness of intermittent lighting is the use of such schedules in the commercial control of flowering time, especially in chrysanthemum, where short exposures to TF lamps do not inhibit flowering (Table 4.4); intermittent lighting schedules considerably reduce the cost of night-break lighting when this must be continued over several hours. Flowering is inhibited in many SDP by only a few minutes of low intensity light from TF lamps; examples are *Xanthium* and *Glycine*. In chrysanthemum, however, a 12 min night-break did not prevent flowering even at 28.9 W m^{-2}, whereas, when a total of 12 min was given intermittently over 4 h, flowering was completely prevented at 2.7 W m^{-2} (see Table 4.4). It was suggested that the different spectral properties of the leaves of *Xanthium* and chrysanthemum might account for these differences. Light from TF lamps consists of a mixture of R and FR wavelengths. In chrysanthemum, most of the R is absorbed at the upper leaf surface and only FR penetrates through the leaves; consequently only about 0.02–0.05 Pfr/Ptot would be established in much of the leaf tissue, even at high irradiances, and this would quickly reach the threshold level following transfer to darkness. *Xanthium* leaves are more transparent to R and a higher percentage of Pfr would be established: Pfr would, therefore, remain above the threshold for a longer time in darkness. Measurements have confirmed that the leaves of chrysanthemum are considerably more opaque to R than are *Xanthium* leaves because of anatomical differences (Cathey and Borthwick, 1964). When high-intensity fluorescent light was used (18.7 W m^{-2}), a 1 min night-break completely prevented flowering in chrysanthemum confirming that, if a sufficiently high percentage of Pfr is established throughout the leaf tissue, it remains above the threshold long

TABLE 4.4. The effect of night-break treatments with fluorescent or tungsten filament lamps on flowering in *Dendranthema grandiflora*

Lamp type	Irradiance (W m^{-2})	Duration of night-break	Response
Fluorescent[a]	18.7	1 min	Vegetative
Fluorescent[a]	6.2	1 min	Flowering
Fluorescent[a]	1.2	12 min	Vegetative
Tungsten filament[a]	1.8	12 min	Flowering
Tungsten filament[b]	28.9	12 min	Flowering
Tungsten filament[b]	2.7	4 h	Vegetative
Tungsten filament[b]	2.7	12 min (9 × 1.3 min in 4 h)	Vegetative
Control (no night-break)[a'b]			Flowering

[a], cv Honeysweet; [b], cv White Pink Chief.
Table compiled from Borthwick and Cathey (1962) and Cathey and Borthwick (1964).

enough to inhibit the flowering response (see Table 4.4). If this interpretation is correct, it not only implies a threshold mechanism of Pfr action, but indicates that most of the leaf tissue is involved in the response and not just the upper leaf surface as has been suggested (Schwabe, 1968). The results emphasise the need to exercise caution in interpreting results with R and FR light; FR can inhibit flowering in SDP by raising Pfr above a threshold (at φ_i), or by reducing Pfr (the Pfr-requiring reaction); additionally, the tissue properties can exert a major effect on the percentage of Pfr established by light which contains both R and FR wavelengths, such as TF lamps and sunlight.

When a long dark period is scanned with an R night-break, a circadian rhythmicity is often observed in the response (Fig. 4.4), although sometimes this rhythmicity is seen only under certain conditions (see Chapter 2). The first NBmax often occurs at or after the 7–9th hour of darkness (depending on species) and, thereafter, inhibition occurs at circadian intervals. With shorter nights, such as would occur naturally, NBmax also occurs several hours after transfer to darkness (see Figs 1.8, 1.10). Since (on the basis of the FR reversibility of the response) the inhibitory effect depends on the formation of Pfr, it is evident that the amount of Pfr in the leaf is inadequate to inhibit flowering after several hours have elapsed in darkness. At the end of the photoperiod a considerable fraction of the phytochrome would be in the Pfr form (Pfr/Ptot = 0.8 in R or white fluorescent light as used in many experiments, and about 0.6 in natural daylight). Thus the night-break response seems to depend on phytochrome in which Pfr is relatively unstable in darkness, reaching an ineffective level within a few hours. This point is returned to later in relation to the more stable Pfr involved in the Pfr-requiring reaction.

PHASE SETTING

It has been proposed that the timekeeping mechanism in SDP is a circadian rhythm for which the phase is controlled by the action of the photoperiod. In *Chenopodium,* the rhythmic response to the duration of darkness has been used as an assay for the

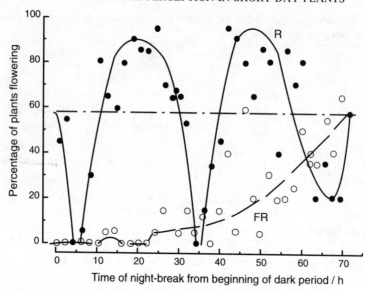

FIG. 4.4. The flowering response of *Chenopodium rubrum* to night-breaks with R or FR. Plants were given a 72-h dark period interrupted by 4 min R, 8.5 W m^{-2} (●) or 10 s FR, 16.8 W m^{-2} (○) at the time indicated on the abscissa. After Cumming *et al.*, 1965.

rephasing action of light (King and Cumming, 1972a, b). A 1 min exposure to R early in the night (after 4–6 h dark) delayed timing in the first cycle but there was no apparent phase-shift of the second or third oscillation (King and Cumming, 1972b). Thus, although an exposure of only 1 min transiently delayed the rhythm, it did not rephase it. The rhythm was advanced by light given at the 9th hour of darkness (see Fig. 2.5). At this time, a 2–6 h exposure was required to rephase the rhythm; a 6 h skeleton was not effective (Fig. 4.5) but partial rephasing was obtained by giving 5 min R every 1.5 h. The involvement of phytochrome is indicated by the effectiveness of repeated R pulses but FR reversibility could not be demonstrated. The transient phase delay obtained by 1 min R given earlier in the night also supports the involvement of phytochrome.

The strongest evidence that phytochrome is the sensor for the phase-shifting action of light in photoperiodism comes from experiments with *Pharbitis*. In light-grown seedlings, the phase of the rhythm as determined by the time of the first NBmax after light-off, could be altered by a single brief exposure to R; after 6 h in the dark, as little as 54 μmol m^{-2} was sufficient to cause a phase delay of 3 h (Lumsden and Furuya, 1986). The effectiveness of such a small exposure to R light strongly indicates that phytochrome is the sensor pigment. Reversibility by FR, which would confirm this conclusion, is difficult to demonstrate because an FR pulse at this time is strongly inhibitory to flowering (Saji *et al.*, 1983). However, reversibility has been reported by Lumsden (1991), although no experimental details are available. Experiments with *Xanthium* also used the time of NBmax as the main indicator of the rhythm phase. Here, a 1 min exposure to white light after 6 h dark was sufficient to elicit a marked phase delay (Papenfuss and Salisbury, 1967). In *Lemna paucicostata*, a new light-on (dawn) signal was given by a 15 min exposure to either R or FR light, whereas in the

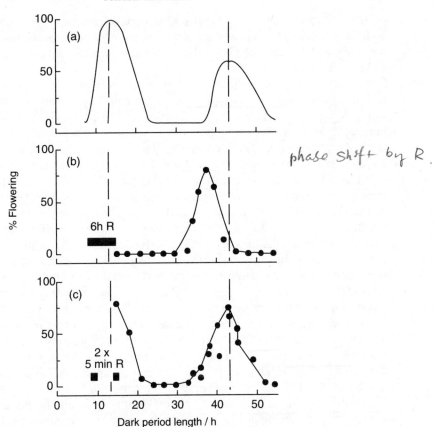

phase shift by R.

FIG. 4.5. The effect of continuous and skeleton photoperiods to phase shift the photoperiodic flowering rhythm in *Chenopodium rubrum*. Beginning at the 9th hour of darkness, plants were exposed (b) to 6 h of R (1.15 W m^{-2}) or (c) to a 6 h skeleton beginning and ending with a 5 or 15 min R pulse at the same irradiance. Control plants (a) received no light interruption. The phase of the nightlength rhythm was shifted (advanced) by the continuous light exposure but not by the skeleton. After King and Cumming, 1972b.

LDP *L. gibba*, the phasing action of R light was found to be reversible by FR (Oota, 1984).

In dark-grown seedlings of *Pharbitis*, a short exposure to R was sufficient to start the flowering rhythm (see Fig. 2.7) but attempts to obtain reversal by FR were inconclusive because of the Pfr requirement for flowering under these conditions. When given during a rhythm which is phased from light-off, a brief light pulse appeared to act as a dawn signal in *Xanthium* only after several hours of darkness had elapsed (Papenfuss and Salisbury, 1967). In dark-grown *Pharbitis*, however, a brief exposure to R acted as a new dawn signal when given at any time in the light-on rhythm started by the initial R pulse (Lee *et al.*, 1987).

Despite the failure to show FR reversibility in most cases, there seems little doubt that a single photoconversion of phytochrome to Pfr can affect the phase of the photoperiodic rhythm, even if only transiently. Effects on the time of NBmax and/ or on CNL have been obtained with at least three species of SDP and with both light-on and light-off rhythms. As expected from the behaviour of overt circadian rhythms, both the sensitivity to light and the direction and magnitude of the resultant phase-shift may vary with the time of giving the exposure (Lumsden and Furuya, 1986; Lee *et al.*, 1987).

With relatively short light periods timing seems to begin from a light-on signal but, with longer photoperiods such as would occur naturally, timing begins from a light-off signal (Chapter 2). Therefore, the function of the photoperiod must not only be to initiate, or re-phase the rhythm at dawn but also subsequently to 'suspend' it at a particular phase point. The identity and functioning of the photoreceptor involved in the apparent suspension of the rhythm in continuous light has hardly been investigated. A few workers have approached the question by examining the light requirement for maintaining the rhythm in suspension and preventing the onset of dark time measurement. The usual protocol has been to give a photoperiod in continuous light which is long enough to suspend the rhythm, and to follow this with an experimental light period. When the photoperiod was extended with light of different wavelengths, the onset of dark timing in *Pharbitis* was prevented most effectively (i.e. at the lowest irradiance) with R compared with other wavelengths (Takimoto, 1967). Red was also more effective that a mixture of R + FR in *Xanthium*, indicating the involvement of Pfr rather than any turnover between Pr and Pfr (Salisbury, 1981).

The photoperiod has also been extended with pulses of R, in order to examine the possible involvement of Pfr. The rationale was that treatments perceived as continuous light would delay the beginning of dark timing, giving NBmax 8–9 h from the end of the extension period. On the other hand, treatments perceived as darkness would result in NBmax 8–9 h from the beginning of the extension period. In such experiments, a brief pulse of R given every hour was sufficient to maintain the rhythm in suspension, with NBmax occurring at the same time as after 6 h of continuous light (Fig. 4.6). It was suggested that the pulses would be sufficient to maintain [Pfr] above a threshold and, consequently, the involvement of Pfr in rhythm suspension was proposed (Lumsden and Vince-Prue, 1984). However, the results of later experiments questioned this conclusion (Lumsden *et al.*, 1986). An analysis of the phase-shifting responses to a single R pulse given at various times in the dark period indicated that the apparent effect of hourly R pulses to maintain the rhythm in suspension could be explained by the effect of each pulse to phase-shift the rhythm which had already been released by the initial transfer from continuous light to darkness. The phase shift produced by a pulse of R after 1 h of darkness would effectively return the pacemaker, and hence the rhythm, to the position occupied at the end of the photoperiod. Subsequent pulses would simply repeat the sequence.

A more direct approach is to examine the effect on rhythm suspension of light conditions given during the entire photoperiod. An irradiance of 0.03 W m^{-2} of continuous white light was insufficient to give a light-off signal in *Pharbitis* but at 0.2 W m^{-2}, or above, the generation of a light-off signal was determined by the duration of the photoperiod (Hoshizaki and Hamner, 1969). Following a 24 h photoperiod in R, FR, B or white light, the first NBmax occurred at the 9th hour of darkness in all cases (Fig. 4.7). Thus, at the irradiances used, all wavelengths appeared to be equally effective in suspending the rhythm and causing timing to begin from a light-off signal. This is not incompatible with phytochrome being the photoreceptor as all wavelengths used would maintain at least a low [Pfr]. Unfortunately, no action

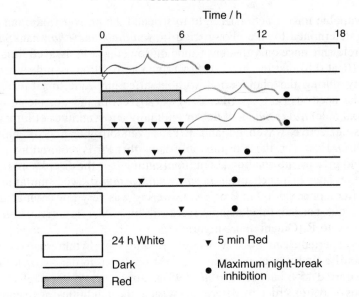

FIG. 4.6. Effect of day extension treatments on the time of maximum sensitivity to a night-break in *Pharbitis nil*. Dark-grown seedlings were exposed to a single photoperiod of 24 h white light followed by a day-extension for 6 h, before transfer to a 48 h inductive dark period. Treatments during the 6 h extension period are indicated on the figure. Time measurement was examined by exposing plants to a 10 min night-break at various times during the dark period; the time at which the night-break had the maximum inhibitory effect on flowering is shown. The R exposures were given at 55 µmol m^{-2} s^{-1} (day-extension pulses and night-breaks) or 12 µmol m^{-2} s^{-1} (continuous 6-h extension). After Vince-Prue, 1983b.

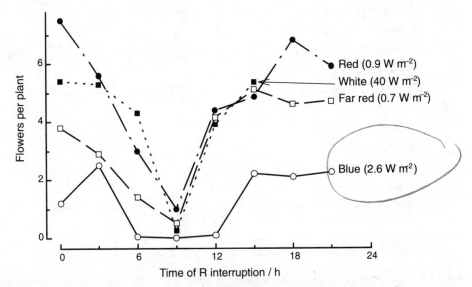

FIG. 4.7. The effect of light quality during the photoperiod on the time and pattern of response to a night-break in *Pharbitis nil*. Dark-grown seedlings received a single 24 h photoperiod in light of different spectral quality at the irradiance indicated; this was terminated by 10 min R in order to establish the same Pfr/P ratio at the time of transfer to darkness. The 48 h inductive dark period was interrupted by a 10 min night-break with R at the time indicated on the abscissa; at the end of the dark period plants were transferred to continuous W. Although the magnitude of flowering was affected (especially by B), night-break timing was essentially unchanged. After Vince-Prue, 1981.

spectrum is available for the action of light to suspend the photoperiodic rhythm; the action spectra determined for the photoperiod in *Xanthium* and *Pharbitis* (Salisbury, 1965) are not relevant since only the magnitude of flowering was assayed. It is evident from Fig. 4.7 (R at 0 h) that the level of flowering can be strongly influenced by the spectral quality during the photoperiod without affecting dark time-measurement, emphasising the need to use a specific assay for effects of the treatments on time-keeping. For example, *Kalanchoë* will flower with only a few minutes of light per day; R is the most effective waveband and R/FR reversibility can be demonstrated (Fredericq, 1963). However, these results only show that Pfr is required for a flowering response and give no information about timekeeping (see the discussion of the Pfr-requiring reaction later in this chapter). *Lemna paucicostata* showed similar responses to *Pharbitis*; after a photoperiod in B or FR, flowering was very low compared with R or white light but normal timekeeping (assayed by CNL) occurred following a terminal exposure to R (Ohtani and Ishiguri, 1979).

Skeleton photoperiods can also be useful in studying rhythm suspension. The second R pulse of a 24 h skeleton acted as a light-on signal in dark-grown *Pharbitis* and NBmax occurred 15 h later (Lee *et al.*, 1987). Consequently, the skeleton can be filled in various ways in order to determine what light conditions are necessary to generate a light-off signal. In one series of experiments (Fig. 4.8), timekeeping was assayed by comparing the flowering response to night-breaks given at the 9th hour of the inductive dark period (NBmax expected for light-off timing) or at the 15th hour

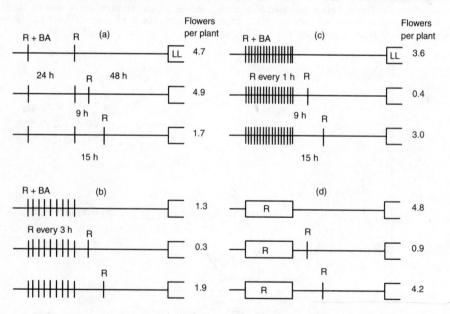

FIG. 4.8. The effect of different pulse frequencies given during the photoperiod on timekeeping in *Pharbitis nil*. After a 24 h continuous photoperiod (d) the time of maximum inhibition of flowering by a night-break (NBmax) occurred at the 9th hour of darkness, with no effect at the 15th hour. After a skeleton 24 h photoperiod (a), NBmax occurred 15 h after Pulse 2 of the skeleton and there was no effect at the 9th hour. When the skeleton was filled with a R pulse every hour or every 3 h, plants responded as if to a continuous photoperiod; a night-break was inhibitory at the 9th hour but had no effect at the 15th hour of darkness. Plants were sprayed with benzyladenine (BA) during the first R pulse in order to maximise flowering. Saji and Vince-Prue, previously unpublished.

(NBmax expected for light-on timing). Following a filled skeleton with a R pulse every 1–3 h, flowering was inhibited (compared with control plants without a night-break) at the 9th but not at the 15th hour of darkness indicating that this treatment was sufficient to suspend the rhythm and initiate light-off timing. Filled skeletons with R pulses at longer intervals gave ambiguous results because of low flowering in control plants. These results support the hypothesis that an intermittent input to the circadian system (presumably from Pfr, although FR reversibility was not demonstrated) is sufficient to initiate light-off timing in *Pharbitis*.

Overall, the experimental results indicate that phytochrome is the most likely photoreceptor for all actions of light to control the phase of the flowering rhythm. A single pulse of R can initiate a new rhythm, and advance or delay the phase of an existing rhythm indicating the action of Pfr. In some species (such as *Chenopodium*; King and Cumming, 1972a, b) several hours of light are apparently required to rephase the rhythm; however, it is possible to replace the continuous exposure with repeated pulses, at least in part. The greater effectiveness of R in maintaining the rhythm in suspension and the effect of R pulses to initiate dark timekeeping indicate that phytochrome is also the photoreceptor for this action of light and that Pfr is the active form. In general, FR reversibility of the phasing action of R pulses has not been demonstrated. This may, in part, be due to the fact that the majority of the studies have been made with dark-grown seedlings of *Pharbitis*, where FR strongly inhibits flowering in the Pfr-requiring reaction discussed later in the chapter. It could also be due to an extremely rapid escape from reversibility.

THE DUSK SIGNAL

Exposure to a nightlength longer than a certain critical duration has been shown to be the overriding factor in the control of flowering in many SDP and timekeeping runs from the end of the photoperiod, provided that this is sufficiently long. Thus a specific function of the photoreceptor must be to sense this light/dark transition. Under natural conditions, the transition between day and night occurs during a period of twilight when the irradiance gradually decreases. In the tropics, twilight is short and the transition between day and night is abrupt, with a rapid decrease in irradiance. At high latitudes in the summer, twilight is very long and the irradiance decreases only slowly. The changes in irradiance are often accompanied by a change in spectral quality, twilight spectra being relatively rich in blue and FR and relatively poor in orange-red light. The pattern of spectral distribution becomes more exaggerated as the solar elevation declines and, during evening twilight, the ratio of R to FR light decreases from a daylight value of about 1.1 to values in the region of 0.7–0.8 (Smith, 1982). Somewhat higher reported values for daylight and twilight are due mainly to the chosen bandwidths and the presentation of data in terms of energy rather than quantum ratios; nevertheless the trend towards a reduced R/FR ratio during twilight is clearly evident (Fig. 4.9). Similar changes in FR/R ratio during twilight also occur underwater (Chambers and Spence, 1984). The decrease in either irradiance or in R/FR ratio could, therefore, theoretically be the environmental dusk signal which initiates dark time measurement. The blue/green or blue/red ratio could also provide

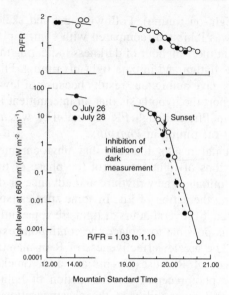

FIG. 4.9. Changes in irradiance and in red:far red ratio during evening twilight. Measurements were made on two separate days in Logan, Utah. Redrawn from Salisbury, 1981.

an index of the progression through twilight (Smith, 1982), but such a signal does not relate to the properties of any known plant photoreceptor.

It has been shown that changes in R:FR during twilight are, in fact, very variable because of cloud conditions on the horizon (see Fig. 4.9) and they may be absent completely, especially at high latitudes (Smith, 1982). Irradiance values can also vary considerably from day to day because of clouds. Such variations could lead to significant errors in timing and, indeed, the effective photoperiod in *Pharbitis* has been shown to be 20–30 min longer on clear days than on cloudy ones (Takimoto and Ikeda, 1960). The error would become very small, however, for plants with a low threshold irradiance. It could also be reduced by averaging when repeated cycles are necessary for induction. Changes in the quality and quantity of twilight occur together under natural conditions, and both can vary considerably with cloud cover, latitude and season. Consequently, experimental approaches with artificial light sources that vary each factor independently are necessary to determine whether either the quality or quantity of light, or both, are important natural cues for the perception of the light/dark transition at dusk.

Most experimental approaches to end-of-day light quality have considered only its effects on the magnitude of the overall flowering response. It is abundantly clear that, under certain conditions, exposing plants to FR before entry to darkness can substantially reduce or even eliminate flowering (see Fig. 4.14). However, such an end-of-day treatment with 5 min FR (Fig. 4.10) appeared to have almost·no effect on the time of NBmax in *Pharbitis*, when compared with plants transferred to darkness from fluorescent light (which would establish a Pfr/Ptot ratio of approximately 0.8). Similarly, there was little or no effect on CNL in *Chenopodium* (King and Cumming, 1972b), *Xanthium* (Salisbury, 1981) or *Pharbitis* (see Fig. 4.10) when the photoperiod was

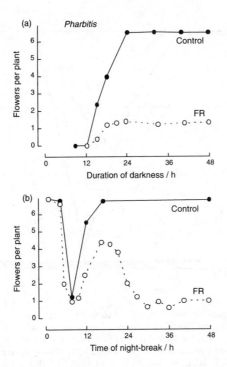

FIG. 4.10. The effect of exposure to FR at the end of the photoperiod on the critical nightlength and night-break timing in *Pharbitis nil*. Seedlings were transferred from continuous W fluorescent light to darkness either directly (controls) or after exposure to FR (6.0 W m^{-2}) for 5 min (FR). In the night-break treatments (b), plants received an inductive dark period of 48 h at 19°C, with a 5-min R night-break (3.3 W m^{-2}) given at the time indicated on the abscissa. In (a), plants received a single dark period at 20°C for the duration shown. Neither the time of night-break sensitivity not the critical nightlength were significantly altered by the end-of day FR, although the magnitude of flowering was depressed. After Takimoto and Hamner 1965.

terminated with FR or FR-enriched light. Thus, reducing [Pfr] at the end of the photoperiod much more rapidly and to a greater extent than would ever occur under natural conditions appears to have little effect on dark timekeeping when compared with plants which have a high Pfr/Ptot ratio when transferred to darkness.

It has been shown that dark time measurement occurs at the same rate as in darkness when plants are transferred to a sufficiently low irradiance of light, irrespective of its R/FR ratio. In *Pharbitis*, where plants remained under fluorescent lamps throughout, time measurement (determined by the time of NBmax) began when plants were transferred from 15.6 W m^{-2} to 0.04 W m^{-2} but was completely prevented by light at 4.0 W m^{-2} (Fig. 4.11). In an attempt to approximate more closely to the natural conditions of a gradually changing irradiance, plants were transferred to darkness through several stepped gradients of irradiance without changing light quality (Fig. 4.12). In all cases, timekeeping (as estimated by the time to NBmax) began when plants were transferred to 2.0 W m^{-2}. Since the decrease in irradiance began from different initial levels, the results also demonstrated that attaining a critical irradiance value was the determining factor, not the rate of change. Although more experiments are needed with other plant species, there seems little doubt that measurement of dark

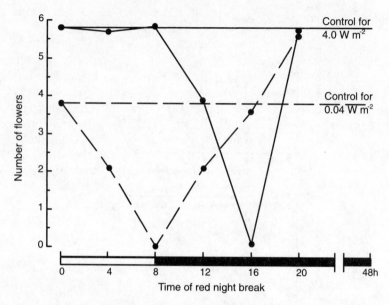

FIG. 4.11. The effect of irradiance on the initiation of dark time-measurement in *Pharbitis nil*. Plants were grown in continuous fluorescent W at 15.6 W m^{-2} and then given 8 h at either 4.0 W m^{-2} or 0.04 W m^{-2} before transfer to darkness for 48 h. Time measurement was examined by exposing plants to a 15 min R night-break. Normal dark timekeeping began when plants were transferred to 0.04 W m^{-2} with maximum inhibition occurring 8 h later. At 4.0 W m^{-2} timekeeping did not begin until plants were transferred to darkness. After Takimoto, 1967.

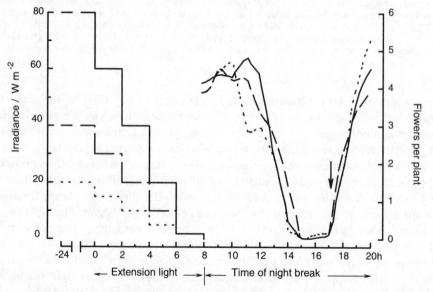

FIG. 4.12. The effect of stepped gradients of irradiance at the end of the photoperiod on dark time measurement in *Pharbitis nil*. Plants received a 24 h photoperiod at 20, 40 or 80 W m^{-2} in W followed by a stepped decrease in irradiance at 2-hourly intervals as shown; during the final 2 h, all plants were maintained at 2 W m^{-2} before transfer to an inductive 72 h dark period. Timekeeping was examined by giving plants a 10 min night-break with R (55 µmol m^{-2} s^{-1}. The arrow shows the predicted time of maximum inhibition if dark timing were coupled to the end of the light period. After Lumsden and Vince-Prue, 1984.

time can begin when the irradiance decreases below a threshold value in the absence of any change in light quality.

Very few investigators have attempted to follow dark timekeeping under natural conditions but, where this has been done, the results indicate that a critical irradiance value is the natural dusk signal which begins dark time measurement. The most pertinent experiments are those of Salisbury (1981) who found that the transition from above to below the range of irradiance values effective in inhibiting dark time measurement in *Xanthium* occurred within 5.5 to 11.5 min, on two separate occasions in Utah (see Fig. 4.9) and was not much affected by clouds. This is remarkably close to the theoretical value of 10 min for the precision of a time signal given by a change in irradiance in an unshaded habitat in the UK (Hughes *et al.*, 1984). Other workers have transferred plants to darkness at different times during evening twilight to determine when dark time-measurement began (Takimoto and Ikeda, 1961). Under these conditions, the inductive dark period for *Pharbitis* began when the natural irradiance had fallen to between 1–2 W m^{-2}, which is in good agreement with the values obtained in artificial light (see Fig. 4.12). Considerable differences between species have been recorded; for example, approximately 0.16 W m^{-2} was found to be the threshold for photoperiodic darkness in sugar cane (Clements, 1968) and <0.04 W m^{-2} in *Xanthium* (Takimoto and Ikeda, 1961).

Although changes in the R/FR ratio were occurring together with changes in irradiance in these experiments, it is unlikely that the former constituted the major environmental cue. The spectral changes are small in relation to the simultaneous changes in irradiance (Hughes *et al.*, 1984, Fig. 4.9) and it has been calculated that the timing precision for the latter would be considerably greater (10 min) than for R/FR (30 min). Moreover, a simple threshold time signal related to R/FR would be obliterated by heavy canopy shading and, even in a relatively unshaded habitat, R/FR would be an unreliable signal because of sunflecks. Considerable day-to-day variation has also been recorded for R/FR changes during twilight. In contrast, irradiance during twilight was influenced to a much smaller extent by cloud cover and canopy shading (Hughes *et al.*, 1984). Overall, therefore, it seems that a threshold irradiance is the most precise environmental time cue and so is likely to be the most important under natural conditions, a conclusion which is supported by the experimental evidence.

Considering the importance of the dusk signal, it is surprising to find that rather few experiments have directly addressed the question of whether phytochrome is the photosensor. The observation that, in dark-grown seedlings, a reduction in spectrophotometrically measurable Pfr occurred following transfer to darkness led Hendricks (1960) to suggest that Pfr might be involved in dark timekeeping; it was proposed that the time taken to fall below a critical inhibitory threshold value for [Pfr] was the way in which a critical nightlength was measured. We now know that this hypothesis was incorrect. For example, the loss of Pfr in thermal reactions has a Q_{10} close to 2, whereas the critical nightlength and especially the time to NBmax are often essentially independent of temperature (King, 1979). Moreover, a brief exposure to FR at the end of the photoperiod has little or no effect on dark timekeeping (for both CNL and NBmax) in several SDP (King and Cumming, 1972b; Salisbury, 1981; see Fig. 4.10), even though [Pfr] would rapidly be reduced photochemically. Nevertheless, Hendricks was the first to propose that a reduction in Pfr is associated with the dusk signal that begins dark timekeeping and this subsequently became the generally accepted concept. It was challenged by Vince-Prue (1983b) who pointed out that much of the physiological evidence for light-grown plants indicated that their Pfr was highly stable in darkness and, therefore, that Pfr loss on transfer to darkness appeared to be an unlikely candidate for the dusk signal. The discovery that there are different phyto-chromes with different kinetics for loss in darkness (Chapter 3) has re-opened the question.

Perhaps the strongest evidence that a reduction in [Pfr] is indeed the light-off signal comes from the experiments with end-of day FR. Although it appears to have no effect on dark timekeeping in some experiments, other, more detailed investigations have revealed that end-of day FR does advance dark timing by a small amount. In *Pharbitis*, the time of NBmax was advanced by about 30–40 min by an end-of day-exposure to FR, compared with plants transferred to darkness from fluorescent light (Fig. 4.13). A rather greater effect on CNL (shortened by 1–1.25 h) has been reported for rice (Ikeda, 1985), *Pharbitis* (Evans and King, 1969) and, with repeated cycles, for *Chenopodium* (Cumming, 1963). This slight advance in timing by end-of-day FR supports the conclusion that the beginning of dark timekeeping (assumed to be the release of the circadian rhythm) is initiated by a reduction in [Pfr] in the tissue. Exposing *Xanthium* plants to low temperature (10°C) for 2 h at the beginning of the dark period lengthened the CNL by 45 min, also suggesting that a thermal reaction (such as a decrease in [Pfr] in darkness) is involved in the initiation of dark timekeeping (Salisbury, 1990). It may be that a particular critical threshold concentration of Pfr must be reached, but this has yet to be established.

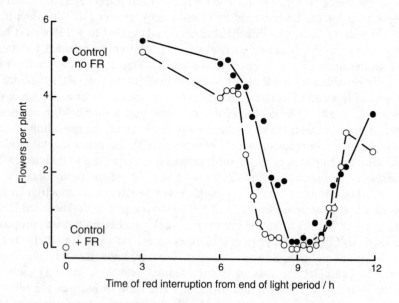

FIG. 4.13. The effect of an end-of-day exposure to FR light on the time of sensitivity to a night-break in *Pharbitis nil*. Seedlings received 24 hours of W fluorescent light (80 W m^{-2}) and then transferred to an inductive 72 h dark period either directly (●) or after exposure to FR (9.2 μmol m^{-2} s^{-1} for 5 min) (○). A 10-min R night-break (55 μmol m^{-2} s^{-1}) was given at the time indicated; control plants received uninterrupted darkness. Following an end-of-day FR exposure, sensitivity to a night-break occurred slightly earlier than when plants were transferred directly to darkness from W fluorescent light. After Lumsden & Vince-Prue, 1984.

If we are to account for the perception of the light-off signal entirely in terms of [Pfr], it is necessary to assume that Pfr is lost rapidly through thermal reactions on transfer to darkness since photochemical lowering of Pfr only advances timekeeping by some 30–40 min (see Fig. 4.13). The loss of Pfr must also become significant at low irradiance, since dark timekeeping is initiated and appears to proceed normally under these conditions (see Fig. 4.11). The small effect of temperature on NBmax time also indicates that the thermal loss reaction must be a rapid one, such that doubling or halving the rate has rather little effect.

THE Pfr-REQUIRING REACTION

One problem that is repeatedly encountered in studies of photoreceptor action in SDP (and often complicates their interpretation) is an additional component of the overall flowering response for which Pfr is required (see Fig. 4.1). The promoting effect of Pfr on the magnitude of flowering in SDP was first shown for *Pharbitis*. In young seedlings (with expanded cotyledons but no foliage leaves), flowering was depressed when a short exposure to FR was given at the end of the day before transferring the plants to an inductive dark period (Fig. 4.14). The inhibition of flowering by FR was prevented by a subsequent brief exposure to R, demonstrating that the effect is associated with a requirement for Pfr; the action spectra for inhibition by FR and re-induction by R were found to be the same as those for other phytochrome-mediated responses (Nakayama *et al.*, 1960). The effect of chlorophyll to shift the R action maximum in *Lemna* from 660 nm (etiolated plants) to 612 nm (green plants) was later demonstrated (Fig. 4.15). The sensitivity of etiolated plants was some 10–30 times greater than for green ones.

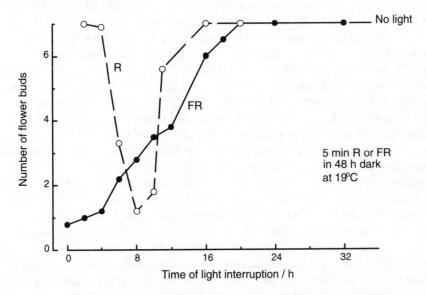

FIG. 4.14. Flowering response of *Pharbitis nil* to night-breaks with R or FR. Plants received a 48 h inductive dark period at 19°C, interrupted with 5 min of R or FR at the time indicated. Light conditions as in Figure 4.11. After Takimoto and Hamner, 1965.

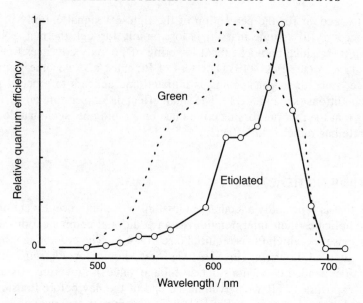

FIG. 4.15. Action spectra for reversal of the inhibition of flowering by FR in etiolated and green plants of *Lemna paucicostata* T-101. Plants received a saturating exposure to FR at 730 nm (80 s at 1.4 W m^{-2}) at the beginning of a 72 h dark period in order to inhibit flowering. The action spectra for the reversal of the FR inhibition were obtained by giving monochromatic light immediately after the exposure to FR. The relative quantum efficiency for the reversal of the FR inhibition is shown. After Ohtani and Kumagai, 1980.

Inhibition of flowering by FR given at the beginning of an inductive dark period has also been reported for *Chenopodium* (see Fig. 4.4), *Begonia boweri* (Krebs and Zimmer, 1983), *Lemna paucicostata* (Mori, 1979), *Xanthium*, sorghum (Borthwick, 1964), rice (Ikeda, 1985), *Kalanchoë* and chrysanthemum (Vince-Prue, 1975); it thus appears to be a common feature of the control of flowering in SDP. Usually the inhibitory effect is seen only after very short photoperiods. For example, in 24 h cycles, FR was most inhibitory to flowering in *Pharbitis* when given after photoperiods of 2-4 h. After intermediate ones (8 h) it had no obvious effect and, after longer ones (12 h), FR promoted flowering (Table 4.5). The promotion of flowering with 12 h photoperiods was probably the result of the slight acceleration of dark time measurement and consequent increase in the effectiveness of the 12 h dark period which, from the control results, was close to the CNL. Similar results were obtained for *Chenopodium rubrum* (Cumming, 1963) and rice (Ikeda, 1985). After suboptimal photoperiods, a terminal treatment with FR depressed flowering, but after supra-optimal photoperiods, flowering was promoted. In both cases, flowering was promoted by end-of day FR when the dark duration was close to the CNL.

From these experiments using 24 h cycles in which both light and dark duration were simultaneously varied, it appeared that the inhibitory effect of FR light given at the beginning of the night was associated with short photoperiods. Later experiments with non-24 h cycles, however, indicated that the inhibition was associated with long dark periods rather than with the very short days *per se*. For example, when a single inductive dark period was given to *Pharbitis* plants which were otherwise kept in continuous light, end-of-day FR depressed flowering when the dark period exceeded

TABLE 4.5. Effects of brief irradiation with FR at the end of photoperiods of different durations and irradiance on flowering of *Pharbitis nil*.

Photoperiod	Flowers per plant	
	No FR	Plus FR
Series A: 62 W m^{-2}		
1 h	4.7	0.0
2 h	5.6	0.2
3 h	5.4	2.1
4 h	6.1	4.2
5 h	6.5	6.4
8 h	6.1	6.5
12 h	0.3	2.6
13 h	0.0	0.0
Series B		
4 h at 53 W m^{-2}	6.3	5.8
4 h at 28 W m^{-2}	6.2	3.3
4 h at 12 W m^{-2}	6.3	0.0

Plants received 3 SD under white fluorescent lamps as indicated and were exposed to FR for 4 min at the end of each of the three photoperiods before transfer to darkness for the rest of the 24 h cycle. Plants without FR were transferred directly to darkness.
After Fredericq (1964).

14–15 h and promoted it when the dark period was less than 12 h (Evans and King, 1969). This promotion response is, again, probably a consequence of increased effectiveness of a dark period which is close to the CNL. In subsequent experiments, the duration of the photoperiod was varied and followed by a very long dark period; the results showed that, even with a dark period of 48 h, end-of-day FR did not inhibit flowering provided that the photoperiod was sufficiently long (Fig. 4.16). Thus the photoperiod duration is important for the Pfr-requiring response but the effect may only be seen when a short photoperiod is combined with a long dark period. These results indicate that it is the absence of Pfr over a long period of time which is inhibitory to flowering; therefore, since Pfr is always present in the light, end-of day inhibition by FR would occur only when long dark periods are associated with relatively short photoperiods. It is also evident that, whatever the relationship of the Pfr-requiring reaction to other components of the photoperiodic mechanism in SDP, there is no obligatory requirement for Pfr to be present during the inductive dark period since many experiments have shown that the Pfr-requiring reaction can be satisfied during the photoperiod (see Fig. 4.16, Table 4.5).

The timing and duration of Pfr action on the magnitude of flowering can be studied by giving a brief exposure to FR at different times in the night. Several early experiments indicated that Pfr is required for a shorter period in the light (Vince-Prue, 1983a). For example, in *Xanthium*, flowering was depressed by giving FR at any time during the first 8 h of darkness following a 2 h photoperiod (in 24 h cycles) but had no effect after an 8 h photoperiod. These early experiments with *Xanthium* confounded the duration of the photoperiod with the duration of darkness and so their interpretation is not entirely straightforward. However, other experiments have also indicated that light conditions can influence the time over which Pfr is required.

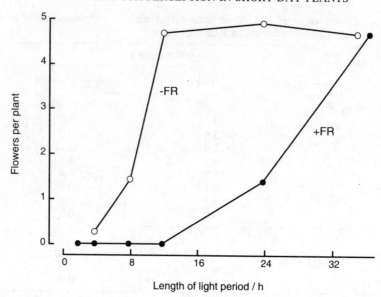

FIG. 4.16. The effect of the duration of the preceding photoperiod on the response to FR given immediately before an inductive dark period. Dark grown seedlings of *Pharbitis nil* were given a single photoperiod in W at 20 W m^{-2} for various durations as shown. They were then transferred to a 48 h inductive dark period either directly (−FR) or after a 10 min exposure to broad-band FR at 9.2 μmol m^{-2} s^{-1} (+FR). Plants were returned to continuous W at the end of the dark-period. After Lumsden 1984.

In *Pharbitis*, for example, flowering was completely prevented by end-of-day FR after 4 h photoperiods only at low irradiances; after a photoperiod at 53 W m^{-2}, a terminal FR treatment had no significant effect (Table 4.5). Similarly, in *Kalanchoë* and *Xanthium*, end-of-day FR inhibited flowering in short photoperiods only when the irradiance was low (Borthwick, 1964; Fredericq, 1965). Thus the Pfr-requiring reaction may require substrates that are generated more rapidly in the light and are irradiance dependent.

Under natural conditions with longer photoperiods at high irradiances, the Pfr requirement is satisfied in the light and does not continue into the dark period. However, if it has not been completed during the preceding photoperiod, the Pfr-dependent process continues during the inductive night and, in some cases, may take a very long time for completion when plants are maintained in darkness. For example, in *Chenopodium rubrum*, FR given even after 50 or 60 h of darkness substantially decreased the flowering response (see Fig. 4.4). In *Pharbitis*, FR completely prevented flowering when given up to 12 h into the dark period following an 11 h photoperiod and up to 18 h following a 10-min photoperiod (Fig 4.17). A further feature of this particular action of phytochrome is that Pfr does not have to be present continuously in the dark period, nor necessarily at the beginning of the dark period. Following an end of day exposure to FR, flowering in *Pharbitis* was re-promoted by forming Pfr with a pulse of R after many hours in darkness (see Figs 4.10, 4.13); this is also so in *Xanthium*, where the inhibitory effect of FR could be reversed by R for up to 30 h in darkness (Reid *et al.*, 1967).

The Pfr-requiring reaction does not appear to be associated with a circadian rhythm.

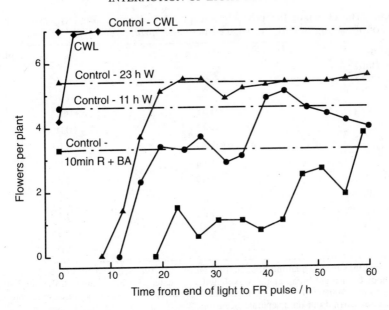

FIG. 4.17. The effect on flowering of the time of giving a FR pulse at various times during a single inductive dark period following photoperiods of different durations. Dark-grown seedlings of *Pharbitis nil* were given a single photoperiod of 10 min R (■), or 11 h (●) or 23 h (▲) of W from Warm White fluorescent lamps. This was followed by a long inductive dark period during which plants were given a brief (5–10 min) exposure to broad-band FR (9.2 µmol m^{-2} s^{-1}) at various times as shown on the abscissa. Other seedlings (CWL) were grown in continuous W from germination (4 d) before transfer to darkness. After Vince-Prue, 1983a.

In general, the inhibitory effect of FR is greatest at the beginning of the night and gradually decreases with time. This contrasts markedly with the inhibitory effect of a night-break, which shows a circadian rhythm. This was elegantly demonstrated in the early *Chenopodium* experiment (see Fig. 4.4) and has also been reported for other SDP, including *Pharbitis* (see Fig. 4.14), *Glycine* (Carpenter and Hamner, 1963) and *Lemna paucicostata* (Mori, 1979). Moreover, although flowering is strongly depressed, removing Pfr at the beginning of the inductive dark period does not substantially alter the time of NBmax, nor the CNL. Thus the Pfr-requiring reaction does not seem to be part of the circadian time-measuring process of photoperiodism, although it is clearly necessary for floral induction and strongly influences the magnitude of the flowering response. By using optical fibres, the Pfr-requiring reaction was shown to take place in the cotyledons of *Pharbitis*, i.e., in the organs where photoperiodic induction occurs (Knapp *et al.*, 1986); no effect of end-of-day FR was obtained when the apex alone was irradiated.

INTERACTION OF LIGHT EFFECTS

The Pfr which is associated with the Pfr-requiring reaction appears to be extremely stable in darkness. For example, in *Chenopodium*, it was apparently still present and active after 50 h in darkness at physiological temperatures (see Fig. 4.4). This contrasts with the more unstable Pfr involved in the night-break response, where

TABLE 4.6. The effect of the phytochrome photoequilibrium (Pfr/Ptot) established by a night-break on flowering in dark-grown seedlings of *Pharbitis nil*.

Wavelength of night-break (nm)	Pfr/Ptot	Flowers per plant
660	0.75	0.0
680	0.58	<0.2
688	0.47	0.0
699	0.24	0.0
708	0.13	0.0
720	0.04[b]	0.0
730	0.03[b]	0.0
750[a]	<0.01[b]	0.1
Control		2.5
Control[a]		1.0

Dark-grown seedlings received an initial saturating pulse of R followed by a 48 h dark period interrupted at the 14th hour by a night-break at the wavelengths indicated. Pfr/Ptot values in the cotyledons were measured spectrophotometically after the night-break or, where too low to measure, were estimated (b). Experiment a was carried out on a separate occasion.
After Vince-Prue (1983a).

the amount of remaining Pfr is no longer effective after a few hours in the dark. A significant point to emerge from these experiments is that, under some conditions, flowering can be inhibited at certain times either by removing Pfr with FR (active Pfr is present) or by the formation of Pfr with R (Pfr is absent or below a threshold); this has been demonstrated in *Chenopodium* (Fig. 4.4), *Pharbitis* (Fig. 4.14), *Xanthium* (Borthwick and Downs, 1964; Reid *et al.*, 1967), *Kalanchoë* (Fredericq, 1965), *Glycine* (Carpenter and Hamner, 1963) and *Lemna* (Saji *et al.*, 1982). A consequence of this is that the inhibition of flowering by end-of-day FR is often associated with loss of reversibility in the night-break inhibition of flowering, because a brief exposure to FR itself inhibits flowering at the time of sensitivity to a night-break with R light. The paradox that, at certain times in the inductive night, flowering may be inhibited by a brief exposure to either FR or R has been recognised for many years and, because of it, several workers have advanced the theory that two different pools of phytochrome are involved in the control of flowering in SDP and that these have different kinetics for the loss of Pfr by thermal reactions (Vince-Prue, 1983a; Takimoto and Saji, 1984). The simpler hypothesis that a single phytochrome pool is involved with different thresholds for the Pfr promotion and Pfr inhibition reactions seems to be eliminated by the results of experiments with dark-grown seedlings of *Pharbitis*, where all wavelengths between 660 nm and 750 nm were inhibitory to flowering at certain times (Table 4.6). Direct spectrophotometric measurements confirm this conclusion; in dark-grown *Pharbitis*, Pfr/Ptot values were approximately 0.08 at the time when both a brief exposure to FR (which would lower the Pfr/Ptot to 0.03) or to R (which would increase Pfr/Ptot to 0.8) were inhibitory to flowering (King *et al.*, 1982). Moreover, the FR inhibition was readily reversed by a subsequent exposure to R at any time throughout the dark period, except at the time of sensitivity to a night-break (Table 4.7).

An action spectrum for the inhibition of flowering in *Lemna paucicostata* at the

TABLE 4.7. The effect of R and FR given at different times during an inductive dark period on flowering in *Pharbitis nil*.

Time of night-break	Night-break treatment	Flowers per plant
5th hour of darkness	R	4.2
	FR–R	4.7
	FR	0
	R–FR	0.1
10th hour of darkness	R	0.2
	FR–R	0.3
	FR	0
	R–FR	0
15th hour of darkness	R	0
	FR–R	0
	FR	0
	R–FR	0
24th hour of darkness	R	5.3
	FR–R	4.0
	FR	0.7
	R–FR	1.1

Dark-grown seedlings received an initial saturating pulse of R (10 minutes at 2.4 W m^{-2}) followed by a 48 h inductive dark period interrupted by a night-break at the times indicated. Night-break treatments were given for 10 min of R (1.8 W m^{-2}) or FR (1.6 W m^{-2}), or 10 min of each wavelength in the sequence indicated.
After Saji *et al.* (1983).

beginning of the night corresponded to the absorption spectrum of Pfr. However, at the time of NBmax, the action spectrum extended throughout R and FR wavebands, with the FR tail beyond 720 nm corresponding almost exactly with the action spectrum, for the inhibition of flowering at the beginning of the night (Fig. 4.18). It was concluded that both the removal of Pfr (the Pfr-requiring reaction) and the formation of Pfr (the night-break reaction at φ_i) inhibited flowering, supporting the concept that two pools of phytochrome with differently stable Pfr are involved in the control of flowering in *Lemna*. Using similar techniques, the same conclusion was reached for *Pharbitis* (Saji *et al.*, 1983). However, here the action spectrum at the time of NBmax corresponded to the absorption spectrum of Pr and showed no FR tail (see Fig. 4.18). The complex exposure-response curves for the control of flowering in *Pharbitis* at certain times in the night (King *et al.*, 1982; Lee *et al.*, 1987) also indicate that more than one pool of phytochrome is involved in the two different responses to light; for example, at the 4th hour, a simple promotion response was obtained (the Pfr-requiring reaction) whereas, at the 16th h (the time of NBmax), both promotion and inhibition responses were obtained as the duration of the R exposure was increased (Fig. 4.19).

The concept of two simultaneous actions of light has important consequences in the interpretation of **null-point experiments**. The null-point method is based on the assumption that a light source will have no physiological effect if it establishes the same Pfr/Ptot ratio as that already present in the tissue at the time of illumination. It was originally devised by Cumming *et al.* (1965) and subsequently used by several

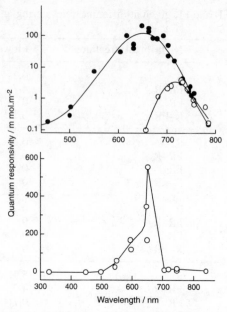

FIG. 4.18. Action spectra for the night-break inhibition of flowering in *Pharbitis nil* and *Lemna paucicostata* 441. *Lemna* (upper); plants were grown in continuous W before receiving a single 16 h dark period; the night-break was given 8 h into the dark period. *Pharbitis* (lower); dark-grown seedlings were given a 10 min photoperiod in R followed by a 48-h inductive dark period; the night-break was given 15 h into the dark period. For both plants, the action spectra for the night-break response (solid lines) show the reciprocal of the exposure required to inhibit flowering to 50% of the dark control values. For *Lemna*, the action spectrum for the inhibition of flowering at the beginning of the dark period is superimposed (open circles). Data from Saji *et al.*, 1983 (*Pharbitis*), Lumsden *et al.*, 1987 (*Lemna*).

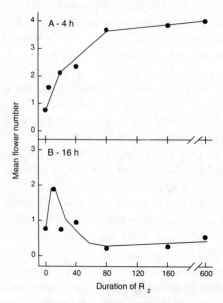

FIG. 4.19. The sensitivity of *Pharbitis nil* to R given at different times in an inductive dark period. Dark-grown seedlings were given a 10 min photoperiod in broad-band R (33 µmol m^{-2} s^{-1}) followed by a 40 h inductive dark period. The night-break with various durations of R (R$_2$ at 3.9 µmol m^{-2} s^{-1}, 660 nm) was given after 4 h or 16 h of darkness. Plants were returned to continuous light at the end of the dark period. After Lee *et al.*, 1987.

other workers to follow Pfr/Ptot ratios in green plants during an inductive dark period, using flowering as the response. However, if Pfr is having two separate and opposing effects, the null test will be misleading. At times when both Pfr input and removal inhibits flowering one can predict that a null value would be unobtainable (see for example Table 4.6). In their original paper, Cumming *et al.* (1965) observed such times but were unable to explain why they should occur. Interpetation of null-point experiments in terms of changes in the internal Pfr/Ptot ratio should, therefore, be made with the greatest care. For example, based on known phytochrome properties, it is essentially impossible that Pfr is re-formed in darkness following an end-of-day exposure to FR as indicated by null measurements in *Chenopodium* (King and Cumming, 1972b).

DO DIFFERENT PHYTOCHROMES CONTROL FLOWERING IN SDP?

A brief summary at this stage may be helpful. The night-break inhibition of flowering in SDP is clearly dependent on Pfr; thus it seems that the Pfr present at the end of the photoperiod declines to a level which is not effective in the night-break reaction before the light-sensitive phase of the photoperiodic rhythm is reached. The Pfr-requiring reaction, which is essential for flowering, is also clearly dependent on Pfr. In this case, however, the Pfr is highly stable in darkness. The dusk signal which initiates dark time measurement is usually attributed to a thermochemical reduction in Pfr, although this has not unequivocally been proved. If Pfr is involved in the dusk signal, it must be highly unstable in darkness since, at physiological temperatures, dark time measurement appears to begin within less than an hour after transfer to darkness. Thus the phytochrome that is associated directly with the photoperiodic mechanism (dusk signal perception and night-break inhibition) and that controlling the Pfr-requiring reaction must be different at least in the stability of Pfr.

Phytochrome Destruction

Phytochromes differ in the stability of Pfr (see Chapter 3). Following exposure to light, type I phytochrome undergoes rapid proteolytic degradation from Pfr with $t_{\frac{1}{2}}$ = 1–2 h. In contrast, type II phytochrome is only slowly degraded from Pfr with $t_{\frac{1}{2}}$ = 7–8 h. There is, at present, no direct evidence concerning the involvement of either type of phytochrome in the photoperiodic responses of SDP. Nevertheless, the stability of the Pfr in the Pfr-requiring reaction (especially in *Chenopodium*, where Pfr apparently remains active for 60 h or more in darkness; see Fig. 4.4) strongly suggests that a type II phytochrome is involved. This is supported by studies with *phyB* mutants of *Arabidopsis*, which have been shown to lack reversible end-of day growth responses to FR (see Chapter 3). For the night-break and dusk signals the situation is ambiguous. Pfr destruction from type I phytochrome is sufficiently rapid to account for loss of activity by the time of night-break sensitivity. Based on physiological evidence, the dusk signal appears to occur within some 30–60 min after transfer to darkness from white fluorescent light. Some Pfr destruction from type I phytochrome would certainly have occurred within this time and, since we do not know the threshold [Pfr] at which time-measurement is coupled, the involvement of type I phytochrome (phytochrome A) is not excluded solely on the basis of the kinetics of Pfr destruction. Probably the

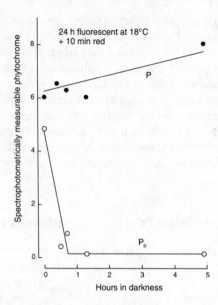

FIG. 4.20. Phytochrome changes during darkness in light-grown plants of *Pharbitis nil*. Dark-grown seedlings were exposed to 24 h W fluorescent light, followed by 10 min R before transfer to darkness at 25°C. The temperature during the light period was maintained at 18°C in order to inhibit chlorophyll accumulation. Changes in total spectrophotometrically-measurable phytochrome (P) and Pfr are shown as $\Delta(\Delta A) \times 10^4$. After Vince-Prue *et al.*, 1978.

strongest argument against the involvement of phytochrome A is the fact that the photoperiodic mechanism continues to operate normally under conditions when little or no unstable phytochrome is present. For example, normal photoperiodic responses are seen in *Pharbitis* when spectrophotometric measurements in light-grown, bleached seedlings indicate that little or no destructible phytochrome remains in the cotyledons, since total phytochrome did not decrease in darkness (Fig. 4.20). *Pharbitis* also shows strong down-regulation of phytochrome A mRNA in the light (see Chapter 5) and a balance between rapid re-synthesis and loss seems unlikely.

Operation of the photoperiodic mechanism through Pfr destruction would require the continued re-synthesis of sufficient Pr phytochrome to maintain sensitivity to the night-break and dawn signals. There was little evidence for re-synthesis in darkness even after some 50 h in dark-grown seedlings of *Pharbitis* following a single brief exposure to R (Fig. 4.21).

Phytochrome Reversion

The relative stabilities of phytochromes I and II concern the extent to which their Pfr is subject to destruction, with consequent loss of the protein moiety. In contrast, the dark reversion process converts Pfr to Pr without any loss of total phytochrome; the need for resynthesis to maintain sensitivity is thus obviated. Partly for this reason, it has often been suggested that reversion of Pfr might account for the dusk signal and it could also account for the need for light in the night-break reaction. One obvious

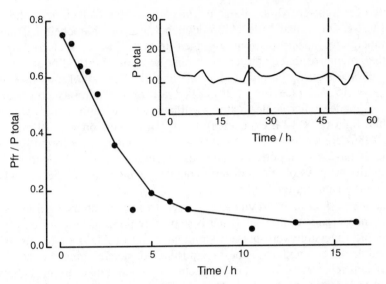

FIG. 4.21. Phytochrome changes during darkness in dark-grown seedlings of *Pharbitis nil*. Dark-grown seedlings were given a saturating exposure to R (664 nm, 10 μmol m^{-2} s^{-1}) before transfer to darkness at 25°C. Total phytochrome (inset) is shown as $\Delta(\Delta A) \times 10^4$.

problem is that, in the early studies, reversion was not observed *in vivo* in dark-grown seedlings of the Gramineae, some members of which are SDP. For example, it is not found in *Zea mays*, which is a SDP for the photoperiodic control of maleness/femaleness. However, based on spectrophotometric measurements, it has been claimed that the reversion process develops during de-etiolation since it was not observed in dark-grown seedlings of *Pharbitis* until they had been exposed to several hours of light (Rombach *et al.*, 1982); possibly seedlings of the Gramineae are similar. It was suggested that the development of photoperiodic sensitivity during de-etiolation was linked to the capacity for phytochrome reversion in *Pharbitis* (Heyde and Rombach, 1988). This introduces a further problem, however, since other workers have found that dark-grown seedlings of *Pharbitis* are responsive to a night-break following a single minute exposure to light (see Figs 2.7, 2.11) when, according to the spectrophotometric measurements, reversion does not take place. Although this night-break effect could be due to residual Pr which was not converted to Pfr by the initial brief exposure to R, several spectrophotometric studies have shown that light is not necessarily required for the development of the capacity for reversion. Reversion has been demonstrated in dark-grown seedlings of *Sinapis alba* (Kendrick and Hillman, 1970) and, even in *Pharbitis*, some 10% of the total phytochrome appeared to undergo reversion following a single pulse of R (Jabben *et al.*, 1980). It is likely that some of the reported inconsistencies are due to the inherent problems of making accurate measurements in very small populations of phytochrome.

Compared with Pfr destruction, dark reversion may be rapid at physiological temperatures with, for example, a $t_{1/2}$ of approximately 8 min being recorded for etiolated seedlings of *Sinapis alba* (Schäfer *et al.*, 1973). Rapid dark reversion has also been reported for light-grown seedlings of *Pharbitis*. After 24 h in white fluorescent light at 18°C, seedlings had little chlorophyll because of photochemical bleaching at this temperature; spectrophotometric measurements were, therefore, possible. Although the amount of detectable phytochrome was low, the total pool appeared to be stable in darkness over a period of at least 5 h at 27°C and all of the Pfr reverted to Pr, with $t_{1/2}$ = 15 min (see Fig. 4.20). This rate of reversion is consistent with measurements made on herbicide-treated seedlings (without chlorophyll) where, after 36 h of white light (3 × 12 h exposures), the $t_{1/2}$ for reversion was 19 min at 22°C (Rombach, 1986). Reversion rates sufficiently fast to counteract the formation of Pfr at low irradiances have been reported for *Sinapis* (Brockman *et al.*, 1987); this would be consistent with the physiological observation that a dusk signal occurs when the irradiance value reaches a threshold.

In cotyledons of *Pharbitis* previously exposed to several hours of light, spectro-photometric measurements indicated that a stable Pfr pool and a pool of Pfr which undergoes rapid reversion were present simultaneously (Table 4.8); it was proposed that these might be equated with the two postulated physiological pools in photo-periodism (Rombach, 1986). So far, only spectrophotometric measurements have been made and so the identity of the phytochrome species in these two pools is unknown; however, neither appears to undergo rapid destruction from Pfr indicating the involve-ment of a type II phytochrome. It is not yet possible to say whether the same phytochrome is present in both the stable and unstable pool, perhaps with differences in their cellular environment, or whether they are different molecules. At least two different type II phytochromes have been recognized immunochemically and one of the genes is characterized by an extended N-terminal domain (Sharrock and Quail, 1989). It is known that, for type I phytochrome, the N-terminal domain is crucial for the thermal stability of Pfr and its deletion in genes used for transformation leads to phytochrome that undergoes rapid dark reversion in the resultant transformed plants. It is possible, therefore, that the modification to the N terminus of one of the type II phytochromes could alter its dark reversion characteristics compared with the other. Thus different type II phytochromes are possible candidates for the processes where fast reversion (photoperiodic timekeeping and the night-break response) and slow reversion (Pfr-requiring reaction) appear to be important *in vivo* factors (Thomas, 1991).

TABLE 4.8. Spectrophotometrically measurable pools of phytochrome in cotyledons of dark-grown *Pharbitis nil* seedlings following different light treatments.

Light treatment	Stable Pfr (Δ (ΔA) × 10^5))	ΔPr in 30 min
10 min red	25 ± 4	−7 ± 12
4 h white; 10 min red	37 ± 4	7 ± 7
6 h white; 10 min red	64 ± 4	18 ± 7
12 h white; 12 h dark; 10 min red	34 ± 2	20 ± 7
12 h white; 2 × (12 h dark; 12 h white); 10 min red	25 ± 3	37 ± 3

Stable Pfr was measured as the Pfr remaining after several hours in darkness. Dark reversion of Pfr to Pr was measured as the increase in Pr (ΔPr) in 30 min of darkness.
After Rombach (1986).

The possibility that the two postulated pools might consist of the same molecular type of phytochrome with different reversion characteristics is not excluded, however. For example, the rate of Pfr reversion of native pea phytochrome *in vitro* can be differentially altered by monoclonal antibodies which react with different epitopes on the protein (Lumsden *et al.*, 1985). The antibody MAP9, which binds near the N-terminus, accelerated the rate of reversion *in vitro*, without proteolysis. Although the relationship between *in vitro* reversion and events *in vivo* is unknown, the fact that reversion may be affected in this way indicates that differently-conjugated pools of phytochrome might show different reversion characteristics *in vivo*. Another pertinent observation concerns the behaviour of phytochrome dimers (see Chapter 3). Because of differences in their dark reversion rates, Thomas (1991) has suggested that different dimers of phytochrome might account for the two physiological pools in the control of flowering, with the night-break and dusk signal being mediated by the faster reverting heterodimer (PrPfr) and the Pfr-requiring reaction by the more stable homodimer. Models involving dimers have also been proposed for other phytochrome-mediated physiological responses (Van der Woude, 1985).

Whatever the identity of the phytochrome molecules undergoing fast and slow reversion as assayed spectrophotometrically in *Pharbitis* (Rombach, 1986), there is as yet no direct evidence linking these two spectrophotometric pools with the two postulated physiological pools in photoperiodism. The best correlation appears to be with the dusk signal. A number of physiological experiments have indicated that dark timing appears to be coupled within 30–60 min after transfer to darkness; by this time, dark reversion would be essentially complete (Vince-Prue *et al.*, 1978; Rombach, 1986). However, the apparent link between the light requirement of 6 h for the development of phytochrome reversion in dark-grown seedlings of *Pharbitis* (Rombach, 1986) and for the generation of a dusk signal (see Fig. 2.11) probably has no physiological significance, since light-grown plants also require several hours of light in order for timekeeping to begin from light-off (see Fig. 2.13).

The dusk signal has not unequivocally been established as being associated with a reduction in the amount of Pfr. Some time ago, the concept of loss of sensitivity to a stable Pfr was proposed to explain the observation that repeated exposures to R were needed to maintain chloroplast movement in *Mesotaenium caldariorum*, even though Pfr was still present in the tissue (Haupt and Reif, 1979). It has been suggested that the dusk signal could be explained in a similar way if, for example, the action of Pfr in the perception of the light signal requires association with a receptor and this Pfr–receptor complex is highly unstable (Thomas and Vince-Prue, 1987). The slight acceleration of dark timekeeping by end-of-day FR, which seems to indicate that a reduction in Pfr itself is the controlling event, would be explained by the immediate loss of effectiveness of the phytochrome–receptor complex when the Pfr component is removed photochemically. There is, however, no direct evidence for the existence of such complexes whereas reversion is a well documented characteristic of some phytochrome molecules, with kinetics that are consistent with the physiological observations.

CONCLUSIONS

In this chapter we have attempted to demonstrate that at least three distinct photo-reactions are relevant to the photoperiodic regulation of flowering in SDP. They are

- the phase-setting effects of light
- the inhibitory action of light given as a night-break
- the requirement of Pfr for a strong flowering response as shown by end-of-day FR effects.

On the basis of present evidence, phytochrome appears to be the photoreceptor for all of these actions of light. However, it is evident that much is yet uncertain regarding the identity of the phytochrome(s) involved and there is even uncertainty about the precise nature of some of the photoperiodic signals. Although physiological experiments have established clearly that there are differences in the apparent stability of the Pfr which is involved in the perception of photoperiodic signals (control of the photoperiodic rhythm and interaction with this rhythm at the inducible phase (the night-break)) and the Pfr which is involved in the Pfr-requiring reaction, it is still not clear how these differences arise. Since much of the physiological evidence seems to exclude the participation of phytochrome A, the most likely suggestion is that a sub-population of a type II phytochrome undergoes rapid dark reversion. Given the multiplicity of phytochromes, this may be a separate species, perhaps uniquely associated with the perception of photoperiodic signals. Alternatively the same type II molecule may behave differently according to the cellular environment, or different dimers could have specific physiological roles. In the future, with the availability of new mutants or transgenic plants, it should be possible to determine whether or not the two sub-populations are distinct molecular species. To date most of the work has been done with mutants of the LDP *Arabidopsis* and there are few useful mutants yet available for SDP (see Chapter 9). However, a presumed phytochrome B-deficient mutant of the SDP *Sorghum bicolor* has been investigated with respect to its photo-periodic responses and timekeeping behaviour (Childs *et al.*, 1995). The mutant (ma_3^R ma_3^R) lacks a light-stable 123 kDa phytochrome and has similarities with other known phytochrome B mutants. In continuous light (but not in darkness), a circadian rhythm in the expression of Lhch and RbcS mRNAs was observed in both mutant and WT genotypes, with no obvious change in phasing. Thus the presumed B phytochrome was evidently not required for entrainment of the underlying oscillator. Unfortunately the NBmax rhythm during an inductive dark period was not investigated in these studies to determine if the oscillator controlling the photoperiodic response also continued to operate normally in the absence of this phytochrome. The ma_3^R mutation also leads to early flowering and relative photoperiod insensitivity; however, when grown in photoperiods ranging from 12 to 24 h, there was still a small quantitative delay in the time to floral initiation when compared to SD. Thus the absence of (presumed) phytochrome B does modify the photoperiodic response in *Sorghum* in some way but some responsivity to daylength still remains. However, further studies are needed on the effect of this mutation on the timing of and sensitivity to a R night-break before any conclusion can be drawn regarding the role of the 123 kDa stable phytochrome in the photoperiodic SD mechanism.

    The discussion has centred on the reactions of higher SDP, where phytochrome appears to be the primary photoreceptor. Blue light is also known to interact with biological clocks in several ways, including the initiation, suppression and phase shifting of rhythms and, from time to time, it has been proposed that a blue-absorbing photoreceptor (BAP) may also be involved in the control of flowering in SDP. However, analyses of the effects of blue light on flowering in *Lemna paucicostata* have led to the conclusion that they are explained by the action of blue light to set an intermediate level of Pfr/Ptot (Hillman, 1967a). This explanation may also be applicable to other reported effects of blue light on flowering in SDP. In contrast, there is good evidence for participation of a BAP in photoperiodic responses of lower plants (Dring, 1988) and also in some species of higher LDP (see Chapters 3 and 5). In cases where the operation of the photoperiodic mechanism depends on light absorbed by a BAP, the problem of stability of a biologically-active ground state molecule (Pfr) does not arise, since the perception of light/dark transitions operates via a short-lived excited state. However, the identity of the BAP remains uncertain and its precise functioning in molecular terms is less well understood than that of phytochrome (see Chapter 3).

The chapter has concentrated on the light-dependent components of the photoperiodic mechanism as deduced from studies of the flowering responses of a relatively small number of SDP, most of which can be induced by a single short-day cycle. To a considerable extent, this avoids the complication of photosynthesis, which undoubtedly contributes to the overall flowering response. This is a restrictive approach as light undoubtedly influences flowering in a number of other ways but the objective has been to analyse those light-dependent reactions which specifically relate to photoperiodism, rather than to flowering in general. Photoperiodism is a particular mechanism for measuring time and, therefore, season and latitude and it may be appropriate to consider it as a 'gating' device which allows induction to proceed (or prevents it from taking place). The subsequent intensity of induction and the later flowering can be influenced by numerous other factors, including light. Much of the confusion in the extensive literature relating to light and photoperiodism probably arises from the fact that many studies have considered only the overall magnitude of flowering responses and have not specifically addressed the question of the role(s) of light in the photoperiodic mechanism itself. For example, many workers have stressed the importance of irradiance and suggested the involvement of one or more irradiance-dependent reactions in the photoperiodic control of flowering. However, while the magnitude of flowering can often be shown to be irradiance dependent during the period of photoperiodic induction (Chapter 1), an analysis of the photoperiodic light reactions in a small number of single-cycle SDP and under particular conditions has shown that these reactions can proceed with only the small quantities of light necessary to input into the photoreversible reaction of phytochrome. It remains to be determined whether this is generally true.

5 Daylength Perception in Long-Day Plants

INTRODUCTION

In Chapter 2, we looked at the mechanisms that are involved in the timekeeping aspects of photoperiodism. Most of the evidence and discussion centred on SDP. This reflects the fact that, in terms of photoperiodic mechanisms, SDP have proved to be the most accessible and rewarding and have therefore received the most attention. Nevertheless the history of research into LDP extends as far back as for SDP, the first experiments on LDP (Klebs, 1910) being performed at about the same time as the first studies on SDP were being undertaken by Tournois (1912). In the initial systematic studies of Garner and Allard (1920), the subjects were Biloxi soybeans and Maryland Mammoth tobacco, both of which were SDP. However, in their follow-up studies they included a number of plants, such as radish, in which flowering occurred in long days but not in short days. Based on the results, they classified species as SDP or LDP and defined LDP as those plants which flower and fruit in a wide range of photoperiods above a critical daylength. This definition, although often attributed to Garner and Allard, appeared in a single-author paper by Garner (1933).

The preference for working with SDP rather than LDP was evident from early studies. When Borthwick and Parker began their investigations into the spectral responses of photoperiodic plants, they consulted with Allard who recommended using SDP. He said that LDP were much less precise in their daylength responses and therefore less suitable for Borthwick and Parker's purposes (Borthwick, 1972). However, where it was possible to compare the night-break responses in LDP with those of SDP, a similar action spectrum, with maximum effectiveness in the red, was obtained (Borthwick *et al.*, 1948). Furthermore, under the appropriate experimental conditions, FR reversibility of the night-break could be demonstrated, confirming the action of phytochrome (Downs, 1956). It was already known (Knott, 1934) that the site of daylength perception in LDP is located in the leaves, as it is in SDP. Thus, it might be argued on the basis of early comparative data that separate consideration of

LDP is unnecessary inasmuch as the mechanisms involved are the same for both response types, but with the responses of LDP being mirror images of those in SDP. Alternatively it could be argued that the imprecise responses of LDP, in experimental protocols which reveal precision in SDP, indicate in themselves that the underlying mechanisms are not the same. It is evident that the situation is not entirely clear cut. There are undoubtedly elements of daylength perception in LDP which are directly comparable to components of the SDP response. On the other hand there are certain characteristics which are highly distinctive for LDP. We will now consider components of the daylength responses of LDP in more detail.

NIGHT-BREAKS

Interruption by light of a non-inductive long dark period in LDP can, under the right conditions lead to floral promotion. However, only a few species of LDP can be induced by a single light break of under 30 min. In this respect they differ significantly from SDP. The initial action spectra for induction in LDP were based on the night-break response of barley and *Hyoscyamus* (Borthwick *et al.*, 1948; Parker *et al.*, 1950). These were obtained by growing the plants on marginally inductive photoperiods of 11.5 or 12 h and induction did not occur when night-breaks were given in combination with 8 h short days. Even with marginal daylengths repeated cycles of treatment were needed to obtain a reasonable flowering response. Evans (1964a), working with the Ceres strain of *Lolium temulentum* which required only one long day for induction, was unable to show inflorescence initiation in a single plant out of 550 which were exposed to a 2 h night-break in the middle of a 16 h dark period. However, 5 min R given 10 h after the end of the main light period did cause some induction when the photoperiod was extended with 4 or 6 h tungsten filament (TF) light which was, in itself non-inductive. Under conditions which allow induction of flowering by night-breaks, 1–2 h is usually sufficient to induce flowering but may not saturate the response (Hughes and Cockshull, 1969; Vince, 1965). The responses of LDP to a night-break are frequently semiquantitative in nature (see Fig. 1.7). Reversibility of the night-break response is also problematical; only partial FR reversal of the promotion of flowering by R could be demonstrated in barley and *Hyoscyamus* (Downs, 1956; Lourtioux, 1961) and to the knowledge of the authors no action spectrum for night-break reversal in LDP has been published. From the limited reversibility studies carried out it can be concluded that phytochrome is the photoreceptor for the night-break response in most LDP, although blue light night-breaks are the most effective for *Sinapis* and maybe other cruciferous plants (Hanke *et al.*, 1969).

LIGHT QUALITY IN THE PHOTOPERIOD

Early studies on the floral responses of a limited range of LDP to light of different spectral quality indicated that orange or red light was the most effective spectral region and that blue and green were relatively ineffective (Rasumov, 1933; Withrow and Benedict, 1936). Most of these studies used fairly rudimentary light sources based on TF lamps in which the likely contamination by far red (near infrared) wavelengths was not assessed. In more rigorous studies by Stolwijk and colleagues in Wageningen

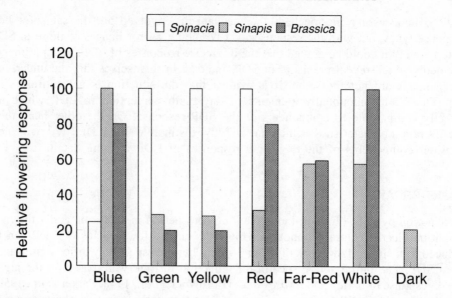

FIG. 5.1. Effects of day extensions with light from different spectral regions on the flowering responses of the LDP *Sinapis alba, Brassica rapa* and *Spinacia oleracea.* (From data of Wassink *et al.*, 1950, 1951; Stolwijk, 1952a, b).

(Wassink *et al.*, 1950, 1951; Stolwijk, 1952a, b) effects of day extensions on the flowering responses of the LDP *Sinapis alba, Brassica rapa* and *Spinacia oleracea* were determined. As shown in Fig. 5.1 the pattern for *Spinacia* was somewhat different from that of *Sinapis* and *Brassica rapa*. In *Spinacia*, day extensions with green, yellow and red were as effective as with white light but blue and infrared had little effect on flowering. In *Sinapis* and *Brassica*, blue and infrared had a strong promoting effect. This observation supported earlier work of Funke (1948) who had noted that members of the Cruciferae appeared to show unusual sensitivity to blue light for floral promotion. This response to blue light is probably mediated by a specific blue photoreceptor. The more puzzling observation was the sensitivity to near infrared (now termed FR) light. The action spectra for night-breaks had indicated that R and hence Pfr promoted flowering. Therefore, it would be expected that FR should inhibit flowering. However, for *Sinapis* in particular, giving FR as a day-extension was much more effective than R in promoting the floral response. Another feature was the contrast with *Spinacia* which was virtually insensitive to FR extensions. Did this indicate that a completely different mechanism was being used by these LDP?

In a series of studies in the 1950s (Wassink *et al.*, 1951; Downs *et al.*, 1958, 1959; Piringer and Cathey, 1960; Friend *et al.*, 1961), researchers consistently found that day extensions with mixtures of R plus FR, or with TF light (which contains both R and FR light), were more effective than R light alone or light from white fluorescent lamps (which contains R but not FR). Acceleration of flowering by the addition of FR light to a day extension with R was first examined in detail in *Hyoscyamus* and was subsequently observed in many other species of LDP, including *Petunia*, dill, barley, sugar beet, wheat, lettuce, carnation, *Lolium* and *Silene* (Takimoto, 1957; Friend, 1963; Vince *et al.*, 1964; Lane *et al.*, 1965; Vince, 1965). The inhibition of flowering in the

SDP strawberry and *Portulaca* has also been shown to be more effective with a mixture of R plus FR than with R alone, indicating that this pattern is not exclusive to LDP; in contrast, the LDP, *Fuchsia* cv Lord Byron showed little acceleration of flowering with added FR indicating control more like that of a SDP (Vince-Prue, 1976).

Vince *et al.* (1964) working on the response to light quality, found that day extensions with R were unable to promote flowering in carnation and lettuce unless FR was added. Vince also carried out experiments with *Lolium temulentum* (Ba 3081), which required four LD cycles for induction. Initial experiments showed that light sources producing R or FR alone were rather ineffective as day extension treatments but that, in combination, they were strongly promotive. In order to determine the relative importance of the R and FR components, further experiments were carried out in which daylength extensions were given with

a)   8 h R
b)   8 h R+FR
c)   7 h R followed by 1 h FR
d)   7 h FR followed by 1 h R.

Of these, treatments (b) and (d) strongly promoted flowering but plants in (a) and (c) remained vegetative (Vince, 1965; Fig. 5.2). Thus, even where a FR extension had only a small effect it could be very strongly promotive if followed by a R treatment before the start of the dark period. Evans *et al.* (1965) also studied the effects of R and FR red light during day extensions using the Ceres strain of *Lolium temulentum*. They found that extending an 8 h photoperiod to 16 h with either R or FR resulted in induction. However, the strongest response was obtained with TF light, which is a

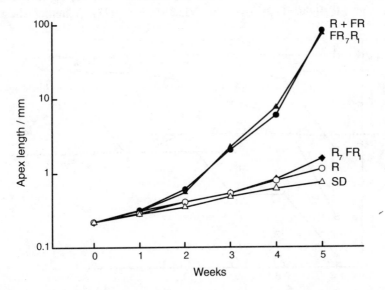

FIG. 5.2.   The effect of R and FR on flowering in *Lolium temulentum* Ba3081. Plants were grown in natural daylight for 8 h and supplementary treatments were given over the next 8 h. Treatments: R, 8 h R; R+FR, 8 h R+FR; FR_7R_1, 7 h FR followed by 1 h R; R_7FR_1, 7 h R followed by 1 h R; S.D., no extension. Plants were transferred to LD and dissected at intervals. After Vince (1965).

TABLE 5.1   Effects of night-breaks on flowering in *Fachsia hybrida* cv Lord Byron.

Night-break	Duration of night-break/h					
	0.5	1	2	4	8	16
Red	43.4	43.2	42.7	42.1	41.5	39.2
Red plus far-red	44.5	43.7	42.4	42.0	38.9	37.5

Results are given as the number of days to anthesis.

mixture of both R and FR. It should be noted that the FR source used by Vince had a longer wavelength cutoff and would have established very low Pfr levels, which could explain the differences in results between the two groups. The marked effect of changing the spectral distribution of the light during the photoperiod, especially altering the FR component, contrasts strongly with the behaviour of most SDP. For this reason it was proposed by Hillman (see Vince-Prue, 1979) that these plants be designated as **light dominant** to distinguish them from plants in which the main criterion for floral induction is the occurrence of a sufficiently long, uninterrupted dark period. These latter plants were then **dark dominant**. Dark dominant and light dominant responses correspond largely, but not exactly to the classification of species as SDP and LDP. However, there are a few LDP (e.g. *Fuchsia hybrida* Lord Byron) where flowering is responsive to a brief night-break, as in SDP, and there is little further effect of increasing the duration beyond 30 min (Table 5.1). There are also a few SDP (e.g. strawberry) which respond poorly, or not at all, to a brief night-break and where the inhibition of flowering requires exposures to long photoperiods containing both R and FR light (Guttridge and Vince-Prue, 1973). Whether these two

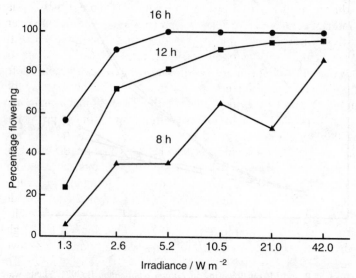

FIG. 5.3.   Interaction of irradiance and duration of a single period of supplementary light on flowering in *Brassica campestris* cv Ceres. Seedlings were given one 8 h SD extended with 8, 12 or 16 h TF light at a range of intensities. After Friend (1968a).

categories really represent different basic mechanisms is unknown, but light-dominant plants show a number of features which appear to be characteristic and which are different from those exhibited by dark-dominant species (Fig. 5.6).

RESPONSES TO LIGHT QUANTITY

The responses of light-dominant plants are frequently of a semiquantitative nature over a wide range of irradiance and duration of light. There are, however, considerable differences between species in the way in which they respond to light. Carnation shows some response to a 2 h night-break in a 16 h dark period, but not to a 2 h extension following an 8 h photoperiod. In this species, acceleration of flowering time continues to increase with increase in the duration of light until it is given continuously throughout the 24 h cycle (Harris, 1968). Irrespective of whether light was given intermittently or continuously, the flowering response was found to be a function of the light integral. However, maximum response was obtained only when the light was continued throughout the 24 h cycle. Thus, here there seems to be evidence for a photon-counting device and also for a photoperiodic timekeeping mechanism since a 2 h exposure to light was effective only as a night-break. Flowering in *Brassica campestris* is also quantitatively related to the duration of light given as a night-break (see Fig. 1.7). When plants were grown in 12 h photoperiods, the percentage of plants flowering increased with increasing irradiance. This seems to be primarily a requirement for assimilates since the addition of sucrose to the medium in SD removed the response to irradiance in green plants and resulted in flowering even in bleached plants (Fig. 5.4). However, although duration can partly be compensated by irradiance (Friend, 1968a), there is not a simple reciprocal relationship as equal amounts of radiation given over different durations did not give the same flowering response. For equal total irradiance a long period of low intensity was more effective than a short period at high intensity (Fig. 5.3). In *Sinapis alba* (Bodson, 1985b), high intensity light is required for flowering but only during part of the day (the first 8 h of the photoperiod); this is thought to be a photosynthetic effect because carbon dioxide is required. However, neither high-intensity light nor sucrose can substitute for the LD effect and plants remained vegetative when given a single SD at the same total energy as during a single LD. It is also evident that the timekeeping mechanism in *Sinapis* is not dependent on exposure to high intensity light, since only low intensities are required during the day extension period.

Thus, even in cases where total exposure may be important there is still evidence for an underlying timing mechanism. In most light-dominant plants, several hours of light are usually required for maximum acceleration of flowering. In general, the same quantity of light is more effective when given as a night-break than when added to the photoperiod indicating that there is a basic timekeeping mechanism which measures the time to NBmax as in dark-dominant plants. Although some 2 h of light may be required, the NBmax time has been shown to be quite clearly defined in several light-dominant species (e.g. *Lolium*; see Fig. 1.10). The existence of a rhythm in time of sensitivity to light (see Fig. 2.6) also demonstrates that the timekeeping mechanism may have similarities to that in SDP.

It might be argued that where an effect of irradiance is found, it could simply be

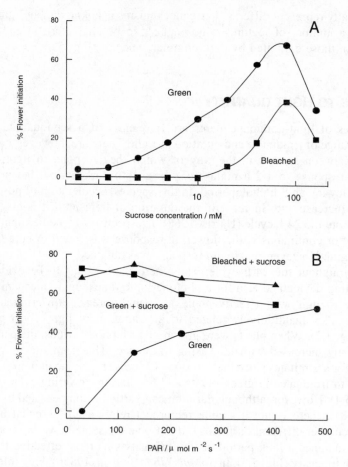

FIG. 5.4. The effect of photon flux density and sucrose concentration on flower initiation in *Brassica campestris* cv Ceres. Plants were grown in 8 h (A) or 12 h (B) photoperiods for 12 days. Bleached plants were cultured on Norflurazon (5 μM). The sucrose concentration was as indicated on the abscissa (A) or 80 mM (B). Friend *et al.*, 1984.

accounted for by the involvement of photosynthesis. As discussed in Chapter 1, it seems likely that the realisation of flowering requires a supply of assimilates, either from concurrent photosynthesis or stored assimilate. Thus, where flowering is measured by the number of flowers or time to anthesis, a photosynthetic effect could be significant. In *Brassica*, the irradiance dependence is lost when plants are supplied with an exogenous source of sucrose (see Fig. 5.4) supporting the suggestion that the irradiance effect is photosynthetic. *Anagallis arvensis*, an absolute LDP with a CDL of 12 h, also has a marked requirement for high-intensity light during the photoperiod; assimilates are evidently necessary for flowering in this plant since removal of carbon dioxide during the day extension period inhibited initiation and the effect could be reversed by the application of glucose. However, even in this case, the requirement for high-intensity light may not be entirely photosynthetic since flowering (flower number) continued to increase at irradiances higher than the saturation value for carbon dioxide uptake (Brulfert *et al.*, 1985).

Not all light-dominant plants show irradiance dependence. In *Lolium temulentum* Ceres only the duration of exposure seems important and, for a day extension with TF light, the response was saturated at only 1.0 W m^{-2} (Evans, 1976). This contrasts with other related species such as wheat, where the flowering response increased with increase in the irradiance of TF light between 0 and 84 W m^{-2} (Friend, 1968b) and barley where the response to 6 h FR light added to a white fluorescent background was irradiance-dependent up to about 10 W m^{-2} at the most sensitive time and up to 50 times that range at the least sensitive period (Deitzer, 1983). However, other aspects of the photoperiodic behaviour of these species has many similarities, so that the differences in irradiance responses of different species may be of secondary importance.

In order to be effective for promoting flowering in LDP lighting does not have to be continuous. Kasperbauer *et al.* (1963a) found that 6 min of TF light in every 60, or 1.5 min in every 15, was equivalent to continuous light in *Melitotus*. This indicated the involvement of a photoreceptor such as phytochrome rather than the photosynthesis system. However, some other results with cyclic lighting suggest that the light and dark reactions of phytochrome are not the only factors involved. In *Hyoscyamus*, the most effective cycle of 6 s/min was still not as effective as continuous light (Schneider *et al.*, 1967). This implies that the decoupling of the photoreceptor from the response takes place in less than a minute, which is much faster than any of the known thermal (i.e. dark) reactions of the phytochrome molecule. Also, in carnation, flowering appears to be related only to total light exposure irrespective of whether this is given as a continuous or intermittent treatment (Harris, 1972). In *Lolium*, cyclic treatments with both R and FR (in any sequence or simultaneously) were always more effective than R (Evans *et al.*, 1965; see Fig. 1.15), indicating that the effect of intermittent exposures is not simply due to maintaining a threshold amount of Pfr during the intervening dark periods.

ACTION SPECTRA FOR LDP — *but missing blue region.*

Action spectra for night-breaks in LDP, where they were possible, showed action maxima at about 660 nm, consistent with light absorption by Pr and hence Pfr formation. Given that the light quality played a significant part in determining the extent of the floral response, a number of workers attempted to obtain action spectra for the effects of extended light treatments either as night-breaks or as day extensions. In these extended treatments reciprocity did not hold. For the most part, spectra have been obtained by comparing the response to a standard light dose at a range of wavelengths. Recently a true action spectrum for the inductive effect of day-extension has been obtained for wheat (Fig. 5.5) which confirms the general features of previous spectral response curves. A feature of all of these spectra is the presence of peaks of effectiveness at wavelengths well beyond the 660 nm absorbance peak for Pr. Action maxima in the 700–710 nm region were found for *Lolium temulentum* (Blondon and Jacques, 1970; Evans *et al.*, 1965) and *Anagallis arvensis* (Imhoff *et al.*, 1971). Peaks at even longer wavelengths, close to 720 nm, were obtained in wheat and *Brassica campestris* (Friend, 1968b; Carr-Smith *et al.*, 1989) and in sugar beet, *Hyoscyamus* and spinach (Schneider *et al.*, 1967; Borthwick *et al.*, 1969). In some

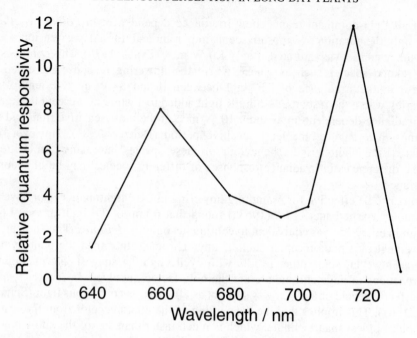

FIG. 5.5.   Action spectrum for floral promotion in wheat. After Carr-Smith *et al.* (1989).

cases a peak in the red part of the spectrum was found together with one at wavelengths greater than 700 nm (Fig. 5.5). In *Hyoscyamus*, the peak of activity at 710 nm was retained even when given against a background of red light (Borthwick *et al.*, 1969). In *Lolium*, there was some evidence in a shift of the action maximum to longer wavelengths when a red background was introduced, consistent with the action of phytochrome, although comparative data for only four different wavelengths were presented (Blondon and Jacques, 1970). (Exceptionally, the LDP *Calamintha nepetoides* shows maximum sensitivity in red light (as observed for SDP) with no evidence of action beyond 700 nm (Jacques and Jacques, 1989).

RESPONSE TO THE TIMING OF LIGHT TREATMENTS

The time in the photoperiod when treatments with different light qualities are given considerably influences their effectiveness. Vince (1965) found that giving a short day from 08.00 to 16.00 h in daylight and extending it with R from 16.00 h to midnight resulted in little induction in *Lolium temulentum* Ba3081. However, a pre-photoperiod extension with R from midnight to 08.00 h resulted in a very strong flowering response. Evans (1976) compared the spectral response for floral promotion in the first and second part of a 16 h period following an 8 h short day in the cultivar Ceres; he also found that maximum effectiveness occurred at wavelengths above 700 nm during the first part of the 16 h but was shifted to between 660 and 680 nm in the latter half of this period. In these experiments, as with those of Vince, light treatments at the

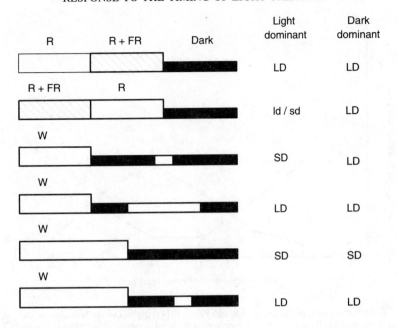

FIG. 5.6. Schematic representation of the typical responses of light dominant and dark dominant plants to combinations of light and darkness. SD, treatment perceived as a short day; LD, treatment perceived as a long day.

end of a 16 h night period can also be considered as being at the beginning of the subsequent photoperiod. Thus R appears to be effective during the early part of a long photoperiod but becomes less so later in the day. Light containing FR, on the other hand, is not required in the early part of the photoperiod but in the latter part of the photoperiod becomes required for the full inductive response in most LDP and, in some cases, is required for any floral response at all. Overall, a more or less consistent pattern of response to day extensions and night-breaks seems to hold for light-dominant LDP; this is summarised in Fig. 5.6.

Vince (1965) looked at the timing of the response to R and FR in more detail. In *Lolium*, extending an 8 h natural photoperiod in daylight with 8 h of low intensity R had little effect on flowering but flowering was markedly promoted when the R day-extension was interrupted for a few hours with FR. There was a marked optimum for the FR interpolation which coincided with the time when R was most inhibitory when inserted into a day extension with FR (Fig. 5.7). Similar results were obtained with several other LDP where flowering was accelerated by inserting treatments rich in FR into day extensions with fluorescent lamps which have virtually no FR output. Maximal promotion occurred when FR was given in the early part of the day-extension and this coincided with the time when fluorescent light was most inhibitory (Lane *et al.*, 1965). Thus, although R (and presumably a high Pfr level) promoted flowering at certain times, as evidenced by the promoting effects of night-breaks, at other times, especially in the early part of a night following a SD, high Pfr levels inhibited flowering. There was thus an apparent requirement for a low level of Pfr at these times. As sunlight contains a mixture of R and FR it was proposed that both the

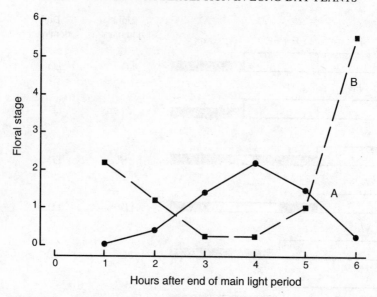

FIG. 5.7.   The effect of time of exposure to R and FR on flowering in *Lolium temulentum* Ba 3081. Plants were grown in natural daylight for 8 h and supplementary treatments were given over the next 7 h. In A, plants were given 2 h FR centred at the times indicated and R for the remainder of the time. In B, they were given R for 2 h centred at the times indicated and FR for the remainder of the period. All plants were then given R for 1 h and returned to darkness for the remaining 8 h of the daily cycle. Ten treatment cycles were given. After Vince (1965).

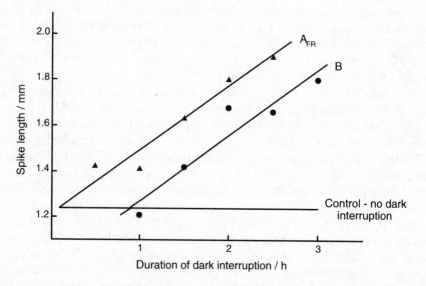

FIG. 5.8.   Effect of a dark interruption during a day extension with R on flowering in *Lolium temulentum* Ba 3081. Plants were grown in natural daylight for 8 h and R was given over the next 9 h. A_{FR} received 10 min FR prior to darkness while B received only the dark interruption. After Holland and Vince (1971).

high and low Pfr reactions would be satisfied. An alternative explanation could be that the response to FR represents a positive action of phytochrome, such as an HIR type of response (see Chapter 3), which requires a low Pfr/Ptot level and thus cannot be elicited by R on its own. The high Pfr response (i.e. to R) would then be due to the action of phytochrome in a simple Pfr-dependent mode and FR would promote flowering in the period following a SD because, at this time, the plants would be sensitive to phytochrome acting through an HIR-type mode. The crucial physiological experiments to distinguish between these two alternative proposals should aim to establish whether the promotion of flowering by FR at certain times is positive (an HIR type of response) or whether merely lowering Pfr levels is enough. The best evidence for the latter comes from experiments with *Lolium temulentum* (Holland and Vince, 1971). They interrupted a day extension of R with varying durations of darkness. This, they reasoned, would allow some of the Pfr to revert or decay thermally. Such dark interruptions promoted flowering and the effect was proportional to the duration of the dark period at least up to 3 h. When a dark interruption was preceded by a brief exposure to FR in order to lower the Pfr level instantaneously, the effect was increased by an amount equal to that resulting from about 45 min of darkness (Fig. 5.8). Furthermore, the promoting effect of FR could be reversed by a subsequent exposure to R. These results are consistent with the idea that a low Pfr reaction is required for flowering in *Lolium* (i.e. that Pfr is inhibitory at this time). However, in the same study it was found that the response to FR increased with irradiance and almost linearly with duration over a 2 h treatment, even though photoconversion from Pfr to Pr was saturated in less than 1 h in etiolated mung bean seedlings given the same FR treatment. They were thus unable to rule out a positive action of FR in their experiments.

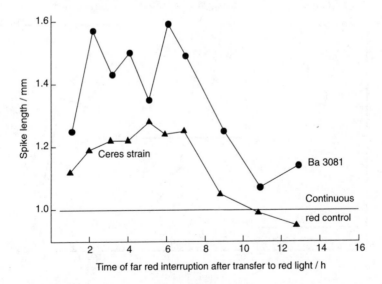

FIG. 5.9. The response to FR light given at different times in two strains of *Lolium temulentum*. After an 8 h day in sunlight, plants were transferred to R for the remainder of the 24 h cycle, interrupted with 2 h FR centred at the times indicated. There were four treatment cycles and plants otherwise remained in SD. After Holland (1969).

What is without question is that the responses of LDP to R and FR change during the course of the daily cycle. This is borne out by experiments carried out using a range of different methodological approaches. For example, when R was given to *Lolium* throughout the whole of a 16 h 'night', interruption with 2 h of FR clearly had a much greater promotive effect in the first part as opposed to the second (Holland, 1969; Fig. 5.9). Similar patterns of response to a 4 h interruption with FR also occurred during the course of the night in *Petunia*, *Fuchsia* and *Lolium* (Fig. 5.10). However, although the pattern of the changing response to FR was similar in all three plants examined, the absolute response varied considerably. In *Petunia*, FR strongly promoted flowering at any time in the 16 h period, whereas in *Lolium*, flowering was actually inhibited by adding FR during the final 4 h of the treatment period. Perhaps the most interesting results are those with *Fuchsia*. This has been classified as a dark-dominant LDP and, compared with continuous R, the substitution of FR at any time depressed flowering. Nevertheless, the pattern of response to FR was similar to that in the other plants indicating that, despite their characteristic differences, the underlying mechanisms may not be so different in the dark-dominant and light-dominant control of flowering. Different approaches with carnation (Harris, 1972) and *Hyoscyamus* (El Hattab, 1968) revealed similar patterns of change, R becoming more promotive and FR becoming less so as the night proceeded. As the relative effectiveness of R and FR alters during the day there can be a change in the apparent optimum photoequilibrium which is revealed by experiments with mixtures of R and FR (Fig. 5.11). In *Lolium* when light was given in the early part of the night the optimum lay towards low R/FR but, when given throughout the night, the optimum was shifted towards higher R/FR.

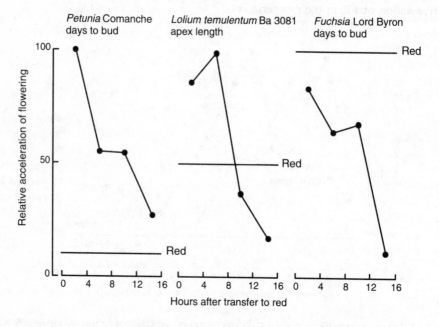

FIG. 5.10. Effect on flowering of interpolating a 4 h period of FR into a 16 h R day extension following 8 h sunlight in the daily cycle. Flowering values are plotted at the mid point of the FR interruption. Red, relative response for plants not receiving a FR interruption. After Holland (1969).

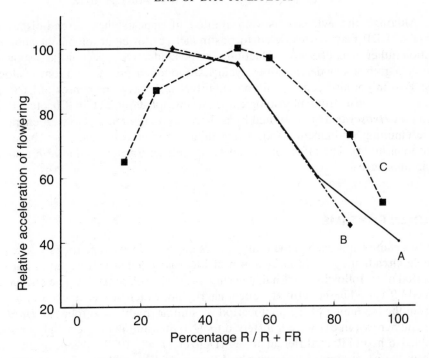

FIG. 5.11. Change in optimum R/FR ratio for maximum acceleration of flowering in *Lolium temulentum*. All values are shown relative to SD controls. Curve A: *L. temulentum* Ba 3081. R/FR mixtures given for 2 h from 19:00 h to 21:00 h, plants otherwise received a R day extension from 16:00 h to 01:00 h for four cycles. Curve B: *L. temulentum* Ceres. R/FR mixtures given as day extensions from 16:00 h to 24:00 h for 21 cycles. Curve C: *L. temulentum* Ceres. R/FR mixtures given throughout the night from 16:00 h to 08:00 h for 1 cycle. After Vince-Prue (1975). Data from Holland and Vince (1971), Lane *et al.* (1965), Evans *et al.* (1965).

END-OF-DAY FR EFFECTS

As shown in Chapter 4, a FR treatment given at the beginning of an inductive dark period can reduce the magnitude of the flowering response in SDP. The effect is more pronounced following a relatively short photoperiod suggesting that a full flowering response requires the presence of Pfr over a period of time which can include either the photoperiod and/or the dark period. Some evidence for an equivalent effect of Pfr can be found in the responses of LDP. Downs *et al.* (1958) found that an end-of-day FR treatment given to dill, a quantitative LDP, enhanced flowering and stem extension, and that this was reversible by a subsequent R treatment. In a later study with *Hyoscyamus*, Downs and Thomas (1982) also found that, under inductive conditions, a low Pfr /Ptot ratio obtained by irradiating with an end-of-day FR promoted flowering. The promoting effect of FR on *Hysoscyamus* was found to be much less during the first half of a 20 day experimental period than in the second. As induction is largely complete after the first 10 LD, they concluded that the main effect of removing Pfr at the end of the daily photoperiod was on floral stem extension and flower development. This is consistent with the known effects of end-of-day FR to promote stem elongation in a number of plants where flowering is not daylength dependent (Downs *et al.*,

1957). Although the evidence is very limited, it appears that an end-of-day FR response in LDP may be restricted to postinductive flower development and stem elongation rather to an effect on induction. The postinductive aspect of the response is obviously significant under multiple cycle LD-induction protocols. From analogies with SDP, it might also conceivably be important in SD-type responses of LDP. For example, it is clear from physiological experiments described in Chapter 1 that flowering in *Hyoscyamus* is inhibited by a short day-long night cycle with characteristics resembling SD-induction (e.g. some light is required for inhibition by a subsequent long night). The effect of an end-of-day FR treatment has not been tested in such circumstances.

CIRCADIAN RHYTHMS

The most detailed approach to the timing of FR has been to examine the response over the whole circadian cycle. In *Lolium*, 4 h of FR was added at various times during a 40 h period of R, following an 8 h day in sunlight (Vince-Prue, 1975). The response to the added FR varied in the form of a circadian rhythm which peaked at about 8–10 and 35 h from the beginning of the photoperiod. A similar rhythm was found in *Hordeum* when plants grown in SD were transferred to 72 h of white fluorescent light to which was added 6 h of FR at different times (Deitzer *et al.*, 1979). Again, a circadian rhythm in the promotion of flowering by FR was evident, with maxima occurring at about 16 and 40 hours from the beginning of the photoperiod. In both plants, the rhythm clearly continued in the light over at least two circadian cycles (Fig. 5.12) and the peaks were separated by about 24 h. Deitzer (1984) also showed that a similar

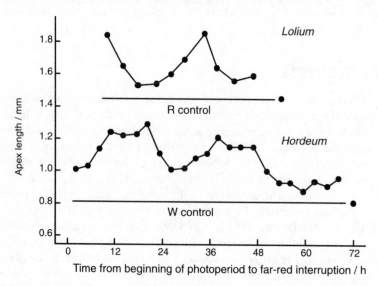

FIG. 5.12. Rhythmic responses to added FR light in the LDP *Lolium temulentum* Ba 3081 and *Hordeum vulgar* Wintex. *Lolium* was given four cycles of 8 h sunlight extended with R. *Hordeum* received a 72 h photoperiod in W fluorescent light. FR, 4 h for *Lolium* and 6 h for *Hordeum*, was added at various times. After Thomas and Vince-Prue (1984).

rhythm could be observed in the response of *Arabidopsis* to added FR light. It seems most likely that this rhythm in the response to FR accounts for many of the observations that the effect of added FR changes with the time of day although, so far, only these three species have been examined in sufficient detail to reveal the circadian rhythm to FR.

It has also been reported that there is a circadian rhythm in response to a night-break in the LDP *Sinapis*, *Hyoscyamus* (Hsu and Hamner, 1967) and *Lolium* (Kinet, 1972; Kinet *et al.*, 1973; Vince-Prue, 1975; Périlleux *et al.*, 1994). Initial studies with *Sinapis* used an 8 h fluorescent W treatment, which the authors considered to be a short day, given in a 40 h or 64 h dark period (Kinet, 1972; Kinet *et al.*, 1973). In the 40 h experiment two peaks at about 10 and 34 h were obtained. With 64 h of darkness, three peaks at 6, 30 and 58 h were found but the middle peak was much smaller than the other two (see Fig. 2.6). When Vince-Prue (1975) used a 2 h R night-break with *Lolium* in a 40 h dark period, she obtained two peaks in similar position to those in the *Sinapis* experiments (see Fig. 2.6). More recently, the effect of 8 h light given in a 40 h or 64 h dark period was investigated in *Lolium*. While the results with the 40 h dark period were comparable to the earlier work with *Lolium* and *Sinapis*, only two peaks at about 4 h and 20–24 h were found when a 64 h dark period was used, but light at the end of the dark period beginning at 56 h was also strongly promotive (Périlleux et al., 1994) (Fig. 5.13). The results indicate that the response to an 8 h light break is similar to that of a 2 h night-break but they also show that the pattern of

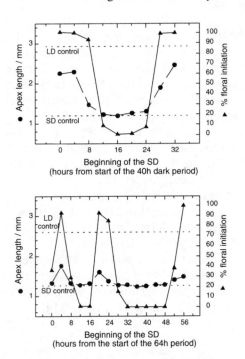

FIG. 5.13. Flowering response of plants of *Lolium temulentum* Ceres to an 8 h DSD (displaced short day) interruption of a 40 h (upper) or 64 h (lower) dark period. Flowering response in terms of apex length or % floral initiation is plotted against the time of the beginning of the interruption. Redrawn from Périlleux *et al.* (1994).

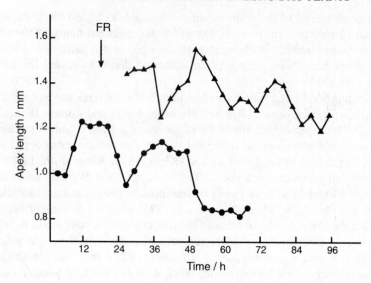

FIG. 5.14.   Phase shifting effect of a 6 h FR treatment on the rhythmic response to FR light in *Hordeum vulgare* Wintex. Plants were given 72 or 96 h continuous daylight fluorescent light to which was added 6 h FR at various times. The lower curve shows the response to single FR treatments. The upper curve shows the effect of an additional 6 h FR treatment at the time indicated by the arrow on the subsequent response to added FR. After Deitzer *et al.* (1982).

response varies with the duration of the experimental dark period. Experiments with *Arabidopsis* also revealed a rhythm in response to a night-break but, again, the results were complex and indicated interactions with both the preceding and subsequent light periods (Mozley and Thomas, unpublished). In this respect the night-break rhythm is significantly different from the FR response which is clearly under the control of a single circadian rhythm. It appears that night-breaks not only interact with a rhythm of light sensitivity that continues in darkness (as in SDP), but can also interact with the subsequent photoperiod. In this, LDP differ from SDP where the NB rhythm is independent of the length of the dark period, the phase normally being set at the end of the preceding photoperiod (see Chapter 2).

As far as the authors are aware, there is no published work on the phase control of the rhythm in sensitivity to night-break light in LDP such as *Hyoscyamus* and *Lolium*. However, for *Hordeum*, there is some information on the phase control of the rhythmic flowering response to 6 h FR added to white light (Deitzer *et al.*, 1982). In this case, it was found that an exposure to FR between 15 and 21 h after the beginning of the 72 h fluorescent light treatment gave about 50% of the promotion gained by adding FR for the whole of the experiment. Subsequent additions of 6 h FR also showed a rhythmic response but with the phase altered by about 12 h (Fig. 5.14).

POSSIBLE MECHANISMS

There is a good deal of general evidence that a circadian clock is involved in the perception of light and/or dark duration in LDP. As discussed in Chapter 2, cycle length experiments and night-break experiments have both revealed a circadian oscillation in the flowering response to R or W and the underlying rhythm seems to be out of phase with that in SDP (as would be predicted from Bünning's original hypothesis). For example a light exposure promoted flowering in the LDP *Lolium* at times when it inhibited flowering in the SDP *Glycine*, while cycle lengths of 24 and 48 h were most inhibitory to flowering in the LDP *Hyoscyamus* but were optimal for flowering in *Glycine* (Hsu and Hamner, 1967). There is, in addition, the circadian rhythm to FR added to R or W fluorescent light that has been observed in the LDP, *Lolium*, *Hordeum* and *Arabidopsis*.

Responses of LDP are much less well understood than those of SDP, both in terms of the action of the photoreceptor and in the way in which the observed circadian rhythms may operate to control the flowering response under natural conditions. Early experiments with non-24 h cycles and the responsiveness to brief R night-breaks demonstrated that LDP measure a critical nightlength which, if exceeded, prevents flowering. Unlike SDP, however, the critical nightlength is not the overriding factor in light-dominant LDP. Experiments with 16 h days and 8 h nights, much shorter than the critical nightlength, show that flowering often does not occur if conditions during light period are incorrect as, for example, if 8 h R plus FR is followed by 8 h R (see Fig. 5.6). Thus, whatever the basis is for timekeeping in LDP, it is not concerned only with measurement of a critical nightlength.

The night-break rhythm experiments (although with only a few species) indicate similarities between the mechanism for time-measurement in LDP and SDP with flowering dependent on whether or not light is given at an inducible phase of the rhythm. Results with *Lemna gibba*, in which a light-on rhythm results in flowering only when a second pulse of light is given at a particular circadian time (Oota and Nakashima, 1978) support this conclusion. This behaviour is very similar to the SDP *Pharbitis*, where a light-on rhythm results in flowering unless a second pulse of light is given at a particular circadian time. The only difference is whether the second pulse of light promotes (LDP) or inhibits (SDP) flowering. However, the way in which the observed night-break rhythm might measure a critical duration of darkness under natural conditions remains to be determined. It is possible that this occurs in the same way as in SDP, with the rhythm being released at dusk to run from a phase point that is established during the preceding photoperiod. However, it is extremely difficult to carry out experiments to examine this possibility as, unlike SDP, altering the duration of the photoperiod can itself alter the flowering response in LDP. Attempts to investigate whether the rhythm runs from a fixed phase point at light-off in *Lolium* were unsuccessful and yielded ambiguous results (Vince-Prue, unpublished results).

Response to Far Red Light

The presence of an action maximum beyond 700 nm and a strong dependence of the response on the irradiance of the experimental treatment have led to the suggestion that the effect might be related to the high-irradiance response (HIR) of photomorphogenesis (see Chapter 3). The HIR, sometimes referred to as the HER (high-energy response) characteristically requires extended irradiation for effect, has action maxima in FR and blue, and shows irradiance dependency; it can also be difficult to demonstrate reciprocity. One argument against the flowering response being the same HIR as that

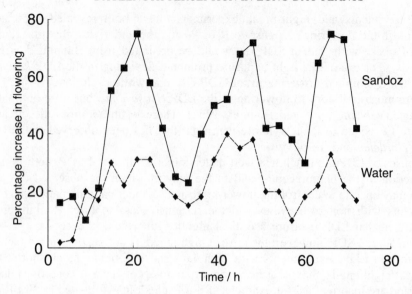

FIG. 5.15.   Effect of adding 6 h FR at different times to a background of white fluorescent light on the flowering of *Arabidopsis thaliana*. Plants were grown on agar or with 5×10^{-7} M Norflurazon (Sandoz) to inhibit chlorophyll formation and hence photosynthesis. After Deitzer (1984).

operating in etiolated seedlings is that a blue action maximum is only found in a few species of LDP (Evans *et al.*, 1965). However it is now known that the blue and FR peaks of the HIR may be mediated by separate photoreceptors. The FR peak is mediated by phytochrome A (Parks and Quail, 1993; Nagatani *et al.*, 1993; Whitelam *et al.*, 1993) whereas the response to blue is mediated by a specific blue-absorbing photoreceptor (Thomas and Dickinson, 1979). An action maximum between 700 and 720 nm might also arise through an interaction with the reaction centre chlorophyll in photosystem I (Schneider and Stimson, 1971, 1972). This possibility was tested by comparing the flowering response of green *Arabidopsis* with that of plants treated with the herbicide Norflurazon, which inhibits carotenoid synthesis leading to photobleaching and lack of chlorophyll (Deitzer, 1984). When 6 h FR was added to a white fluorescent background at different times during a 72 h period, both sets of plants showed a similar rhythmic pattern of stimulation by FR with a periodicity of about 24 h (Fig. 5.15). Thus the response to FR could be demonstrated in the absence of active photosystem I indicating the involvement of a separate photoreceptor; photosynthesis was at most interactive and exerting a modulating effect.

Recent evidence that the promotion of flowering by light containing a high proportion of FR is mediated through phytochrome comes from the finding by Whitelam and colleagues (Johnson *et al.*, 1994) that phytochrome A-deficient mutants of *Arabidopsis* are relatively insensitive to floral promotion by day extensions with TF light when grown from germination to flowering under these conditions. The inductive response to extensions with FR (Fig. 5.16) or R plus FR mixtures (Thomas, unpublished data) was also lost when only a single inductive photoperiod was given; thus the effect of phytochrome A deficiency to prevent the flowering response to day extensions with R plus FR light could not be explained by an effect on pre- or postinductive development

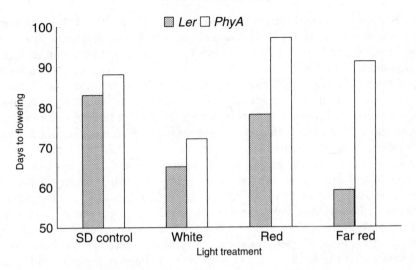

FIG. 5.16. Induction by a single long day in phyA mutants of *Arabidopsis* Thomas, unpublished.

in the seedlings. However, the response to FR was retained in other mutants which contained the photoreceptor but were unable to show the FR HIR for hypocotyl inhibition in young seedlings (Thomas, unpublished data). These are presumed to be transduction mutants and the conclusion is that, whereas the response to FR for both hypocotyl extension and flowering is mediated by phytochrome A, the signal transduction pathways are not the same for both.

A corollary of the suggestion that phytochrome A mediates the strong floral response to light containing FR in LDP is that phytochrome A must be expressed in the photoperceptive tissues. There have been relatively few studies of phytochrome A expression in light-grown tissues and these have suggested that phytochrome A mRNA levels are down-regulated to different extents in light-grown tissues of different species. In monocotyledons such as *Avena*, maize and rice, down-regulation of mRNA is severe, in contrast with the situation in most dicotyledons where phytochrome is either partially down-regulated or not down-regulated at all (Quail, 1994). Interestingly, the SDP dicotyledonous plant *Pharbitis nil* is an exception, showing strong down-regulation of phytochrome A mRNA by light (Robertson, 1995). Additionally, in monocotyledons, phytochrome A destruction is generally believed to be a zero-order event. Therefore it would be predicted that phytochrome A levels would be very much lower in monocotyledons than in most dicotyledons under the same irradiation conditions, such as in daylight. However, many LDP are graminaceous monocotyledons and are as responsive to FR-enriched light as those which are dicotyledons. To address this apparent paradox Carr-Smith *et al.* (1994) used immunoassays to follow the kinetics of phytochrome A synthesis and destruction in a daylength-sensitive cultivar of wheat. Although some evidence for down-regulation of photoreceptor synthesis by light was found, this was much less than in *Avena* which was used for comparison. A major difference was, however, that Pfr destruction in both dark-grown and light-grown tissues was approximately first order in wheat but not in *Avena*. Taken together, these results predicted that phytochrome A levels in wheat should be wavelength-dependent and higher in FR than in R. Measurements of phytochrome A were made by ELISA at a range of wavelengths and compared under the day extension conditions which had been used to construct an action spectrum for floral promotion (Carr-Smith *et al.*, 1989). The measured levels correlated well with those predicted from the kinetic experiments and the flowering response (Carr-Smith, 1990). The results from this one species are consistent with phytochrome A being the photoreceptor for the FR promotion of flowering in LDP but further studies are needed to confirm this as a general feature of graminaceous LDP. It must be noted, however, that *Avena* (where down-regulation is very strong) is also a LDP although, as far as the authors are aware, nothing is known about its requirement for FR light.

The mutant and expression data, while providing compelling evidence that phytochrome A mediates the response to continuous light rich in FR, remain difficult to reconcile with the action of phytochrome A in seedling development. Action spectra showing a peak at about 720 nm are common for responses such as inhibition of hypocotyl extension or pigment synthesis in etiolated seedlings. A prerequisite for such spectra is that the seedlings previously be maintained in darkness so that a high level of phytochrome A is present at the beginning of the experiment. The effectiveness of FR is then considered to be due partly to the fact that phytochrome destruction is minimised and a large pool of phytochrome is maintained. To retain a large pool of phytochrome A, a low photoequilibrium (e.g. 3–5% Pfr/Ptot) is needed. Pretreatment with red light, which depletes the pool, leads to a loss of the FR peak in etiolated mustard seedlings (Beggs *et al.*,1980). Yet the effect of FR in long day floral induction takes place in light-grown tissues where relatively low levels of phytochrome A will exist at the beginning of the treatment. Furthermore the most effective treatments for LD induction are mixtures of R and FR which give photoequilibria much higher than those which maintain high levels of phytochrome A, whereas simultaneous irradiation with red light abolishes sensitivity to FR in etiolated seedlings. If phytochrome A mediates the FR response in LDP it appears to do so in a way which does not require high phytochrome levels and in a way which is not directly related to photoequilibrium. In other words, phytochrome A appears to act specifically as a detector of FR, either alone or in R plus FR mixtures. One possible mechanism for this is that the addition of FR to white fluorescent light, while decreasing the proportion of phytochrome as Pfr, would also lead to an increase in the Pfr –Pr heterodimer (see Chapter 3). A model by which Pr–Pfr heterodimers can account for HIR-like responses has been advanced by Van DerWoude (1987). A changing sensitivity to added FR would then be observed if there were a circadian rhythm of response to the heterodimer (Thomas, 1991). Other possibilities can also be advanced; for example, the response might be coupled to the cycling of phytochrome between the Pr and Pfr forms.

There is now good evidence that a circadian rhythm in sensitivity to FR during the photoperiod accounts for the response of LDP to added FR and that the response to FR is mediated by phytochrome A. Measurements on phytochrome A mRNA levels in *Sinapis* have failed to show any evidence for a circadian rhythm in continuous light or in darkness (Robertson, 1995). The rhythm in sensitivity to FR seems, therefore, not to arise from a rhythm in photoreceptor synthesis but rather seems to be a rhythm in the cellular response to the photoreceptor signal. This could arise through the circadian control of a specific phytochrome receptor or transduction chain component.

Although *Lolium temulentum* resembles barley and *Arabidopsis* in showing a rhythmic response to FR, it is difficult to explain this through a positive action of phytochrome. A detailed series of experiments showed that, at the time when FR promoted flowering most strongly, there was no optimum R/FR ratio and indicated that flowering was promoted when Pfr was lowered below a certain threshold value, equivalent to that established by a mixture of 25% R and 75% FR light (see Fig. 5.11, curve A). This conclusion was further strengthened by the fact that a period of darkness also strongly promoted flowering at this time and, moreover, that the promotion was increased by a brief exposure to FR given at the beginning of the dark period and decreased by a brief pulse of R given near the middle. Reversibility was obtained in both cases, strongly suggesting that the promotion of flowering by FR was largely, if not entirely due to a reduction in Pfr and not to heterodimer formation. Detailed studies showed that, when a dark interruption was preceded by a brief FR treatment in order to lower Pfr immediately, its effect was increased by an amount equal to that resulting from about 45 min of darkness (see Fig. 5.8). Pfr thus appeared to fall to a non-inhibitory level within about 45 min of darkness following transfer from R; this is consistent with the estimates for Pfr loss associated with the dusk signal in SDP (see Chapter 4). At this time of day, darkness appeared to be more promoting than R for at least 3 h (see Fig. 5.8).

One question raised by the *Lolium* experiments is whether a dark interruption of a day-extension and the day-extension itself act on the same part of the photoperiodic mechanism. By introducing a dark period there is a possibility of generating false dusk and dawn signals within the treatment cycle, which could have unexpected effects on any endogenous rhythms in light sensitivity. It was shown, for example in *Hyoscyamus* that a 9 h photoperiod combined with a 15 h dark period (i.e. in a 24 h cycle) is non-inductive but flowering occurs if the cycle length (and the duration of the dark period) is either increased or decreased (Finn and Hamner, 1960; see Fig. 2.8). It very difficult to rule out the possibility that the effects of dark interruptions involve some complex interaction with cycle length. Another problem with the Pfr-removal hypothesis is that the response to FR in, for example, wheat and barley and even in *Lolium* is irradiance dependent at light dosages beyond those which saturate the conversion of Pfr to Pr. This suggests that the FR has effects in addition to those on the photoequilibrium. If part or all of the effect of adding FR is merely to lower the Pfr level, it would be expected that there should be a circadian rhythm in response to a dark interruption of an extended light period without FR in barley and *Lolium*. Such experiments have not to our knowledge been performed.

Rhythms have been found in the responses to both R and FR suggesting that for whatever reason low Pfr/Ptot inhibits flowering at some phases of a circadian rhythm and promotes it at others. More than one rhythm may be involved as the time of NBmax (R night-break) in darkness (see Fig. 2.6) did not precisely coincide with the time that adding FR light (in 40 h R) had least effect (see Fig. 5.12). For the FR rhythm, the phase must be established at light-on because the rhythm runs in continuous light. If a single rhythm were rephased at light-off, this could provide an explanation for the non-coincidence of R maxima in darkness (rhythm rephased) and FR minima when added to continuous R (rhythm rephased from light-on).

One other type of approach to the problem of the FR enhancement of flowering has been used with *Lolium*, but not apparently with other plants. Intermittent treatments with 1 min R and 1 min FR were used in various sequences and continued throughout a 16 h experimental period following an 8 h day in sunlight. In early experiments with the cultivar Ceres, cycles containing both R and FR were always more effective than R pulses alone; there was no evidence for reversibility and cycles of R/FR promoted flowering as strongly as cycles of FR/R (Evans et al., 1965). A problem with these experiments is that the low irradiances used may not have been sufficient to saturate the photoconversion, so that an intermediate level of Pfr may have been established in all treatments. More recent experiments, using higher irradiances, have confirmed that any sequence of R and FR, or a mixture, is more effective that R or FR alone (see Fig. 1.15). However, there was a degree of reversibility and sequences ending in R were considerably more effective than sequences ending in FR, pointing to the involvement of Pfr.

CONCLUSIONS

Two different explanations have been given for the observation that some FR light is required for optimal flowering in light-dominant plants. These are an HIR operating through phytochrome A (Deitzer, 1984; Thomas, 1991; Johnson et al., 1994) or an inhibitory reaction of Pfr which is prevented (Vince-Prue and Takimoto, 1987). At present, it is not possible to say with certainty which of these explanations is correct. Nor can we rule out the possibility that different mechanisms may operate in different plants or that, even in the same plant, FR may enhance the flowering response in more than one way. The recent availability of mutants in *Arabidopsis* offers a real opportunity for evaluating the different possibilities but there is need for detailed physiological studies which are so far lacking in this species. In addition, more information is required about phytochrome levels and the kinetics of different phytochrome pools in a range of species before the FR enhancement of flowering in LDP can be fully understood. The availability of conventional or transgenic mutants in other LDP would aid these investigations.

Timekeeping in light-dominant LDP is much less well understood than in SDP. Although circadian rhythms in sensitivity to FR and R light have been observed, it is not known how they operate in relation to photoperiodic timekeeping in these plants. A critical nightlength is measured, but it does not appear to be the overriding factor as it is in dark-dominant SDP since the spectral conditions during the photoperiod may prevent flowering even when the nightlength is less than the critical duration when measured under natural daylight conditions. Any model for the control of photoperiodic flowering in such LDP needs to account for the following observations:

- circadian rhythms in the responses to cycle lengths, night-breaks during the dark period, and FR supplements during the photoperiod
- the changing sensitivity to R and FR light during the daily cycle
- the requirement for long exposures to light for photoperiodic induction.

Bünning's hypothesis proposed that light promotes flowering in LDP when it falls in the skotophile phase. Although this is consistent with the promotion of flowering by a night-break, it has been difficult to reconcile with the apparent requirement for FR (low Pfr) during a long photoperiod, especially where this can apparently be replaced by a dark interruption, as in *Lolium*. Vince-Prue and Takimoto (1987) suggested that flowering in LDP may be controlled by two rhythms, one in which Pfr promotes flowering and the other in which Pfr is inhibitory. Sunlight (or any mixture of R and FR which establishes an intermediate photoequilibrium) would generate sufficient Pfr to interact with the promoting response but insufficient to cause inhibition and thus is the most effective for flowering. This proposal would account for the responses of *Lolium* but does not predict the loss of response to FR of phytochrome A deficient mutants in *Arabidopsis*. Based on studies with mutants and transgenic plants, responses to R are now ascribed to phytochrome B and maybe other type II phytochromes, while the FR-HIR is attributed to phytochrome A. If this holds true for flowering, the changing sensitivity to R and FR in LDP could be considered as changing sensitivity to phytochromes A and B respectively. Thus phytochrome B would be required and effective during the early part of the photoperiod, whereas phytochrome A activation would be required in the latter part of the photoperiod. Red light, acting through phytochrome B, would set the phase of a rhythm in sensitivity to a HIR acting through phytochrome A (Fig. 5.17). This model has much in common with the SDP model proposed in Chapter 2 (see Fig. 2.12). In both cases light-on sets the phase of a rhythm in light sensitivity. In LDP, this continues to run in light whereas, in SDP, the rhythm is suspended. The rhythm (in LDP) establishes a phase of sensitivity to phytochrome A which requires extended periods of light containing FR for a significant photoactivation. In this type of model, extended night-breaks could either interact with the rhythm or rephase the endogenous rhythm so as to potentiate its interaction with the subsequent photoperiod. In practice it would probably do both, thus leading to the quite complex patterns of response described in the literature.

Overall, the photoperiodic control of flowering in LDP has proved more difficult to understand than that in SDP. In part, this is due to the requirement for long exposures

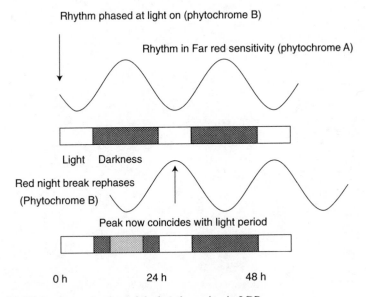

FIG. 5.17. Model for the mechanism of daylength sensing in LDP.

to light, which are less easy to work with than the brief exposures that have been found to be effective in many SDP. However, whatever the mechanism (and it may not be the same in all species), it is clear that LDP are well adapted to flowering under long days at the R/FR ratio present in natural daylight.

6 The Physiology of Photoperiodic Floral Induction

Flowering involves a dramatic change in the pattern of differentiation at the shoot apex, or in the axillary buds close to the apex. The idea that this is under the control of a specific flower-forming substance dates back to Julius Sachs who, in the 1880s, concluded that leaves in the light produce small quantities of substances which direct assimilates into the formation of flowers. This idea was later opposed by other workers who thought that only sugars were involved in determining whether or not flowering would occur. The idea of a specific floral hormone was perhaps first clearly enunciated by Chailakhyan (1936), who proposed the term **florigen** for the hypothetical flowering substance. The idea that the differentiation of floral organs is under the control of a hormone applies to day-neutral plants as well as those where flowering depends on exposure to appropriate daylengths but, in the latter, it is supposed that the hormone is synthesised only when the correct daylength is given. It has also been proposed that substances inhibiting flowering may be involved; the appropriate daylength would then lead to removal of an **antiflorigen** rather than (or in addition to) synthesis of a floral hormone.

Much of the evidence for both flower-inducing substances and floral inhibitors comes from physiological experiments of various kinds and attempts to extract and identify such substances have, at best, met with limited success. This chapter, therefore, will first review the physiology of photoperiodic floral induction before considering what is known about its biochemistry in Chapters 7 and 8.

THE NATURE OF PHOTOPERIODIC FLORAL INDUCTION

The perception of the daylength signal occurs primarily in the leaves and, from grafting experiments, it is known that leaves are independently capable of generating a transmissible signal which can bring about flowering at receptive shoot apices. Leaves which are capable of causing flowering are said to be **induced** (see Chapter 1).

After a sufficient number of favourable cycles, flowering will often occur when plants are returned to non-inductive cycles (Table 6.1); this is true for both LDP and

TABLE 6.1   Minimum numbers of photoinductive cycles required for floral initiation.

Response type	Cycles for induction
Short-day plants	
Chenopodium rubrum	1
Lemna paucicostata	1
Pharbitis nil cv Violet	1
Wolffia microscopica	1
Xanthium strumarium	1
Kalanchoë blossfeldiana	2
Begonia boweri	2–3
Glycine max cv Biloxi	2–3
Cannabis sativa	4
Perilla (red)	3–5
Fragaria × *ananassa* cv Blakemore	6
Perilla (green)	7
Dendranthema grandiflora	12
Long-day plants	
Anagallis arvensis	1
Anthriscus cerefolium	1
Brassica campestris	1
Lemna gibba	1
Lolium temulentum	1
Sinapis alba	1
Hyoscyamus niger	2–3
Arabidopsis thaliana	4
Silene armeria	6
Long-short-day plants	
Cestrum nocturnum	5 LD: 2 SD
Bryophyllum daigremontianum	60 LD: 15 SD
Bryophyllum crenatum	20 LD: 9–12 SD
Short-long-day plants	
Echeveria harmsii	20 SD: 10 LD

Different cultivars and plants of a different age and/or under different environmental conditions may have different requirements.

SDP. In the most extreme cases a single favourable (inductive) cycle is able to bring about floral initiation. The number of plants for which a single cycle is sufficient is not great but, because of ease of experimentation particularly in relation to the timing of events, they have been studied in far greater detail than other species. In these and other plants requiring only a few inductive cycles, the photoperiodic control of flowering is an inductive phenomenon in the strict sense and no morphological changes in the shoot apex can be observed at the end of the induction period. At the other extreme, favourable cycles must be continued until the apex has become recognisably floral (Battey and Lyndon, 1990). In *Impatiens balsamina*, for example, returning the plants to non-inductive LD causes the tip of the placenta to resume vegetative growth, even after the flower has formed fertile anthers and an ovary with abortive ovules. Similarly, in *Anagallis arvensis*, flower reversion occurs on return to non-inductive SD after a sustained period of flowering in LD; as in *Impatiens*, the latest reversion was when a terminal leafy shoot was formed from the placenta of the ovary. Since floral organs are formed in rapid succession, it appears that the threshold between flower formation and reversion is crossed within a few hours in these plants, indicating the need for a continual supply of a floral stimulus (presumably from the leaf) if flower development is to be sustained. The production (or supply) of such a stimulus must also cease extremely rapidly following transfer to non-inductive conditions.

In contrast, it is evident that in some species a leaf can continue to supply floral stimulus over many days, or even weeks. In these plants, induction in the leaves involves a (largely) irreversible change, such that they continue to generate a floral stimulus for a long time even after they have been returned to inappropriate day-lengths. The existence of a more or less permanent induced state in the leaf is evident from the grafting experiments with *Perilla* (red), already discussed in Chapter 1. A leaf removed from a plant exposed to SD is able to cause flowering when grafted to a receptor plant; subsequently, this leaf can be detached and regrafted to a second receptor plant. Following 28 inductive SD cycles to the original leaf and 7 regrafts, flowering of the receptor took place in 27 days at the 7th graft compared with 21 days at the time of the first graft (Zeevaart, 1958). It was concluded that the induced leaf continued to export a flowering stimulus during its entire life, despite various treatments including high temperature and the application of auxins and enzyme poisons such as azide. The strict localisation of induction in *Perilla* was demonstrated by exposing part of a leaf to SD. Only that part exposed to a sufficient number of SD could function as a donor of flowering to receptor plants maintained in LD (Zeevaart, 1962a); thus, in *Perilla,* the induced and non-induced state can coexist in the same leaf without affecting each other. The induced state persisted during the whole lifetime of the leaf in the experiments of Zeevaart (1958). However, it was shown later that the duration of the induced state appeared to depend on the extent of induction; after fewer than 20 SD, plants produced vegetative shoots within 6 weeks whereas no reversion was observed when more than 27 SD were given (Lam and Leopold, 1961). Repeated decapitation and debudding increased the number of plants which reverted to vegetative growth; the number of debudding treatments required to cause reversion also depended on the degree of induction, being greater with an increase in the number of SD given. After 50 SD, no reversion occurred in any treatment. Similar results have been obtained for other SDP. *Xanthium* normally fruits and dies when given more than three SD cycles, but with fewer than three cycles, some of the plants eventually reverted to vegetative growth; if repeatedly decapitated, all of the plants reverted (Lam and Leopold, 1960). For both *Xanthium* and *Perilla*, therefore, the evidence suggests that induction in the leaf is not an all-or-none process, but develops progressively and becomes more stable with an increasing number of inductive cycles.

The need for several inductive cycles in order to achieve flowering raises the question of how these cycles are summated by the plant. Summation in the leaf is clearly evident in some cases. Since an excised *Perilla* leaf requires several SD cycles before it can be an effective donor of flowering, the summation of the photoperiodic signal must occur in the leaf itself. In *Xanthium,* the magnitude of the flowering response is increased with an increase in the number of inductive cycles; this effect also seems to be associated with the leaves rather than with the apex as 4 SD given to a single leaf were more effective than 2 SD given consecutively to two leaves (Lincoln *et al.*, 1958), although it should be noted that all leaves are not necessarily equally sensitive to induction.

The requirement for multiple cycles may not depend entirely on events in the leaf. It is possible that a true 'induced state' does not develop and that each cycle produces some stimulus. Flowering might then require the accumulation of a certain threshold amount at the apex. A number of factors might be associated with the requirement for different numbers of inductive cycles: the threshold could vary from plant to plant; the apex could be more or less sensitive; the amount produced in each cycle might vary; destructive or inhibitory mechanisms could operate to a greater or lesser extent. An earlier suggestion that the minimum number of photoinductive cycles corresponds with the number of days in the plastochron was found not to be the case in several plants (including *Pisum*, *Lolium* and *Sinapis*; Reid and Murfet, 1977). However, the possibility of summation at the apex has not received a great deal of attention. A positive effect was obtained in the grass, *Rottboellia exaltata*, where a minimum of 6 SD are necessary for flowering; these could be given separately to two leaves, clearly indicating summation at the apex (Evans, 1962a). In this experiment, successful summation was obtained when the uppermost leaf was wrapped to receive 1 SD and was then excised before all of the lower leaves were exposed to the remaining 5 SD. However, summation was not obtained by giving 1 or 2 SD consecutively to individual leaves; giving 2 or 4 SD to the uppermost leaf and the remaining 4 or 2 SD to the leaves below also failed. In a similar experiment, flowering in *Perilla* (green) was observed when 2–3 SD were given to each leaf in succession, whereas plants remained vegetative when 2 SD were given to all leaves concurrently. However, these results were not confirmed in subsequent experiments with red *Perilla* and there was no evidence for summation at the apex when 2–3 SD were given consecutively to separate leaves (Table 6.2); this

TABLE 6.2 Failure of summation of photoperiodic induction in red *Perilla* when the inductive SD treatment was given successively to different leaves.

Treatment	Flowering response	Days
Each plant with 8 leaves		
All leaves receive 3 SD simultaneously	V	80
3 SD consecutively to each leaf starting at the tip	V	80
3 SD consecutively to each leaf starting at the base	V	80
Control: 1 leaf pair received 24 SD	Fl.b	17
Each plant with 6 leaves		
All leaves received 3 SD simultaneously	V	80
All leaves received 3 SD, leaves not removed	V	80
3 SD consecutively to each leaf, starting at the tip	V	80
3 SD consecutively to each leaf, starting at the base	V	80
3 SD to each leaf; leaves removed after the last SD	V	80
Control: 1 leaf pair received 18 SD	Fl.b	17

Individual leaves received 16 h dark periods daily by wrapping in lightproof bags. SD-treated leaves were excised 8 h after the end of the dark period unless otherwise stated. V, vegetative; Fl.b, flower buds; days indicate when flower buds were visible, or experiment discontinued.
Data of Zeevaart (1971).

finding is clearly more consistent with the fact that SD cycles are summmated by excised leaves in *Perilla*. The single confirmed example of summation at the apex occurred when the number of successive inductive cycles given to a single leaf was close to the threshold and, overall, the evidence strongly points to the conclusion that multiple cycles are summated by the leaf leading to a progressively stable degree of induction. This is an important conclusion with respect to the biochemistry of events occurring in the leaf during photoperiodic induction.

Fractional Induction

Fractional induction is the summation of inductive cycles despite the intercalation of non-inductive cycles. Plants reported to show fractional induction include the LDP sugar beet, *Hyoscyamus niger, Plantago lanceolata, Lolium temulentum* and the SDP *Bidens tripartita, Glycine* max cv 'Biloxi', *Chenopodium amaranticolor, Impatiens balsamina, Caryopteris clandonensis* (Lang, 1965, 1980; Vince-Prue, 1975) and *Lemna paucicostata* (Oota, 1984).

In *Hyoscyamus*, summation occurred when 1 LD was alternated with 1 SD and flower initiation was delayed, at most, by the number of days corresponding to the total number of intercalated SD (Lang, 1980). However, summation frequently fails unless more than a certain number of consecutive inductive cycles is given (Vince-Prue, 1975). In *Kalanchoë*, cycles of 1 SD/1 LD were not effective in inducing flowering, but cycles of 2 SD/2 LD were. Similarly in *Glycine* 'Biloxi', cycles of 1 SD/1 LD were ineffective, but pairs of SD could be partly summated even with as many as 12 intercalated LD; the biggest delay occurred with the first intercalated LD, and subsequent ones had little further effect (Fig. 6.1). Although summation is possible in both LDP and SDP, it has been suggested that, in general, the inhibitory effect of intercalated non-inductive photoperiods is greater in the latter (Lang, 1980);

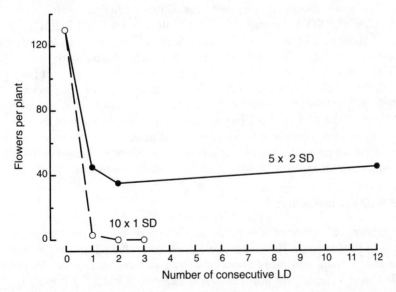

FIG. 6.1.   Effect of the number of consecutive intercalated LD on flowering in *Glycine max* Biloxi. After Carr, 1955.

in *Hyoscyamus*, for example, six cycles of 1 SD/1 LD gave flowering about as rapidly as 6 LD whereas, in the SDP *Kalanchoë*, a single LD interrupting 2 × 6 SD reduced flowering by more than 50%. Where summation occurs, the number of intervening non-inductive cycles does not seem to be very important provided that an adequate number of consecutive inductive cycles is given (see Fig. 6.1). Other examples include *Impatiens balsamina*, where SD cycles were summated with as many as 16 intercalated LD and *Caryopteris* (which is day-neutral for initiation but requires 11 SD for development to anthesis), where up to 30 LD could interrupt the SD treatment without altering the number of inductive cycles required. In okra, summation requires three consecutive SD; this could be carried over during four intercalated LD but the summation effect was partially lost when more than four consecutive LD cycles were given (Nwoke, 1986). Thus, it seems that, after an adequate number of inductive cycles, the resulting change is relatively stable and can be maintained for some time (the precise duration depending on the plant); it can then be added to when subsequent inductive cycles are experienced.

Partial Induction

The results from fractional induction treatments give further support to the concept that induction is not an 'all or none' process. It appears that, once a sufficient number of inductive cycles has been given, a condition is attained which is stable but is insufficient to elicit flowering. The existence of degrees of induction is also evident from other types of experiment. In red *Perilla*, for example, the minimum duration between the donor leaf and receptor plant decreased as the number of SD given to the donor leaf was increased, indicating that the rate of export of a flowering stimulus from the leaf is proportional to the duration of the inductive treatment. The number of days before flower buds appeared on the receptor also decreased with increasing numbers of SD given to the donor, and was similar whether grafts were made immediately after the SD treatment or after the donors received a further 28 LD. The degree of induction of a particular leaf, therefore, appeared to be maintained during non-inductive LD (Zeevaart, 1958). Other experiments with intact plants support the idea that induction varies in degree. For example, in *Xanthium*, plants given multiple SD cycles produced mature flowers in about 2 weeks whereas, when given a single SD, plants progressed only slowly to flowering over several months (Salisbury, 1963b). The first visible signs of flowering occurred at about the same time in both. The rate of development was not accelerated by reducing the number of buds; it seems, therefore, to be a function of a progressive degree of induction in the leaf and not to the amount of stimulus available at each receptive meristem.

Direct and Indirect Induction

The demonstration of a more or less permanent induced state in the leaves has been demonstrated in *Perilla* (red). However, while the leaves which are exposed directly to SD become induced and can bring about flowering in a receptor plant, leaves taken from this flowering receptor are not themselves capable of inducing flowering. Thus, in *Perilla* (red), leaves can only be induced by exposing them directly to inductive SD cycles (Fig. 6.2). The presence of buds seems unimportant for maintenance of the

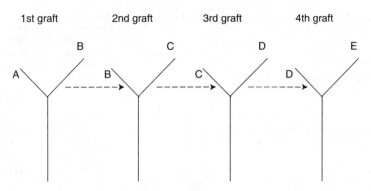

FIG. 6.2. Direct and indirect induction of flowering. Plant A was induced to flower by exposure to SD; subsequently all graft combinations were kept in LD. In *Xanthium strumarium*, all receptor plants B to E flowered. In *Perilla* (red), only plant B flowered.

induced state and an induced leaf continues to produce floral stimulus over a long period of time.

Xanthium behaves differently from *Perilla* in many respects. As with *Perilla*, although far less readily, an excised leaf exposed to SD can cause flowering when grafted to a receptor in LD. Thus signal perception and stimulus formation can be completed in the leaf alone. However, it appears that the generation of a stimulus by the leaf continues for only a relatively short time and flowering does not occur unless a developing bud or young leaves are present within this period (Salisbury, 1955). A major difference between the two plants is that, in *Xanthium*, leaves can become induced as a consequence of a stimulus received from a donor plant (see Fig. 6.2). After removing the donor plant originally exposed to SD, the receptor was grafted to another vegetative plant which, in turn, became reproductive. A flowering condition was transferred by indirect induction in this way through many successive grafts with no evidence of a diminished response (Zeevaart, 1958). It seems most unlikely that these results can be explained in terms of the stimulus exported from the initial, directly-induced plant. Thus, in *Xanthium*, a stimulus from the induced donor appears able to bring about induction in a receptor plant; this occurs irrespective of whether the stimulus comes from the initial, directly-induced plant or from the subsequent, indirectly-induced receptor. From these experiments, it is not possible to say that the stimulus is necessarily the same in both cases; only that it has the same physiological effect.

Any general consideration of photoperiodic induction must take into account the marked difference between *Xanthium*, where leaves can be induced by the arrival of a stimulus from other leaves and *Perilla* (red), where only leaves directly exposed to SD can be induced. Little is known about the behaviour of other plants in this respect. *Silene* (LDP), *Bryophyllum* (LSDP) and green *Perilla* (SDP) have been shown to be capable of indirect induction (Zeevaart, 1984); thus, the property of indirect induction is not associated with any particular photoperiodic response group. The floral stimulus here seems to be self-perpetuating and it has been suggested that some type of positive feedback may occur, with the floral stimulus causing synthesis of more floral stimulus in the young leaves and buds. To date, only in red *Perilla* has it been shown that

TABLE 6.3. Indirect and direct induction in green and red *Perilla*.

Donor in first graft	Receptor in second graft	Flowering quotient	Days to visible flower bud
A. Indirect induction in green *Perilla*			
SD green	Green	9/9	51.6
	Red	7/10	78.6
LD green	Green	10/10	78.9
	Red	4/10	109.3
SD red	Green	10/10	71.4
	Red	7/10	98.1
LD red	Green	10/10	83.3
	Red	8/10	116.5
B. No indirect induction in red *Perilla*			
SD green	Green	0/10	–
	Red	0/6	–
LD green	Green	0/10	–
	Red	0/6	–
SD red	Green	0/4	–
	Red	0/4	–
LD red	Green	0/4	–
	Red	0/4	–

All donor leaves in first grafts received 26 SD. In A, leaves on green receptor shoots of the first grafts were used as donor leaves in the second grafting. In B, leaves on red receptor shoots of the first grafts were used as donor leaves in the second grafting. The second grafts were carried out 20–23 days after the first grafting.
Data of Zeevaart and Boyer (1987).

receptor shoots are themselves unable to function as donors of flowering. However, a similar situation has been demonstrated in two species of *Echeveria*, *E. harmsii* (SLDP) and *E. pulv-oliver* (LDP), where vegetative shoots taken from induced plants failed to cause flowering in receptor plants of *Kalanchoë blossfeldiana*. In contrast, flowering shoots from the same plants were effective donors (Zeevaart, 1978).

Despite the differences in their induction characteristics, the floral stimulus is readily transmitted between green and red *Perilla* and, although it cannot itself be indirectly induced, red *Perilla* is able to bring about indirect induction in green receptors (Zeevaart and Boyer, 1987). Moreover, leaves taken from indirectly induced plants of green *Perilla* were able to cause flowering in both red and green receptors, irrespective of whether the original donors were green or red (Table 6.3). In contrast, leaves from indirectly-induced plants of red *Perilla* never caused flowering in either green or red receptors. The difference between direct and indirect induction thus appears to be associated with cellular conditions within the young leaves and buds, which influence their response to the arrival of the floral stimulus. It does not appear to be associated with the identity of the floral stimulus, which is interchangeable between the two types of induction. The underlying mechanism for the apparent self-perpetuation of the stimulus in plants that are capable of indirect induction has yet to be elucidated. An earlier suggestion that the immediate product of a directly induced leaf might differ from that produced in leaves and buds following its arrival did not receive any support from subsequent experiments (Zeevaart, 1971).

EFFECTS OF AGE

Plants often vary with age in their sensitivity to daylength, frequently becoming more easily induced as they grow older. While this may, in part, be due to a larger leaf area giving rise to more stimulus, this is by no means the only factor; for example, the same number of cycles is required to induce an intact plant as one defoliated to a single leaf. In many cases, fewer inductive cycles are required for flowering as the plants continue to grow in non-inductive conditions. *Lolium temulentum* Ba 3081 will flower with only a single LD when the plant has developed 6–7 leaves but requires at least 4 LD for a minimal flowering response at the 2–3 leaf stage (Fig. 6.3). A number of similar cases has been recorded. In some plants, the photoperiodic requirement may be lost entirely as, for example, in the facultative SDP, *Amaranthus retroflexus*, where it was concluded that progress towards reproduction could occur in both long and short days, but that a single SD was equivalent to approximately 4 LD (Koller *et al.*, 1977). Somewhat surprisingly, a few plants have been shown to become more difficult to induce as they become older. These include the LDP *Anagallis arvensis*, where the sensitivity of leaves decreases with ontogenetic rank (Ballard and Grant-Lipp, 1954), and the SDP *Pharbitis nil*, where seedlings grown in continuous light reach maximum photo-periodic sensitivity 5–6 days after germination (Vince-Prue and Gressel, 1985), and *Chenopodium rubrum*, where 3-day-old seedlings require fewer inductive cycles than older plants (Seidlova and Krekule, 1973).

Individual leaves vary in their sensitivity to daylength as they develop; in many cases, maximum sensitivity is reached as leaves approach full expansion and then declines until the oldest leaves show low sensitivity. The youngest, fully expanded leaf was found to be most sensitive in *Perilla*, *Glycine*, chrysanthemum (Vince-Prue, 1975) and *Begonia* (Zimmer and Krebs, 1980). In the SDP *Xanthium* (Fig. 6.4) and

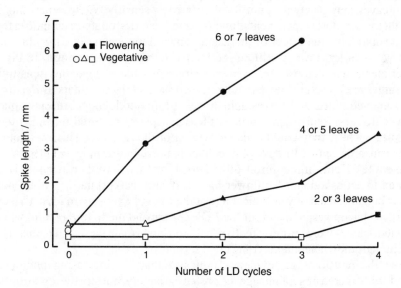

FIG. 6.3.   Effect of plant age on sensitivity to photoperiodic induction in the LDP *Lolium temulentum* Ba 3081. Data of Vince-Prue.

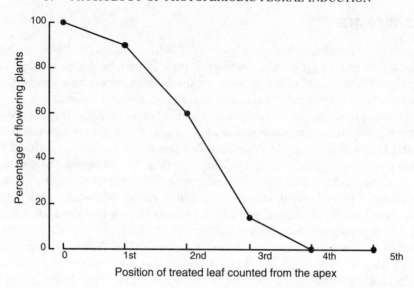

FIG. 6.4.   The effect of leaf age and position on flowering in *Xanthium strumarium*. The plants were defoliated to a single leaf and exposed to one SD. The first to fifth leaves were fully expanded (40–45 cm^2 area) and the expanding leaf, 0, was 10–15 cm^2 area. After Vince-Prue 1975 (Data of Khudairi and Hamner, 1954).

Rotboellia (Evans, 1962a), maximum sensitivity was reached in leaves which were most rapidly expanding, whereas fully expanded leaves retained their full inductive capacity for a considerable period in the LDP *Lolium* (Evans, 1969a).

Since the photoperiodic stimulus has been shown to be translocated in the phloem, the fact that the carbohydrate supply to the apex is mainly derived from the uppermost expanded leaves may partly account for their greater sensitivity. However, this is not the only factor. In most experiments, plants have been de-foliated except for the test leaf. This would have the effect of increasing upward movement of assimilates to the apex and yet such leaves may still be ineffective in causing flowering (see Fig. 6.4). The experiments which most clearly demonstrate that the endogenous sensitivity of leaves changes are those of Zeevaart (1958), who induced intact plants of *Perilla* (red) and then removed individual leaves, which were grafted to identical receptor plants in LD. Leaves from nodes located lower on the donor plants required more SD to bring them to a state where they could cause flowering of the receptor. This was not due to differences in area since 30 cm^2 of the fifth leaf from the base resulted in 100% flowering while 61 cm^2 of the third leaf had no effect. It was further demonstrated that the capacity to act as a donor for flowering was determined primarily by leaf position (ontogeny) and not by its physiological age at the time of exposure to inductive cycles. A similar conclusion was reached for the LDP *Lolium* and the SDP *Begonia*, where the later formed leaves were more effective per unit area than the earlier ones (Evans, 1969a; Zimmer and Bahnemann, 1981).

Much of the physiological evidence indicates that the increasing photoperiodic sensitivity with increasing plant age is probably largely due to the contribution of leaves of higher ontogenetic rank. Nevertheless, other factors are likely to be important in some cases. For example, the seedling apices of tobacco are less competent to respond to a given level of floral stimulus than the apices from mature plants (Singer *et al.*, 1992). The presence of roots may also influence the overall response and contribute to the effects of age. In tobacco (a day-neutral cultivar), roots were inhibitory to flowering and, as the seedlings aged, the increasing number of internodes between the roots and apex reduced their inhibitory effect (Gebhart and McDaniel, 1991).

Plants which show quantitative photoperiodic responses eventually flower without exposure to inductive cycles. Similarly, day-neutral plants flower in the absence of photoperiodic induction. In both cases, the age of the plant appears to be a major factor in the attainment of reproductive growth but rather little consideration has been given as to whether this ageing affect is associated with the leaves or apex. It is known that the leaves of such plants are induced and capable of causing flowering when grafted to appropriate receptor plants (see Table 6.8; Heinze *et al.*, 1942). However, there seems to be little information from grafting experiments concerning the development of the induced state in the leaves. For example, do all the leaves of quantitative photoperiodic plants become induced, or only those of higher ontogenetic rank which develop as the plants become reproductive? Are all leaves of day-neutral plants equally capable of causing flowering in receptors or do they become more effective with increasing ontogenetic rank? We do know that donor leaves from young vegetative plants of the DNP *Coleus blumei* caused flowering in receptors of the SDP *C. frederici* just as effectively as leaves from older flowering plants (Jacobs, 1980). However, no systematic study of the effect of ontogenetic rank on the induced state in leaves of DNP appears to have been undertaken.

At least some of the effects of age on autonomous induction have been directly attributed to events occurring in the leaf. In the quantitative LDP *Pisum sativum*, inhibitor production in the leaf appears to decline as the plant ages, leading to a reduction in the number of LD needed to induce flowering (Reid and Murfet, 1977). However, physiological changes in the meristem itself may also be involved since gradual changes in the morphology and histology of the apex have been observed in plants grown for long periods under non-inductive photoperiods (Cockshull, 1979). In the SDP poinsettia (*Euphorbia pulcherrima*), the time of initiation in LD appeared to be a function of the ontogenetic age of the meristem since leaf-removal did not affect the leaf-number to flowering (Evans *et al.*, 1992). A further possibility is that inhibitory factors from the roots decrease with age, since the distance between roots and apex increases as the plants continue to grow. There is some evidence for this in the day-neutral tobacco Wisconsin 38, where floral initiation was considerably delayed by air-layering to produce roots on the stem and, therefore, closer to the apex. However, air-layering had no effect on the long-day leaf number in poinsettia. Without further experimental evidence it is not possible to determine the extent to which effects of age on flowering in autonomously-induced plants is due to changes in the supply of promoters and/or inhibitors from the leaf (and perhaps also the roots) or to changes in the sensitivity of the apex to these substances. It is probable that both are involved and are to a greater or lesser extent important in different plants. One can state with certainty, however, that flower-promoting substances are produced in the leaves of plants which have been autonomously induced and that these are able to induce flowering in photoperiodically-sensitive plants under non-inductive conditions (Table 6.8).

JUVENILITY

Juvenility is the name usually given to an early phase of growth during which flowering cannot be induced by any treatment. In this way, sexual reproduction is delayed until plants reach a size sufficient to maintain the energetic demands of flowering and seed production. In many plants the transition to reproduction occurs after the juvenile phase has been completed without exposure to any particular stimulus; in others appropriate environmental treatments (such as daylength) are necessary. Sometimes the term **ripe-to-flower** is used for plants which have completed the juvenile phase but have not yet experienced the correct conditions for flowering. The duration of the juvenile period varies widely. In most woody plants it lasts for several years but in herbaceous plants it is rarely more than a few days or weeks. *Xanthium* cannot be photoperiodically induced during the first week after germination, while *Perilla* does not become sensitive for 15–20 days. In *Glycine*, the duration of the photoperiod insensitive phase varied from 11–33 days in a range of four cultivars of diverse origin and was found to be a strong determinant of the time to first flowering (Collinson *et al.*, 1993). In extreme cases, there is no juvenile phase since flower primordia are found in the seed.

Size appears to be important in the transition to maturity and, in general, conditions that promote growth reduce the duration of the juvenile period. Two possible explanations for the effect of size have been considered. One is that a plant of sufficient size transmits one or more signals to the apex, which then undergoes a phase change from juvenile to adult. The second is that the apical meristem behaves independently and undergoes the phase transition at a particular time. There is some evidence for both views (Thomas and Vince-Prue, 1984). Once attained, the adult phase is usually highly stable and persists until sexual reproduction occurs.

The existence of a juvenile phase is obviously one of the factors associated with the lower sensitivity of younger plants to photoperiodic induction, since the daylength cycles given to juvenile plants are not effective. That this is a property of the leaves rather than of the apex is indicated by the results of experiments with *Bryophyllum*, where growing points from juvenile plants were able to flower when grafted to photoperiodically-induced mature plants (Zeevaart, 1962b). A similar result has been obtained in the SDP *Ipomaea batatas* (Takeno, 1991); here, 13-day-old seedlings flowered when grafted to *Pharbitis nil* in SD, but not when they were themselves exposed to inductive cycles. The hypothesis that the limitations of juvenile plants reside mainly in their inability to produce a floral stimulus is supported by the behaviour of obligatory viviperous species of grasses. Viviperous plantlets may develop a new generation of proliferated inflorescences while still attached to their parent plants but detached plantlets cannot be induced to form such inflorescences and remain vegetative (Heide, 1994). Apparently the attached plantlets are able to respond to a floral stimulus from the parent plant, but are not themselves able to produce it. Additional evidence that juvenility effects may be associated with the leaves comes from grafting experiments with *Glycine*, where the long-juvenile genotype appeared to be associated with one or more transmissible compounds from the leaves, which delayed flowering (Sinclair and Hinton, 1992). In contrast, seedling apices of tobacco have been shown to be less competent to respond to a given level of floral stimulus than apices from mature plants (Singer *et al.*, 1992). Biochemical and morphological changes in the meristems have been observed to occur with increasing plant age and could confer increased sensitivity to the floral stimulus; for example, a number of polypeptides were unique to either young or aged apices in the LDP *Silene coeli-rosa* growing in non-inductive SD and there were also differences in morphology and mitotic index (Nougarede *et al.*, 1989). Thus changes in both leaves and apices may contribute to the development of competence to respond to photoperiodic induction.

A few experiments have been made with woody plants in which induction is autonomous. When juvenile apices of Japanese larch were grafted to mature trees, only one (out of 56) flowered in the first year (Robinson and Wareing, 1969); when nearly-mature apices were similarly grafted, 36 of the grafts flowered. A similar situation has been reported for mango, where graft induction failed in 1–4-year-old seedlings but was successful when the receptors were 6 years old and nearly mature (Kulkarni, 1988). Thus, in these two examples, it seems that the transition from juvenile to mature in woody plants is not primarily determined by signals reaching the apex from the rest of the plant.

INTERACTIONS WITH TEMPERATURE

The requirement for exposure to a particular photoperiod in order to effect induction can be profoundly modified by temperature. Examples include the SDP strawberry, which is strictly photoperiodic only at temperatures above about 15°C (Guttridge, 1985) and *Clarkia amoena*, which is day-neutral at 20–24°C but is an LDP at lower temperatures (Halevy and Weiss, 1991). Many other similar interactions have been recorded (Vince-Prue, 1975 and see Appendix I). In a few cases, temperature has been shown to influence the induction process in the leaf. *Perilla* (green) is a SDP at 22°C but will flower in continuous light when maintained at 5°C; a single leaf from a plant induced to flower by exposure to low temperature acted as a donor of flowering when grafted to a receptor plant maintained in continuous light at 22°C (Deronne and Blondon, 1977). Similarly, in *Pharbitis*, there was little flowering response when cotyledons were removed after 10 days in continuous light at 13/14°C but the number of flower buds was the same as in intact plants when they remained for a further 2 days at 23°C before excision (Shinozaki and Takimoto, 1982). In the majority of cases, however, it is not known whether changes in temperature modify the induction process in the leaf or other components of the overall flowering response. Interactions between temperature and photoperiod are not always found, however, even when flowering is responsive to both environmental factors (Roberts *et al.*, 1985).

Insensitivity to temperature is a characteristic of circadian rhythms and temperature usually has little effect on the time of maximum sensitivity to a night-break (Chapter 2). In contrast, the critical nightlength has been shown to be markedly influenced by temperature, although there is considerable variation between species in the extent to which this occurs (Fig. 6.5). Little effect was observed in the SDP *Chenopodium*, *Lemna* and *Xanthium*, but both the SDP *Pharbitis* and the LDP *Hyoscyamus* showed a strong dependence of CNL on temperature. The effect is presumed to occur in the leaf, although this has not been strictly demonstrated.

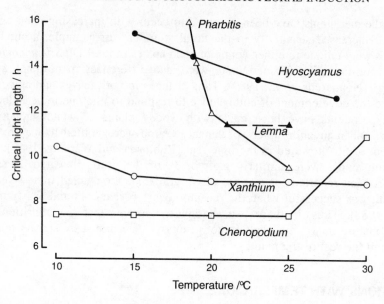

FIG. 6.5.   Effects of temperature on the length of the critical dark period for flowering. Adapted from King, 1979.

Vernalisation

Not all workers agree on the precise definition of vernalisation but the term is probably best restricted to the specific promotion of flowering by a cold treatment given to the imbibed seed or young plant (Thomas and Vince-Prue, 1984). There are also direct effects of low temperature on floral initiation in some plants (for example in brussels sprout, *Brassica oleracea gemmifera*, and in some bulbous plants such as *Iris* Wedgwood and onion). In these cases, floral initials differentiate during the cold or cool treatments and so they can be distinguished from vernalisation, which is an inductive phenomenon; after vernalisation is complete, flower initials are not yet present and they differentiate only later when the plant is returned to a higher temperature and, in many cases, also to particular photoperiodic regimes. However, the biochemical changes that occur in response to cold are not necessarily different in the two response types.

The phenomenon of vernalisation is particularly associated with photoperiodism. A cold requirement for flowering is often linked with a requirement for a particular photoperiod and, in some plants, one process can substitute for the other. The requirement for vernalisation is most commonly, but not only, found in LDP and many plants requiring low temperature vernalisation also require subsequent exposure to LD in order to induce flowering (Appendix I). These may be annuals (e.g. winter cereals), biennials (e.g. carrot, sugar beet) or perennials (e.g. *Lolium perenne*). There is, however, no necessary relationship between a vernalisation requirement and a particular photoperiodic response group; many day-neutral species also require vernalisation as do some SDP (Table 6.4). Many cold-requiring plants grow as rosettes in the absence of a chilling treatment and exposure to low temperature results both in the

TABLE 6.4   Examples of daylength requirements following vernalisation.

Plants day-neutral after vernalisation

Apium graveolens cv Prazsky obrovsky
Brassica napus
Cardamine pratensis
Daucus carota
Dendranthema grandiflora several cvs
Geum urbanum

Plants requiring LD following vernalisation

Obligate LD requirement
  Beta (sugar beet, cv Kleinwanzleben)
  Blitum capitatum
  Campanula pyramidalis
  Hyoscyamus niger
  Lotus spp.
  Oenothera biennis
  Ornithopus sativus
Facultative LD requirement
  Apium graveolens var. dulce cv Florida
  Arabidopsis thaliana Stockholm strain
  Calceolaria × *herbeohybrida*
  Digitalis purpurea
  Dianthus barbatus
  Lunaria annua
  Pisum sativum late genotypes
  Secale cereale winter strains

Plants requiring SD following vernalisation

Obligate SD requirement
  Dendranthema grandiflora several cultivars
Facultative SD requirement
  Allium sativum

From Napp-Zinn (1984) and Halevy (1985).

elongation of a flowering stem and in the formation of flower primordia. In this they resemble the many LDP which grow as rosettes without stem elongation in short photoperiods.

Cold-requiring perennials pose a special problem. As with juvenility, the vernalisation requirement is normally re-established during sexual reproduction. Thus the association of the perennial habit with a vernalisation requirement appears, at first sight, to be contradictory because the vernalised state is perceived by the shoot apex and is normally transmitted through all subsequent mitotic divisions. All vernalised buds should, therefore, produce flowers and the plants would be monocarpic. This does not occur for a number of reasons (Thomas and Vince-Prue, 1984). **Devernalisation** (i.e. reversal of vernalisation) can occur, especially in high-temperatures and at the low light intensities that would be experienced by perennating buds at the base of the flowering shoot. An alternative strategy is seen in some perennial grasses where the vernalisation effect is not perpetuated indefinitely so that tillers which develop late in the summer are not vernalised and overwinter without flowering. In other cases, the sensitivity to low temperature appears to be transient so that vernalisation is not achieved in all of the axillary buds.

In the context of photoperiodism, perhaps the most interesting interactions between the two processes are the abolition or modification of a photoperiodic requirement by exposure to low temperature and its converse, the replacement or modification of a vernalisation requirement by a daylength treatment. There are many cases where vernalisation can modify or abolish a photoperiodic response, either reducing the critical daylength or causing plants to become indifferent to daylength (Lang, 1965). Biennial sugar beet normally requires both vernalisation and LD but, with an extended cold treatment, becomes able to flower in SD. Similarly, in certain strains of spinach which normally require daylengths longer than 14 h, vernalisation reduced the critical daylength to about 10 h. A reduction in critical daylength following vernalisation has also been recorded for a number of other LD species (Lang, 1965; Napp-Zinn, 1984).

A photoperiodic treatment can also substitute for vernalisation. In particular, exposure to SD can often substitute either partly or entirely for cold. All plants where SD can substitute for low temperature are LDP. The SD or low-temperature treatment must precede the LD exposure, the reverse order being ineffective; thus they could be classified either as SLDP or as cold-requiring LDP. A detailed study of *Campanula medium* by Wellensiek (1960) revealed several differences between the two alternative pre-LD treatments. For example, plants became sensitive to SD one month sooner than to low temperature and fewer LD were necessary after SD than after cold. Consequently, it was concluded that the low temperature and cold effects involved different mechanisms leading to the same result. In Petkus rye, as with *Campanula,* SD seems to be an alternative to vernalisation in hastening the flowering response to subsequent LD (Purvis, 1961). In rye, the SD effect has been shown to be a true photoperiodic one and is annulled by a night-break treatment; however, SD are much less effective than cold in this case. A similar SD effect has been demonstrated in a range of wheat cultivars (McKinney and Sando, 1935), in oat (Sampson and Burrows, 1972), in several grasses (Heide, 1994) and in a number of dicotyledons (Napp-Zinn, 1984). In most temperate perennial grasses with dual induction requirements, vernalisation and SD are essentially interchangeable factors for primary induction, while secondary induction requires exposure to LD (Heide, 1994). Although SD and low temperature are independently able to bring about the primary induction response, their effects are additive when applied sequentially. They are, however, highly interactive in the primary induction process. At low temperatures (0–6°C), primary induction appears to be identical with vernalisation and is independent of photoperiod but, as the temperature rises, primary induction occurs only in SD. Above a certain threshold temperature primary induction cannot take place.

The possible substitution of SD for low temperature seems not to be uncommon and may have implications for the mechanism of vernalisation. It is interesting to note that SD can also abolish the requirement for cold in *Trifolium repens* (a New Zealand strain), a plant in which the low-temperature treatment leads directly to floral initiation with no inductive (vernalisation) effect (Thomas, 1979). The SD effect is, however, inductive and must be followed by LD if flowering is to occur. The ecological significance of the substitution of SD for low-temperature may be to allow some flowering to occur after mild winters when vernalisation is incomplete. In most cases,

TABLE 6.5 Effect of daylength during and after vernalisation on flowering in celery, *Apium graveolens* L. cv Florida.

	Plants flowering (%)
A. Daylength during vernalisation	
Short days (9.75–12.5 h)	87
Long days (16 h)	37
B. Daylength after vernalisation	
Short days (8 h)	0
Long days (16 h)	73

A, maintained in natural conditions between 7 and 10°C; B, vernalised for 3 months at 10/5°C in 20 h photoperiods. Temperature after vernalisation was 27/22°C.
Data of Pressman and Negbi (1980).

SD are considerably less effective than low temperature, as would be expected if this explanation is correct. However, the interchangeability of SD and vernalisation in high-latitude grasses indicates that the relationship is also advantageous in plants which grow in regions with long winters at low temperatures. In some arctic-alpine species, initiation of inflorescence primordia takes place during the SD primary induction so that primordia are initiated in the autumn; this seems to be an important adaptive strategy to a short and cool growing season (Heide, 1994).

Although SD cannot substitute for a cold treatment in some other LDP, they may promote flowering when given during or before an exposure to low temperature. Examples of this type of response are found in some perennial grasses and in table beet. It is also seen in celery (*Apium graveolens* cv Florida) where LD during vernalisation delayed flowering whereas, after vernalisation, LD promoted flowering (Table 6.5). This has become of some importance in growing early celery as a glasshouse crop in the UK; in order to avoid bolting, plants can be given LD during the period of winter cold in an unheated glasshouse. However, daylength during chilling had no effect on flowering in a different cultivar of celery (Dulce), although vernalisation did not occur when plants were maintained in darkness; this light requirement appeared not to be photosynthetic since irradiances between 6 and 85 W m^{-2} were equally effective (Ramin and Atherton, 1994). In contrast, the effectiveness of a vernalisation treatment may be increased by exposure to LD (Napp-Zinn, 1984). This occurs in *Pisum sativum* containing the Sn gene, as would be expected since both low temperature and LD are thought to depress the action of this gene or destroy its product.

LD can also substitute for vernalisation. For example, in the LDP *Arabidopsis*, vernalisation has no effect on flowering in the early flowering strains Columbia when plants are subsequently exposed to LD, but markedly accelerates flowering in SD. In this case LD can be considered to abolish the vernalisation requirement. The introduction of the FRIGIDA gene (Lee and Amasino, 1995) resulted in a strong vernalisation effect, even when plants were subsequently grown in LD. However, the low temperature requirement for maximum acceleration of flowering was less in LD (30 d) than in SD (90 d), again indicating that one process (LD) is substituting for the other (vernalisation).

Physiology of Vernalisation

A detailed consideration of the physiology of vernalisation (Thomas and Vince-Prue, 1984) is outside the scope of this book but some aspects that particularly relate to the interaction with photoperiodism can be briefly summarised.

The characteristic feature of vernalisation is that the optimum temperature lies between about 1° and 7°C, with the effective temperature ranging from just below freezing to about 10°C. A broad optimum temperature curve of the type actually seen in many vernalisation responses could be generated by assuming that vernalisation consists of an inhibitory and a promoting process with very different Q_{10} values (Salisbury, 1963b) but this remains a purely theoretical concept. The effect of low temperature increases with the duration of exposure until the response is saturated at a duration which varies widely with species and cultivar (< 10 to > 100 days). Just as with photoperiodic induction, vernalisation is a progressive process and the effect becomes increasingly stable as the duration of cold is increased (Table 6.6).

Vernalisation has been achieved by applying localised cooling treatments to the stem apex and the effect seems to be largely independent of the temperature experienced by the rest of the plant. Excised shoot tips have also been successfully vernalised (Metzger, 1988) and, where seed vernalisation is possible, fragments of embryos consisting essentially of the shoot tip are sensitive to chilling (Purvis, 1940). Callus cultures showed little evidence of vernalisation in *Lolium* but somatic embryoids formed before and during the cold-treatment gave rise to vernalised plants; however, those developing after exposure to cold did not (Arumuganathan *et al.*, 1991). Thus, in contrast to photoperiodic induction, the effect of low temperature appears to be largely confined to the shoot apical meristems. Once vernalisation has occurred, it is perpetuated through mitosis and seems to be a property of all subsequent daughter cells, although the nature of the changes which lead to this semi-permanent vernalised state are unknown. Based on the effects of DNA-demethylating agents to accelerate flowering in non-vernalised plants, it has been proposed that low temperature may demethylate DNA and so release a block to floral initiation; such thermoregulation of an apex-specific gene which regulates production of an enzyme of the GA-biosynthetic pathway may be involved in *Thlaspi* (Burn *et al.*, 1993). It is unlikely that the GA-biosythetic pathway is involved in all vernalisation responses, however, since GA does not substitute for the low-temperature promotion of flowering in many plants (e.g. *Campanula pyramidalis* (LDP) and chrysanthemum (SDP)).

It is possible that not only apical meristems but all dividing shoot cells are potential sites for vernalisation. For example, excised leaves of *Lunaria annua* taken from unvernalised plants regenerate new plants which must be vernalised before they can flower. If the cut leaves were first held at 5°C, however, the regenerated plant was able

TABLE 6.6   Progressive stabilisation of vernalisation with increasing duration of exposure to cold in winter rye (cv Petkus).

Cold treatment (weeks)	2	3	4	5	6	8
Plants remaining vernalised after 2 d at 25°C (%)	0	42	44	75	84	97

Adapted from Purvis and Gregory, 1952; Thomas and Vince-Prue, 1984.

TABLE 6.7 The effect of embryo age at excision on sensitivity to vernalisation in *Lolium temulentum*.

Age of embryo at excision	Leaf number to flower		
(days after anthesis)	0 weeks cold	4 weeks cold	8 weeks cold
5	14.3	13.7	8.7
10	13.1	13.8	8.7
15	13.8	14.2	8.9
20	13.9	13.5	8.6
25	13.9	14	8.3
30	13.8	12.5	8.3

Embryos were exposed to 2°C for 0, 4 or 8 weeks, following excision at different times. Embryos excised as early as 5 days after anthesis were sensitive to vernalisation and flowered with a reduced leaf number. Data of Arumuganathan *et al.* (1991).

to flower. The action of cold is limited to the base of the petiole where cells are dividing since, when this was removed after the cold treatment, the regenerated plant was not vernalised (Wellensiek, 1964). Although these results point to the need for dividing cells, vernalisation is possible in winter rye and in *Cheiranthus allionii* under conditions when mitosis is not occurring. Cell division is clearly not a prerequisite for vernalisation in *Thlaspi arvense*, since flowering shoots were regenerated from leaves that were fully grown before the cold treatment was imposed (Metzger, 1988). However, these thermoinduced cells subsequently became organised into shoot meristems. Thus, although cell division *per se* during vernalisation seems not to be essential, vernalisation appears to be restricted to the meristematic zones of shoots or to cells which are capable of being organised into shoot meristems.

In winter annuals the cold requirement can be satisfied by treating imbibed or germinating seeds. In *Lolium temulentum*, excised embryos could be vernalised as early as 5 days after anthesis and there was no increase in effect with increasing age up to 30 days after anthesis (Table 6.7). However, chilling the developing ear for up to 30 days after anthesis did not result in vernalisation. Some biennial plants can also be vernalised as seeds (e.g. beet, *Oenothera)* but, as with the response to daylength, many must first undergo a period of growth before the cold treatment is effective. This juvenile phase lasted for about 10 days in *Hyoscyamus* and for more than 7 weeks in *Lunaria*. Cultivars may also vary; for example the juvenile phase ended after the production of 17–21 leaves in four different cultivars of cauliflower (Wurr *et al.*, 1994). The duration of the juvenile phase is particularly important in the production of biennial root crops; a long juvenile phase can allow earlier sowing and enables growers to take advantage of a longer growing season without the problem of bolting in the first year. Seed vernalisation is also possible in some perennial grasses but, as with biennial plants, cold treatment is without effect in many cases until some growth has occurred (Heide, 1994).

Devernalisation

The possibility of devernalisation, or loss of the vernalised condition, seems to be widespread and is probably ecologically significant. The most usual devernalising agent is high temperature (about 30°C) but devernalisation by SD has also been reported. High temperature is usually most effective immediately following a vernalising treatment, especially when the latter is suboptimal (Table 6.6). Partial or suboptimal low temperature treatments can sometimes be stabilised by maintaining seeds or plants at a neutral temperature of about 12–15°C before exposing them to the devernalising temperatures. There are, however, many degrees of response ranging from no stabilisation to no high-temperature devernalisation. There are also complex interactions with light which, when given during the chilling treatment itself, appears to have a stabilising effect in some plants; for example, *Lactuca serriola*, can be devernalised only before germination and in the absence of light (Marks and Prince, 1979). It appears that seeds can be devernalised in all species investigated, whereas many plants cannot.

The devernalisation process also seems to interact with daylength and, in some plants, devernalisation is induced by short days rather than by high temperature. Exposing plants of *Beta* to SD caused devernalisation at any stage and, even after stem extension had already commenced, transfer to SD conditions resulted in the formation of a perched, vegetative rosette (Margara, 1960). Devernalisation by SD has also been reported for *Oenothera biennis* and *Cheiranthus allionii* (Wellensiek, 1965).

Although high temperatures may lead to devernalisation in the early stages of chilling, the vernalised condition is normally extremely stable once it is fully established and persists throughout the vegetative phase until the plants finally flower. In biennial *Hyoscyamus* (which requires both vernalisation and LD for flowering), the vernalised state has been shown to persist for several months in SD and flowering occurred promptly when plants were returned to LD.

Transmissibility of the Vernalisation Effect

One of the still unresolved questions about low-temperature vernalisation is whether, like photoperiodic treatments, it leads to the production of a transmissible stimulus. As with photoperiodism, the main approach has been grafting experiments, using non-cold-requiring or vernalised plants as donors and cold-requiring, non-vernalised plants as receptors. Both successes and failures have been reported (Lang, 1965). Transmission of a flowering effect has been observed between different species and genera, and from both non-cold-requiring and vernalised donors. Failures have been reported in SDP (cultivars of chrysanthemum), LDP (*Oenothera*) and DNP (*Thlaspi*).

There are many parallels with the transmission of the stimulus resulting from photoperiodic induction. Some tissue union between grafted plants is essential and treatments which favour the movement of assimilates from donor to receptor (e.g. leaf removal on the receptor) also favour the transmission of the flowering effect from plant to plant. The key question is whether the vernalisation treatment leads directly to the production of a transmissible stimulus which is different from the flowering stimulus produced in favourable daylengths. Alternatively, vernalisation could lead to a change in the condition of the cells such that they are able to produce a floral stimulus when subsequently returned to favourable conditions. Most of the grafting studies are consistent with the latter hypothesis, for the donor plants have usually been in conditions in which they flower; in these circumstances, the vernalisation requirement in the receptor could have simply been by-passed by the transmission of a floral stimulus from the flowering donor. This certainly seems to be the case in *Thlaspi arvensis* where non-vernalised shoot tips remained vegetative when grafted on to cold-treated donors, but flowered when grafted (in LD) to the LDP *Sinapis alba* (Metzger, 1988). Vernalisation may, therefore, initially be a strictly localised state, which is perpetuated by cell division and enables the plant to produce a transmissible substance under favourable conditions at a later stage. This conclusion is supported by results with biennial *Hyoscyamus,* where photoperiodic induction by LD cannot be consummated in the absence of vernalisation and plants remain vegetative indefinitely. In this plant, therefore, vernalisation seems to bring about changes which lead to photoperiodic sensitivity. In those cases where SD can substitute for vernalisation (as in many grasses) it appears that their effect on primary induction is the same, both treatments enabling the plant to respond to and flower in subsequent LD, as in *Hyoscyamus*. No transmissible stimulus appears to be produced as a consequence of primary induction in dual-inductive grasses (Heide, 1994) and it has been suggested that the effect may be to increase the responsiveness of the shoot apex to substances translocated from the leaves during secondary induction in LD. However, the grafting experiments with *Thlaspi* suggest that it is the subsequent ability of the leaves to export a flowering stimulus in inductive photoperiods which is influenced by vernalisation (and perhaps also by primary induction in SD) rather than the sensitivity of the apex.

A TRANSMISSIBLE FLORAL STIMULUS

Circumstantial evidence from physiological experiments strongly suggests that leaves in inductive photoperiods generate a stimulus which evokes flowering at receptive shoot meristems. Three kinds of experiments have given rise to the concept of a floral stimulus and all have in common that photoperiodic induction acts over a distance, implying the movement of a signal from the site of production to the site of action. Firstly, only the leaves have to be exposed to the inductive treatment in order to obtain flowering. Exposing a single young leaf of *Xanthium strumarium* to one SD cycle was sufficient to lead to the development of submacroscopic flowers after 39 days, even when the remainder of the plant (including all the other leaves) remained in LD (Naylor, 1941a). It seems most unlikely that nutritional factors or the absence of inhibitor produced by a single leaf for a single day can explain these results; production of a flower-promoting stimulus by the induced leaf is strongly indicated. Other examples are less dramatic but there are many cases where exposing one, or even part of a leaf to photoinductive cycles is sufficient to cause flowering. Secondly, in species that require only one inductive cycle, formation of a flower-promoting stimulus in the leaves and its subsequent movement to the shoot tips can be demonstrated by removal of the leaves at intervals (Fig. 6.6). Thirdly, the flowering condition can be transmitted via a graft union from an induced shoot or leaf (donor) to a non-induced partner (receptor). A single leaf of *Perilla* (red) can be excised, exposed to photoinductive SD cycles and then grafted to a receptor plant which subsequently flowers. It has been demonstrated that roots are not necessary for induction in the excised donor leaf, nor for the evocation and initiation of floral primordia in the receptor (see Table 1.1). In *Perilla*, an induced leaf can be regrafted several times and retains its capacity to cause flowering in a series of receptors (Zeevaart, 1958). Once again, it is difficult to explain these results except in terms of export of a flower-promoting substance from the

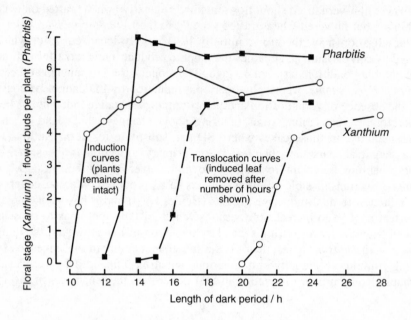

FIG. 6.6.   Induction and translocation of the floral stimulus in *Pharbitis* and *Xanthium*. Solid lines show the effect of increasing dark period duration on the flowering responses. Dashed lines show the effect of removing the induced cotyledons (*Pharbitis*) after various durations of darkness and returning the plants to light, or of removing the induced leaf (*Xanthium*) at various times, the plants remaining in darkness. In both cases, flowering was prevented if the induced organ was removed before translocation of the floral stimulus had occurred. After Vince-Prue, 1975.

induced leaf (other explanations – such as an inhibitor-free carbohydrate supply – are possible, but unlikely). The existence of a chemical signal acting in a concentration-dependent fashion to elicit flowering is also indicated by a different type of approach, using excised pieces of tobacco stem taken from flowering plants (McDaniel and Hartnett, 1993). Floral activity was assessed by the number of leaves formed on an axillary bud before a terminal flower was formed and was found to be determined by the position of the associated internodes rather than the bud itself. For example, when bud 10 and bud 5 were associated with the same internodes, their flowering performance was similar.

Translocation of the Floral Stimulus

Leaving aside for the moment the nature of the floral stimulus, it is possible to ask how it is transmitted from an induced leaf to the apex. Treatments which restrict phloem transport, such as localised heat, cold treatments, removing a ring of tissue external to the xylem, or narcotic treatments with chemicals such as chloroform, prevent the movement of the floral stimulus. Most experiments have been carried out with SDP (chrysanthemum, *Perilla*, *Kalanchoë*, *Xanthium*) but at least one LDP, *Hyoscyamus*, has been shown to behave in the same way (Lang, 1965). From this and similar evidence it has generally been concluded that the floral stimulus moves in the phloem with assimilates. However, all of these treatments would be expected to limit any form of symplastic movement and indicate only that the stimulus does not move in the xylem with the transpiration stream. Perhaps the best evidence for transport in the phloem is that grafted donor leaves or shoots can cause flowering only when a functional phloem connection has been established. This was first demonstrated for *Perilla* (Zeevaart, 1958). Other workers have also found that stimulus transport across a graft required tissue union, although phloem connection was not strictly shown to be the limiting factor.

Assuming that the floral stimulus moves primarily in the phloem, is it co-transported with the mass flow of assimilates? In some cases, good correlations have been reported between the translocation of assimilates and the movement of floral stimulus. For example, a high flowering response in *Perilla* was associated with the presence in the bud of a large amount of ^{14}C label from an induced leaf, while low flowering was associated with label from a non-induced leaf (Chailakhyan and Butenko, 1957). In contrast, movement against the expected flow of assimilate has been demonstrated in young, predominantly-importing leaves of *Lolium*, which were highly active in causing flowering (Evans and Wardlaw, 1966). This experiment is not conclusive, however, since 5–7% of the labelled carbon was exported and the floral stimulus could have accompanied this fraction (Evans and King, 1985). However, induced leaves of *Pharbitis* held in darkness exported floral stimulus at a rate comparable with that from leaves in the light, although no concurrent movement of labelled assimilate was detected (King *et al.*, 1968).

Simultaneous studies of the rates of assimilate movement and the movement of the floral stimulus have been carried out in only a few cases. Needless to say, one of the problems here is the accuracy with which the rate of movement of an unknown substance(s) can be determined. Another complication arises in the interpretation of experiments in which leaf excision has been carried out; while this has been necessary in order to estimate the rate of movement from induced leaf to responding apex, removing a leaf must have affected the supply of substances other than the floral stimulus (Vince-Prue and Gressel, 1985). In *Pharbitis* (Fig. 6.7), the calculated rates for the movement of floral stimulus (240–370 mm h^{-1}) and labelled assimilates (330–370 mm h^{-1}) through the stem were similar, indicating that both were moving by mass flow in the phloem. In contrast, experiments with the LDP *Lolium* indicated that the transport rate through the leaf blade was only 10–24 mm h^{-1} for the floral stimulus, compared with approximately 1000 mm h^{-1} for sucrose (Evans and Wardlaw, 1966). Based on these apparent differences in the rate of transport it was suggested that the stimulus in the LDP *Lolium* might be different from that in the SDP *Pharbitis* and transported via a different mechanism. However, a detailed analysis of ^{14}C profiles in *Lolium* revealed that, although the major component moved at 400–840 mm h^{-1}, there was a slower component which moved along the leaf blade at 10 mm h^{-1} and which might be associated with the movement of the floral stimulus (Evans and King, 1985).

Based on the rather limited evidence presently available, it appears that the floral stimulus moves from the induced leaf to the apex by a symplastic route which is the phloem in most, but perhaps not all instances. The transport mechanism has not yet been established. Movement together with assimilates has been demonstrated in some

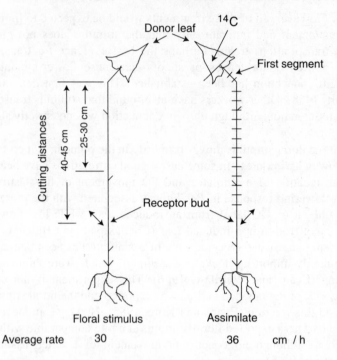

FIG. 6.7. Velocities of translocation in mature *Pharbitis* plants. Assimilate movement was established by cutting 2 cm long segments between the donor leaf and receptor bud at various times after exposure to $^{14}CO_2$. Floral stimulus movement was estimated by removing the donor leaf at various times; the excisions were made below the petiole of the donor leaf, or about 25 or 40 cm down the stem as shown. In this way it was possible to estimate the velocity of movement specifically through the stem. All other leaves were removed prior to the experiment. After Vince-Prue and Gressel, 1985 (data of King *et al.*, 1968).

cases; however, movement against the main assimilate flow can also occur and so a co-transport mechanism with sugars seems unlikely.

The Nature of the Floral Stimulus

Grafting experiments indicate that the final product of photoperiodic induction is the same in plants of different photoperiodic classes and is interchangeable with a floral stimulus in day-neutral plants. That this stimulus might be common to (or at least physiologically equivalent for) many plants is indicated from the results of inter-specific and intergeneric grafts. As the stimulus is only transported when a successful graft union with a functioning phloem has been established, exchange between plants can only be demonstrated where this occurs. This means that no work has been possible with monocotyledons. It also means that interchangeability can only be studied between closely related plants.

Successful transfers of flowering between grafts of photoperiodically sensitive plants have been tabulated by Lang (1965) and more recently by Grayling (1988). A large number has been recorded and a selection of examples is given in Table 6.8. The majority of experiments have been carried out with herbaceous plants but graft transmission of flowering has also been recorded for woody species. In mango

TABLE 6.8 Successful transfers of flowering by grafting.

Donor	Response type	Receptor	Response type
Intraspecific grafts			
Glycine max cv Agate	DNP	*G. max* cv Biloxi	SDP
Chenopodium rubrum 60° 47′ N	SDP[a]	*C. rubrum* 34° 90′ N	SDP
Pisum sativum various genetic lines	DNP or LDP[a]	*P. sativum* line G	LDP
Interspecific grafts			
Gossypium hirsutum	DNP	*G. davidsonii*	SDP
Nicotiana tabacum Delcrest	DNP	*N. sylvestris*	LDP
Intergeneric grafts			
Blitum virgatum	LDP	*Chenopodium rubrum*	SDP
Chenopodium polyspermum	SDP	*Blitum capitatum*	LDP
Cucumis sativus	DNP	*Sicyos angulatus*	SDP
Centaurea cyanus	LDP	*Xanthium strumarium*	SDP

[a] Facultative photoperiodic response.
Examples taken from Grayling (1988).

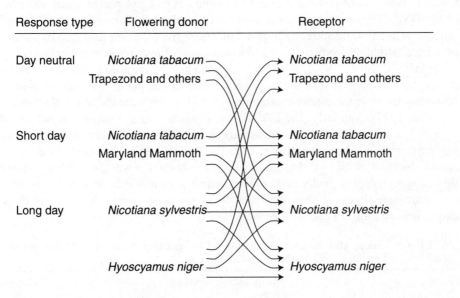

FIG. 6.8. Diagrammatic representation of the successful graft transmission of flowering from donors to receptors in various members of the Solanaceae. After Lang, 1987.

(*Mangifera indica*), the cultivar Royal Special (which was capable of flowering) was a successful donor for defoliated scions of cultivars which were in the off-season (non-flowering) condition (Kulkarni, 1986). Successful intergeneric and interspecific grafts indicate that the stimulus is interchangeable between plants of different genera and species. Flowering has also been transferred between all major photoperiodic classes.

In the Solanaceae, LDP, SDP and DNP have been shown to cause flowering in receptor plants of all three photoperiodic categories and successful inter-generic grafts between *Hyoscyamus* and *Nicotiana* have been carried out (Fig. 6.8). In some, though not all combinations, a single leaf is sufficient as the donor (Lang, 1987). Acceleration of flowering was obtained in the DNP *N. tabacum*, Trapezond when plants were grafted to the SDP, Maryland Mammoth or to the LDP, *N. sylvestris* and maintained in photoperiods inductive for the donors; thus transmission of a floral stimulus occurs from photoperiodically-sensitive plants to DNP as well as *vice versa* (Lang *et al.*, 1977). Work with the Crassulaceae (Table 6.9) has extended the range of photoperiodic categories in which interchangeability has been demonstrated to include dual daylength plants; for example, both the SLDP *Echeveria harmsii* and the LSDP, *Bryophyllum daigremontianum*, caused flowering in receptors of *Kalanchoë blossfeldiana* (SDP) and flowering plants of the LSDP *B. daigremontianum* were successful donors of flowering for vegetative plants of the SLDP *E. harmsii,* maintained in either LD or SD. Other examples of transfer of flowering between plants of different photoperiodic classes are given in Table 6.8.

It is also evident that a flowering stimulus can be transferred from plants where induction is achieved by means other than exposure to favourable photoperiodic cycles (Table 6.10). Day-neutral plants, which achieve flowering through autonomous induction, can bring about flowering in both LD and SD receptor plants; even a single leaf of the DNP *Glycine* cv Agate was sufficient to induce flowering in the SD cv Biloxi (Heinze *et al.*, 1942). Indirectly induced plants of green *Perilla* (induced by the arrival of a floral stimulus from flowering donors) cause flowering in red *Perilla* receptors (Zeevaart and Boyer, 1987). *Perilla* (green) can also be induced to flower by exposure to low temperature in LD; a single leaf from these plants can act as donors of flowering to receptor plants maintained in LD at high temperature (Deronne and Blondon, 1977). Similarly, the LDP *Silene armeria* can be induced in SD by either high or low temperature, or by the application of GA_3; all can act as donors to cause flowering in vegetative receptors (Lang, 1987). The LSDP *Bryophyllum daigremontianum* induced in SD by the application of gibberellin was an effective donor of flowering to receptor plants of *Echeveria harmsii* in non-inductive SD; this was not due to transmission of gibberellin from the donor plant since GA treatment does not cause flowering in *Echeveria* (Zeevaart, 1982).

TABLE 6.9 Successful transfers of flowering by grafting between different genera and photoperiodic response types in the Crassulaceae.

Donor (flowering)	Response type	Receptor	Response type
Kalanchoë blossfeldiana	SDP	*Sedum* spp.	LDP
Sedum spp.	LDP	*Kalanchoë blossfeldiana*	SDP
Bryophyllum daigremontianum	LSDP	*Kalanchoë blossfeldiana*	SDP
Kalanchoë blossfeldiana	SDP	*Bryophyllum daigremontianum*	LSDP
Echeveria harmsii	SLDP	*Kalanchoë blossfeldiana*	SDP
Echeveria × pulv-oliver	LDP	*Kalanchoë blossfeldiana*	SDP
Bryophyllum daigremontianum	LSDP	*Echeveria harmsii*	SLDP

From Zeevaart (1978, 1982).

TABLE 6.10 Successful transfers of flowering by grafting with donor plants induced to flower by treatments other than exposure to appropriate photoperiods.

Inductive treatment	Donor	Receptor
Autonomous	*Nicotiana tabacum* Trapezond (DNP)	*Nicotiana tabacum* Maryland Mammoth (SDP)
	Nicotiana tabacum Trapezond (DNP)	*Nicotiniana sylvestris* (LDP)
	Glycine max Agate *(DNP)*	*Glycine max* cv Biloxi (SDP)
SD + GA₃	*Bryophyllum daigremontianum* (LSDP)	*Echeveria harmsii* (SLDP)
SD + high temperature	*Silene armeria*	*Silene armeria* (LDP)
SD + low temperature	*Silene armeria*	*Silene armeria*
SD + GA₃	*Silene armeria*	*Silene armeria*
LD + low temperature	*Perilla* (green)	*Perilla* (green) (SDP)
Indirect, by grafting to flowering donor	*Perilla* (green)	*Perilla* (red) (SDP)

For details see text.

Based on the results of many grafting experiments, it appears that the floral stimulus of plants in different photoperiodic classes is identical, or at least interchangeable. Moreover, plants which flower in response to triggers other than daylength produce a stimulus which is able to cause flowering in photoperiodically-responsive receptor plants. Thus, the differences between them appear to lie in the mechanisms leading to induction and production of the stimulus, rather than in the end product itself.

The many successful transfers of flowering between different species and genera (Tables 6.8, 6.9, Fig. 6.8) has sometimes led to the conclusion that there is a specific and universally effective flowering hormone, or **florigen**. However, it must be emphasised that neither the specificity for flowering nor the universality of florigen has been demonstrated. Interchangeability of the stimulus is limited by the need to establish a graft union between donor and receptor plant. This means that interchangeability can be shown only between plants of the same family. A reported inter-family transfer of flowering from *Xanthium strumarium* (SDP, Compositae) to *Silene armeria* (LDP, Caryophyllaceae) has not been repeated and was probably due to removal of the roots, which results in flowering of *Silene* in SD (Zeevaart, 1984). Nothing is known about interchangeability between species and genera in monocotyledons although, as in dicotyledons, their leaves produce a transmissible flowering stimulus as a result of photoperiodic induction. Much of this evidence comes from work with *Lolium temulentum* (Evans and King, 1985). In some cases, the floral stimulus does not appear to be interchangeable even between closely-related plants (Table 6.11). One of the most striking examples is the failure to transfer flowering between the LSDP *Cestrum diurnum* and *C. nocturnum*. This does not appear to be due to the failure of graft union or transport because, when an intergraft of *C. nocturnum* was placed between two shoots of *C. diurnum* (one of which was defoliated to act as receptor), both shoots flowered, although defoliated shoots normally remain vegetative (Griesel, 1963). A further example of failure to obtain graft transmission in the Solanaceae is from the DNP *Nicotiana tabacum* cv Delcrest to the SD cultivar Maryland Mammoth; however, Delcrest was able to cause flowering in *N. sylvestris* and other day-neutral

TABLE 6.11   Examples of failure to transfer flowering by grafting.

Donor	Receptor	Comments
Sedum spectabile (LDP)	*Bryophyllum daigremontianum (LSDP)*	Reciprocal graft effective. *S. spectabile* is effective donor for *Brassica crenatum*
Cestrum diurnum (LSDP)	*C. nocturnum* (LDP)	
Cestrum nocturnum	*C. diurnum*	
Ipomoea repens (SDP)	*I. batatas* non-flowering cv	Flowering of a non-flowering cv of *I. batatas* achieved by graft with other early flowering cvs
I. batatas early blooming cv	*I. batatas* non-flowering cv	
Kleinia articulata	*K. articulata* (SDP)	
K. repens (LSDP)	*K. articulata* (SDP)	Reciprocal graft succcessful
Nicotiana tabacum Maryland Mammoth	*N. tabacum* Maryland Mammoth (SDP)	No known success for this combination
Silene cucubalis	*S. cucubalis* (LDP)	
Silene armeria (LDP)	*S. gallica* (LDP)	
Aster savatieri (LDP)	*Xanthium strumarium* (SDP)	
Coreopsis lanceolata (LDP)	*Xanthium strumarium* (SDP)	

Examples taken from Grayling (1988).

tobacco cultivars were successful donors of flowering to Maryland Mammoth (Lang, 1987). Within the family Chenopodiaceae, transmission from the SDP *Chenopodium* to the LDP *Blitum* occurred without defoliation of the receptor shoot whereas, to obtain the reciprocal transfer, it was necessary both to remove mature leaves from the *Chenopodium* receptor and to remove flowers from the *Blitum* donor (Jacques and Leroux, 1979). Another case involves members of the Compositae. The SDP *Kleinia articulata* caused flowering in the LSDP *K. repens* but not vice versa (Kulkarni and Schwabe, 1984). There are, of course, a number of possible explanations for the failure to obtain graft transmission of flowering, other than differences in the identity of the stimulus. Assimilate movement from donor to receptor may not have been assured (in *Impatiens balsamina*, the flowering effect was not transmitted between branches when the receptor shoot was defoliated, unless the apical bud was also removed (Sawhney *et al.*, 1978); tissue union may not have been established; there may have been inhibitory effects within the receptor plant; the amount of stimulus produced by the donor leaves may differ between species, as could the sensitivity of the apical meristems of the receptor to the arriving stimulus (donor plants of *Pharbitis nil* required 4 SD in order to induce flowering in receptors of *Ipomaea batatas*, whereas *Pharbitis* itself requires only a single SD (Takeno, 1991)).

An important observation with reference to the specificity for flowering of the putative floral stimulus comes from grafting experiments in which tuber-forming plants have been used as receptors. The potato, *Solanum andigena*, forms tubers only when the leaves are exposed to SD; however, tuberisation also occurred when receptor plants were grafted to flowering donors of either the LDP *Nicotiana sylvestris* or the SDP *N. tabacum* Maryland Mammoth (Martin *et al.*, 1982). Non-flowering plants did not cause tuberisation. Although these experiments do not prove that the stimuli for flowering and tuberisation are interchangeable, it is evident that exposing leaves to photoperiodic cycles that are inductive for flowering also leads to the production of a transmissible stimulus which can evoke photoperiodically-controlled tuberisation at the apical meristems of shoots in plants of another species. Neither species of tobacco is, itself, capable of forming tubers. A similar situation has been reported (Nitsch, 1965) for grafts between *Helianthus annuus* (sunflower) and *H. tuberosus* (artichoke). At least some of the events in the induced leaf appear to be common to both the tuber-forming and flowering pathways as a gene expressed in the leaves of potato plants that are beginning to initiate tubers is also expressed in tobacco plants during flowering (Jackson *et al.*, 1993). However, since this occurs fairly late, it may only be associated with a secondary event such as the increased export of substances to the newly developing sinks. Nevertheless, it remains possible that the specificity of the response to the arrival of a stimulus from induced leaves lies in the target cells rather than in the identity of the substance itself. This important question requires further investigation.

Overall, a considerable body of physiological evidence points to the existence of transmissible flowering stimulus which appears to be common to (or, at least, interchangeable between) plants of all photoperiodic classes. From such studies, the idea of a florigen, i.e. a specific floral-forming hormone, has developed. Nevertheless graft transfer of flowering is only possible between closely related plants where a graft union is possible; thus there is no evidence for a universally effective florigen, as has sometimes been suggested. Attempts to extract and identify a photoperiodic floral stimulus are discussed in Chapter 8.

THE INHIBITION OF FLORAL INITIATION IN NON-INDUCTIVE DAYLENGTHS

It has been argued that 'the problem may quite as properly be considered as one of "failure to flower" as of promoting flowering' (Gregory, 1948). Certainly, much of the evidence for a floral stimulus could also be interpreted in terms of a floral inhibitor. Leaves in inductive daylengths may, for example, only be needed to provide an inhibitor-free supply of carbohydrate. The arguments for a floral promoter are strong and control only by floral inhibitors seems unlikely. There is, however, a substantial amount of evidence showing that leaves in non-inductive cycles may exert a positive inhibitory effect on flowering and are not just neutral. Three major interpretations have been proposed:

- Leaves in non-inductive cycles may affect the pattern of solute translocation and interfere with movement of the floral stimulus.
- Non-inductive cycles may act in the leaves to interfere with the synthesis or export of a floral stimulus, or cause its destruction.
- Non-inductive cycles lead to the production of a transmissible inhibitor of flowering which interferes with the action of the floral stimulus at the apex.

Interference with Translocation of the Floral Stimulus

Although flower initiation can be brought about by exposing one leaf, or part of a leaf to appropriate photoperiods, this frequently occurs only when some, or all of the other leaves are removed. To induce transfer of the floral stimulus from one branch to another of a two-branched plant, or from donor to receptor in grafting experiments, it is often necessary to de-foliate the receptor branch or plants. Several lines of evidence suggest that the non-induced leaves may, at least partly, act by interfering with the transport of the floral stimulus to the receptive site.

Inhibition by leaves in non-inductive daylengths is usually not observed unless they are between the induced leaf and the receptive meristem (Lang, 1980; Jacobs, 1980). This is explicable if it is assumed that the main bulk of the hormone moves with the carbohydrate stream, since the shoot meristem receives carbohydrate from the nearest photosynthesising leaf; if this is in non-inductive cycles the flowering stimulus would arrive from a more distant leaf. The inhibitory effect has sometimes been shown to be more pronounced for a mature leaf as would be expected from carbohydrate translocation patterns and immature leaves, which import rather than export carbohydrate, have been found to inhibit flowering less (Lincoln et al., 1956). Treatments which would be expected to reduce or prevent the flow of sugars out of a leaf, such as a lower light intensity (Lincoln et al., 1956) or killing a section of the petiole by heat (Gibby and Salisbury, 1971), have also been found to decrease the inhibitory effect of the leaf on flowering. The most direct piece of evidence for action via interference with stimulus transport was a report that the inhibitory effect of a long-day leaf in Perilla could be largely replaced by feeding sucrose. However, subsequent experiments with Xanthium (Jacobs and Eisinger, 1990) showed that supplying water to a cut petiole also resulted in a delay of flowering which was not significantly different from that obtained with sucrose; it was concluded that the inhibitory effect was probably due to wounding, rather than to sucrose per se.

Although some of the inhibitory effect on flowering by leaves in non-inductive cycles may result from interference with the transport of a floral stimulus due to changes in assimilate movement, many experiments are difficult to interpret in these terms. In Kalanchoë, modifying the photosynthetic capacity of the interposed leaf by maintaining it in low, or high carbon dioxide concentration, or applying DCMU, had little effect on the inhibition of flowering (Papafotiou and Schwabe, 1990). Night-breaks, which would have little effect on photosynthetic capacity, were also inhibitory when given to the interposed leaf in both Kalanchoë and Xanthium. When the apical and basal parts of the same leaf were exposed to different photoperiods, flowering in the SDP Perilla and Xanthium was decreased when the basal part of the leaf was in non-inductive daylengths, whereas LD given to the apical part had no effect (Table 6.12). The total movement of

TABLE 6.12 Effects of exposing apical or basal halves of leaves to long and short days on flowering in Perilla and Xanthium.

Daylength treatment		Days to bud	Floral stage
Basal	Apical	(Perilla)	(Xanthium)
SD	SD	36	6
SD	LD	38	5.2
LD	SD	61	0
Dark	SD	40	–
SD	Dark	36	–

Data of Chailakhyan (Perilla) and Gibby and Salisbury (Xanthium); from Vince-Prue (1975).

floral stimulus with carbohydrate from the leaf would be expected to be similar in both cases. In the SDP *Rottboellia exaltata* and in the LDP *Lolium temulentum* (Evans, 1960, 1962a), leaves in non-inductive daylengths inserted below the induced leaf were found to reduce the flowering response. Carbon labelling experiments with *Lolium* showed that most of the carbohydrate from these lower leaves was exported to the roots and their presence did not reduce the movement of carbohydrate from the upper, induced leaf to the apex (Evans and Wardlaw, 1966).

Thus, although the inhibition of flowering by leaves in non-inductive daylengths may arise, at least in part, as a consequence of interference with stimulus movement to the apex, there also appears to be a more specific inhibitory process associated with leaves which are exposed to non-inductive photoperiods. Moreover, it appears that more than one kind of inhibition can occur in the same plant (Ogawa and King, 1990). In *Pharbitis nil*, the presence of one of the cotyledons in continuous light reduces the floral response to a single inductive dark period given to the other. Parallel inhibitory effects on flowering and on the amount of assimilate reaching the apex from the induced cotyledon point to interference with co-transport. Moreover, there was a clear relationship between flowering and assimilate imported to the apex from the non-induced cotyledon, as affected by the area of the non-induced cotyledon (Table 6.13A). In addition, chemical treatments which restricted the production or export of assimilates promoted flowering when applied to the non-induced cotyledon, but were inhibitory when applied to the induced cotyledon. Thus effects on assimilate/ stimulus co-transport are able to account for part of the inhibitory effect of non-induced coptyledons in *Pharbitis*. However, when the non-induced cotyledon was exposed to a night-break rather than to continuous light, flowering was reduced without changing the pattern of assimilate import into the apex from the induced cotyledon (Table 6.13B). The night-break treatment reduced flowering well below that

TABLE 6.13 Relationship between import of ^{14}C into the apex from an induced cotyledon and flowering in seedlings of *Pharbitis nil*.

	^{14}C activity in apex from induced cotyledon (dpm g^{-1})	Flowers per plant
A. Area of non-induced cotyledon (%)		
100	932	0.8
50	1667	1.8
25	2069	1.75
0	2489	1.25
B. Treatment of non-induced cotyledon		
Dark (14 h)	120	5.9
Night-break	116	0.9
Removed		4

In both experiments, one cotyledon was given an inductive dark period by wrapping it in foil for 15 or 16 h. In A, the other cotyledon was exposed to continuous light at 45 µmol m^{-2} s^{-1} from white fluorescent lamps and its area was varied. In B, the other cotyledon was maintained in darkness, given a 10 min night-break at the 8th hour after beginning the dark period, or removed. After Ogawa and King (1990).

brought about by cotyledon removal, demonstrating that the non-inductive cycle was actually inhibitory and the reduced flowering was not just due to the lack of induction. Interestingly, the characteristics of the inhibitory effect were similar to those previously observed for the inhibition of induction (see Chapter 4); NB_{max} occurred near the 8th hour of darkness and the threshold irradiance was between 0.08 and 3.0 W m^{-2}, a value close to that for the suppression of dark time measurement. A photoperiod-dependent switch between induction and inhibition is indicated. However, in *Xanthium*, a somewhat longer night-break and a slightly higher irradiance were required to establish inhibition than to prevent induction (Papafotiou and Schwabe, 1990).

Inhibitory Processes in the Leaf

If an inhibitory process occurs in leaves exposed to non-inductive daylengths, it may be antagonistic in some way to the induction process in the leaf or it may lead to the production of a substance that inhibits floral initiation at the apex.

The existence of inhibition of the induction process in the leaf has largely been deduced from the results of fractional induction experiments. However, a careful examination of these results shows that other interpretations are possible in some cases. A quantitative study of the effect of intercalated LD in the SDP *Kalanchoë blossfeldiana* (Schwabe, 1956) showed that the number of buds was progressively reduced with a decrease in the number of *consecutive* SD, when the *total* number of SD given remained constant at 12. Even a single LD inserted between two groups of 6 SD reduced the flower number per plant by 50%. Increasing the number of consecutive LD did not increase their effect and 3 LD given singly between groups of 3 SD were much more inhibitory than three consecutive LD. Using a sequence including 24 h of darkness, it appeared that the LD cycle had an inhibitory effect only on the subsequent SD cycle, since inhibition of flowering occurred when a LD immediately preceded a SD but not when it preceded a 24 h dark period, which largely neutralised its inhibitory effect (Table 6.14). Similar results were obtained with the SDP *Perilla* and *Glycine* (Schwabe, 1959) and it was proposed that the LD-inhibitory process acted to prevent induction in the leaves during the subsequent SD cycle. However, an analysis of the treatments actually given in Table 6.14 suggested an alternative explanation, namely that the number of inductive cycles was not the same in all treatments (Grayling, 1990). The cycle SD/24 h dark/LD actually consists of 8 h light/40 h dark/16 h light/8 h dark; this sequence includes only one inductive (longer

TABLE 6.14   Effect of intercalated long days and periods of darkness on flowering response to short days in *Perilla* and *Kalanchoë*.

Treatment cycle	Flowering quotient (*Perilla*)	Flowers per plant (*Kalanchoë*)
SD–DD–LD	0/10	2.5
SD–LD–DD	12/12	150.1
SD–DD	8/8	174.5
SD–LD	0/10	0.75

SD, 8 h light–16 h dark; LD, 16 h light–8 h dark; DD, 24 h darkness. Treatment cycle repeated 12 times. Data of Schwabe (1956, 1959).

than critical) dark period. In contrast the cycle SD/LD/24 h dark consists of 8 h light/ 16 h dark/16 h light 30 h dark and so includes two inductive dark periods. Since each cycle was repeated 12 times, plants in the first treatment received 12 long nights, whereas those in the second received 24 long nights, accounting for their greater flowering response. This explanation is very plausible, but does not account for the high level of flowering in the SD/24 h dark treatment which would also have received only 12 inductive dark periods. Although the possibility that a non-inductive cycle acts in the leaf to inhibit induction in the subsequent cycle cannot be completely ruled out, the evidence for such a process is not very strong and most attempts to determine the timing of the inhibitory effect of leaves have concluded that this occurs after induction (Ogawa and King, 1990). Thus interference with the movement, or action of a floral stimulus is indicated rather than with its production in the leaf.

Production of a Transmissible Inhibitor

Several types of physiological experiment suggest that one or more transmissible inhibitors may be produced in the leaves of some plants when they are in non-inductive cycles. A few plants have been found to flower when the leaves are removed; examples are known for both LDP (*Hyoscyamus*; Lang, 1980) and SDP (strawberry; Guttridge, 1985). In *Hyoscyamus*, the presence of even one mature leaf in non-inductive cycles restores the inhibition. It is evident that, in these plants, the inhibition of flowering in non-inductive cycles could not result from an interference with transport of a floral stimulus, nor could the inhibitory effect be due to an antagonistic process in the leaves which prevents stimulus production.

More direct evidence for a transmissible inhibitor of flowering comes from strawberry, where an inhibitory effect is transmitted from a mother plant in LD to a daughter joined to it by a stolon. Mature leaves were found to be most inhibitory and transmission of the LD inhibition was increased by treatments which would be expected to increase the flow of carbohydrates from donor to receptor; for example, by restricting the number of daylength hours given to the daughter plants, or reducing the light intensity (Guttridge, 1985).

The strongest evidence for the existence of transmissible inhibitors of flowering comes from grafting experiments in the Solanaceae (Lang, 1987). These are summarised in Fig. 6.9. In grafts between the LDP *Nicotiana sylvestris* and the day-neutral tobacco Trapezond, flowering was inhibited in Trapezond when the grafts were maintained in SD. The growth habit of the grafted Trapezond was also modified; the internodes were much shorter and became thickened, approaching to some extent the rosetted habit of the LD partner when maintained in SD. Similar results were obtained when a SD tobacco cultivar was used as the receptor plant and the LDP, *Hyoscyamus niger* was used as the donor in SD. A single leaf of the LD partner was sufficient to cause these effects. In contrast, flowering in the day-neutral tobacco was not affected by grafting to the SDP, Maryland Mammoth in long days. In similar grafts between Maryland Mammoth (as donor) and Trapezond, *Hyoscyamus* or *Nicotiana sylvestris*, no consistent inhibitory effects on flowering were obtained. Thus, whereas the LDP *N. sylvestris* and *H. niger* appear to produce a transmissible inhibitor of flowering in non-inductive photoperiods, Maryland Mammoth does not, or at most produces it in much lower quantities. The putative *antiflorigen* is apparently inter-

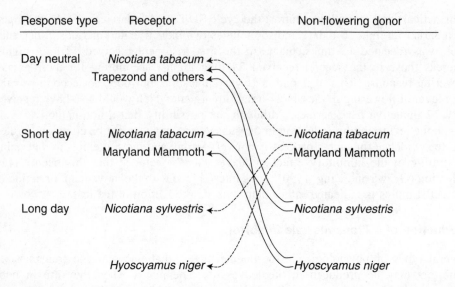

Response type	Receptor	Non-flowering donor

Day neutral   *Nicotiana tabacum*
                Trapezond and others

Short day   *Nicotiana tabacum*   *Nicotiana tabacum*
                Maryland Mammoth   Maryland Mammoth

Long day   *Nicotiana sylvestris*   *Nicotiana sylvestris*

                Hyoscyamus niger   Hyoscyamus niger

FIG. 6.9. Diagrammatic representation of experiments demonstrating the existence of graft-transmissible floral inhibitors in some members of the Solanaceae. Solid lines indicate that flowering was inhibited in receptors capable of flowering, by donors maintained in non-inductive photoperiods. Dashed lines indicate that the donor had no inhibitory effect on flowering in the receptor. After Lang 1987.

changeable between species and photoperiodic categories. For example, the inhibitor from the LDP *Hyoscyamus* affects flowering in both the SDP *Nicotiana tabacum* Maryland Mammoth and the day-neutral tobacco, Trapezond. A simple stoichiometric relationship appears to exist between the flowering promoter and inhibitor in these Solanaceae; when a scion of Maryland Mammoth and one of *N. sylvestris* were grafted to Maryland Mammoth, flowering in the receptor became earlier as the number of induced leaves on the Maryland Mamoth donor was increased and later as the number of non-induced *N. sylvestris* leaves was increased (Table 6.15).

Grafting experiments have also been carried out in the LDP *Pisum sativum* and these have been coupled with a detailed genetic analysis. It has been shown that the *Sn* gene controls the production of a graft transmissible inhibitor which begins within 4 h of the beginning of darkness (Reid and Murfet, 1977). Graft-transmissible promoters of flowering also appear to be produced, however, since leaves of wild-type shoots can supply a substance necessary for flowering in mutant scions (Taylor and Murfet, 1994). It is thought that this floral stimulus is transported with assimilates within the phloem, which would bring its movement under the influence of the photoperiod gene system (*Sn Dne Ppd*) which controls the production of the inhibitor. It is suggested that the delayed flowering in late photoperiodic genotypes under SD conditions arises because the inhibitor directs an increase in the basipetal flow of assimilate, leading to transport of the floral stimulus away from its site of action in the apical bud (see also Chapter 9). Thus, in this plant, the inhibitory action of leaves in non-inductive daylengths appears to depend on assimilate movement, even though a transmissible inhibitor is produced. It is interesting that the floral promoter has a specific role in floral initiation in pea, while the floral inhibitor has generalised effects on both reproductive and vegetative characters.

TABLE 6.15 Flowering of indicator shoots of *Nicotiana tabacum* Maryland Mammoth in double grafts with a Maryland Mammoth and a *N. sylvestris* scion.

No. of leaves on MM donor	No. of leaves on NS donor	MM indicator shoot	
		Days to first flower bud	No. of leaves to flower bud
10	0	29	27
10	2	47	34
10	5	89	41
10	10	122	46
5	0	37	29
5	2	75	42
5	5	113	44
5	10	136	46
2	0	82	41
2	2	127	47
2	5	136	49
2	10	145	54

Grafts consisted of a shoot of the LDP *Nicotiana sylvestris* (NS) and one of the SDP Maryland Mammoth tobacco (MM) both grafted onto a MM receptor. The amount of induced and non-induced tissue was varied by altering the numbers of leaves on both the induced (MM) and non-induced (NS) shoots. From Lang (1980).

In SDP, grafting experiments which indicate the production of a transmissible inhibitor have only been reported in detail for *Coleus* (Jacobs, 1980). A single leaf of the SDP *Coleus frederici* was grafted into the third internode of a *C. frederici* stock plant with two leaves retained; flowering was recorded at the node below the grafted leaf. A leaf taken from a flowering plant was an effective donor of flowering to plants maintained in LD, whereas one taken from a vegetative plant (in LD) was found to delay flowering compared with plants without a grafted donor leaf. There is also some evidence for the production of a transmissible inhibitor in SD tobacco. When grafted to the day-neutral cultivar, Trapezond, there was litle evidence for the production of an inhibitor by Maryland Mammoth in LD; however, a marked inhibition of flowering was obtained in grafts with *N. sylvestris* (Lang, 1980). As with a flower-promoting stimulus, therefore, differences between success and failure to transmit an inhibitory effect may be due to differences in the sensitivity of the receptor plants.

Transmissible inhibitors also appear to participate in the control of flowering in plants which respond to a single inductive cycle. For example, in *Lolium*, 10 cm² of the sixth (uppermost) leaf was sufficient for the induction of flowering in the absence of the other leaves, but flowering failed if the remaining leaves were exposed to SD (Evans, 1960); these leaves were inserted below the LD leaf and did not affect the movements of assimilates from it. The number of plants which initiated flowers depended on the length of time that the SD leaves remained on the plant during induction of the uppermost leaf by a single LD (Fig. 6.10). It was assumed that the inhibition of flowering resulted from action of a transmissible inhibitor at the apex. Like other inhibitory effects (and unlike induction) it is not cumulative since the threshold number of LD for flowering decreased, rather than increased as the plants continued to grow in SD (Fig. 6.3). An interesting experimental observation was that

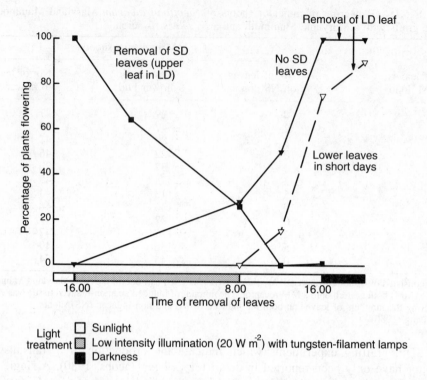

FIG. 6.10. Flower promotion by a LD leaf and flower inhibition by SD leaves in *Lolium temulentum* as influenced by the time that the leaves were allowed to remain on the plant. The uppermost leaf was given a single inductive exposure of 24 h light; the other leaves were either removed at the beginning of the inductive LD or remained in SD. At the time indicated, either the LD leaf or the SD leaves were removed. After Vince-Prue, 1975 (data of Evans, 1960).

the production of the putative inhibitor in SD leaves was inhibited by anaerobic condition (nitrogen gas) but not the production of the floral stimulus in LD (Evans, 1962b).

It is evident that leaves in non-inductive cycle antagonise, to a greater or lesser extent, the effect of leaves in favourable cycles. While this may, in part, be due to an alteration in the pattern of assimilate translocation affecting the movement of the floral stimulus from induced leaves to the apex, there is evidence in both LDP and SDP that processes antagonistic to flowering may occur in leaves exposed to non-inductive cycles. In some cases, the results indicate the production of a transmissible inhibitor which presumably acts at the apex. In other cases, it has been suggested that it is the inductive process in the leaves which is affected. A characteristic of the inhibitory effect is that, in contrast to the promoting effect of favourable cycles, it is not cumulative and several unfavourable cycles usually have no greater effect than a single one (Fig. 6.1). However, a cumulative inhibitory effect of LD has been reported in the SDP *Hibiscus cannabinus* (Ren *et al.*, 1982).

Even in those plants where an inhibitory effect of unfavourable cycles is most clearly seen, a flower-promoting effect has also been demonstrated. Although *Hyoscyamus* flowers readily in any daylength when defoliated, it will flower more quickly with one leaf in LD (Lang, 1965). Similarly, *Lolium* initiated floral primordia when the inhibitory effect of SD was removed by maintaining the leaves in nitrogen during the long dark period, but the characteristic development of the spike required the inductive effect of LD (Evans, 1962b).

CONCLUSIONS

It seems likely that the photoperiodic control of flowering is regulated by more than one transmissible substance from the leaves. Whether or not flowering occurs may depend on a balance of promoters and inhibitors (involving possibly also nutritional factors) rather than on the accumulation at the apex of a threshold amount of a single floral stimulus. The demonstrated stoichiometry between promotion by leaves in inductive cycles and inhibition by leaves in non-inductive cycles (Table 6.15) points to control of evocation by a balance between promoters and inhibitors in tobacco but one should not, perhaps, expect to find a single regulating mechanism. In plants, such as *Xanthium*, the requirement for a positive stimulus may be limiting; in others, such as *Lolium*, promotive and inhibitory effects may be balanced; while in others, such as strawberry, the inhibitory effect may be dominant. In pea, although both inhibitors and promoters appear to be present, the results of grafting experiments between different genotypes indicates that an absolute quantity of the floral stimulus is required at the shoot apex before floral transition can occur, rather than a simple balance of two opposing substances (Taylor and Murfet, 1994). Even in closely related plants, the balance between inhibition and promotion may differ as indicated by the results of grafting experiments between *N. sylvestris* and *N. tabacum* Maryland Mammoth. However, the flowering response can be linked to specific substances only when they have been isolated, identified and shown to be essential. The nature and possible identity of the substances involved in the photoperiodic control of flowering is discussed in the next chapter.

7 The Nature and Identity of Photoperiodic Signals

There is a spatial separation between the generation of a stimulus in leaves exposed to a particular daylength and the action of that stimulus to evoke flowering at the shoot apex. The perceptive mechanism for many of the vegetative responses of plants to photoperiod, such as tuber formation and the onset of dormancy, has also been shown to be located in the leaves. Thus, in addition to any stimuli specifically associated with flowering, substances which can bring about a variety of vegetative responses at widely differing sites in the plant must also be produced in the leaves. It is, therefore, not surprising to find that many chemical differences can be detected in leaves exposed to different daylengths. However, in view of the physiological evidence from grafting experiments for the interchangeability of the floral stimulus in plants of different photoperiodic classes, it is perhaps more surprising to find that many different substances and treatments can bring about flowering (Table 7.1).

Experiments aimed at discovering the chemical identity of the floral stimulus have been of two kinds. One is to make extractions or obtain exudates from induced and non-induced plants and to examine these for their ability to evoke flowering in receptor plants. The second is to examine the effects of known substances (especially

TABLE 7.1 Some factors that modify flowering.

Environmental factors

Daylength; light integral; temperature; carbon dioxide; water stress; nutrient supply

Chemicals

Growth regulators: auxins, ethylene, cytokinins, abscisic acid, gibberellins, growth retardants, polyamines

Membrane perturbers: valinomycin, gramicidin, filipin, salicylic acid

Sucrose, ATP, DCMU, calcium, chelators, cell wall framents, vitamin K, arsenate

Plant manipulation

De-rooting, de-leafing, bending, bark ringing

growth regulators) on flowering and to correlate these with changes in their concentration and/or turnover under inductive and non-inductive conditions. Similar methods have been used in attempts to identify the transmissible floral inhibitors that have been shown to be produced by non-induced leaves in some plants. Neither approach has yielded simple answers.

ATTEMPTS TO EXTRACT A FLORAL STIMULUS

Many attempts have been made to extract a chemical fraction that will evoke flowering when introduced into vegetative plants growing in strictly non-inductive conditions. Most extracts have been made with organic solvents and these have generally yielded only negative results or, if positive, they have been hard to duplicate (Zeevaart, 1979; Cleland and Ben-Tal, 1983; Vince-Prue, 1985). The crude extracts have also not been amenable to further purification for identification of the active fraction. Among early experiments were those of Hamner and Bonner (1938) who tested some 246 different kinds of extracts from SDP, LDP and day-neutral plants; none had any flower-evoking effect. The most promising early extracts were from freeze-dried tissue of flowering plants of *Xanthium*; these caused some flowering when applied to the leaves of *Xanthium* plants in LD (Lincoln *et al.*, 1961) or to the SDP *Lemna paucicostata* (Hodson and Hamner, 1970). The flowering response was, however, discouragingly low; in one reported experiment, 70 plants flowered out of a total number of 12 350 treated (Zeevaart, 1979). Extracts from vegetative plants were not active but some activity was obtained with similar extracts from flowering plants of the day-neutral species *Helianthus annuus* (Lincoln *et al.*, 1962) and from the fungus *Calonectria rigidiuscula*, a flowering gall of cacao (Lincoln *et al.*, 1966). A considerable improvement in the flowering response of *Xanthium* was obtained when GA$_3$ was added, but when applied to *Lemna* the extracts alone were more effective (Hodson and Hamner, 1970). However, the extracts did not appear to contain all of the necessary materials for flower development in *Xanthium* since most plants did not produce normal male inflorescences; instead there were minute, sessile, perianth-like structures in some of the bracts. A few normal male flowers developed after several months but no female inflorescences were produced at any time. No identification of the active ingredients of the flower-evoking extract was achieved, although it was shown to be water-soluble and acidic (Lincoln *et al.*, 1964). Attempts to purify it further resulted in loss of activity.

The results of these early experiments were generally not encouraging. Although some degree of activity was often obtained, the results were difficult to reproduce and identification of the active ingredients proved impossible. In part, the variability may have been due to the method of applying the test substances during bioassay. In more recent experiments, the methods of feeding through a leaf or stem flap as used by Lincoln *et al.* were unsuccessful and the plants suffered major stress (Paré *et al.*, 1989); an alternative procedure of feeding the test solution through split petioles was developed and found to have no inhibitory effect on primordia initiation in induced control plants, nor any primordia-inducing effects in vegetative control plants. Nevertheless, when extracts from *Xanthium* plants were tested using this bioassay, none were capable of inducing flower primordia formation in vegetative (LD) test plants.

There was, however, evidence for the presence of a substance promoting floral development, since the neutral-basic fraction from a methanolic extract of shoots that had received 4 SD was found to stimulate primordia development when fed to test plants that had been induced by exposure to a single SD; the active substance was not identified.

florigen

Chailakhyan and co-workers have reported successful results with extracts from lyophilised Maryland Mammoth tobacco into boiling ethanol (Chailakhyan *et al.*, 1977; Chailakhyan, 1982; Chailakhyan *et al.*, 1989). When a purified active fraction from the leaves of flowering plants of Maryland Mammoth was applied to the plumules of 12-day-old seedlings of *Chenopodium rubrum* in continuous light, all of the plants produced flower buds within 7 days; these flower buds were apparently quite normal in appearance (Chailakhyan *et al.*, 1989). The pattern of growth responses did not resemble the effects of known phytohormones but there were some similarities with the growth changes resulting from photoperiodic induction. Younger plants responded less, although the earliest stages of terminal flower formation occurred. Control plants in continuous light without the tobacco extract remained fully vegetative and extracts from non-flowering plants of Maryland Mammoth tobacco had little or no flower-evoking effect. Some of the partially purified fractions were tested for their effect on flowering in stem explants from the DNP Trapezond tobacco (Chailakhyan, 1982). Two of the fractions increased the number of flowering buds, while samples from leaves of vegetative plants depressed flowering. Although these results appear striking and are claimed to be repeatedly reproducible, no information is yet available about the possible chemical nature of the active florigenic ingredient. There are also a number of problems in interpreting the results. As Zeevaart (1979) has pointed out, it is crucial that an extract is active in the plant from which it was obtained, while a similar extract from vegetative plants is not active. The latter criterion has been met but there are no reports that the extract is active on the SDP Maryland Mammoth. With respect to the reported effect on the tobacco cultivar, Trapezond, this is a day-neutral species and the explants were already in a partially flowering condition.

It is evident that extracts can be prepared which are able to elicit at least some flowering response in test plants maintained under non-inductive conditions. However, the results have often been variable and hard to repeat; in some cases, the original laboratories have not been able to repeat even their own earlier results. Without better reproducibility and/or further identification of the active components the significance of the results remains in doubt.

Phloem Analyses

The long-distance transport of the floral stimulus is thought to take place in the phloem and one approach to the problem of the identity of the floral stimulus has been to analyse phloem contents. A commonly used technique is to collect EDTA-enhanced phloem exudates (King and Zeevaart, 1984) from induced and non-induced leaves. In *Perilla* (green), over 80% of the dry weight of the exudate consisted of carbohydrates but the relative composition did not change significantly with different photoperiods, nor were significant differences observed in the acidic and basic fractions (Zeevaart and Boyer, 1987). Although there was no change in the relative amounts of different

TABLE 7.2 Effect of daylength on the carbohydrate content of the EDTA-enhanced phloem exudate obtained from green *Perilla* leaves.

	Short days	Night-break	Long days
Dry weight (mg)	17.8	11.1	9.6
Carbohydrate (mg)	16.5	8.6	8.1
Carbohydrate (% dry weight)	92	77	84

From Zeevaart and Boyer (1987).

sugars, the absolute amounts varied with photoperiod and almost twice as much carbohydrate was present in exudates from plants in SD, compared with those in LD or with a night-break (Table 7.2). However, this change in sugar level takes place independently of the ability of the leaves to cause a flowering response. Plants grown in SD and then returned to night-break conditions showed a gradual decline in the amount of carbohydrate present in the phloem exudate even though such leaves remain in the induced state and are still able to elicit flowering.

The persistence of the induced state in *Perilla* leaves is a very useful control for determining whether a particular change associated with daylength conditions is also associated with the ability of the leaf to bring about a flowering response. This point is highly relevant to studies of phloem carbohydrates since it is evident that these increase under inductive daylengths in other photoperiodically-sensitive plants, including both LDP and SDP. In *Xanthium*, the sucrose content of EDTA-enhanced leaf exudates increased (compared with the LD control) after the end of an inductive dark period, the maximum increase being observed during the first 24 h (Fig. 7.1). However, the enhanced movement of sucrose out of the leaf also occurred in vegetative control plants exposed to a night-break. It is evident that sucrose alone is not an adequate trigger for the switch to reproductive development of the apex in *Xanthium*, although an increase in the supply of carbohydrate may be required by apices undergoing floral transition.

An increase in carbohydrate flux from the leaves of induced plants has also been observed in the LDP, *Sinapis alba* (Lejeune *et al.*, 1991). Sucrose was the main sugar present and the amount exported from leaves increased during the latter part of a 22 h inductive LD (Fig. 7.2A), whereas the SD controls showed little variation during the same period. When the phloem exudate was collected close to the presumed target site, the apex, the amounts in SD increased during the light period and remained low during the long night; in contrast, sucrose levels remained high during the LD extension in high intensity light (Fig. 7.2B). However, once again the behaviour of control plants indicates that the increased flux of carbohydrate is not the trigger for floral evocation. *Sinapis* can be induced to flower by a single LD or by a displaced SD and the timing of a number of early evocational events has been found to be the same when computed from the beginning of the LD or displaced SD treatment. In contrast, the timing of the increase in carbohydrate flux does not appear to be the same when computed in this way (see Fig. 7.2); the patterns are similar but the peaks for the displaced SD treatment occur at different times. Thus, the increased flux of carbohydrate is not temporally linked to apical events that are typical of early floral transition, indicating that this is dependent on some trigger other than the increased supply of

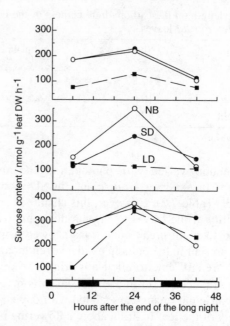

FIG. 7.1.   Effect of daylength on the sucrose content of exudate from leaves of *Xanthium strumarium*. Results are from 3 independent experiments in which exudate was collected during a 12 h period beginning 0, 18 and 36 h after the end of the inductive 16 h night. Values are plotted at the middle of the exudation period. SD, induced plants given a single 16 h night; NB, vegetative control plants given a 5 min night-break; LD, vegetative control plants which remained in 18 h days throughout. After Houssa *et al.*, 1991.

sucrose. A similar conclusion was reached from studies with a mutant of clover which does not flower with a normally inductive LD, unless supplied with exogenous GA (Jones, 1990). Transfer to inductive LD (with the addition of GA_3 to mutant plants) resulted in an increase in soluble carbohydrate in the apices within 1–2 days; however, a very similar pattern of changes occurred in SD (+GA_3) when the plants remained vegetative. The same situation is seen in *Sinapis*, where exposure to a single SD cycle at high irradiance resulted in an increase in the soluble carbohydrate content of the apex but did not trigger floral initiation (Bodson and Bernier, 1985).

   In both LDP and SDP, an increase in the export of soluble carbohydrates from leaves and/or in soluble carbohydrates in the shoot apex has been recorded during or immediately following photoperiodic induction. However, a careful examination of the behaviour of different types of control plants eliminates this increase as being the photoperiodic trigger for evocation, although it may be required during floral transition. For example, in *Lolium temulentum*, there was no evidence of an increase in sucrose in the shoot apex during an inductive LD (16 h low-intensity day-extension) or the following high-intensity light period when evocation occurs (King and Evans, 1991). Increasing the sucrose supply by increasing the irradiance during the daily light period did not result in flowering in SD, nor did the high concentration of sugars in apices cultured *in vitro* on 5% sucrose. However, following arrival of the photoperiodic stimulus from LD leaves, increasing the sucrose concentration enhanced floral development *in vitro*, especially after initiation. Direct effects of

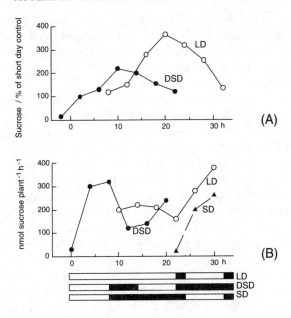

FIG. 7.2. Time course of changes in the sucrose content of phloem exudate from plants of *Sinapis alba* induced to flower by a LD or by a displaced SD. Phloem exudate was collected during periods of 4 h from excised leaves (A) or at the apical part of the shoot from plants decapitated just below the 3rd or 4th leaf (B). The data for the LD and SD treatments are plotted from the beginning of the 22 h LD; those for the displaced SD (DSD) are plotted from the beginning of the displaced 8 h SD, which was delayed by 10 h compared with the normal SD. The periods of light and darkness are shown below the figure. After Lejeune *et al.*, 1991, 1993.

daylength on assimilate partitioning have also been observed in a number of cases where floral induction is clearly not involved (see Chapter 13).

Possible changes in phloem sap proteins have been investigated in EDTA-enhanced exudates from *Pharbitis* cotyledons (Friedman *et al.*, 1987). No differences were detected between induced and control plants, indicating that proteins do not form part of the phloem translocatable stimulus; however, this does not exclude a possible role for lower molecular weight polypeptides. In this context, it is worth noting that a polypeptide (with a molecular mass of $\leq$22 kDa) present in buffer extracts has been found to have florigenic activity in a weakly-responsive SD strain of *Lemna pauci-costata* tested in continuous light (Kozaki *et al.*, 1991); its relevance to any photo-periodic stimulus is, however, obscure since the extract was made from plants growing in non-inductive daylengths.

Phloem exudates from *Perilla* leaves have been tested for their ability to evoke flowering but the results have been negative (Cleland and Ben-Tal, 1983) or inconsistent (Zeevaart and Boyer, 1987). When the entire exudate from induced red *Perilla* leaves was incorporated into the medium on which shoot tips of the same species were cultured, flowering was promoted in a few cases, but not always (Purse, 1984). However, phytotoxicity limited the concentration that could be used. The active constituent was not identified but ABA and sucrose were excluded as being inadequate to account for the flowering response. Substances diffusing from shoot explants taken from stems of vegetative or flowering plants of the DNP tobacco Trapezond have also been tested for florigenic activity (Chailakhyan, 1982). In this case, donor plants were cultivated on agar for 3–7 days and receptor explants were then placed on the same medium. There was no stimulation of flowering in vegetative explants, although the number of flower buds was increased when flowering receptors were cultivated after flowering donors.

A promising result has been obtained with phloem exudates from cotyledons of *Pharbitis nil* tested on excised apices taken from seedlings growing in 16 h LD. The apex cultures were also maintained in LD and the exudate from a single cotyledon was added to 10 ml of the medium. A considerable degree of flowering was obtained (Fig. 7.3) and the response was clearly related to the degree of photoperiodic induction, since there was a critical night length and the magnitude of flowering in the test apices increased with increasing duration of the dark period given to the cotyledon. A 15 min night break also reduced the florigenic activity of the exudate. The florigenic ingredient was not, however, identified. Nor could the results be reproduced elsewhere (Takeno, 1994), although a low and variable degree of flowering was obtained when the phloem exudate was introduced into the seedlings using a perfusion technique; phloem exudate from non-induced plants was also active in some assays.

A different approach has been to examine the effect of honeydew from aphids feeding on vegetative or flowering plants. Aphid honeydew is known to be qualitatively quite similar to phloem sap and a number of plant growth regulators have been obtained from honeydew in active form. Following thin layer chromatography of an acidic ethyl acetate fraction of honeydew from aphids feeding on flowering plants of the SDP *Xanthium strumarium*, one of the zones obtained was found to cause flower-

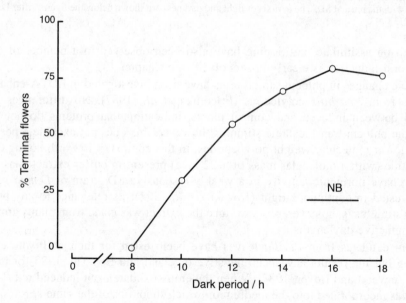

FIG. 7.3. The effect of the duration of the dark period on the flower-inducing activity of phloem exudates from *Pharbitis nil*. Apex explants were cultured under 16 h LD on a medium that contained phloem exudate from cotyledons of plants exposed to a single dark period of various durations. The exudate from 1 cotyledon was added to 10 ml of medium. The horizontal bar (NB) show the response to exudate from plants given a 15 min night-break in the middle of a 16 h dark period. After Ishioka *et al.*, 1991a.

ing when applied to the LDP *Lemna gibba* G3 (Cleland and Ajami, 1974). The promoting substance was identified as salicylic acid (SA), and it was also shown that authentic salicylic acid could induce a good flowering response in *Lemna gibba* G3 and in at least two other species of *Lemna*. However, SA had no effect on flowering in *Xanthium* when given alone or in combination with GA and/or kinetin, nor was there any difference in the SA content of honeydew from aphids feeding on flowering or vegetative plants of *Xanthium*. Consequently, it must be concluded that SA is not associated with florigenic activity in *Xanthium* itself. However, flowering was induced under non-inductive conditions by the application of SA to the SDP *Impatiens balsamina*; it was suggested that SA and other phenols may modify IAA levels through an effect on IAA oxidase (Kumar and Nanda, 1981). SA has also been reported to promote flowering in a few other species, including *Pistia stratioides* (Araceae) and the LDP *Arabidopsis thaliana* (Raskin, 1992).

Does SA play a role in the endogenous control of flowering in *Lemna*? It has been shown to be effective in evoking flowering in many different members of the Lemnaceae and is able to induce flowering in LD, SD and day-neutral species, one of the properties expected of a true florigen. The effect is also quite specific in one member of the Lemnaceae, *Wolffiella hyalina*, which flowered in SD only with the addition of 10^{-5} M SA to the medium; in this case no other tested compound was able to substitute for the SA effect (Tamot *et al.*, 1987). However, the relationship between photoperiodic induction and the benzoic acids remains unclear. The general conclusion is that SA and related compounds probably act in the frond to influence induction in some way (see Chapter 8) but the fact that the amounts present do not change with daylength and are not correlated with flowering (Fujioka *et al.*, 1985; Raskin, 1992) casts doubt on any endogenous role in the photoperiodic control of flowering in *Lemna*.

Metabolic Studies

Attempts to isolate and characterise a floral stimulus have mostly been based on bioassays, in which extracts or diffusates from induced plants have been tested for their ability to cause flowering when applied to vegetative test plants. An alternative, indirect approach is to feed potential precursors of the floral stimulus prior to photoperiodic induction and to follow their fate during and after exposure to photo-inductive cycles. When radioactive precursors were applied to cotyledons of *Pharbitis* before a long night, there were no differences between induced and non-induced plants with labelled mevalonate and glyoxylate, but consistent changes were observed with acetate (Aharoni *et al.*, 1985). Induction appeared to cause the disappearance of a major metabolite soon after the end of the critical dark period; this occurred simultaneously in both cotyledons and plumules but not in the roots. Treatments which prevented induction (e.g. night-break or exposure to ethylene) also prevented the loss of this metabolite. Interpretation is complicated by the fact that the labelled metabolite (which was not identified) underwent chemical modification during the prolonged methanolic extract. In an earlier study, mevalonate was applied to leaves of *Xanthium* prior to an inductive night, in order to investigate the possibility that the floral stimulus might be a steroid. In this study, no consistent differences in steroid composition of induced and non-induced plants could be detected (Bledsoe and Ross,

1978). However, changes in the relative amounts of cholesterol and sitosterol in the apices of *Lolium* were recorded on the day after photoperiodic induction and continued during the following 5 days (Garg and Paleg, 1986). The reported inhibition of flowering in *Xanthium* and *Pharbitis* by the application of substances known to prevent the synthesis of steroids (Bonner *et al.*, 1963) was probably due to an effect on photosynthesis (Zeevaart, 1979).

Earlier reports that flowering in non-inductive photoperiods could be evoked by the application of certain steroidal oestrogens or androgens in both LDP (*Callistephus chinensis*) and SDP (chrysanthemum cv Princess Anne) do not appear to have been followed up. An increase in unidentified estrogen-like substances under photoinductive cycles has been reported in both LDP and SDP; however, this increase was correlated with the period of flower development rather than with early evocation. Although the involvement of steroidal oestrogens and androgens in reproductive development cannot be ruled out, unreliable methods were used in many of the earlier experiments and the evidence is so far unconvincing (Jones and Roddick, 1988).

ASSAYS FOR FLORIGENIC ACTIVITY

One of the problems associated with the identification of a floral stimulus is undoubtedly that of developing a suitable assay. The response of intact plants may be influenced by inhibitors coming from non-induced leaves and there is the further problem of introducing the test material in sufficient quantities. Leaf removal may affect the supply of substances which, while not themselves florigenic, are necessary for the response to a floral stimulus. Wounding itself might also modify the flowering response of the test plant (Jacobs and Eisinger, 1990). Grafting experiments with *Perilla* indicate that the floral stimulus must be supplied over a period of time; this point is often not considered, although it is obviously not relevant to plants where initiation occurs in response to a single photo-inductive cycle followed by removal of the induced leaf (see Figs 6.6, 6.10).

One assay which overcomes the problems of inhibition from non-induced leaves is to culture the shoot apex on substrates containing the test material. Some success has been reported with *Perilla* apices cultured on phloem extracts from flowering plants (Purse, 1984). However, the possibility of inhibition from non-induced leaves is not entirely excluded since young leaves capable of daylength perception are likely to develop during the test period; in *Xanthium*, cultured shoot apices were able to respond to daylength even if all leaves were removed at the beginning of the culture period (Jacobs and Suthers, 1974). Nevertheless, explants offer a useful way of introducing test materials into the plant and several different kinds have been used as assays for examining the florigenic activity of plant extracts, as well as of known substances such as growth regulators. Examples of such *in vitro* systems include callus from stem segments of tobacco (Chailakhyan *et al.*, 1975), stem segments of *Torenia* (Tanimoto *et al.*, 1985) and *Plumbago* (Nitsch and Nitsch, 1967), leaves of *Streptocarpus* (Simmonds, 1982), roots of *Cichorium* (Margara and Touraud, 1967), node plus leaf of *Glycine*, and apical meristems of several species (Dickens and van Staden, 1985). A shortcoming of some of these assays is the fact that explants were derived from parent material which was already flowering, or had been induced to flower. In such cases, the regeneration of floral or vegetative buds can frequently be modified by various treatments but the overiding factor seems to be the origin of the explant material; for example, stem callus from flowering plants of tobacco Trapezond (DNP) produced flower buds, while those from young vegetative plants formed only vegetative buds. Stem callus from both flowering and vegetative plants of Maryland Mammoth (SDP) and *Nicotiana sylvestris* (LDP) formed only vegetative buds.

The *Lemna* System

The floating pondweed *Lemna* has often been used to test for florigenic activity. The system has both advantages and disadvantages. Substances can readily be supplied to intact plants without manipulation via the culture solution and several species are strongly photoperiodic with both LDP (e.g. Lemna gibba G3) and SDP (e.g. strains of *L. paucicostata*) being available. Positive results have often been obtained when extracts have been tested on *Lemna* but most of these have not fulfilled Zeevaart's criterion that extracts should be active in the species from which they were originally obtained (Zeevaart, 1979). A major disadvantage is the wide range of substances that have been shown to modify the flowering response in *Lemna* (Vince-Prue, 1975; Kandeler, 1984). In many cases, it is clear that the substances in question are not acting as a floral stimulus to evoke flowering at the shoot meristems. For example, SA and related compounds cause a flowering response in most of the plants belonging to the Lemnaceae, including both LD and SD species, but the action appears to be on the process of floral induction in the fronds. Other substances that modify flowering behaviour in plants of the Lemnaceae include iron- and/or copper-chelating agents (Khurana and Maheshwari, 1984), cytokinins (Gupta and Maheshwari, 1969), plant cell wall fragments (Gollin et al., 1984), vitamin K (Kaihara and Takimoto, 1985), lithium ions (Kandeler, 1984), nicotinic acid (Fujioka et al., 1986a) and some amino acids (Fujioka and Sakurai, 1992; Tanaka et al., 1994). The site of action of many of these substances is not known. It appears that vitamin K acts cooperatively with benzoic acid, copper and iron (at least in low concentrations), although the dependence of their action on photoperiodic conditions differs. For example, both benzoic acid and SA shortened the CNL in the SDP *L. paucicostata*, whereas vitamin K_5, copper and iron promoted flowering independently of daylength (Kaihara and Takimoto, 1985). In general, it appears that most, or all of these widely different compounds may act at the level of floral induction (see Chapter 8) rather than being components of a transmissible floral stimulus.

Crude water extracts from a wide range of plants have been found to cause flowering in the weakly responsive SDP, *Lemna paucicostata* 151 when tested under continuous light; control plants remained vegetative (Kaihara et al., 1989). Plants from which active extracts were obtained included *Spinacia* (LDP) and *Xanthium* (SDP), while activity was not observed in extracts from *Silene* (LDP) or *Perilla (SDP)*. Extracts and phloem exudates from *Pharbitis nil* were also active when tested on *Lemna*, but neither the activity of the water extract nor that of the exudate could be correlated with photoperiodic induction. The florigenic component has not been identified but activity (tested on strain 151) was obtained when the tyrosine metabolite norepinephrine was incubated with the pellet from a centrifuged homogenate of *Lemna* (Takimoto et al., 1991). There is no evidence that the florigenic activity of these extracts is associated with the endogenous induction of flowering since it occurred equally in extracts from induced and non-induced plants. It has been suggested, however, that it may be an endogenous regulator which is unavailable until released (perhaps by a change in compartmentation) by an appropriate stimulus such as photoperiod. Whether the florigenic activity occurs at the level of induction or evocation is unknown.

The Tobacco Thin-Cell-Layer System (TCL)

Explants consisting of a few of the outermost cell layers of the stem have been used by several groups to study the action of various substances on development. When taken from stems of flowering tobacco plants, these TCLs produced roots, shoots or flowers with no apparent intermediate callus (Tranh Than Van *et al.*, 1974). Four types of organ-inducing media (distinguished by auxin:cytokinin concentrations) have been defined, namely root, vegetative shoot, and flower media and a transition medium on which few organs develop (Mohnen *et al.*, 1990). Apart from variations in the IBA and kinetin concentrations in the medium, *de novo* flower formation without callus has been shown to be influenced by pectic cell-wall fragments (Eberhard *et al.*, 1989) and by polyamines (Galston and Kaur-Sawhney, 1990). The flowering response is also strongly dependent on the developmental state of the source tissue (Rajeevan and Lang, 1993).

Extracellular matrix carbohydrate molecules have been shown to play a role in the development of plants (Ryan, 1987) and it has been suggested that the cell wall may be a repository of signal molecules which are released by appropriate environmental and/ or developmental signals. Fragments from sycamore, tobacco and citrus cell walls induced *de novo* flower formation in tobacco TCLs cultured on transition medium. The flower-inducing activity was found to be stable to protease and mild acid and base treatments, but was sensitive to subsequent treatment with endopolygalacturonase suggesting that the active component is an oligogalacturonide-containing carbohydrate (Mohnen *et al.*, 1990). Flowering was also induced in explants cultured on a vegetative medium by the addition of the polyamine, spermidine (Table 7.3), while inhibitors of polyamine biosynthesis inhibited floral initiation in explants on a flowering medium. Neither polyamines nor pectic fragments were effective in causing

TABLE 7.3   Effects of spermidine on bud differentiation in thin layer tobacco tissue cultures.

Spermidine concentration (mM)	Buds per explant	
	Vegetative	Floral
0	15	0
0.5	9	3.4
1.0	9	2.3
5.0	6	2.2

Explants were grown on a medium containing 1 μM NAA and 10 μM kinetin. From Kaur-Sawhney *et al.* (1988).

flowering when applied to TCLs derived from vegetative tobacco plants, indicating that they are not truly florigenic in the tobacco TCL tissue system and may only permit expression of an already established floral state. A somewhat similar conclusion was reached for the action of polyamines on flowering in the quantitative LD duckweed, *Spirodela punctata* (Bendeck de Cantu and Kandeler, 1989). Specific inhibitors of polyamine biosynthesis depressed flowering when plants were grown under LD in the presence of EDDHA. The addition of spermidine reversed this inhibition but could not replace the requirement for EDDHA. Thus, spermidine appeared to be a necessary but not sufficient factor for flower initiation in this plant.

At present the results with TCLs are difficult to relate to the intact plant where a daylength-dependent floral stimulus changes an existing vegetative meristem to a reproductive one. In the majority of cases, flower buds have only been obtained when explants were taken from the flowering regions (e.g. inflorescence branches, pedicels, flower parts) of plants that were already in an advanced stage of floral development and daylength appears to have no effect in such explants (Rajeevan and Lang, 1993). However, following the discovery that a small percentage (about 0.2%) of shoots developed on stem explants taken from young vegetative plants were floral in a day-neutral tobacco (McDaniel *et al.*, 1989), a system has been developed using the strict SD cultivar Maryland Mammoth, in which explants taken from the apical regions of plants in the earliest stages of floral transition showed an absolute SD requirement for the production of floral buds (Altamura and Tomassi, 1994). It will be of interest to determine whether any substance(s) can replace the SD requirement in such explants.

Polyamines in Intact Plants

Exposure to a photoinductive dark period has been shown to result in changes in polyamine metabolism in the leaves of intact *Xanthium* plants (Hamasaki and Galston, 1990). Conjugated polyamines (putrescine, spermidine and spermine) increased sharply during the first or second photoinductive dark period, especially in expanding leaves, which are the most sensitive to photoperiodic induction (Fig. 7.4). Buds from the apical region also showed sharply elevated levels of conjugated polyamines (especially spermine), when these were exposed to 4 SD and harvested after a further 4 LD (Table 7.4). From these results it seems that conjugated polyamines may follow the pattern expected of a floral stimulus, namely an early increase in leaves exposed to inductive cycles, followed by possible transport to developing buds. In the SDP, *Glycine max* cv Williams, SD was found to enhance the transport of label from leaf-applied spermidine, especially to the apical bud and the youngest leaf (Caffaro *et al.*, 1994). In this cultivar, 2 SD are required to induce flowering, with floral transition in all plants occurring within 3–5 days after exposure. It was reported that free putrescine, spermidine and spermine decreased in the leaf at the end of the first inductive long night, again suggesting that inductive cycles might increase the export of these free polyamines from the leaf (Caffaro and Vicente, 1995). However, no changes in the amount or composition of free polyamines were detected in the stem apex during the first 3 SD cycles (Caffaro and Vicente, 1994). Moreover, the significance of the changes in leaf polyamines is difficult to assess since there was considerable variability from day to day, even in plants that remained in LD condi-

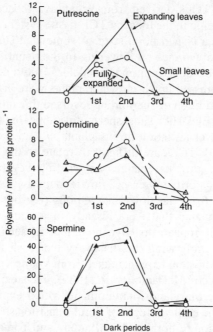

FIG. 7.4.   Effect of daylength on conjugated polyamines in leaves of *Xanthium strumarium*. Plants were exposed to 4 successive cycles of inductive SD (8 h L/16 h D) and otherwise remained in non-inductive LD. After Hamasaki and Galston, 1990.

TABLE 7.4   Effect of exposure to short-day cycles on polyamine content of *Xanthium* buds.

	Free polyamines (nmol mg^{-1} protein)		Soluble, conjugated polyamines (nmol mg^{-1} protein)	
	0 time	After 4 SD	0 time	After 4 SD
Putrescine	58	1360	39	388
Spermidine	155	262	19	456
Spermine	214	689	49	3720

Plants were exposed to 4×8 h light: 16 h dark cycles and the buds were harvested 4 days later. After Hamasaki and Galston, 1990.

tions. Particularly with spermidine and spermine, the difference between SD and LD treatments at the end of the first long night was due to a sharp increase in leaves given a 3 h night-break, rather than to any marked decrease in those exposed to the inductive dark period. At the end of the second cycle, there was no peak following the night-break and no difference between this and the uninterrupted long night. Only with putrescine was the LD pattern reasonably consistent over the first three cycles and, in this case, there was a marked decrease at the end of the first long night.

A rise in free polyamines after photoperiodic induction has also been observed in the LDP, *Rudbeckia hirta*, where an increase in free putrescine and spermidine began after 4 LD cycles and essentially coincided with the beginning of the events of floral initiation (Harkness *et al.*, 1992). Polyamines may also be involved in *Sinapis alba* since inhibitors of putrescine biosynthesis dramatically reduced the flowering response to one LD and induced leaves were found to export an early pulse of putrescine in the phloem sap (Bernier *et al.*, 1993). This finding is consistent with the transient decrease in leaf putrescine during SD induction in *Glycine*, as discussed above. Changes in polyamine composition during reproductive development have also been reported in other plants (Kakkar and Rai, 1993).

Despite the several observed correlations between polyamines and floral initiation, there have been few reported studies of their effects when applied to intact plants or other test systems. Both putrescine and cadaverine resulted in more than 90% flowering in seedlings of *Pharbitis nil* (dwarf cv Kidachi) when applied for 24 days through the roots under conditions where all control plants remained vegetative (Wada *et al.*, 1994); putrescine appeared to be the effective molecule in this case and applied spermidine had little effect. In apple, floral initiation in young trees was increased by the infusion of polyamines, as well as by fertilisation with ammonia which resulted in an increase in the endogenous levels of the polyamine precursor, arginine (Edwards, 1986). In tobacco TCLs, as already discussed, the application of spermidine promoted flowering, but this effect was elicited only when the tissue explants came from flowering plants and were already in a florally determined state. Thus the results with polyamines are proving interesting but further investigations are needed to determine whether they constitute part of a transmissible signal and whether they are associated with the switch from a vegetative apex to a floral one or, as appears more likely, with the expression of floral development.

ATTEMPTS TO EXTRACT FLORAL INHIBITORS

The idea of a specific floral inhibitor(s) was proposed almost as early as that of a floral promoter (Lang, 1980; Jacobs, 1980). The identification of such substances, or **antiflorigens** has, however, met with no more success than the identification of **florigens**. One problem in evaluating the physiological significance of flower-inhibiting chemicals is the notorious lack of specificity common to inhibitory effects. The inhibition of flowering by a particular chemical is not strong evidence for its being a natural **antiflorigen** and there are few examples of the extraction of inhibitory compounds from plants. Many of these are known plant growth regulators and are discussed later under the appropriate heading. A different type of naturally-occurring inhibitor has been identified in leaves of the SDP *Kalanchoë blossfeldiana*. Here, crude sap expressed from LD leaves was inhibitory to flowering when injected into SD leaves, while the injection of sap from SD leaves had little effect (Schwabe, 1972). It was later suggested that 2-flavan-3-ol fractions of relatively low molecular weight might be involved (Schwabe, 1984). A report that gallic acid was the endogenous inhibitor of *Kalanchoë* could not be confirmed, nor did it have any inhibitory effect on flowering in nodal explants (Dickens and van Staden, 1990). In *Kalanchoë*, the inhibitor appeared to act in the leaves to prevent production or export of the floral stimulus and would presumably be different from the graft-transmissible anti-florigen deduced from the results of grafting experiments. There are no reports of the isolation and chemical identification of such antiflorigens (Lang, 1980, 1987). A compound identified as *bis*(2-ethylhexyl)hexane dioate (BEHD) was found to decrease transiently towards the end of an inductive dark period in the leaves of several unrelated SDP and also inhibited flowering in a range of test plants, including LDP, SDP and DNP (Jaffe *et al.*, 1987). However, BEHD was later shown to originate from the plastic containers in which the plants were growing; consequently, it is highly unlikely to have any physiological relevance.

GROWTH REGULATORS AND FLOWERING

All classes of plant growth regulators (PGRs) have some effect on flowering when applied to a particular species under certain conditions (Vince-Prue, 1985). The problem is to determine what part they play, if any, in the endogenous control of the flowering process and their relevance to photoperiodism. All, except ethylene, are found in the phloem, i.e. they are transported along the same pathway as the floral stimulus. Although the application of PGRs often modifies the flowering response, in the majority of cases the component of the overall flowering process which is being affected is not known. For example, application of a PGR could affect the induction process in the leaf, export from the leaf or transport of a floral stimulus; the PGR might be part of the floral stimulus itself or could affect sensitivity to such a stimulus when it arrives at the apex. Since most PGRs are readily transported around the plant, it is not always easy to determine even the site of action. Timing of action is also a problem since the growth substance has to be taken up and possibly transported. It may also have to be metabolised to an active molecule. It is not surprising therefore, that it has not proved easy to determine the precise roles that natural growth regulators play in the photoperiodic control of flowering.

GIBBERELLINS

Of the major groups of naturally-occurring growth substances, gibberellins are the most interesting with respect to the photoperiodic control of flowering. It has been known for a long time that applied gibberellin (GA) can substitute for photoperiodic induction in many plants (Lang, 1965; Vince-Prue, 1975). In particular, GAs cause flowering in many LDP which grow as rosettes in SD; in these plants, the flowering response (either to GA or LD) is accompanied by elongation of the flowering stem. There are also a few qualitative SDP where GA application can evoke flowering in non-inductive conditions (Sawhney and Sawhney, 1985). GA application can also substitute, either partially or completely, for a low-temperature (vernalisation) signal in several biennials, and also in other cold-requiring plants, such as tulip bulbs. In addition striking effects to accelerate flowering of juvenile plants have been observed in several Gymnosperm families. Thus, in an impressive list of plants, GAs have been found to be able to substitute for the primary environmental signals of daylength and low temperature, as well as for the endogenous trigger of 'age' in the autonomous induction of flowering. Is the floral stimulus, therefore, one or more of the many known GAs? Unfortunately, any simple interpretation of the action of GA in the control of flowering is ruled out because exceptions, or opposite responses, can be listed for all of the generalisations made above. The application of GA inhibits flowering in some LDP and is ineffective in others, including some (e.g. *Hieracium* spp.; Peterson and Yeung, 1972) which grow as rosettes in SD and elongate in response to GA application. Similarly GAs are inhibitory to flowering in some SDP and ineffective in others. For most woody angiosperms, GAs are strongly inhibitory (Pharis *et al.*, 1989) while, with respect to flowering, GAs are not always an effective substitute for low-temperature vernalisation. Examples of some of the effects of the application of GA (usually GA_3) on flowering are given in Table 7.5.

Site of GA Action

The photoperiodic signal is perceived in the leaf and a major question has been whether the GA effect on flowering occurs there or at the apex, as would be expected if GA is replacing the floral stimulus. There is some evidence for both. Action in the leaves is best documented for the LSDP *Bryophyllum*, where the application of GA to leaves in SD (but not in LD) promoted flowering. Grafting studies indicated that when GA is applied in SD, leaves produce the floral stimulus (Zeevaart, 1969a) but the GA itself is not exported and has no effect at the apex. For example, when *Bryophyllum* plants were induced to flower in SD by the application of GA, they were effective donors of a floral stimulus to the SLDP *Echeveria harmsii* maintained in non-inductive conditions (see Table 6.10); since GA treatment does not cause flowering in this plant in either LD or SD (Zeevaart, 1982), it is evident that flowering in the *Echeveria* receptor is not the result of the transmission of GA from the *Bryophyllum* donor. The conclusion is that GA is not part of the floral stimulus in *Bryophyllum*, but substitutes for the action of LD in the leaves and results in production of the floral stimulus when the treated leaves are subsequently exposed to SD; this also explains why LD must precede SD in this dual-daylength plant. SLDP, in contrast, appear to have a quite different response to gibberellins. The primary induction of flowering by SD in several perennial temperate grasses (e.g. *Bromus, Poa*) is prevented by the application of GA_3 and a reduction in GA levels seems to be required in order for the SD induction process to take place (Heide *et al.*, 1986a). The GA biosynthesis inhibitor CCC was found to enhance the primary induction of flowering, especially under marginal conditions so that, in these plants, the effect of SD may be to remove the inhibitory effect of LD which, in turn, is a consequence of the higher GA levels under long photoperiods (Heide, 1994). Moreover the application of GA_3 to plants already induced by SD was not able to substitute for the transition to LD in *Bromus inermis*. In complete contrast, GA can substitute for SD in the SLDP *Scabiosa succisa*, where the application of GA promotes initiation in LD (Wilkins and Halevy, 1985). Thus, there appears to be no parallel in these SLDP to the situation in the LSDP, *Bryophyllum* although, in the dual daylength grasses, the action of gibberellin clearly also occurs in the leaves where it prevents the primary induction process.

If GAs are associated with a floral stimulus, they would be expected to act at the apex. This has been demonstrated for the SDP *Impatiens* (Sawnhey *et al.*, 1978) where the photoperiodic induction of flowering by SD requires the presence of leaves, whereas the application of GA promotes flowering in LD in the absence of leaves; it is clear that, in this case, the mode of GA action cannot be to promote the production

TABLE 7.5 Some examples of the flowering response to gibberellin application in plants with different environmental requirements for flowering.

Environment for flowering	Response to GA	Species
Long-day plants	Promotes in SD	*Lolium temulentum*
		Hyoscyamus niger (annual)
	Inhibits in LD	*Fuchsia hybrida*
		Lemna gibba G3
	No effect	*Sinapis alba*
		Hieracium aurantiacum
Short-day plants	Promotes in LD	*Zinnia elegans*
		Impatiens balsamina
	Inhibits in SD	*Fragaria* × *ananassa*
		Begonia × *cheimantha*
	No effect	*Xanthium strumarium*[a]
		Glycine max
Long–short-day plants	Promotes in SD	*Bryophyllum daigremontianum*
		B. crenatum
	Inhibits in LSD	*Cestrum nocturnum*
Short–long-day plants	Promotes in SD	*Coreopsis grandiflora*
	Promotes in LD	*Scabiosa succisa*
	Inhibits in SD	*Bromus inermis*
	No effect	*Campanula medium*
		Echeveria harmsii
Day-neutral plants	Promotes	Taxodiaceae, Cupressaceae (precocious flowering)
	Inhibits	*Citrus latifolia* (lime)
		Malus (apple)
	No effect	*Campanula pyramidalis*
		Oenothera biennis
Require vernalisation		
LDP	Promotes in LD	*Avena sativa* (winter strains)
		Hyoscyamus niger (biennial)
SDP	No effect	Chrysanthemum (SD cvs)
	Promotes in SD	None known
Day-neutral	No effect	*Eryngium variifolium*
		Saxifraga rotundifolia
	Promotes	*Brassica napus*
		Daucus carota

Gibberellin application can promote, inhibit or have no effect on flowering, even in plants with the same environmental requirements. Although the application of GA_3 almost invariably causes stem elongation, especially in plants which grow as rosettes in the vegetative state, this is not always accompanied by flowering.

[a] Slight promotion in SD.

Examples from Lang (1965), Vince-Prue (1985), Sawhney and Sawhney (1985) and Halevy (1985).

of the floral stimulus in the leaf. Action at the apex is also indicated in several LDP which grow as rosettes in SD (e.g. *Rudbeckia*, Chailakhyan, 1958) but here the interpretation of the role of endogenous GAs in flowering is complicated by the fact that such plants respond to LD induction by elongation of the stem. In *Hyoscyamus*, applied GA is thought to have two different effects; to cause production of the floral stimulus in the leaves and to act at the shoot apex to cause stem elongation (Warm, 1980). In neither case is GA considered to be the equivalent of the floral stimulus itself. The inhibition of flowering by GA application in the LDP *Fuchsia* also appears to result from its action at the apex (Sachs *et al.*, 1967).

Endogenous Gibberellins

Both the amount and type of endogenous GA may be influenced by photoperiod (Vince-Prue, 1985). The content is often higher in LD and this seems to be independent of the photoperiodic class for flowering (Appendix II). The content of both diffusible and extractable GA may increase, but not always in parallel; a higher content in LD has also been reported for phloem exudates. Probably more important is the fact that there are considerable differences in the metabolism of GAs in different photoperiods.

The 13-hydroxylation sequence (GA_{53}, GA_{44}, GA_{19}, GA_{17}, GA_{20}, GA_{29},) is a major biosynthetic pathway leading to the bioactive molecule, GA_1, and at least two steps in this pathway have been shown to be photoperiodically regulated in several plants. In the LDP *Silene armeria*, the main effect of daylength was to control the activity of GA_{53} oxidase, which increased when plants were transferred from SD to LD, and decreased when plants were returned to SD (Talon and Zeevaart, 1990); this led to a transient increase in later members of the pathway (e.g. GA_{19}, GA_{20} and GA_1). In spinach, GA_{19} accumulates in SD. Following transfer to LD, the conversion of GA_{19} to GA_{20} was increased, as well as the further metabolism of GA_{20} to GA_{29} (Metzger and Zeevaart, 1982). Studies with cell-free extracts (Gilmour *et al.*, 1986) indicated that daylength acted at two steps in the 13-hydroxylation pathway; the activities of the enzymes oxidising GA_{53} and GA_{19} were higher in extracts from LD leaves and lower in those from SD leaves. The 53 $\rightarrow$ 44 and 19 $\rightarrow$ 20 conversions decreased rapidly following transfer to darkness and there was essentially no activity after 16 h; conversely, both increased in the light (Fig. 7.5). The presence of a SD-dependent inhibitor of enzyme activity was ruled out. Similar changes have been observed in another LD rosette plant, *Agrostemma githago* (Jones and Zeevaart, 1980). Biosynthesis of the GA precursor, *ent*-kaurene was also enhanced by LD in *Agrostemma* and *Spinacia* (Zeevaart and Gage, 1993). It appears that, in spinach, flow through the GA biosynthetic pathway is much increased during the high-intensity light period but GA turnover also occurs during the low-intensity day-extension, both being responsible for the LD effect (Talon *et al.*, 1991b). Thus, in several rosette LDP, the GA content may increase in LD, even if only transiently, and there is often an increase in GA metabolism together with the formation and/or accumulation of new GAs.

Similar effects of daylength on GA metabolism have also been observed in at least one SDP, *Begonia* $\times$ *cheimantha*. Transfer to SD resulted in a rapid fall in the content of several GAs in the leaves (Fig. 7.6). The amounts of GA_4, GA_9, GA_{19} and GA_{20} decreased to about half of that in LD and most of this change had already occurred

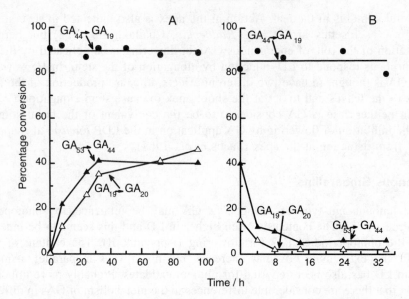

FIG. 7.5.   Gibberellin metabolism in cell-free extracts from spinach leaves. In A, plants were transferred from SD to continuous light. In B, plants were transferred to darkness following 8 LD (continuous light). Enzyme extracts were made at different times from leaves of equal fresh weight. After Gilmour *et al.*, 1986.

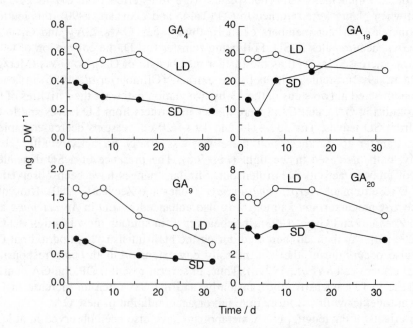

FIG. 7.6.   Effects of photoperiod on the content of several gibberellins in the leaves of the SDP, *Begonia* × cheimantha. Plants were maintained in non-inductive LD (○) or transferred to inductive SD (●); the temperature was 24°C in both photoperiods. Oden and Heide, 1989.

TABLE 7.6 Effect of exposure to a single inductive long-day cycle on endogenous GA-like substances in shoot apices of *Lolium temulentum*.

Treatment	GA-like substances (μg g^{-1} dry wt)		
	Polyhydroxylated GAs[a]	Polar GAs[b]	Less polar GAs[c]
Control	2.6	12.6	0.3
Induced (day II)	12.5	11.6	11.3
Induced (day III)	3.0	1.2	11.7

GA-like substances at Rts of [a]GA$_{32}$ and GA$_8$; [b]GA$_1$; [c]GA$_{20}$, GA$_4$, GA$_9$ and GA$_{12}$.
From Pharis *et al.* (1989).

within 2 SD cycles. After the initial reduction, the amount of GA$_{19}$ in SD increased slowly to levels higher than the controls in LD, indicating that a photoperiodic block earlier in the pathway may become leaky allowing GA$_{19}$ to accumulate because its further conversion to GA$_{20}$ is blocked in SD as observed in *Spinacia* and *Silene*. Since both the GA$_9$ → GA$_4$ and GA$_{19}$ → GA$_{20}$ pathways can give rise to GA$_1$, which has been identified in *Begonia* leaves, it was suggested that this GA may be an active inhibitor of flowering in LD.

Photoperiod has also been shown to influence the content of GAs in the shoot tips where floral evocation occurs. In *Silene*, the levels of several 13-hydroxy GAs (44, 19, 20, 1 and 8) were higher in the tips of LD-induced plants than in SD, especially in the meristematic zone approximately 0.5–1.5 mm below the apical meristem. Compared with SD, the highest increase in LD was observed for GA$_1$ (Talon *et al.*, 1991a). In *Lolium temulentum*, exposure to a single photoinductive LD increased the content of bioactive free GAs in the true shoot apex above the leaf primordia, most consistently for GA-like substances at retention times where polyhydroxylated GAs elute (Table 7.6). As discussed below, these polyhydroxylated GAs may have a specific flower-promoting effect in *Lolium*.

Numerous studies have demonstrated that daylength-dependent changes in the content and/or metabolism of gibberellins can occur both in the leaves and in the apical parts of the shoot. Leaves may also export more GA in the phloem in LD (Hoad and Bowen, 1968), or in SD (Takeno, 1994; Yang *et al.*, 1995). In *Silene*, as in other plants, LD treatment of the mature leaves alone results in elongation and flowering; exposure of mature leaves to LD also resulted in an increase of GA$_1$ in the shoot tips which was independent of the daylength to which the latter were exposed (Talon and Zeevaart, 1992). Thus, daylength-dependent changes in GA might modify flowering in a number of ways. In the leaf, such changes might modify floral induction, GAs might be a component of the photoperiodic stimulus exported from the leaf, and changes in GA at the apex might be a consequence of the arrival of a floral stimulus from the leaves. It is also possible that the observed daylength-dependent changes in GA metabolism are not causally related to flowering, although they could influence other processes such as stem elongation.

The question as to whether endogenous GAs have a function in the flowering process in LDP other than to cause stem elongation is not easily resolved. GAs are primarily stem growth factors and their effect on flowering may be only secondary.

The role of endogenous GA in LD-induced stem elongation is discussed in Chapter 13. A role in flowering is less certain. Caulescent LDP do not flower in response to GA application and the situation in rosette LDP is not straightforward. In *Silene*, the application of the growth retardant AMO 1618 completely suppressed stem elongation, yet flower initiation occurred normally in the absence of detectable GAs (using the d5 corn assay, Cleland and Zeevaart, 1970); nor do exogenous GAs have any florigenic activity in *Silene*. A similar situation has been reported in spinach, for the growth retardant tetcyclasis (Talon and Zeevaart, 1990). Thus, in these two rosette LDP, GA seems to be critical for bolting but not for flowering. In *Silene armeria*, the effects of GA_3 on cell division at the apex were quite different from those of LD after the first 24 h; although both treatments resulted in an early mitotic activation, the mitotic indices in the shoot apex of the GA_3-treated plants reverted to values similar to those of the SD controls, while a new mitotic wave (typically associated with the transition to flowering) occurred in the photoinduced apices (Besnard-Wibaut *et al.*, 1989). On the other hand, the application of growth retardants suppressed both flowering and elongation in the LDP *Samolus* and GA_3 reversed these effects; hence the higher GA content in LD appears to be necessary for initiation in this case (Baldev and Lang, 1965). A comparison of the pattern of differentiation at the shoot apex of *Rudbeckia* indicated that the immediate responses to LD and GA application were different; GA increased the mitotic activity in the medullary zone of the apex, whereas transfer to LD stimulated mitotic activity in the central zone. These and other results led to the conclusion that the sites of action of GA and the floral stimulus in the shoot apex are different, although both may be required for floral evocation (Chailakhyan, 1975).

One problem in interpreting results of experiments with growth redardants is that some of them (e.g. cycocel) may not be particularly specific as inhibitors of GA biosynthesis and may affect other aspects of metabolism; endogenous levels of GA may even increase with cycocel treatment (Pharis and King, 1985). Such effects may explain the synergistic promotion of flowering by GA and cycocel in *Lolium temulentum* and *Blitum capitatum*. In the case of *Lolium*, it is highly unlikely that the increased content of GAs in the shoot apical meristem is associated with stem elongation; a single LD results in only a very small increase in stem length during the subsequent 3 weeks of growth in SD, whereas the increase in GA occurred within a few hours after the end of the inductive LD (Pharis *et al.*, 1987a). Structure activity studies have also made it clear that, in *Lolium*, the promotion of inflorescence initiation and development by GA does not result indirectly from an effect on stem elongation since the order of effectiveness of different GAs for floral transition of apices cultured *in vitro* (where stem elongation does not occur) was the same as in intact plants (King *et al.*, 1993).

In vitro experiments have confirmed that GAs act at the shoot apex to influence floral transition in *Lolium* (King *et al.*, 1993). When apices were excised from plants in SD, none reached the earliest stage of floral differentiation (double-ridges) in the absence of GA, even after 11 weeks in high-irradiance light. When GA_3 was added to the medium, however, all apices reached the double ridge stage when excised from plants that were at least 8 weeks old. In *Lolium*, floral evocation approaches completion by the end of the day following induction by a single LD (day II); however, in the absence of GA in the medium, there was virtually no progress beyond the stage of

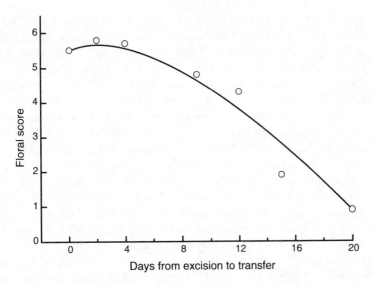

FIG. 7.7. The effect on flowering of time of transfer of excised shoot apices of *Lolium temulentum* from a medium with no gibberellin to a medium containing gibberellin. Intact plants were induced by exposure to a single 24 h LD (day I) and shoot apices were excised on day III. They were cultured on a medium without GA for the number of days shown on the abscissa and then transferred to 10^{-6} M GA_3. The floral stage was assessed 3 weeks after the initial LD. After King *et al.*, 1993.

advanced double-ridges even when apices were excised several days later. Moreover, inflorescence development of shoot apices from LD-induced plants was unaffected by the absence of GA from the medium for the first 4–6 days after excision on day III, although after that its absence severely limited inflorescence development (Fig. 7.7). Based on experiments with inhibitors of GA synthesis, it appears unlikely that the apparent indifference to the presence of GA in the medium immediately following excision is due to the carry-over in excised apices of sufficient endogenous GA. The conclusion that GA exerts its greatest effect after the double ridge stage is supported by experiments with intact plants, where the maximum flowering response occurred when GA was applied between day IV and day VI.

A role other than on elongation is also evident in those instances where GA action in the leaf leads to the production of a floral stimulus. Even where elongation is part of the overall response of LDP to GA application, an examination of the dose relationships has shown that a flowering response can sometimes occurs at doses below those needed to effect stem elongation. It seems reasonable to conclude that endogenous GAs are associated with the transition to flowering in many LDP, although they do not always act in the same way, nor at the same site.

Endogenous gibberellin may also be necessary for the switch to flowering in some SDP. In most SDP the application of a bioactive GA generally depresses flowering in inductive photoperiods and does not promote flowering in non-inductive cycles. However, although GA application does not evoke flowering in the SDP *Xanthium*, the response to extracts from flowering plants was more pronounced, or was only successful when GA was added; the flowering response to thermocycles (4/23 °C) was also increased by GA (Mirolo *et al.*, 1990). The effect of GA application in *Xanthium* thus seems to be associated with the promotion of floral expression. In *Pharbitis nil*, GA had no effect on flowering in tall strains (cv Violet) nor in isogenic tall lines (King *et al.*, 1987), although some increase in the magnitude of flowering has been reported for the cv Violet when plants were treated with GA_3 under marginally-inductive conditions (Takeno, 1994). When tested on the somewhat GA-deficient dwarf variety, Kidachi, GAs were found to have either a promoting or inhibitory effect on flowering depending on the dose and type of GA, and when it was applied. When applied prior to a marginally inductive dark period, flowering was inhibited only at rather high doses of GA; in contrast, lower doses promoted flowering (Fig. 7.8). The proposed effector for vegetative elongation, GA_1, promoted flowering only at very high doses. When applied after an inductive dark period, the effect on flowering was always inhibitory. These results demonstrate that GAs can play a positive role in the induction of flowering in this qualitative SDP, but apparently the promotive effect is only seen in a GA-deficient dwarf and/or under marginally inductive photoperiods. Even under these limiting conditions, GA can be highly inhibitory at high doses, or when applied immediately after the inductive dark period, or when the GA applied is long-lived (e.g. GA_3, King *et al.*, 1987). In contrast, GA treatment can completely substitute for SD induction to promote flowering in a few qualitative SDP (see Table 7.5). The best documented species is *Impatiens balsamina*, where gibberellins A_3, A_{13}, A_{4+7} were found to be effective in inducing flowering in strictly non-inductive 24 h photoperiods (Nanda *et al.*, 1969a). Sub-threshold photoinduction (one SD) could be added to sub-threshold GA but, for the additive effect to occur, it was necessary to expose plants to a SD before making the GA application, although several LD could be intercalated without losing the effect of the subsequent GA treatment (Nanda and Jindal, 1975). Thus the effect of GA may be to increase the response to a sub-threshold

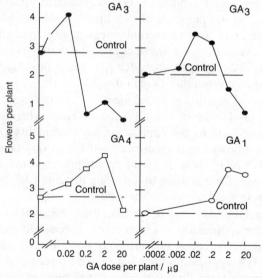

FIG. 7.8. Effect of gibberellins on flowering in *Pharbitis nil* cv Kidachi. Varying amounts of gibberellins were given as a single application to the petiole either 5 h (left) or 16 h (right) before exposure to an inductive 13.25 h night. After Pharis *et al.*, 1989.

amount of floral stimulus. Unfortunately, there seems to be no information about the effect of photoperiod on endogenous GAs in *Impatiens*, nor on the effect of lowering the endogenous GA content by the application of growth retardants.

In some SDP, the increased amount of GA in LD may be associated with the inhibition of flowering. The application of GA_3 has been found to inhibit or delay flowering in several different SDP, when applied in inductive photoperiods (Vince-Prue, 1975). In strawberry, a large dose of GA_3 completely prevented floral initiation in SD although, once initiation was complete, GA_3 accelerated development of the flower (Tafazoli and Vince-Prue, 1978). However, the role of endogenous GAs and their relationship with a transmissible inhibitor of flowering (which is produced in LD in strawberry) remain in doubt, since the application of growth retardants in LD was only marginally promotive of flowering (Guttridge, 1985). Moreover, although a wide spectrum of GAs inhibited flowering, they also caused elongation of the main stem; in LD alone, the main stem remains as a rosette. However, apart from the aberrant elongation of the stem, the effects of GA were very similar to those of LD for petiole growth, stolon formation and promotion of flower growth, suggesting that endogenous GAs are probably involved in these responses.

The application of GA_3 inhibited the induction of flowering in the SDP *Begonia* × *cheimantha* and transfer to inductive SD cycles decreased the content of GA_4 and GA_{20} (see Fig. 7.6). Since both of these GAs can give rise to GA_1 (which was also present), it was suggested that this GA may be an active inhibitor of flowering in LD. The fact that the content of GA_{20} was also decreased by transfer from LD at 24 °C to LD at 15 °C, conditions which are also inductive for flowering in *Begonia*, supports the hypothesis that endogenous GAs are inhibitory to flowering in this plant. However, the changes were recorded in the leaf and their relevance to evocation at the apex is not known. The regeneration of adventitious buds in detached leaves (see Chapter 13) is also influenced by daylength and inhibited by exogenous GA in *Begonia*.

There are a few LDP in which flowering is inhibited by GA application (see Table 7.5). Evidence has been presented that an endogenous GA is inhibitory for flowering in LD strains of *Pisum sativum*. In the strain G_2, a late-flowering line, ^3H-GA_9 was metabolised in SD to more polar GAs but this was largely blocked by exposure to a single LD, which also induced flowering. Moreover, when reproductive plants were reverting to vegetative growth, they regained the ability to synthesise polar GAs from ^3H-GA_9 (Proebsting and Heftmann, 1980). It was concluded that flowering in this strain of pea is inhibited by a GA which is produced in SD (Davies *et al.*, 1982). However, expression of the flowering response to daylength was not prevented in GA-deficient mutants (Murfet and Reid, 1987) and it was concluded that the flowering genes *Sn* and *Dne* (which confer photoperiodic responsivity and appear to control synthesis of the flower inhibitor) probably do not act by influencing GA metabolism. Other major flowering genes in peas (e.g. *Lf*) appear to be specifically concerned with events at the shoot apex; however, allelic differences at the *Lf* locus were also clearly expressed in GA-deficient plants. In *Fuchsia*, the inhibitory effect of exogenous GA occurred only when it was applied to the apical bud before, or at least concurrently with the presumed arrival of the floral stimulus from the leaves. Later applications were without effect. Repeated application of the growth retardant, daminozide, promoted initiation in SD and also increased the effectiveness of sub-inductive LD treatment, while the simultaneous application of GA and daminozide in

SD prevented the expected enhancement of flowering. Thus in this caulescent LDP, there is evidence for an inhibitory role of endogenous GA in flowering; this may be part of the photoperiod-dependent control since the content of endogenous GA may actually decrease during LD (Sachs *et al.*, 1967).

A Role for Specific GAs

One proposal to account for the variation in the direction of the flowering response, or the ineffectiveness of GA application in some cases, has been that only specific GAs are involved in flowering. Although this has been considered unlikely by some workers on the basis of the metabolic pathways for GA interconversions (Zeevaart, 1984), there is an increasing number of results which lend support to the idea. For example, it has been reported that GA_{13} promoted flowering in three quantitative SD species of *Amaranthus* without increasing stem elongation, whereas GA_3 and $GA_{4/7}$ increased elongation but had no effect on flowering (Kohli and Sawhney, 1979; Sawhney and Sawhney, 1985). No GA completely substituted for SD but GA_{13} application decreased the critical nightlength by 1–2 h. GA was applied to the apex and the authors assumed that its action occurred there. Based on the flowering behaviour of crosses between tall and dwarf genotypes, it was concluded that the gibberellins which influence stem elongation are not important in determining the differences in critical night length in the dwarf Kidachi and tall Violet strains of *Pharbitis* (King *et al.*, 1994).

In the Pinaceae, only the less polar GAs ($GA_{4/7}$) are reliably effective in causing flowering in juvenile seedlings and older propagules (Pharis *et al.*, 1989); success has been reported for at least twenty species in five different genera. These GAs may also be endogenous flowering factors since several diverse cultural treatments that accelerate flowering (e.g. water stress, root pruning) have also been found to increase the amount of less polar GAs within the plant, while there was no effect, or a decrease in more polar GAs (Pharis *et al.* 1987b, 1989; Pharis and King, 1985). Root pruning was also found to retard the metabolism of applied GA_4 in Douglas fir and it is suggested that cultural treatments may interfere with the metabolism of native GA, leading to an increase in specific GAs, most notably those without, or with only one hydroxyl group. However, conifers have not been reported to be sensitive to photoperiod for the induction of flowering, although there are some effects of daylength on sex expression which is also influenced by GA (see Chapter 10). It was also observed that flowering of apple trees in the 'off' year was increased by the application of GA_4 or C-3 *epi*-GA_4 in the previous season (Looney *et al.*, 1985). In view of the generally strongly inhibitory effect of applied GAs on flowering in woody angiosperms including apple, this response also indicates the possibility that specific GAs may be associated with the regulation of flowering, although the effect of GA_4 was very variable and flowering was sometimes inhibited (Greene, 1993). In conifers, the effect on flowering does not appear to be dependent on assimilate movement; in *Pinus radiata*, GA_3 and $GA_{4/7}$ enhanced assimilate accumulation into differentiating primordia to the same extent, yet only $GA_{4/7}$ significantly promoted flowering (Pharis *et al.*, 1989).

In contrast to the promotion of flowering in the Pinaceae by less polar GAs, the highly polar polyhydroxylated GA_{32} (four hydroxyl groups) was much more effective

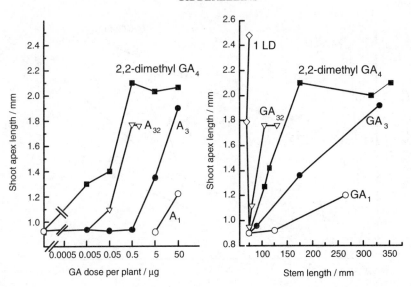

FIG. 7.9. Effect of several gibberellins on the flowering response (shoot apex length) and elongation growth (stem length) of *Lolium temulentum* plants in non-inductive short days. Re-drawn from Pharis *et al.*, 1987a.

than either GA_3 or GA_1 in causing flowering in the LDP *Lolium temulentum* (Fig. 7.9), while the addition of a ß hydroxyl group at C-15 increased the florigenic activity of GA_3 by 10–20-fold (Fig. 7.10). Moreover, of the many GA-like compounds present in the shoot, the putative polyhydroxylated components showed the most consistent increase in the shoot apex after exposure to a single inductive LD (Table 7.6). It has also been shown for this plant that some GAs are more effective in promoting flowering, while others cause stem elongation (see Fig. 7.9). A single LD resulted in an increase in flowering (as measured by apex length) with little effect on elongation at the time of dissection while GA_1 caused stem elongation with a relatively small effect on flowering. In contrast, the polyhydroxylated GA_{32} gave a flowering response approaching that of a LD, with only a small increase in stem length. When different GA structures were compared, the results indicated that 'building towards the GA_{32} structure also builds towards the effect obtained with a single LD' (Pharis *et al.*, 1989). Thus, GA_1 (hydroxyls at C-3 and C-13 but no double bond in ring A) gave poor flowering but good vegetative growth; adding a double bond in ring A (GA_3) increased the florigenic effect but also promoted stem elongation; adding a third hydroxyl group at C-15 (15ß-OH-GA_3) increased the florigenic effect at lower doses, at which there was only a small effect on stem length (Fig. 7.10). There is good evidence that a double bond in the A ring at either C-1,2 or C-2,3 is essential for florigenic activity, although not for stem elongation (Evans *et al.* 1990). A notable exception is 2,2-dimethyl GA_4 with no double bond in ring A; this was highly florigenic at low doses (see Fig. 7.9) and its relative effect on stem elongation and flowering resembled that of GA_{32} rather than GA_3 (Pharis *et al.*, 1987a). A steric effect, such as a flattening of the A-ring, may explain the anomalous responses to this GA (Evans *et al.*, 1990). High florigenic activity is also associated with the presence of hydroxyl groups; GA_{32}, the

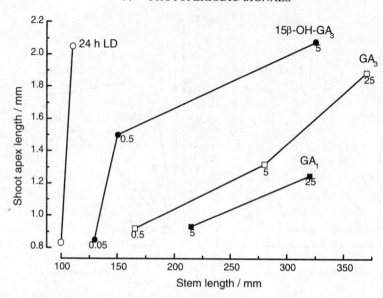

FIG. 7.10. Relative effects of several gibberellins on flowering (shoot apex length) and elongation growth (stem length) of *Lolium temulentum* plants. 24 h LD, plants given a single LD without gibberellin; GA-treated plants were grown in non-inductive 8 h SD. Gibberellins were applied to a single leaf at the doses indicated on the figure (μg per plant). From Pharis *et al.*, 1989.

most active of the natural GAs for inflorescence initiation in *Lolium*, has four hydroxyls (at C-3, C-12, C-13 and C-15). Hydroxylation at C-12, 13 and 15 has also been shown to enhance the florigenic activity of other GAs, whereas 3ß-hydroxylation enhanced stem elongation but reduced the promotion of flowering. Hydroxyls at C-3 appear to be of much greater importance for stem elongation than for flowering in *Lolium* since the application of inhibitors of 3ß-hydroxylation reduced stem growth but increased the flowering response to LD (Evans *et al.*, 1994a); however, the inhibitor studies also indicated that subsequent inflorescence development may require 3ß-hydroxylated GAs. In contrast, the 3α-hydroxy epimer of 2,2-dimethyl GA$_4$ promoted flowering with almost no effect on stem elongation. The structure activity studies in *Lolium* have made it clear that GAs differ in their effectiveness for stem elongation and flowering. An extreme situation is seen with 16,17-dihydro GA$_5$ and GA$_{20}$, which actually inhibit stem elongation, whereas loss of the C-16,17 double bond generally enhanced the promotive effect on inflorescence development (Evans *et al.*, 1994b). Different GAs have also been shown to vary in their effectiveness in a tall line of the LDP wheat, the most florigenic being 15ß-OH GA$_3$, GA$_3$ and 2,2-dimethyl GA$_4$ and the least effective being GA$_{13}$, GA$_{16}$, GA$_{19}$ and GA$_{54}$ (Evans *et al.*, 1995); these rankings are similar to those for *Lolium*. In an isogenic dwarf line, a wide range of exogenous GAs had no effect on stem and leaf blade elongation when applied in SD but were able to elicit flowering; thus the shoot apex of the dwarf line was responsive to applied GAs for flowering even when leaves and stems were not. However, in contrast to the tall line, the dwarf lines showed little difference in their flowering response to any of the applied GAs.

The fact that flowering in *Lolium* can be evoked in SD by the application of various gibberellins and by no other compounds so far tested, implies some equivalence between LD and GAs in their effects on flowering. However, if the photoperiodic floral stimulus is a specific florigenic GA, it seems unlikely that the ranking of GAs for their flower promoting effect would be the same in SD and after induction by LD, as they are (Evans *et al.*, 1990; King *et al.*, 1993). The increasing responsiveness with age to both LD (see Fig. 6.3) and to GA *in vitro* could reflect a slow accumulation of the floral stimulus in SD; the effect of this stimulus at the apex might be potentiated in some way by GA, which would account both for the effect of applied GA to evoke flowering of older plants in SD and for the parallel florigenic effects of different GAs in SD and following LD-induction.

The possibility that certain specific GAs may be particularly associated with the regulation of flowering is indicated by the results of several different kinds of experiment. In summary these are: the greater effect of less polar GAs (e.g., $GA_{4/7}$) to cause flowering in many members of the Pinaceae, together with the observation that cultural treatments which promoted flowering specifically increased the endogenous content of the less polar GAs; the positive effect of GA_4 to promote return flowering in apple, contrasted with the inhibitory effect of GA_3; the similar effect of GA_3 and $GA_{4/7}$ on assimilate partitioning in *Pinus*, compared with their different effects on flowering; and the effects of different GA structures on elongation and flowering in the LDP *Lolium* and SDP *Amaranthus*. In the LDP *Sinapis*, however, none of the GAs tested (including 2,2-dimethyl GA_4) had any florigenic activity (Bernier *et al.*, 1993).

Only in *Lolium* has photoperiod been shown specifically to increase the content of the highly hydroxylated C-19 GAs that appear to be particularly associated with the promotion of flowering in this plant. This increase occurred in the true shoot apex, where floral differentiation occurs, but the source of these GAs was not demonstrated (Pharis *et al.*, 1987a): if they constitute part of the endogenous photoperiodic signal in Lolium, they would be expected to originate in the leaf. Moreover, if the leaf-generated stimulus to flowering in *Lolium* is a GA, it must be one which promotes inflorescence initiation without causing appreciable stem elongation in the early stages of floral initiation (King *et al.*, 1993; Evans *et al.*, 1994b). On the other hand, based on several types of evidence, it has been suggested that applied GAs may not evoke flowering by replacing the leaf-generated LD-stimulus but rather act indirectly by potentiating the action of that stimulus as it accumulates slowly, with age, in SD (King *et al.*, 1993, Evans *et al.*, 1994b).

Conclusions

It is clear that, while gibberellins influence flowering in a wide range of plants, their endogenous role is not well understood and their involvement varies from plant to plant. GAs may act either in the leaf, or at the apex. In general, there appear to be increased amounts and/or turnover of bioactive GAs in LD, and this may be associated with the regulation of flowering by daylength. There is evidence for inhibitory action in certain SDP (e.g. strawberry, *Begonia*) and in SLDP. On the other hand, the increased GA content in LD may be necessary for synthesis of the floral stimulus in leaves of the LSDP *Bryophyllum* and the LDP *Hyoscyamus*. In rosette LDP, the increased GA content (or turnover) in LD is necessary for stem elongation which normally accompanies flowering in these plants, although (in *Silene* and *Spinacia* at least) it has been established that GAs probably have no direct role in floral initiation. Growth retardant studies with other rosette LDP do, however, suggest a role for endogenous GA. Gibberellin also appears to be required for flowering in *Arabidopsis* since a mutant severely defective in *ent*-kaurene synthesis remained vegetative in SD unless supplied with exogenous GA (Wilson *et al.*, 1992); the relationship with daylength is not clear, however, as the mutant flowered readily in LD, albeit with some delay. The greater sensitivity of the flowering response in several plants also indicates that GA may be having an effect on flowering distinct from that on elongation. For example in both the SDP *Pharbitis* and the LDP *Hyoscyamus*, low doses of GA promoted flowering but did not increase elongation (Warm, 1980). Similarly, flowering in both SDP (*Amaranthus*) and LDP (*Lolium*) is promoted by certain gibberellins at concentrations which have little effect on elongation. There is good evidence that the metabolism of GAs is influenced by daylength and the possibility that particular GAs have a specific role in flowering remains open but appears increasingly likely. Daylength-induced changes in sensitivity to GAs have been demonstrated for stem elongation and may also operate in the control of flowering.

CYTOKININS

The application of cytokinins (CKs) has been found to promote flowering in a few plants but the effect is usually seen only under marginal photoperiodic conditions (Vince-Prue, 1985). Kinetin (6-furfurylamonopurine, a CK not known to occur naturally in plants) reduced the number of SD required for induction in the SDP red *Perilla*. In *Pharbitis*, kinetin significantly enhanced flowering under near threshold dark periods, while benzyladenine (BA) resulted in flowering in dark-grown plants given only a few minutes of R before transfer to an inductive dark period, conditions under which control plants remained vegetative (Ogawa and King, 1979a). SDP which have been induced to flower by the application of CK in non-inductive photoperiods include *Wolffia microscopica* and *Lemna paucicostata* but the wide range of chemicals that can cause flowering in the Lemnaceae has already been discussed. The application of benzyladenine to the SDP chrysanthemum Pink Champagne caused some flowering in LD when controls remained vegetative; in this case, there was a strong synergism between gibberellins and cytokinins, the combination of GA_5 with benzyladenine being highly effective (Pharis, 1972). Synergistic effects between GA and CK have also been reported for other plants (Bernier *et al.*, 1990). Cytokinin application to buds in LD promoted the development of *Bougainvillea* inflorescences that would normally have failed to develop under these conditions (Tse *et al.*, 1974), and promoted the development of *Phaseolus* flowers which normally abscise in long photoperiods (Morgan and Morgan, 1984). Among LDP, *Arabidopsis* appears to be the only example of the induction of flowering by the application of cytokinins in non-inductive photoperiods (Michniewicz and Kamienska, 1965). Thus the effects of exogenous CKs to promote flowering have been observed mainly in SDP and under photoperiodic conditions that are marginal for the initiation and/or development of flowers. In several of the examples given, the application of CK may have modified

TABLE 7.7   Effect of daylength on cytokinin content of phloem exudate from leaves.

Plant	Zeatin		Zeatin riboside		2iP + 2iPA	
	Induced	Control	Induced	Control	Induced	Control
Sinapis alba[a]	0.55	0.15	1.77	0.63	0.89	0.28
Perilla[b] (red)			330	120	240	80
A						
B					160	90

[a] *Sinapis*: exudate collected for 16 h starting 16 h after beginning the single LD (inductive) or SD cycle.
Cytokinin values are means of nine separate experiments, expressed as pmol equivalents of benzyladenine
per plant.
Data of Lejeune *et al.* (1988).
[b] *Perilla*: exudate collected for 24 h from the beginning of the night. Cytokinin values are expressed as
fmol equivalent riboside g^{-1} FWt.
A, plants decapitated above 8th node at start of SD treatment (25 SD); B, not decapitated (30 SD).
Data of Grayling and Hanke (1992).

the movement of assimilates to the shoot apex and the effect on flowering was
probably indirect.

There is evidence that the metabolism and/or transport of endogenous cytokinins is
influenced by daylength (Table 7.7). In *Xanthium*, the decrease in cytokinin levels
after a 16 h inductive dark period was partly prevented by giving a 5 min R night-
break at the 8th hour demonstrating true photoperiodic control (Henson and Wareing,
1977a). A night-break also reduced the effect of a long night to increase the exudation
of CKs from leaves of red *Perilla* (Grayling and Hanke, 1992). There is, however, no
clear cut correlation between daylength effects on CK content and flowering. For
example, CKs (both zeatin and *iso*-pentenyladenine types) were higher in inductive
conditions (SD at 24°C) in leaves of the SDP *Begonia* × *cheimantha* (Hansen *et al.*,
1988) while, in the LDP *Dactylis glomerata*, the CK content in the shoots (mainly
zeatin and zeatin riboside) decreased after two cycles following transfer to inductive
LD (Menhenett and Wareing, 1977). Thus the CK content was higher in SD in both
plants, irrespective of the photoperiodic class for flowering. Similarly, the amount of
CKs present in root exudates was higher in LD in both the LDP *Sinapis alba* and the
SDP *Xanthium strumarium*. Photoperiod has also been shown to control the level of
CKs in *Chenopodium* species, with the amount in leaves, stems and roots decreasing
during the dark period and increasing again during the light period; there were no
significant changes when plants were maintained in continuous light (Macháčková *et
al.*, 1993). In the SDP *C. rubrum*, this fluctuation pattern in SD was coupled with the
photoperiodic regime which induces flowering while, in the LDP *C. murale*, it
occurred in the photoperiod in which plants remain vegetative. Thus there was no
apparent correlation between changes in CK levels in the leaf and photoperiodic floral
induction. In contrast, the CK content in the shoot apices increased, at least transiently, during induction in both LD and SD species and, in the SDP *C. rubrum*, the
increase observed at the end of a 12 h inductive dark period was greatly diminished by
a R night-break at the 6th hour (Fig. 7.11). The correlation with flowering is debatable,

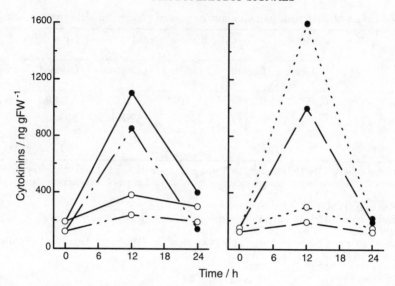

FIG. 7.11. The effect of a red night-break on the content of cytokinins in the apical part of the SDP, *Chenopodium rubrum*. Plants received a cycle consisting of 12 h light/12 h dark which was interrupted at the 6th h of darkness by 15 min R (○); control plants received 12 h uninterrupted darkness (●). The cytokinins assayed were: isopentenyladenine (——), isopentenyl adenosine (–···–), zeatin (— —), and zeatin riboside (. . . .). After Machackova *et al.*, 1993.

however, since flowering was restored by a subsequent exposure to FR, whereas the effect on CK content was not.

Cytokinins have been found in both phloem and xylem sap. In red *Perilla*, the CKs in the phloem sap (with a greater predominance of *iso*-pentenyladenine types) appeared to differ from those in the xylem, where zeatin ribosides predominated (Grayling and Hanke, 1992). In xylem sap, a higher content of CK (zeatin riboside) in SD has been reported for the SDP *Perilla* (Beever and Woolhouse, 1973) and in *Phaseolus vulgaris*, where SD promotes flower development (Morgan and Morgan, 1984). If CKs form part of the floral stimulus, their content would be expected to increase in the phloem sap following transfer to inductive photoperiods. The most detailed studies have been carried out with the SDP *Perilla* and *Xanthium* and with the LDP *Sinapis alba* and, in each case, the CK content of the phloem sap has been found to increase in SD or LD, respectively.

In *Xanthium*, the amount of CKs co-chromatographing with zeatin and zeatin riboside increased in SD in honeydew from phloem-feeding aphids (Phillips and Cleland, 1972). A more detailed study of EDTA-enhanced leaf exudates in red *Perilla* has shown that the amount of *iso*-pentenyladenosine (iPA) and zeatin riboside (ZR) increased 2–5 fold in plants exposed to 30 SD, compared with those that remained in LD (see Table 7.7). However, zeatin-type CKs were only detected in plants that were decapitated prior to the experimental treatment. Intact plants flowered when given 30 SD, while controls in LD remained vegetative. Since the induced state in *Perilla* leaves is known to persist for a considerable time following return to LD conditions, the increased export of CKs would also be expected to continue if it is correlated with the induced state; this was not recorded. The case for a correlation between photo-

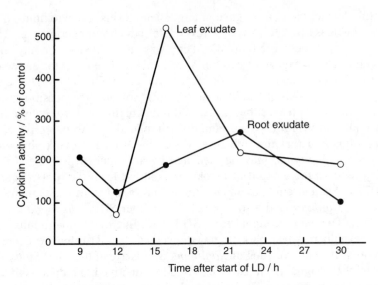

FIG. 7.12. Cytokinin activity in the butanol-soluble fraction of leaf and root exudates from *Sinapis alba* plants exposed to a single inductive 22 h LD. Exudates were collected for 16 h beginning at the times shown on the abscissa. After Bernier *et al.*, 1990.

periodic induction and CK export from the leaves is strengthened by results with the LDP *Sinapis alba*. Exposure to a single LD, which is sufficient to induce flowering, also resulted in an increase in CKs in both leaf and root exudates as determined using the *Amaranthus* bioassay (see Table 7.7). The increase in CK export was detected 16 h after the beginning of the LD and fits well with the time that the first mitotic wave is observed at the shoot apex (Fig. 7.12). A subsequent and more detailed investigation using radioimmunoassay also measured the amount of cytokinins present in phloem sap reaching the apical part of the shoot, close to the target bud (Lejeune *et al.*, 1994). The analyses confirmed the earlier results and also demonstrated an increase in the amount of CKs directed to the upper part of the shoot, although this was smaller than the increase in the root and leaf exudates. There were, however, differences between the leaf and apical exudates in the timing of the increase. In the apical exudate, the maximum increase occurred when the exudate was collected between 9 and 25 h from the beginning of the LD and there was little difference from the SD controls when the exudation began at the 16th hour; in contrast, the maximum increase in CK levels in the leaf exudate occurred during the 16–32 h period of collection (Fig. 7.12). It was suggested that the discrepancy between leaf and apical exudates might be explained by the harsher treatment to which the leaves were exposed, with the apical exudates reflecting more closely the changes occurring in the intact plant.

In contrast to these findings, transfer to SD resulted in a substantial decrease in the amount of cytokinin in leaves, buds and root exudates of *Xanthium*. In this case, there was a strong correlation between the reduction in cytokinin content and the induction of flowering; both were effected by a single SD, an increase in the duration of the dark period increased the response, the effect of a long dark period could be negated by a R night-break, and the leaf was the site of photoperiodic perception for both responses (Henson and Wareing, 1977a). Moreover, conditions which resulted in suboptimal rates of floral development also gave intermediate levels of cytokinins. As with flowering, the decrease in CKs was an irreversible induction phenomenon. Although the cytokinin content decreased in SD, there was no evidence for any effect of daylength on the rate or pattern of cytokinin metabolism in leaves, nor on the content of cytokinin in detached leaves (Henson and Wareing, 1977b). The higher cytokinin content in phloem sap (Phillips and Cleland, 1972) indicates the possibility of more rapid export from the SD leaves, but this would not explain the lower cytokinin levels in the buds, unless utilisation is substantially stimulated by the inductive SD treatment. It was suggested that the influence of daylength on the cytokinin content of leaves and buds in *Xanthium* may be mediated by an unknown SD signal from the leaves, which reduces the amount of cytokinin exported from the roots (Henson and Wareing, 1977a). Such a reduction in cytokinins in root exudates in SD treated plants was observed in *Xanthium* and also reported for another SDP, *Chenopodium rubrum* (Krekule, 1979). The putative signal from SD leaves, like the floral stimulus, appears to move in the phloem since bark-ringing prevented the SD effect to decrease the cytokinin content in root exudates of *Xanthium*. Similarly, in the LDP *Sinapis*, it has been suggested that a signal from leaves exposed to inductive LD increases the amount of CKs exported from the roots (Lejeune *et al.*, 1988). However, there is no information on the possible chemical identity of such a signal. Whether changes in cytokinin export from the root are part of the overall process of photoperiodic floral induction remains open; despite the increase in CK export from roots under inductive conditions in the LDP *Sinapis* and the SDP *Perilla*, both can be induced in the absence of roots (Lejeune *et al.*, 1988; Zeevaart and Boyer, 1987).

The lowered cytokinin content in induced plants of *Xanthium* is not consistent with the reported effects of cytokinins to enhance flowering in some other SDP. However, cytokinins may be responsible for the root-mediated suppression of flowering that has been observed in a number of plants. For example, excision of roots enhanced flowering in the quantitative SDP *Chenopodium polyspermum* in non-inductive conditions and the addition of zeatin to the buds mimicked the presence of roots and reduced flowering; zeatin also counteracted the effects of inductive SD (Sotta, 1978). Similar results have been obtained in the LDP *Scrophularia arguta* (Krekule, 1979) and it has been suggested that, in these plants, flowering may be controlled by a balance between a floral stimulus coming from the leaves and inhibitory cytokinins originating in the roots. However, the effect of exogenous CK has been shown to depend on concentration and time of application (Bismuth and Miginiac, 1984). In *Chenopodium*, flowering was promoted when kinetin was applied after photoinductive SD, but inhibited when it was applied during SD; similarly zeatin was generally inhibitory to flowering in cuttings of the LDP *Anagallis arvensis*, but promoted flowering if applied when root primordia began to develop.

Cytokinins are active in the regulation of cell division and, in many plants, one of the earliest events observed at the shoot apex following photoperiodic induction is a transitory increase in the mitotic index (see Chapter 8). A single application of benzyladenine to the apical bud of the LDP *Sinapis* in SD, resulted in several changes at the apex that are typical of the transition to flowering, including an increase in mitotic activity similar to that seen after exposure to a single LD. However, many other changes normally occurring during floral transition did not occur and, unlike the LD treatment, the application of cytokinin did not cause flowering (Havelange *et al.*,

1986). Defoliation experiments have shown that the stimulus responsible for the early mitotic wave in the meristem comes from the leaves and begins to move out of them about 16 h after the beginning of an inductive LD (Lejeune *et al.*, 1988); this correlates well with the time at which an increase in CK export from leaves in LD begins to increase above the SD controls (see Fig. 7.12), although an increase in phloem CKs was detected considerably earlier when measured close to the apex. All attempts to dissociate this increased mitotic activity from flowering have been unsuccessful and it is thought to be an essential component of evocation in *Sinapis*. Such observations have led to the proposal that the floral stimulus in *Sinapis* consists of at least two components, one of which is a cytokinin that stimulates mitotic activity. However, even if this conclusion is generally true, the lack of response to applied CKs in many plants suggests that the cytokinin component is rarely limiting for flowering.

The site of action of cytokinins appears to be at the shoot apical meristems, as would be expected if they were part of the floral stimulus. Analyses of phloem exudates in which the amount of cytokinin exported from the leaves of *Perilla* is increased by exposure to inductive SD also suggest that cytokinins may be part of a floral stimulus and, in some cases at least, may be a limiting factor for flowering. The increase in CK content of the shoot apices during induction in LD and SD species of *Chenopodium* also indicates a role in evocation although, in this case, the CK contents in leaves, roots and stems do not show any correlation with the daylength conditions that are required for the induction of flowering (Macháčková *et al.*, 1993). Cytokinins are involved in the regulation of the cell-division cycle and, as deduced from the results with *Sinapis*, their mode of action at the apex appears to be to increase mitotic activity. However, alone, cytokinins do not evoke flowering in *Sinapis* nor do they cause production of floral buds in strictly vegetative plants in the tobacco TLC system, although they are required for the initiation of floral buds in florally determined explants. Furthermore, neither in *Arabidopsis* nor tobacco were effects on time of flowering associated with increased CK levels in transgenic plants (Medford *et al.*, 1989), although various morphogenetic effects were observed (e.g. release of axillary buds from dominance). Thus, while they have been shown to be under photoperiodic control in some plants and undoubtedly play a part in the realisation of floral expression, cytokinins cannot be equated with the daylength-dependent stimulus which evokes flowering in vegetative shoot meristems. They may, however, be a component of this stimulus in some cases.

AUXIN

In early studies, the application of auxins was shown to be inhibitory to flowering in a number of SDP but was ineffective, or promoted flowering in LDP. The inhibition appeared to be primarily localised in the leaf and was thought to be due to interference with synthesis of the floral stimulus, to an increase in its breakdown, or perhaps an effect on its transport (Lang, 1965; Jacobs, 1985; Vince-Prue, 1985). In typical experiments with *Pharbitis* and *Xanthium*, flowering was depressed or inhibited when auxin was applied during an inductive dark period, indicating action in the leaf. In *Pharbitis*, the inhibition of flowering was maximal when auxin was applied immediately before a photoinductive dark period or during the first half of the night

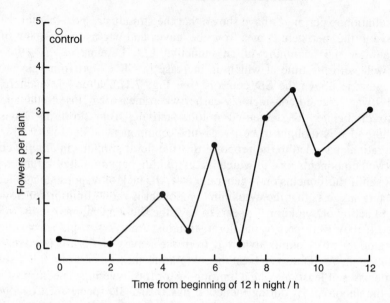

FIG. 7.13. Inhibition of flowering in *Pharbitis nil* by auxin application. Plants were given a single 12 h inductive dark period and sprayed with 1 mM IBA at the time shown on the abscissa. O, unsprayed control plants. After Halevy *et al.*, 1991.

(Fig. 7.13) and the effect was shown to be greatest when the cotyledons were treated. In *Xanthium*, auxin was also inhibitory in the second half of the dark period, or early in the following light period, suggesting possible interference with translocation of the floral stimulus. The application of auxin antagonists to *Xanthium* plants maintained in threshold conditions for induction led to a marginal degree of flowering, suggesting that endogenous auxins may be interfering with the translocation of threshold amounts of the floral stimulus, or may in some way lead to its destruction. In *Pharbitis*, however, the action of auxin may be different, as it was effective earlier in the night and auxin antagonists did not promote flowering. In *Xanthium*, the application of auxin to the growing point promoted flowering; since the promotion was greatest in poor light conditions or when some of the leaves were removed, the effect is probably related to an increase in the flow of materials into the treated bud.

In another SDP, *Chenopodium rubrum*, at least part of the inhibitory effect of auxin application appeared to occur at the apex via the suppression of axillary bud differentiation, but auxin also appeared to exert an inhibitory effect in the leaves (Krekule *et al.*, 1985). In *Chenopodium* and in the LDP *Brassica campestris*, auxin application both inhibited flowering and promoted apical dominance (Krekule and Seidlova, 1977; Seidlova, 1980). Since one of the earliest changes that occurs with the onset of flowering in many plants is the activation of axillary meristems and loss of apical dominance, a decrease in the level of auxins at the shoot apex may be associated with the transition to flowering. Such a decrease has been recorded in the apical bud of *Sinapis* when induced to flower in LD (Bernier *et al.*, 1993). However, although floral events at the apex have been linked with a drop in auxin level, auxin application promotes flowering in some LDP, especially in subthreshold photoperiods (Evans, 1964a).

The endogenous auxin content (usually based on bioassay) has been found to vary in different photoperiods, but no very consistent trends have been observed (see Appendix II) and the role, if any, of endogenous auxins in the photoperiodic regulation of flowering remains obscure. For example, the fluctuations of free IAA in shoots and roots of LD and SD species of *Chenopodium* were similar under 16 h dark periods and were not correlated with the photoperiodic conditions necessary for the induction of flowering (Krekule *et al.*, 1985). In contrast, the content of IAA in leaves of *Begonia* × *cheimantha* decreased rapidly (within 2–4 days) following transfer to inductive conditions (SD at 24 °C or 15 °C in either SD or LD; Oden and Heide, 1989). Whatever the direction of response, the site of action in many cases appears to be in the leaf, to affect either the process of floral induction or the transport of substances out of the leaf. Auxins do not, therefore, appear to be components of the photoperiodic floral stimulus, nor can they be equated with daylength-regulated, transmissible inhibitors of flowering. It has been suggested, however, that auxins may be endogenous inhibitors of flowering providing a general background of inhibition which the stimulating effects of photoperiod can overcome (Jacobs, 1985). A role in floral expression is indicated by the correlation between the inhibition of flowering by auxin and the suppression of axillary bud differentiation. The effects of auxin/CK ratio to switch between vegetative and floral development in TCL explants of tobacco also indicate a role for auxin in the regulation of floral expression; there was, however, no gradient in the content of endogenous IAA, despite the gradient of flower-forming activity at different locations on tobacco stems (Noma *et al.*, 1984).

ETHYLENE

Within recent years, some of the effects of applied auxin have been attributed to an increase in the release of ethylene which, with few exceptions, has inhibitory effects on flowering. Abeles (1967) observed that ethylene prevented flowering in *Xanthium* plants (in 16 h dark periods) and attributed the inhibitory effect of applied auxin in this plant to the release of ethylene. The inhibitory effect of auxin on flowering in the SDP *Chenopodium rubrum* and the LDP *C. murale* has also been attributed to ethylene since aminoethoxyvinyl glycine (AVG, an inhibitor of ethylene biosynthesis) counteracted some of the inhibitory effect of IAA, when applied simultaneously with it (Krekule *et al.*, 1985). On the other hand, auxin appears to act differently in *Pharbitis* (Halevy *et al.*, 1991) since the immediate precursor of ethylene, ACC, stimulated ethylene production to the same or greater degree than auxin but had little or no effect on flowering (Table 7.8). Although exogenous ethylene inhibited flowering in *Pharbitis*, it was only effective when applied in the latter half of the night (Lay-Yee *et al.*, 1987b), whereas the effect of auxin occurred mainly during the early part of the inductive dark period (see Fig. 7.13). Ethylene also inhibited SD-induced flowering in *Perilla* and chrysanthemum, but induced flowering in another SDP, *Plumbago indica* (Vince-Prue, 1985). In the facultative SDP chrysanthemum cv Polaris, the application of an ethylene-releasing compound, 2-chloroethylphosphonic acid (CEPA), delayed flowering in LD as well as in SD, the number of additional leaves formed being the same in both photoperiods (Cockshull and Horridge, 1978); in this plant, therefore, the delaying effect of ethylene is independent of daylength. Ethylene

TABLE 7.8 Effects of auxin and the ethylene precursor 1-aminocyclopropane 1-carboxylic acid (ACC) on ethylene production and flowering in *Pharbitis nil*.

Treatment	Ethylene production (nl h^{-1})	Flowers per plant	% Flowering terminals
Control	2.6	5.2	83.3
0.5 mм IBA	8.9	1.7	5.0
0.5 nм ACC	11.7	4.5	71.4

Indolylbutyric acid (IBA) and ACC were applied to seedlings immediately before a single 12 h night. Ethylene production was measured during the 3rd hour of the dark period. From Halevy *et al.* (1991).

also inhibits flowering in sugar cane and CEPA has been used commercially to prevent flowering and increase the yield of cane.

Ethylene has been shown to promote flowering a few SDP, including the qualitative SDP *Plumbago indica* cv Angkor (Nitsch and Nitsch, 1969). However, the most striking effect of ethylene to promote flowering occurs in pineapple (most cvs are quantitative SDP) and other members of the Bromeliaceae. Auxins also promote flowering in bromeliads, an effect which is almost certainly related to the release of ethylene. As far as is known, this effect of ethylene applies to all bromeliads and it has been exploited commercially by the application of ethylene-releasing compounds to promote flowering in pineapple and ornamental members of the Bromeliaceae such as *Aechmea* and *Billbergia* (Halevy, 1990). Similar treatments using the ethylene precursor ACC have also been used with good results. Control of flowering in these plants also involves prevention when not desired; a single treatment with an inhibitor of ethylene biosynthesis, such as AVG, can prevent flowering for several months and subsequent treatment with ACC will then induce flowering. Ethylene is also the only plant growth regulator with a significant effect on flowering in geophytes (Halevy, 1990) and bulbs of iris and Tazetta narcissus are treated commercially with ethylene to promote forcing.

Although ethylene can either promote or inhibit flowering, it is not known how it acts, nor whether endogenous ethylene is involved in the regulation of flowering. In *Pharbitis*, ethylene may inhibit the induction process in the leaf, since the inhibitory effect occurred during the inductive dark period. In some cases, however, it appears that exogenous ethylene may modify flowering through an effect at the apex. In the SDP chrysanthemum, action at the apex is indicated by the observation that the inhibitory effect of ethephon is generally greatest when applied close to the apical bud (Stanley and Cockshull, 1989). In *Triteleia laxa*, the application of ethylene to corms substantially increased the apex size, an effect which is thought to be associated with the promotion of flowering in this plant (Halevy, 1990); mobilisation of carbohydrate to the apex was, however, unaffected.

Ethylene gas has been shown to influence flowering but there is no evidence that it interacts with photoperiod. There are few studies on the effect of daylength on the rate of endogenous ethylene evolution and there appears to be little correlation between photoperiod-dependent changes in ethylene evolution and physiological responses to daylength (see Appendix II). For example, ethylene evolution increased following

TABLE 7.9 The effect of photoperiod on ethylene evolution in the SDP *Chenopodium rubrum* and the LDP *Spinacia oleracea*.

	Photoperiod	Ethylene	% Flowering
Chenopodium rubrum (intact plants)	Continous light	1516.8[a]	0
	3 SD	880.8[a]	100
	3 SD + night-break	1418.4[a]	0
Spinacia oleracea (leaves)	Continuous light	113.8[b]	
	SD	52.4[b]	

Ethylene evolution was measured in plants receiving LD (continuous light), SD (12 h light/12 h dark, *Chenopodium*; 8 h light /16 h dark, *Spinacia*), or SD with a night-break in the middle of the dark period. Spinach plants flowered only in LD and *Chenopodium* only in SD with uninterrupted dark periods; the effect of LD to increase ethylene evolution was, therefore, not correlated with the flowering response.
[a] ng g^{-1} in 24 h; [b] μg^3 g^{-1} in 24 h.
Data for *Spinacia* from Crevecoeur *et al.* (1986); *Chenopodium* from Macháčková *et al.* (1988).]

transfer to inductive conditions in the LDP *Spinacia* but decreased in the SDP *Chenopodium* (Table 7.9). In both species, the amount of ethylene produced was greater in LD than in SD. The photoperiod-dependent changes in ethylene production in *Spinacia* appeared not to be associated with the onset of flowering since the increase was also observed when previously induced plants were transferred to continuous light (Crevecouer *et al.*, 1986). Similarly, in the quantitative SDP chrysanthemum, there was no interaction between photoperiod and the inhibition of flowering by ethylene treatment. Since ethylene is a gas, it is most unlikely that it is part of any floral stimulus; this does not, however, exclude precursors of ethylene which could be transported from the leaf.

ABSCISIC ACID AND RELATED INHIBITORS

Abscisic acid (ABA) was first isolated as an inhibitor of growth from woody plants and as a substance which accelerated petiole abscission in cotton explants. The observation that ($\pm$) ABA was able to mimic the effect of SD in some daylength-sensitive plants suggested a possible role for endogenous ABA in the photoperiodic control of flowering. The application of ABA stimulated flowering in several SDP (e.g. blackcurrant, strawberry, *Chenopodium rubrum*, *Lemna paucicostata* and *Pharbitis*) and delayed, or inhibited flowering in the LDP *Spinacia*, *Lolium*, carnation and petunia (Vince-Prue, 1985). ABA also induced flowering when applied to excised shoot tips of *Perilla* in LD, although the flowers were abnormal, resembling those produced with threshold induction (Purse, 1984). The fact that the application of ABA promoted flowering of SDP in LD led to the suggestion that ABA might be the floral stimulus produced during SD induction. However, in subsequent experiments under strictly non-inductive conditions ABA was not able to induce flowering in SDP, although it enhanced the flowering response of *Pharbitis* and *Chenopodium* when plants were partly induced. Moreover, the application of ABA was inhibitory to flowering in the SDP *Kalanchoë* when injected into induced leaves (Schwabe, 1972), although it increased flower number when supplied to nodal explants maintained in

SD (Dickens and van Staden, 1990). It seems unlikely, therefore, that endogenous ABA is generally involved in the induction of flowering by SD.

Although ABA was inhibitory to flowering in the LD species mentioned, many others were not affected and ABA seems to have no general function as an inhibitor of flowering in LDP. A role for endogenous ABA in the control of flowering in *Lolium temulentum* was suggested by the fact that water stress both inhibited flowering and increased the content of ABA in the apex (King and Evans, 1977). However, since the contents of ABA and xanthoxin in leaves and shoot apices were not changed in any consistent way by daylength (King *et al.*, 1977), it was concluded that the photoperiod response in *Lolium* (which has an inhibitory component from SD leaves) is not due to either of these two inhibitors. In the LDP *Spinacia*, the endogenous levels of ABA were higher in LD than in SD, while xanthoxin remained unchanged, and there appeared to be no function for either substance in the photoperiodic control of flowering in this plant, despite the inhibitory effect of applied ABA (Zeevaart, 1974). Thus, in general, the presently available information does not suggest that ABA is a component of the **antiflorigen** produced by some LDP under non-inductive SD.

The postulated role of ABA as a promoter of flowering in SDP is also not consistent with physical measurements of endogenous ABA levels, which were higher in LD in the SDP chrysanthemum and strawberry, and did not differ in extracts from induced and non-induced leaves of *Perilla* (see Appendix II). Thus there is little evidence to suggest that changes in endogenous ABA or xanthoxin content are involved in the photoperiodic regulation of floral initiation. It remains possible that ABA might play a role in flowering if it were produced at a constant rate, irrespective of daylength, its effect being overcome in LD by the presence of increased amounts of other compounds. There is some evidence for this in the facultative LDP *Arabidopsis*, where mutants deficient in, or insensitive to ABA flower early in SD but not in LD (Bernier *et al.*, 1993); part of the LD effect may, therefore, be to overcome an inhibitory effect of ABA.

The site of action on flowering has not been investigated in most cases. However, in *Lolium*, where water stress both inhibited flowering and increased the ABA content in the shoot apex, the site of action is presumed to be at the apex, as would be expected if ABA were part of a complex flower-regulating stimulus. The finding that a brief water stress leads to a dramatic increase in the level of ABA in the leaf of *Lolium* some 8–12 h before there is an increase in the apex indicates that the ABA originates in the leaf and is subsequently transported to the apex (King, 1976). Although there is little evidence to suggest that ABA content of leaves is under direct photoperiodic regulation, the elevated levels that are sometimes seen in LD may be associated with the development of thinner leaves (see Chapter 13), leading to a greater likelihood of water stress.

DOES FLORIGEN EXIST?

The idea of **florigen**, a substance which is generated by leaves and is supposed uniquely to evoke flowering at the shoot apex, underlies many considerations of the biochemistry of floral induction and evocation. Much of the experimental evidence in support of such a floral stimulus comes from physiological experiments on plants where flowering is controlled by exposure to appropriate daylengths but the idea has been extended to include plants where flowering is induced by other means. Thus, the original and still sometimes accepted concept of florigen is of a unique chemical substance which evokes flowering in a wide range of species and in all photoperiodic categories. However, florigen continues to remain a physiological concept rather than a chemical entity. Its elusive nature, together with the many effects of naturally-occurring growth regulators has led several workers to question whether it is a single substance (Cleland, 1978; Evans, 1971; Bernier, 1986). Certainly, other morpho-genetic processes, such as the initiation of roots and buds on callus, require the interaction of two or more hormones and, as we have seen, there is evidence that the auxin/cytokinin ratio can control the switch between vegetative and floral buds in florally induced tobacco tissue. Similarly, floral evocation at vegetative shoot meristems may depend on a particular ratio and/or temporal sequence of substances which are not themselves unique for flowering.

Despite the failure to characterise florigen, there is much circumstantial evidence for the existence of a transmissible substance, which is needed for floral evocation in a wide range of different plants. There are several reasons which might account for the failure, so far, to isolate and characterise such a substance; these include lack of suitable assay method, use of the wrong extraction solvents, difficulties of reintroduc-tion into the plant at the right site and in the right quantity, lability, low concentration in the extracts, the presence of inhibitors from non-induced leaves of test plants or rapid breakdown in such leaves, and failure of phloem loading in leaves in non-inductive daylengths. In most attempts to extract a floral stimulus it has been assumed that it is a simple low-molecular-weight compound; however, preliminary attempts to look for larger compounds (such as soluble proteins and polypeptides) have also met with little success. A further possibility is that action of the floral hormone requires the presence of a complementary stimulus, such as a giberellin or cytokinin.

Chailakhyan's original florigen concept (1936) was replaced in 1958 by a hypoth-esis in which he proposed that flowering is regulated by a stimulus consisting of two components, namely gibberellins and a hypothetical substance, or group of substances, called **anthesins** (Chailakhyan, 1982). It was supposed that LDP produce anthesins in all daylengths but GAs only in LD, while SDP produce sufficient GAs in any photoperiod but anthesins only in SD; finally, the formation of florigen in the shoot apex requires both GA and anthesin. This scheme could also account for plants with a dual daylength requirement, if it assumed that LD are needed for GA production and SD for anthesin production, although it does not explain why the obligatory sequence should be different in LSDP and SLDP. Moreover, GA has been shown to substitute for LD in the LSDP *Bryophyllum* and for SD in the SLDP *Scabiosa succisa* (see Table 7.5), while in the SLDP *Bromus* and *Poa*, GA prevents the primary induction of flowering in SD (Heide *et al.*, 1986a). It is also evident that interaction at the apex cannot occur in the LSDP *Cestrum nocturnum* since LD and SD must be given to the same leaf (Sachs, 1985), although this is not the case in *Bryophyllum* (see below). The idea that GA limits the flowering of LDP in SD, while anthesin limits the flowering of SDP in LD, requires that donors of one photoperiodic response type would evoke flowering in receptors of the other type, even when the donor itself is in non-inductive daylengths. However, in the majority of cases that have been examined, donors could only evoke flowering in receptors of other photoperiodic classes when they themselves were induced. There appear to be only two reported exceptions (Lang, 1965; Zeevaart, 1958). Some flowering response was obtained in Maryland Mammoth receptors (SDP maintained in LD) when grafted to vegetative shoots of *N. sylvestris* (LDP maintained in SD), which is surprising in view of the strong evidence for antiflorigens from non-induced leaves of *N. sylvestris* (see Table 6.15). In the LSDP *Bryophyllum*, flowers formed on receptor shoots situated between vegetative parts of the graft which were in SD and LD respectively. It was suggested that, in this case, flowering was evoked by GAs from LD- and anthesins from SD-treated leaves (Chailakhyan, 1975) but another and more likely explanation is that GA from LD leaves moved into SD leaves, thereby resulting in the production of florigen (Zeevaart, 1976). Thus the anthesin/GA hypothesis does not appear to be generally applicable to different photoperiodic response groups and, in any event, there is no information on the chemical identity of the putative anthesins. Even so, the possibility that one or more additional substances may be needed for the synthesis, transport, or action of florigen is not excluded. An appropriate supply of hormones and nutritional factors is necessary for the initiation, differentiation and growth of floral parts and, in many species, these further stages of floral development are also under photoperiodic control (see Chapter 10).

Several hypotheses for the control of floral evocation have been put forward during the past 70 years. These include control by a single specific flower-promoting hormone, a relatively simple stoichiometric interaction between two substances, **florigen** and **antiflorigen**, a complex multicomponent stimulus with different factors being limiting in different conditions, and a nutrient-diversion theory that assumes only a secondary role for hormones. If there is a specific florigen, its identity and even the class of substance remains a mystery. The idea of a multicomponent stimulus is attractive and it has, for example, been suggested that the floral stimulus may consist of a mixture of known growth regulators. However, there is no evidence that any of the known plant growth substances, or any combination of them, can replace the floral stimulus and evoke flowering in a wide range of plants in all photoperiodic categories. Even in *Lemna*, where flowering can readily be modified by several growth regulators, the endogenous levels were not correlated with flowering and it was concluded that changes in hormone levels were of secondary importance in the photoperiodic control of flowering (Fujioka *et al.*, 1986b). No plant growth regulator was able to substitute for photoperiodic induction in test plants of the qualitative SDP *Chenopodium rubrum*; their action on flowering was strictly stage-dependent and rather non-specific growth effects (such as inhibition of cell division in axillary meristems by auxin and enhancement of cell division by CK) resulted in stimulation or inhibition of flowering depending on the stage of floral transition when they were applied (Ullman *et al.*, 1985). In contrast, extracts from induced tobacco plants were able to cause flowering in strictly non-inductive conditions, suggesting the participation of an unknown compound. This does not mean that growth substances do not play a role in the overall flowering process, but indicates that they are probably non-limiting for flowering in most plants. Nevertheless, many are under photoperiodic control and this, together with their numerous effects on flowering, makes it likely that they are at least modifying factors for floral evocation and expression. All classes of known growth regulators or their precursors are transported over long distances and thus could, in theory, be part of a transmissible signal exported from the leaf.

Studies of the changes that occur at the shoot apex in response to the arrival of the floral stimulus following photoperiodic induction have indicated that the initiation and growth of floral primordia is probably controlled by a complex system of interacting factors including, among others, gibberellins and cytokinins (Bernier, 1986). Moreover, the different factors may act in sequence since some of the changes seem to occur before the floral stimulus arrives at the apex. For example, in the SDP *Pharbitis*, a first wave of enhanced uridine incorporation into RNA at the shoot apex (which was 2-fold greater than in the LD controls) occurred even before the end of the critical dark period (Vince-Prue and Gressel, 1985). An even more extreme situation is seen in *Sinapis*, where bark-ringing studies indicated that a leaf signal was exported to the roots before the 12th hour of an LD and an increase in cytokinins was detected in root exudates at the 9th hour (i.e.,within 1 h of extending the 8 h SD, Bernier *et al.*, 1993). The role of these early increases is not immediately evident since they occurred under conditions that are not inductive for flowering. In some cases, changes normally associated with flowering can be caused by treatments which do not themselves bring about floral initiation. For example, in *Sinapis alba*, one of the earliest events at the shoot apex following photoperiodic induction is an increase in mitotic activity and any treatment which abolishes this also abolishes flowering. The application of CK to the shoot apex resulted in an increase in mitotic activity similar to that caused by exposure to a single LD. However, the long day induces flowering while the cytokinin treatment does not. These results have led to the proposal that cytokinin is a component of the floral stimulus, although it is insufficient alone to evoke flowering (Havelange *et al.*, 1986). If the floral stimulus does have more than one component (acting either sequentially or together) then, in a given species, all factors will not necessarily be absent in conditions that do not allow flowering. This could explain the wide range of substances and conditions that can cause flowering in different plants. Bernier (1986) has emphasised that floral induction is achieved by different sets of environmental factors, often in the same plant, and proposed that the idea of a specific leaf-generated stimulus is essentially erroneous in the general context of floral induction. However, the existence of a leaf-generated signal in plants where induction is achieved through exposure to the appropriate photoperiod can hardly be questioned, and the identity of this signal is presently unknown. Moreover, it is evident from grafting experiments that, at least in some cases, the resultant transmissible product from leaves is the same stimulus (or is functionally equivalent), despite the fact that induction may be achieved via different means (Table 6.10). There is, moreover, no *a priori* reason to assume that a floral stimulus is necessarily complex. Although flowering itself is a complex morphogenetic process and is likely to involve a number of regulatory substances, the role of the floral stimulus may only be to control the time at which it begins and the rate at which it progresses; thus, a simple molecule might be sufficient.

The best evidence for a multifactorial stimulus for flowering comes from detailed studies of photoperiodic induction in the quantitative LDP *Sinapis alba*, which is induced to flower by exposure to a single LD. Increases (often transient) in a number of different components exported from the leaf and/or reaching the apical bud have been detected during, or immediately following induction. These include sugars, cytokinins, polyamines and calcium ions, while auxin levels in the apical bud have been shown to decline (Bernier *et al.*, 1993). Given that the application of CK and the elevation of sucrose levels by high-intensity light can bring about a number of the changes in the shoot apical meristem that are normally associated with the LD induction of flowering, it is reasonable to conclude that CK and sugars are components of a multifactorial signal for flowering in this LDP. However, neither sugars nor CKs (nor both together) can evoke flowering in SD; moreover GA is not florigenic for *Sinapis* in SD, nor is any other substance tested. Thus, even if the transition to flowering depends on several regulatory factors, an essential component (florigen?) of the daylength-dependent stimulus generated by leaves remains, as yet, unidentified.

One of the earliest theories for the control of flowering was that it depended on the carbon:nitrogen ratio, flowering being favoured by a high C/N ratio and vegetative growth by a high N/C ratio. The idea is generally attributed to a paper by Kraus and Kraybill in 1918, although their work did not specifically address the question (Cameron and Dennis, 1986). Debate on the role of carbohydrates in the initiation and development of flowers has continued since the early part of the century and several 'nutritional' hypotheses of different kinds have been put forward. For example, in chrysanthemum, the initiation of floral primordia appears to occur when the apex has reached a critical size (Horridge and Cockshull, 1979) and so might be a relatively simple consequence of apex enlargement which, in turn, could arise from the action of a number of factors including nutritional ones. The correlation of apex size with flowering is not invariable, however, and it is evident that apex enlargement alone is not an adequate explanation in many plants; in some cases, the apex may actually decrease in size during floral initiation while, in *Chenopodium rubrum*, older plants with larger apices are less sensitive to SD induction than younger ones (Seidlova and Opatrna, 1978). In excised apices from non-induced *Pharbitis* plants, however, floral bud formation was promoted by a high concentration of sucrose and also by a low nitrogen concentration (Ishioka *et al.*, 1991b) indicating that nutritional conditions can affect the morphogenetic pathway taken by the apex.

A related suggestion has been that the transition to flowering is a consequence of an enhanced flow of nutritional factors to the apex. A specific **nutrient-diversion hypothesis** is that photoperiod controls flowering by altering the pattern of nutrient movement such that, in favourable daylengths, nutrients are diverted to the shoot apex where floral evocation results (Sachs, 1977, 1987). In this context, it is evident that some of the effects of growth regulator application result from enhanced transport of substances to the apex, or of a restriction of transport to the apex; for example, the promotion or inhibition of flowering in *Pharbitis* by CK application to non-induced or induced cotyledons respectively (Ogawa and King, 1979b). Moreover, daylength has been found to influence the partitioning of assimilates in both LDP and SDP (see Chapter 13). Nevertheless, it is evident from the results of many different kinds of experiment that nutritional factors alone cannot account for the daylength-dependent switch to flowering. As discussed earlier in this chapter, studies of the effect of photoperiod on the carbohydrate content of leaf exudates have demonstrated that, although an increased supply of carbohydrate may be crucial for floral development, it is not alone sufficient to trigger the transition from vegetative growth to flowering (Houssa *et al.*, 1991; Lejeune *et al.*, 1991). Furthermore, although the shortening of phases of the cell cycle at the apex of LD-induced plants of *Sinapis* was better mimicked by high-intensity light together with CK application than by either treatment alone, plants remained vegetative (Bernier *et al.*, 1993). In another approach, it was shown that GA_3 and $GA_{4/7}$ have equal effects to mobilise assimilates to potential cone buds but only the latter promoted flowering (Pharis *et al.*, 1989). Moreover, many of the daylength-dependent changes in carbohydrate distribution and concentration observed in normal clover plants also occur in mutants which are not initiating floral development (Jones, 1990). Any nutritional hypothesis also needs to explain why, in *Xanthium*, a single leaf in SD results in flowering whereas all the LD leaves have no effect. Similarly, it is difficult to understand how assimilate supply can account for the fact that a single leaf of *Nicotiana sylvestris* in LD is sufficient to cause flowering in Maryland Mammoth tobacco, whereas all the leaves on the Maryland Mammoth plant (more than 20 in some cases) are ineffective. Such results make a simple nutritional hypothesis unlikely although, once evocation has occurred, the developing flower is a strong sink and nutrients are diverted to it.

Attention has been drawn to the fact that, unlike the situation in animals, developmental processes in plants do not seem to be strictly dependent on the presence or absence of highly specific molecules and it has been argued that there is no reason to suppose that the transition to flowering is different. In support of this point, the effect of auxin/cytokinin ratios to switch the developmental pathways in tobacco TLCs between the production of vegetative and floral buds can be cited (Mohnen *et al.*, 1990). Perhaps one of the most persuasive arguments for a relatively non-specific stimulus is the enormous range of responses, in addition to floral initiation, that may be under daylength control (see Part II), indicating that specificity may be a function of the target site rather than of the leaf-generated stimulus itself. Nevertheless, despite these arguments, it is clear from many physiological experiments that the stimulus exported from a photoperiodically induced leaf differs in some as yet unidentified way from that exported from a non-induced leaf and that this difference is related to the switch between vegetative and reproductive growth at the shoot apex. It is also evident that the stimulus from an induced leaf is effective in evoking flowering in plants of all photoperiodic categories. However, the failure to isolate either **florigen** or **anti-florigen** has led to the concept that floral evocation may depend on a complex stimulus such as a particular ratio and/or temporal sequence of endogenous growth regulators, inhibitors and other factors not specifically associated with flowering, rather than on a specific floral stimulus and/or inhibitor (Bernier *et al.*, 1993). Moreover, many of these have been shown to be regulated by daylength (e.g., gibberellins, cytokinins, carbohydrates, calcium ions) and so could be constituents of a leaf-generated photoperiodic signal. The possibility that some, at least, of these factors act by sensitising the apex to small amounts of a floral stimulus should not be discounted, expecially as the effects are often seen only under marginal conditions for flowering. Unfortunately, 21 years on from the first edition of this book and despite a great deal of research effort, the identity of the daylength-dependent stimulus which is graft-transmissible and evokes flowering in plants of different photoperiodic categories is still unknown; not has its specificity for the flowering process yet been resolved.

8 Biochemical and Molecular Aspects of Photoperiodism

In trying to trace the path between cause and effect in biological systems, it is usually the case that the factors that can act as inputs to the system can be identified through experimentation and the output is identified as the measurable final response. In photoperiodism, the key input functions are the durations of light and darkness and the output functions include, for example, the initiation of flowering, the onset of dormancy, and the initiation of vegetative storage organs. Linking the two are a series of cryptic biochemical and molecular events which physiologists frequently take as a *black box*, the contents of which do not necessarily have to be known in any detail as long as a reasonably consistent relationship between the selected inputs and outputs can be established. The contents of the photoperiodism black box still remain for the large part obscure, despite the combined efforts of many scientists over a long period of time. Almost without exception these efforts have been directed at the photoperiodic regulation of flowering and relatively little is known about the intervening processes involved in the daylength control of other aspects of plant development. When dealing with the photoperiodic control of flowering it is probably more accurate to think of interconnected black boxes, one in the leaves containing the mechanism of induction and the other in the shoot apices where floral evocation is located. In the previous chapter we discussed the possible identity of the transmissible signal(s) which links these two black boxes. In this chapter, we consider the biochemical and molecular aspects of the processes of photoperiodic floral induction in the leaf and of floral evocation in the shoot apical meristems. We would expect that much of the information concerning the operation of the photoperiodic mechanism in the leaf would be relevant to all of the processes controlled by daylength.

BIOLOGICAL TIMEKEEPING

Photoperiodic induction in the leaf occurs as a consequence of an interaction between the photoreceptors which perceive light and darkness and the time-measuring system. The biochemistry of photoperception and possible signal transduction chains have

224

already been discussed in Chapter 3. The molecular basis of the rhythmic phenomena which provide the origins of photoperiodic timekeeping are much less well understood. As detailed in Chapter 2, the core of the timekeeping function of the photoperiodic mechanism appears to be a circadian rhythm in light sensitivity. Circadian rhythms are believed to be coupled to an underlying oscillatory process and entrainment of the observed rhythms occurs by phase shifting as a result of the action of light on the underlying oscillator (Fig. 2.3).

The characteristics of circadian rhythms have already been detailed in Chapter 2 and a number of mathematical and biochemical models have been advanced which describe or predict the behaviour of the underlying oscillator. Mathematical models include harmonic oscillators, relaxation oscillators and limit cycles. Of more intuitive use in a biochemical sense are negative feedback or linear models. These can predict oscillations with a 24 h periodicity and limit-cycle behaviour (Lumsden, 1991). There are, generally speaking, three types of biochemical models for underlying oscillators, based in some way on product feedback, although elements of one type can overlap into another. *Metabolic oscillators* are simple or complex metabolic pathways in which allosteric effects of substrates and products feed forward or backwards to produce intermittent fluxes through the pathway. Examples of such processes include the rhythmic glycolysis in fermenting yeast in response to adding extra substrate to cultures in metabolic steady state and the endogenous rhythms in energy charge and NADPH/NADP ratios in species such as *Chenopodium rubrum* (Kluge, 1982). *Membrane-based oscillators* involve the transport, or diffusion of inorganic ions or other solutes across a membrane barrier in a feedback regulated manner (Njus *et al.*, 1974). Evidence for membrane models includes the observation that compounds which alter membrane permeability, such as ethanol or the ionophore valinomycin, and certain ions such as lithium alter the phase or free-running period of some plant and animal rhythms (Kondo and Tsudzuki, 1980, Jacklett, 1982). *Biosynthesis oscillators* assume that the synthesis of a key enzyme or component is subject to transcriptional control, in which its mRNA level varies, or translational control from a constant mRNA level by the product of its action. There is good evidence that protein synthesis is required for the generation of some circadian rhythms. In *Gonyaulax*, for example, effects of light on advancing or delaying rhythm phase were mimicked by cycloheximide, an inhibitor of protein synthesis (Dunlap *et al.*, 1980). One could also envisage combinations of these basic patterns, such that the synthesis of a membrane transport protein could be regulated by the level of the compound whose transport it mediates. A model of this type, in which the synthesis of proteins which allow potassium ion transport is inhibited when the ion concentration reaches a threshold, was proposed by Burgoyne (1978).

The best characterised mechanism for generating and sustaining biological rhythms in animal systems is a biosynthetic oscillator and involves the period (*per*) gene, which is involved in regulating circadian rhythms in the fruit fly *Drosophila melanogaster*. The *per* gene is expressed in a circadian manner, where fluctuations in *per* mRNA abundance are influenced by its own translation product, which also cycles in abundance (Hardin *et al.*, 1992). Fluctuations in *per* mRNA are primarily controlled by fluctuations in *per* gene transcription. Also, *per* mRNA has a relatively short half-life, and sequences sufficient to drive *per* mRNA cycling are present in 1.3 kilobases of 5′ flanking sequences of the gene. These and other results indicate that the *per* feedback

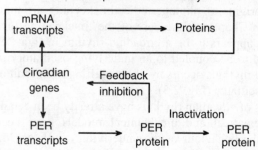

FIG. 8.1.   Schematic representation of control of circadian activity by the PER protein of *Drosophila*. PER transcription and translation leads to accumulation of PER protein. When the level is high enough PER blocks transcription of its own mRNA. Inactivation of the PER protein by folding upon itself provides a competing temperature-sensitive reaction which allows for temperature compensation, a common feature of circadian rhythms. PER acts to control the transcription of other genes, placing them under circadian control.

loop has all of the basic properties necessary to be a component of a circadian oscillator (Fig. 8.1). A similar mechanism appears to operate for the frequency (*frq*) locus of *Neurospora crassa*, which was originally identified in searches for loci encoding components of the circadian clock. The *frq* gene encodes a central component in a molecular feedback loop in which the product of *frq* negatively regulates its own transcript. The result is a daily oscillation in the amount of *frq* transcript. Rhythmic messenger RNA expression is essential for overt rhythmicity in the organism and stepped reductions in the amount of FRQ-encoding transcript sets the clock to a specific and predicted phase (Aronson *et al.*, 1994). Together, these properties define *frq* as encoding a central component in a circadian oscillator in *Neurospora*.

As the precise nature of the underlying oscillator and the process to which it is attached remain unknown in the case of photoperiodism, we can only speculate about how they might be coupled within the cell. However, for certain circadian phenomena, we have some information on the mechanism that leads to the observed rhythm. Circadian rhythms in leaflet movement in species such as *Samanea* and *Albizia* are a result of rhythmic changes in membrane transport properties which modify the lateral transport of potassium ions within the pulvinus, leading to differential alterations of turgor in the flexor and extensor motor cells (Satter and Galston, 1981). It has also been reported that light given in the dark period to *Samanea* produced an increase in the concentration of inositol phosphates and a decrease in the concentration of phosphatidyl inositol bisphosphate (Morse *et al.*, 1988). This suggests that the phosphatidyl inositol cycle, which is known to be an important signal transduction route in animals, might also be important in circadian control of leaflet movement. However, these data have not since been followed up with positive results and there must be some doubt about how important the role of phosphoinositides will turn out to be. Some evidence for the involvement of calcium in this response comes from *Albizia* where leaflet closure in response to red light, which only occurs at a particular point in the cycle, was blocked by inhibitors of calcium transport (Moysset and Simon, 1989). It is, however, difficult to distinguish between possible effects of calcium transport on the transduction of the light response and in coupling the endogenous oscillator to the establishment of light sensitivity.

There is some evidence for a role for calcium ions in either the operation or coupling of the oscillator responsible for the rhythm in light sensitivity for photoperiodic induction. Friedman *et al.* (1989) showed that spraying with EGTA, a chelator of calcium ions, inhibited the flowering response of *Pharbitis nil* only when treatments were given immediately before or early on in the inductive dark period and that the calmodulin antagonist chlorpromazine and lanthanum chloride, a calcium channel blocker, also inhibited flowering. Tretyn *et al.* (1990) also found that EGTA inhibited flowering in *Pharbitis* seedlings but, in their experiments, the time of application was not very important and inhibitors of calcium transport or calmodulin action were relatively ineffective. Consequently, these authors concluded that the action of calcium was non-specific. Takeno (1994) examined the possibility that the differences between the two earlier studies might be due to the degree of penetration of the sprayed-on chemicals through the cuticles of the cotyledons, coupled with confounding effects of EGTA directly on the apex. He developed a perfusion system for delivering the chemicals through the hypocotyls of the seedlings and with this method was able to confirm the findings of Friedman *et al.* that EGTA and other calcium inhibitors only acted before an inductive dark period. The findings of Friedman *et al.* are particularly interesting because they showed that the application of 20 mM EGTA extended the critical nightlength from about 10 h in controls to about 15 h (Fig. 8.2). Thus, calcium appears to act on the timing of responses as well as any effect it may have on the induction process. In a further paper (Friedman *et al.*, 1992) the same authors reported that the effect of EGTA given in the photoperiod varied with the time at which it was applied and was particularly inhibitory either 8 h or 16 h before the beginning of the dark period. These results remain to be independently verified but

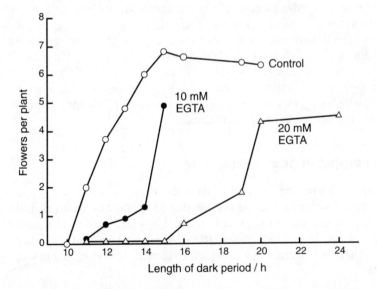

FIG. 8.2. Effect of different concentrations of EGTA applied immediately before the dark period on the flowering response of *Pharbitis nil* to dark periods of different lengths. After Friedman *et al.*, 1989.

again suggest a time-dependent effect of calcium on flowering. As the points of maximum EGTA effectiveness occurred at more or less the same time before the start of the inductive dark period, irrespective of the length of the photoperiod, the authors suggested that the EGTA was acting to initiate or rephase the circadian rhythm in light responsiveness in relation to the start of the dark period, rather than acting on a rhythm running in continuous light. However, it is difficult to understand how EGTA could operate in this way since, following several hours in continuous light, the rhythm would be restarted (or rephased) at the next onset of darkness. A possibility is that EGTA does indeed phase-shift the rhythm running in a light limit cycle in continuous light as such a phase shift would presumably alter the phase in the limit cycle at the time that plants are transferred to darkness and, therefore, the time taken to reach the inducible phase (see Chapter 2). However, since the CNL in *Pharbitis* can be modified without affecting the circadian timekeeping response to a night-break, it is necessary to carry out night-break studies in order to determine whether or not the application of EGTA during the photoperiod actually alters the phase of the rhythm at the onset of darkness.

A particularly attractive hypothesis is that the photoperiodic rhythm in light sensitivity involves the rhythmic synthesis of a phytochrome receptor or components of the signal transduction chain. Circadian rhythms in the transcription of plant genes, including the chlorophyll a/b-binding protein (Cab), Rubisco small subunit and nitrate reductase, are now well established. The first report of a rhythm in the amount of translatable mRNA in plants was by Kloppstech (1985). He found that mRNA levels for Rubisco small subunit and Cab varied in a circadian manner in peas maintained in continuous light. Circadian rhythms in Cab gene mRNA have since been demonstrated in a range of species, including wheat, barley, tomato, soybean, various beans and mustard (Meyer *et al.*, 1989). Fusion of the Cab gene promoter to luciferase and transformation into *Arabidopsis* resulted in circadian light emission which could be monitored remotely. Mutagenesis of the transgenic *Arabidopsis* plants with circadian luciferase expression offers a means of identifying genes and gene products which form part of the underlying oscillator or the oscillator coupling mechanism (Millar *et al.*, 1995a,b and see Chapter 9). While offering a powerful means of dissecting circadian rhythms in plants, it is important to note that the Cab rhythm in *Pharbitis* is not phase-regulated in the same manner as the photoperiodic light sensitivity rhythm (Fig. 8.3), which suggests that Cab gene expression and photoperiodic sensitivity are coupled to separate underlying oscillators.

PHOTOPERIODIC FLORAL INDUCTION

In addition to the biochemical and metabolic changes which form part of the day-length-perception mechanism we assume that there must be changes in the leaf which occur as part of the induction process itself. These are related to modification of the properties of the leaf which allow it to initate and sustain the plant's flowering response. There are three ways in which evidence for these changes can be sought. These are:

- testing for effects on the floral induction response of chemicals or environmental treatments which are known to modify specific biochemical processes

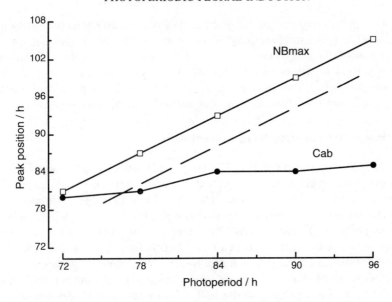

FIG. 8.3. Comparison of the effect of increased photoperiod length on the phasing in the Cab expression and NB sensitivity rhythms in seedlings of *Pharbitis nil*. The broken line shows the time of transfer from light to darkness. Data of Lumsden and Thomas, unpublished.

- comparing the biochemical or molecular fingerprints of induced and non-induced tissues in order to identify specific components whose changes are correlated with inductive treatments
- testing for floral induction responses in mutant plants with lesions in specific biochemical pathways.

Unfortunately, there are major problems in interpreting results from such approaches. One is that, at the present time, induction in the leaf can only be inferred from the observed flowering response, unless grafting experiments are carried out. Since the realisation of flowering can be influenced at many steps in the overall process (induction, loading and/or transport of the stimulus, apical events of evocation and/ or initiation) it may be difficult to determine which step is being affected by applied chemicals or particular biochemical lesions. This problem does not arise when leaf changes are directly investigated but, here, the problem is to ascertain that any observed change is implicated in the induction process.

Biochemical modifications

The idea of treating plants with chemicals which modify some particular aspect of metabolism and then testing whether that treatment affects the photoperiodic flowering response is attractive, at least superficially. However, interpreting any results is fraught with problems. Firstly, there is the question of whether and how efficiently, chemicals are taken up by the plant. Lack of uptake or concentration to unphysiological levels can both give misleading results. Secondly, if compounds taken up are transported within the plant, effects can be remote from the site of application and unrelated to the process of induction in the leaf. Thirdly, even if chemicals can be shown to be acting in the leaf, it is not always possible to determine whether they are directly affecting induction or some other process (e.g. energy supply or transport) necessary for, but not specific to the induction response itself. There is, moreover, the additional problem of distinguishing between effects on timekeeping and effects on induction. It is not surprising, therefore, that no clear picture has yet emerged from the many published studies on the effects of metabolic modification on photoperiodic floral induction.

Photosynthesis and carbohydrate metabolism

As already discussed in more detail in Chapter 1, it is unlikely that photosynthesis is directly involved in the photoperiodic induction mechanism. This conclusion is based on the observation that, in both LDP and SDP, induction is possible in situations where photosynthesis is essentially elimated as, for example, in *Kalanchoë*, where only 1 s of light a day is sufficient to allow induction (Harder *et al.*, 1944) and in seedlings of *Pharbitis nil*, which can be induced prior to de-etiolation (Ogawa and King, 1979). However, photosynthesis clearly contributes to the overall flowering response in many cases, as evidenced by the requirement for carbon dioxide during the inductive photoperiodic cycle. For example, in the SDP *Glycine max* and *Kalanchoë blossfeldiana* short days were reported to be less effective if carbon dioxide was not present during the photoperiods (Fredericq, 1962; Parker and Borthwick, 1940), although carbon dioxide was apparently not required in *Perilla* and *Panicum miliaceum* (Bavrina *et al.*, 1969, Fredericq, 1962). In LDP a supply of carbon dioxide during the photoperiod has been found to be necessary for induction in a number of species (Bavrina *et al.*, 1969). This requirement could be replaced to some extent by sucrose, indicating that photosynthesis is operating primarily to supply an energy source needed for the realisation of flowering. Inhibition of photosynthesis with N′-(3,4-dichlorophenyl)-N,N-dimethylurea (DCMU) also prevented flowering in the LDP *Lolium temulentum* and *Rudbeckia bicolor* and the SDP *Chenopodium rubrum* but had no effect on another SDP, *Xanthium* (Bavrina *et al.*, 1969; Cumming, 1959; Evans *et al.* 1965). Thus, although there is some indication that photosynthesis during inductive cycles may influence flowering in certain cases, it is not universally required and its participation appears to be peripheral to the photoperiodic induction mechanism itself. Photosynthetic products could facilitate export of the floral stimulus (Evans and King, 1985) and/or be involved in the processes of evocation and floral initiation at the apex.

There is a substantial amount of evidence that carbohydrates, and in particular sucrose, may have an important function in floral evocation and initiation. However, while transient increases in the flux of sucrose in phloem sap have been found to occur during or immediately following inductive treatments in both LD and SD plants, this increase has been eliminated as being a sufficient trigger for evocation, although it may be required during floral transition. The possible role of sucrose as a component of the floral stimulus which is transported from leaves to the target tissues was discussed in detail in Chapter 7.

Membrane properties

The involvement of changes in membrane properties in processes related to induction has been only been tenuously established, if at all. The idea was popular in the late

1970s and early 1980s for two reasons. Firstly, the ideas of Njus on the membrane origin of circadian rhythms (Njus *et al.*, 1974) were attractive and secondly it was also believed by many that membranes were the primary site of phytochrome action. It is now clear that phytochrome is a soluble cytosolic protein and that interaction with membranes is not essential for its action; circadian rhythms may also operate in ways that do not involve membranes. Nevertheless, even though the supporting evidence is not very great, the possible involvement of membranes cannot be excluded.

Circumstantial evidence for a role for membranes can be found in the responses of *Lemna* to various chemical treatments. *Lemna* species are small aquatic plants which show examples of both LDP (*Lemna gibba*) and SDP (*Lemna paucicostata*). Plants grow as fronds without recognisable leaves or stems. The apices are located within reproductive pockets close to the photosynthetic tissues which from the point of view of induction are the physiological equivalents of leaves. The aquatic habit and small size of *Lemna* make chemical treatments easy to carry out. However, the difficulty of separating the photosynthetic organs from apices makes the specificity of any treatment for induction rather than action at the apex quite difficult to establish. Transfer of *Lemna paucicostata* from nutrient media to distilled water inhibited flowering, particularly at the time when night-breaks were most effective (Halaban and Hillman, 1970). This could be partially reversed by incubating the plants in calcium nitrate solution or in water in which the plants had been incubated during the dark period (Halaban and Hillman, 1971). In the LDP *Lemna gibba*, transfer to distilled water during the subjective night, but not the subjective day, also inhibited flowering (Oota and Kondo, 1974). These results suggested that membrane leakage at particular times may have influenced induction.

The LD requirement for flowering in *Lemna gibba* G3 can be satisfied by giving only 5–10 min light at two defined times, designated L1 and L2, separated by 12 h darkness (Oota and Nakashima, 1978). The ionophores gramicidin and valinomycin, which enhance potassium movement, substituted for light at L1 while the light requirement at L2 could be satisfied by propranolol or salicylic acid (SA), both of which can bind to or affect the permeability of membranes (Oota and Nakashima, 1978). In further studies with *Lemna gibba* G3, Cleland (1984) found that SA promoted flowering in plants grown under 15 min of light given every 1, 2 or 4 h. The response was shown in the presence of the herbicide Norflurazon, which prevents carotenoid and consequently chlorophyll accumulation, and was observed at levels as low as 0.1 μM SA (Fig. 8.4). Optimal doses of SA caused a shift in the critical daylength in both LD and SD species of *Lemna* (Cleland and Tanaka, 1979). In the LDP *Lemna gibba* G3, the critical daylength was decreased from about 10 to 8 h while, in the SDP *L. paucicostata* it was increased from 14 to 15 h, suggesting that SA may be interacting directly with the photoperiodic induction mechanism (Cleland, 1984). However, an opposite conclusion was reached from comparative studies of photoperiodic and benzoic acid-induced flowering in the weakly responsive SDP *L. paucicostata* 151 (Takimoto *et al.*, 1987); there were no differences in benzoic acid levels in induced and non-induced plants and the effect of benzoic acid on flowering appeared to be independent of, but additive to that of daylength. Based on the fact that SA must be present continuously in the medium and is quickly inactivated once taken up, Cleland (1984) concluded that it must act during uptake into the plant, or shortly thereafter. Although these responses have been interpreted as being effected through

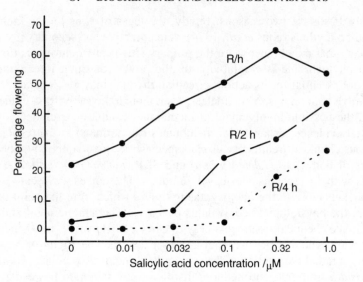

FIG. 8.4.   Effect of salicylic acid on flowering in *Lemna gibba* G3 grown in the presence of 10 μM Norflurazon and given 15 min red light every 4 hours (R/4 h), every 2 hours (R/2 h), or every hour (R/h). After Cleland 1984.

modification of membrane properties, it is now known that SA can mediate the stimulation of transcription of certain defence-related genes by pests or pathogens and, consequently, alternative explanations for its action are possible. SA has been reported to promote flowering in a few other species of both LD and SD plants but, in general, it is ineffective (see Chapter 7).

The idea that membranes act as a site of transduction for the photoperiodic signal is supported by the studies of Greppin *et al.* (1986) on the induction of flowering in the LDP, spinach. This group has identified a wide range of metabolic alterations in the leaves, including changes in bioelectric potential and adenine and pyridine nucleotides, as well as inferred alterations in energy metabolism (Bonzon *et al.*, 1985; Greppin *et al.*, 1986; Montavon and Greppin, 1985). They sought direct evidence for membrane changes by conducting a structural study of the spinach plasma membrane during photoinduction. Densitometric analyses showed that changes in the membrane were confined to an enhancement of the layer adjacent to the cell wall (Penel *et al.*, 1988). These data paint a picture of complex changes in the leaves correlated with photoinduction in spinach, but with so many differences it is almost impossible to estimate their significance, either singly or in concert.

Molecular changes

An induced leaf can be regarded as being in a quasi-stable metabolic state which enables it to direct vegetative apices to be transformed to floral structures. It has been suggested that this altered metabolic state must involve changes in the pattern of gene expression through altered transcription or translation. A simple test of this hypothesis would be to investigate the effects of transcription and translation inhibitors on floral induction. However, many of the problems in evaluating the effects of exogenously applied chemicals on induction apply equally well to protein synthesis inhibitors. Given the potential pitfalls, it is to be expected that experiments may produce inconclusive or contradictory results. Application of cycloheximide to *Xanthium* leaves was found to inhibit flowering without cycloheximide being translocated out of the leaf (Ross, 1970). Kinetic studies showed that inhibition was greatest when cycloheximide was applied either at the beginning or end of the inductive dark period, but there was rather less effect at intermediate times. This suggests a requirement for new protein synthesis at different times during the inductive cycle, but does not prove that any of these new proteins are specific for the induction process itself. Application of actinomycin D, an inhibitor of RNA synthesis, to the cotyledons of *Pharbitis* also suppressed flowering (Arzee *et al.*, 1970). In general, other inhibitors of protein synthesis, such as chloramphenicol or amino acid analogues, are not inhibitory to flowering when applied to leaves during induction. However, two amino acid analogues, ethionine and *p*-fluorophenylalanine, did inhibit flowering when applied to the leaves of some SDP. Ethionine inhibited flowering in *Xanthium* and *Pharbitis* (Collins *et al.*, 1963; Zeevaart and Marushige, 1967) and its effect was reversed by methionine. Ethionine inhibits the incorporation of methionine but could also act by being incorporated to give aberrant proteins or by sequestering ATP to form *S*-adenosyl-ethionine. The application of ethionine to *Pharbitis* cotyledons was inhibitory even at the end of the inductive night, adding further uncertainty to the specificity of its action in induction. Flowering in *Xanthium* was also inhibited by *p*-fluorophenylalanine, its effect being reversed by the corresponding amino acid, phenylalanine (Miller and Ross, 1966). Again there is uncertainty as to the specificity of the effect since *p*-fluorophenylalanine could inhibit protein synthesis or increase the levels of unincorporated phenylalanine. Furthermore, neither ethionine nor *p*-fluorophenylalanine inhibited flowering in *Lolium* (Evans 1964b) and ethionine did not inhibit the photoperiodic induction of *Perilla* (red) leaves nor prevent them acting as donors of the floral stimulus (Zeevaart, 1969c). The infiltration of 2-thiouracil (2-TU), which is incorporated into template RNA and misleads translation, into the leaves of the LDP *Hyoscyamus niger* induced flowering in SD of up to 80% of the treated plants (Eichoff and Rau, 1969). The effect of 2-TU was localised in the leaf and could not be mimicked by defoliation or by inhibitors such as DNP and azide. The authors concluded that 2-TU application led to the synthesis of a floral stimulus.

If induction does involve modified expression of particular genes this might be detected through qualitative or quantitative differences in the polypeptide or mRNA populations of induced versus non-induced leaves. However, the design of experiments to test this hypothesis can be more difficult than first imagined. The common approach has been to compare leaves given an inductive treatment with those which have not. It is then necessary to wait for the appropriate changes to occur, but as the precise nature and timing of the changes are unknown this can be a problem. For example the changes could be transient, in which case delaying sampling could miss them, or they could be cumulative, in which case they might not be detectable until some days after beginning the inductive treatment. This is less of a problem in, for example, single-cycle plants such as *Pharbitis* and *Xanthium*, where there is information from defoliation experiments on the time taken for the inductive signal to leave the leaf. Obviously any change of direct relevance to induction must have occurred before this time.

A further problem is that many plant genes are under circadian control and their expression levels at any one time are dependent on the phase of the rhythm which controls their expression. Almost inevitably, induced and control leaves will have received different light/dark regimes, which may rephase the expression patterns of many genes unrelated to induction. The criteria for identifying a gene as being related to induction must, therefore, be based on more than a change in the level of its expression at a single time point following an inductive treatment. An induction-related gene should exhibit similar changes in expression following alternative induction strategies; for example, in a SDP, the response to a short night and a long night with a short night-break should be similar and the light requirement for the regulation of the gene should be the same as that for induction. Secondly. the time-scale of the response should match the timescale for the production of the floral signal by the leaf, where this is known. It is also to be expected that changes in the expression of equivalent genes would be linked to induction in other photoperiodic species. We suggest that no gene or gene product so far identified meets these criteria.

Some early electrophoretic analyses with *Pharbitis* and *Impatiens* indicated changes in polypeptide composition of leaves or cotyledons in response to inductive treatments (Stiles and Davies, 1976; Sawnhey *et al.*, 1976) but the methods used lacked sensitivity and resolution and the findings were not taken further. The advent of two-dimensional polyacrylamide gel electrophoresis and silver-staining methodologies has increased the potential for detecting small changes in polypeptide composition (and also for generating artifacts). Lay-Yee (1986) reported that the abundance of four polypeptides between 25.5 and 44 kDa and with PI between 6.3 and 6.0 decreased with an inductive treatment in *Pharbitis*. Bassett *et al.* (1991), on the other hand, were not able to repeat this finding nor indeed find any leaf changes that might be linked to floral induction in *Pharbitis*. It has been reported that two peptides of 15 and 16 kDa disappeared from photoperiodically induced leaves of *Xanthium* and it was suggested that these may be related to the loss of inhibitor production under inductive conditions (Kannangara *et al.*, 1990). This remains to be confirmed. It is possible that changes related to induction could be quite subtle. For example, Ono *et al.* (1993) were unable to find any differences in the amount of particular polypeptides between SD-induced and non-induced *Pharbitis* but ^{35}S-methionine labelling of a 22 kDa spot (PI 7.5) was consistently lower in the induced cotyledons. Thus protein turnover may be important for induction.

A comparison of the composition of mRNA populations in induced and control tissues has been carried out in a number of species, using either *in vitro* translation or differential screening of cDNA libraries. The first study based on *in vitro* translation was reported by Warm (1984) working with the LDP *Hyoscyamus niger*. Over 200 translation products were resolved by two-dimensional polyacrylamide gel electro-phoresis and several quantitative and qualitative changes were observed between induced and non-induced leaves. The significance for induction is not yet known, however, nor have the reported changes been verified independently. *In vitro* studies with *Pharbitis* were also reported as showing changes in translation products (Lay-Yee *et al.*, 1987b) but have not been confirmed by others (e.g. Ono *et al.*, 1988). In general, the more sophisticated and reliable the experimental methods used, the fewer changes in mRNA are found. No qualitative changes and either minor or no quantitative changes were found in studies of *Perilla* (red) and *Sinapis alba* (Kimpel and

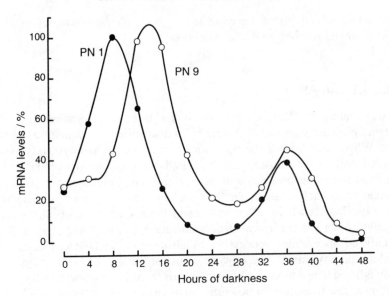

FIG. 8.5. Changes in mRNA levels for PN1 and PN9 genes from *Pharbitis nil* during an extended dark period. Seedlings were grown in continuous light for 6 h before the start of the dark treatment. Data from O'Neill *et al*. 1994.

Doss, 1989). Overall, these studies indicate that changes in mRNA composition related to photoperiodic induction are very difficult to detect, suggesting that if they occur they represent either very rare or subtle events.

The alternative and potentially more powerful approach has been to construct cDNA libraries representing the mRNA populations of induced or non-induced leaves and perform differential screening with cDNA probes derived from these mRNA populations. By plating out the library, tens of thousands of clones can be challenged with cDNA at the same time. If a clone is labelled by cDNA from one mRNA population but not from the other, it is assumed to be differentially expressed in the induced and non-induced tissues. Besides allowing much larger numbers of mRNA species to be tested, this method, if successful, results directly in cloning the gene of interest. Positive results have up until now only been obtained by one group, working with *Pharbitis nil* (O'Neill, 1989; Zheng *et al.*, 1993; O'Neill *et al.*, 1994). Several cDNAs whose abundance alters after transition to darkness were isolated. Two clones, PN1 and PN9, both showed rhythmic patterns of mRNA accumulation in darkness but, whereas PN1 was expressed maximally at about 8 h, PN9 peaked about 6 h later (Fig. 8.5). Both are expressed in leaves and shoots but not in roots and their sequences do not match those of any other genes. A further gene isolated from *Pharbitis* by the same laboratory has high sequence similarity to high-mobility group 1 (HMG1) DNA-binding proteins. This gene is expressed at higher levels in dark-grown tissues such as roots and at lower levels in light-grown tissues such as cotyledons and stems. In prolonged darkness the HMG1 mRNA showed an initial transient increase in abundance with a further peak 24 h later. As yet none of the *Pharbitis* genes can necessarily be assumed to be involved in induction. They do, however, indicate a close linkage between particular light/dark regimes and the circadian regulation of gene expression, which might be expected to be a characteristic of genes whose expression is linked to photoperiodic induction.

CHANGES AT THE APEX

A distinctive feature of plant development is that there is no separate germ cell line and, in plants growing vegetatively, there is no discernible antecedent to reproductive structures. When cells within the apex are marked with phenotypically observable mutations no precise relationship can be found between any particular cell and its developmental fate. The properties of cells within an apex depend heavily on their position relative to each other rather than on their lineage. This is shown by micro-dissection experiments where, after portions of the apex are removed, the remaining cells are able to recreate the apical structure (Sussex, 1989). Vegetative development is potentially an open-ended process in which there is a repeated reiteration of the basic module, or **phytomer**. The phytomer consists of a node with an attached leaf, and a subtending internode at the base of which there is a bud within the axil of the next phytomer. The transition to flowering involves a series of changes at the apex in which the indeterminate pattern of leaf and internode production is modified to produce a determinate floral structure. It is often possible to observe two distinct phases in the transition to flowering. In the first, changes in the pattern of apical organisation lead to an altered three-dimensional distribution of growth points to produce an inflorescence. This frequently involves an increase in apical volume and altered primordium spacing and phyllotaxis, sometimes accompanied by precocious axillary bud growth. Floral differentiation occurs in the second phase. This involves suppression of internode elongation and modification of leaf structures to form whorls of sepals, petals, stamens and carpels. Once an apex becomes fully florally differ-entiated, development becomes determinate and further growth from that apex is not possible.

The floral signal from the induced leaves (see Chapter 7) initiates the series of changes which occur at the apex. In some plants, such as *Pharbitis* and *Xanthium*, once the signal reaches the apex the full processes of floral initiation and differentia-tion occur. This is shown by defoliation experiments where the source of the inductive stimulus is removed within the first 24 h or so and yet floral differentiation still occurs days later. This pattern of development fits the **relay hypothesis** (Heslop-Harrison, 1963) in which flowering is seen as an inevitable series of events inexorably set in motion by the appropriate stimulus, each stage and its associated genes triggering progression to the next. If we accept this interpretation, only the earliest changes at the apex following inductive treatments will be specifically linked to the daylength signal. Although this is probably the case in most instances, it is worth bearing in mind that, in some plants the complete sequence from inflorescence formation through to full floral differentiation requires a continuation of the inductive daylength regime (Battey and Lyndon, 1990). Inflorescence reversion to vegetative growth was investigated in some detail for *Sinapis alba* by Bagnard *et al.* (1972). Following LD induction, plants were returned to SD at low irradiances. Under these conditions, the main shoot produced determinate side shoots with single determinate flowers followed by lateral shoots bearing several flowers with or without subtending bracts. This is the reverse of the normal sequence associated with flower initiation, i.e. suppression of axillary branching followed by the suppression of bracts. Reversion can even be found at late stages of floral differentiation. Reversion to leaf formation in the SDP *Impatiens balsamina* can be brought about at any stage of flower development by returning plants from SD to non-inductive LD (Battey and Lyndon, 1984). Transfer back to LD after different periods of SD treatments produced a range of reversion types. A short treatment with SD resulted in a zone of leaves with no, or modified, axillary structures. Longer exposures gave flower development up to and including stamens and carpels followed by a return to petal or leaf production. In species such as *Impatiens*, commitment to flower only occurs when floral development is complete or almost complete. Clearly, the daylength signal is here required throughout floral differentiation. However, the vast majority of the biochemical or molecular changes which occur will be related to the process of floral differentiation itself rather than to the arrival of the daylength signal. Again, if we wish to identify changes directly related to the floral signal from the leaves, the most likely ones are the earliest that can be detected following the inductive treatment.

Cell structure changes

There have been numerous studies on the structural changes at the apex that take place in response to the assumed arrival of a flowering stimulus. Tissues which initiate floral morphogenesis in response to the flowering stimulus have been called the **target tissues** and become sites of activity in the evoked meristem. In plants where the flowers are not terminally located, the central zone of the plant remains virtually unchanged but there is an increase in activity at the axillary bud sites. In dicotyledons, the target tissues usually include the central and peripheral zones of the apical meristem but, in grasses, the primary target areas are the bud sites in the axils of the leaf primordia (Fig. 8.6). Such structural changes might suggest that the reorganisation of growth patterns represents the primary response of the apex to the daylength signal. However the meristems of some photoperiodic plants also alter their

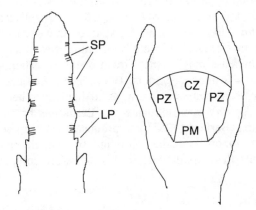

FIG. 8.6. Target tissues in the apical meristem of dicotyledonous species (right) and the shoot tip of grasses (left). CZ, central zone; PZ, peripheral zone; PM, pith-rib meristem; LP, leaf primordia; SP, potential spikelet sites. After Bernier 1971.

organisation when kept for protracted periods under conditions which do not allow flowering (Bernier, 1988). Some of the changes are similar to those occurring during the floral transition, although they are usually slower and less pronounced. Evocation can be considered to be complete at the time at which the meristem is irreversibly committed to flower. In *Sinapis* the point at which inhibitors can no longer prevent flowering occurs at the time when most histological and morphological changes begin and more than 12 h before the first structural signs of flower initiation can be observed (Bernier, 1986; Bernier *et al.*, 1981). The key processes in evocation are, therefore, likely to be the earliest cellular and molecular events rather than structural changes.

Cellular changes

Cellular changes in the apex following an inductive treatment can include increases in the rate of cell division and also its synchronisation (Bernier, 1988). In *Sinapis* exposed to a single LD, cell proliferation is stimulated in two waves, the first occurring about 26–30 h after the start of the LD and the second at about 62 h, coinciding with the initiation of flower primordia (Bernier *et al.*, 1967). The cell doubling time is reduced from 206 to 29 h (Bodson, 1975), caused by both a shortening of cell cycle duration from 86 to 32 h and an increase in the growth fraction (i.e. the proportion of cells cycling rapidly) from 30–40% to 50–60% (Gonthier *et al.*, 1987). The time interval between the two peaks of mitosis is the same as the shortened cell cycling time, indicating that synchronisation occurs. Although synchrony can be observed in *Silene* and *Lolium* when induced by LD (Francis and Lyndon, 1979; Ormrod and Bernier, 1987) and is not observed in shoot meristems other than during the floral transition, it does not seem to be an essential event for floral transition. In *Silene*, suppression of synchrony during floral transition did not prevent flower formation (Francis and Lyndon, 1985) and the authors proposed that synchronisation could be a secondary effect of the abrupt arrival of the inductive floral stimulus at the apex. In *Silene*, a rapid, transient shortening of the cell cycle and accumulation of G2 cells was observed within an hour of the start of the first of seven LD extension treatments (although a day-extension of 1 h would not itself lead to flowering). Interpolation of a 20 min dark period at the start of the extension inhibited both flowering and the G2 increase. Cell cycle shortening was unaffected and it was suggested that the G2 increase may be the critical event for establishing competence for responding to the floral stimulus (Francis, 1987). The differences between *Sinapis* and *Silene*, together with unsuccessful attempts to identify other common features (e.g., changes in apical growth rates) indicate that there may be no essential cellular changes during evocation. Bernier (1988) has proposed a more general requirement, namely that a sufficient proportion of the cell population should be in a critical phase of the cycle because only the cells that are in this phase are competent to respond to the inductive stimulus.

Metabolic changes

The size and complexity of shoot apices make it difficult to make accurate measurements on individual metabolic components during evocation. However, careful studies with several plants, including *Sinapis*, *Spinacia* and *Xanthium* suggest that increased metabolic activity, as indicated by increased levels of sucrose, ATP, mitochondrion number and energy charge, is an early consequence of inductive treatments (Bernier *et al.*, 1981; Bodson, 1985a; Bodson and Bernier, 1985; Bodson and Remacle, 1987; Diomaiuto-Bonnand and Le Saint, 1985; Havelange, 1980; Kanchanapoom and Thomas, 1987). In contrast, ATP levels in *Pharbitis* undergo a transient decrease (Thigpen and Sachs, 1985) and, in *Xanthium*, changes in soluble sugars, invertase and fumarase are not seen after an inductive SD treatment (Bernier, 1988). There is one report of an increase in the activity of pentose phosphate pathway enzymes in spinach (Auderset *et al.*, 1980) but the lack of any additional work in this, or any other species, makes it difficult to assign any significance to this observation. Sucrose accumulates very early in the apical meristems of *Sinapis alba* during and following an inductive treatment (Bodson and Outlaw, 1985) this precedes the activation of mitosis or other energy-dependent processes. However, increases in apical sucrose levels can be brought about by increasing the irradiance of a non-inductive SD and increased sucrose alone is insufficient to cause evocation (Havelange and Bernier, 1983 and see Chapter 7). Bernier's group has also carried out measurements of auxin and cytokinin levels in *Sinapis* apices following inductive treatments. They report that there is a transient increase in cytokinins and a decrease in auxins and, consequently, a significant alteration in the cytokinin/auxin ratio at the apex (Sotta *et al.*, 1992). This may be important for the realisation of flowering since a high cytokinin/auxin ratio has been found to be necessary for floral expression in the thin-cell-layer test system derived from induced tobacco plants (see Chapter 7). However, although some, but not all of the apical changes associated with LD induction in *Sinapis* occur following the application of CK to the apical bud, these are not sufficient to cause flowering.

Molecular changes

Under the right conditions, applying inhibitors of nucleic acid synthesis to the shoot apex inhibits floral evocation with little effect on vegetative growth. In *Sinapis alba* the application of 2-TU to the terminal bud was inhibitory between 12 and 20 h after the beginning of an inductive LD but had no effect after the 28th hour (Fig. 8.7). Injection with 5-fluorouracil (5-FU) near the apex had a similar effect in *Lolium*, although the time of maximum effectiveness was a little later (see Fig. 8.7). In both cases orotic acid, a precursor of both DNA and RNA, reversed the response but thymidine, which is DNA-specific, did not, indicating that RNA and not DNA synthesis was involved. Similar evidence can be found for RNA synthesis being necessary for evocation in SDP. Application of 5-FU during the first part of an inductive night inhibited flowering in *Xanthium* (Bonner and Zeevaart, 1962) and application of 2-TU or actinomycin D before or during an inductive dark period inhibited flowering in *Cannabis sativa*, *Pharbitis nil* and *Chenopodium amaranti-color* (Heslop-Harrison, 1960; Galun *et al.*, 1964; Watson and Matthews, 1966).

Inhibition of DNA synthesis by 5-fluorodeoxyuridine (5-FDU) was also very effective in suppressing flowering in *Xanthium* and *Pharbitis* (Bonner and Zeevaart, 1962; Zeevaart, 1962a). The inhibition in *Pharbitis* was greatest at about or prior to the time of the arrival of the floral stimulus at the apex (Fig. 8.8) indicating that the floral stimulus acts on cells with replicating DNA, which are presumably actively dividing. In

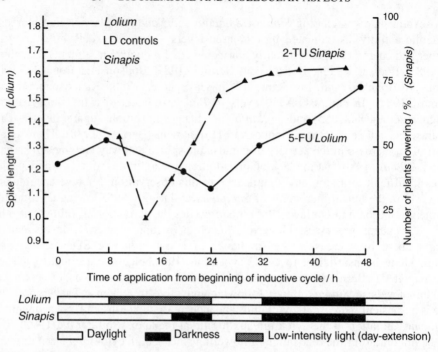

FIG. 8.7.   The inhibition of flowering in LDP by inhibitors of RNA synthesis applied at different times during or after a single inductive LD. 5-FU (0.1 ml 5 mM) applied to *Lolium temulentum* by injection near the apex. 2-TU (0.2 ml 7 mM) applied to the terminal bud of *Sinapis alba*. The irradiation schedules are indicated below the abscissa. After Vince-Prue 1975 (data of Evans, 1964b and Bernier, 1969).

Xanthium, the step requiring RNA synthesis apparently precedes the step requiring DNA synthesis. Reversal of the inhibitory effect of 5-FU at the beginning of the inductive dark period was only possible with orotic acid if it was applied within 8 h. In contrast, inhibition by 5-FDU given at the same time could be reversed by thymidine up to the end of the inductive dark period. Similarly in *Lolium*, events requiring DNA synthesis occur later than those requiring RNA synthesis. Injections of actinomycin D or 5-FU near the apex were most inhibitory to flowering early in the morning of the day following an inductive LD. The injection of 5-FDU had little effect at that time but was inhibitory when injected on the third or fourth day after the inductive treatment (Evans, 1964b).

Taken together the evidence suggests that, while DNA replication and hence cell division is required for the transition of the vegetative apex to reproductive growth, it is not the earliest event. It has been proposed that the requirement for DNA replication is a consequence of a requirement for cell division at the time when the apex is being structurally reorganised (O'Neill, 1993). The key primary events appear, therefore, to be RNA and protein synthesis, which are likely to be tied to an alteration in the pattern of gene expression. Basing conclusions simply on the effects of inhibitors of protein synthesis is, however, unreliable. In *Chenopodium rubrum*, for example, actinomycin D failed to inhibit ^{3}H-uracil incorporation into RNA even though it inhibited floral evocation (Seidlova, 1970). Thus its effect must have been mediated through some process other than RNA synthesis.

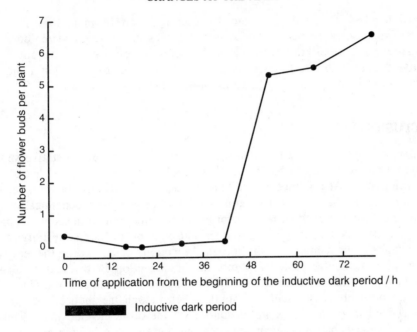

Number of flower buds per plant

Time of application from the beginning of the inductive dark period / h

■■■■■■ Inductive dark period

FIG. 8.8. Inhibition of floral induction in *Pharbitis nil* by FDU application to the plumules at different times during or after a single inductive dark period 2×10^{-5} μmol 5-FDU applied per plant. After Zeevaart 1962a.

Some evidence for the synthesis of new proteins at the apex following an inductive treatment was produced by Lyndon *et al.* (1983). They labelled apices of *Sinapis alba* with ^{35}S-methionine and separated labelled polypeptides by two-dimensional poly-acrylamide gel electrophoresis. Changes in the polypeptides being synthesised were observed after about 50 h of inductive LD treatment. In this sort of experiment it is difficult to distinguish qualitative effects (i.e. the synthesis of new proteins in induced apices) from quantitative effects, such as an increase in the rate of synthesis of particular proteins from levels which were too low to be detected by the labelling method used. Analysis of the total protein composition of *Pharbitis nil* apices following induction did not reveal the presence of any new proteins (Araki and Komeda, 1990) but showed quantitative changes in five polypeptides out of about 1000 which could be resolved.

There have, as yet, been few studies on changes in the mRNA populations of shoot apices following photoperiodic induction. One problem is the small amount of tissue available from apices and hence the low yields of mRNA for cDNA synthesis. Kelly *et al.* (1990) solved this problem by using a polymerase chain reaction protocol for amplifying the amount of cDNA produced from induced tobacco shoot apices. In addition they enriched the amount of apex-specific cDNA through a subtractive hybridisation procedure and, after this, were able to identify some evocation-related cDNA species. Subtractive hybridisation was also used by Melzer *et al.* (1990) to isolate floral initiation-related genes from induced *Sinapis alba*. They placed their cloned genes into two groups based on the patterns of change in transcript abundance. Group I genes were present at low levels in the vegetative apex and increased to maximal levels between days 2 and 10 after the start of induction. They were expressed mainly in the peripheral zone, sometimes in well-defined regions, which is consistent with their involvement in the primary response of the apex to the inductive stimulus. Group II genes, on the other hand, were expressed at a later stage and thought to be primarily related to floral differentiation.

CONCLUSIONS

Photoperiodic induction and the consequent floral evocation of vegetative shoots is mediated through a series of biochemical and molecular changes in the leaves and at the growing apex. At this time the cellular processes which form the components of the underlying mechanisms still remain to be discovered. The biochemical origin of the biological rhythm(s) that forms the basis for photoperiodic timekeeping is not yet clear and neither is it certain how photoperiodic rhythms of light sensitivity relate to other rhythms of biochemical or physiological function which occur in the leaves. The molecular or metabolic basis of photoperiodic induction in the leaf is also still unknown. The evidence suggests that differences between induced and non-induced leaves are probably both small and subtle, even though the induced state may be highly stable, as in *Perilla*. Careful studies on the apex indicate that changes in patterns of gene expression occur and may be the key step in commitment of the apex to flower. However, alterations in patterns of cell division and growth may also be key processes in the vegetative/floral transition and form targets for the floral stimulus. This chapter has dealt with a body of work which is based mainly on the correlative biochemistry of circadian rhythms, induction and evocation. A pervasive theme has been the difficulty of establishing that a particular biochemical event or change is actually relevant to the physiological problem being addressed. Although the information produced will undoubtedly be of great value in gaining a full understanding of the component processes of photoperiodic induction and evocation there is, at the moment, an informational log-jam. As we shall discuss in Chapter 9, the best prospect for breaking through this impasse is offered by the increasing attention that is being given to the genetics and molecular genetics of photoperiodic induction.

9 Genetic Approaches to Photoperiodism

INTRODUCTION

In the face of the complex physiological processes and developmental phenomena which constitute photoperiodic responses, it is not possible to dissect fully the underlying mechanisms using only physiological and biochemical approaches. For example the changes associated with photoperiodic induction in the leaves are likely to involve a modification of only a minute portion of the biochemical behaviour of the leaf and are not discernible against the biological noise within the system. Similarly, isolation of signalling compounds linking inductive changes in the leaf with changes at the apex has defied an analytical approach, for reasons that could include that they are too small to detect, involve compounds which have not been evaluated or involve mixtures of signalling compounds which would not be mimicked by chance in experimental formulations. For this reason, attention has been directed more recently towards genetic approaches in model systems such as *Arabidopsis*. The rationale is that if mutations can be identified which alter a particular response of plants, and can be genetically mapped, the genes responsible can be physically isolated and their sequence may indicate the function of the gene product and highlight the biochemical processes which are involved in the daylength response.

There are a number of potential problems with the use of genetic approaches to photoperiodism. In the first place it is important to distinguish between genes which are linked directly to the target process and those which are related to its regulation by daylength. In the most obvious example, a plant lacking a gene required for the transition from vegetative to floral development would be unable to respond to a daylength signal, but would also be unable to respond to other signals (e.g. vernalisation or applied growth substance). Although knowledge of such genes is obviously important for a full understanding of the overall process, they are not strictly photoperiodism genes which are best defined as 'genes involved in the perception of daylength signals and their transmission to particular developmental processes'. A second point concerns a situation where isolation and sequencing of a gene identifies a product that may be involved in a range of developmental processes. This might include genes for common signal transduction intermediates such as those involved in calcium/calmodulin signalling pathways or maybe putative transcription factors. One would expect that such genes would impinge on photoperiodism as they do on almost every aspect of development. Photoreceptor genes may fall into this category if gene products are involved both in daylength-sensing systems and also mediate changes in flowering time through light-dependent modification of development. A third feature of this approach is that it is restricted to species in which map-based gene cloning is feasible or to those in which genes may be 'tagged' by tDNA or transposable elements. For species where gene isolation is difficult or not feasible, genetic approaches can still provide valuable insights into potential mechanisms by indicating the number of genes and the complexity of their interactions required for particular changes in behaviour.

Genes for photoperiodic responses are of direct relevance to crop improvement. Breeders frequently include a particular requirement for flowering time as targets in their breeding programme. In many cases, early flowering or daylength insensitivity are desired characteristics and modern cultivars have often had strong photoperiodic responses bred out of them. Although plant breeders have, therefore, been active in a huge volume of work based on the genetics of photoperiodism, this has mainly been carried out on an empirical basis. In only a few species has the genetics of photoperiod response been studied with a view to understanding the processes involved in photoperiodism. Two of the best characterised crop species are pea (*Pisum sativum*) and wheat (*Triticum aestivum*) which are both LDP. Recently there has been an explosion of work on flowering-time genes in the LDP *Arabidopsis thaliana*, some of which is related to photoperiod response.

A mutant phenotype is one which differs from the WT line in which the mutation has occurred and may require growth under a defined set of conditions to be observed. Mutants which are of relevance to photoperiodism can be divided roughly into those which were isolated because they flower at different times from the WT in LD or SD (flowering-time mutants) and those which have been isolated for some other type of response, e.g. growth substance or photoreceptor mutants, which show aberrations in their photoperiodic responses. An important class of mutants for photoperiodism would be those in which the ability to measure time is affected. As yet, no photoperiod clock mutants have been identified in *Arabidopsis*, but there are mutants in which the circadian expression of photosynthetic genes is modified. As components of the underlying circadian timekeeping mechanism of photoperiodism may be common or equivalent to those in other circadian rhythms these rhythm mutants will also be included.

FLOWERING-TIME MUTANTS

Arabidopsis: Most of the *Arabidopsis* mutants which are relevant to photoperiodism have been generated by chemical- or irradiation-induced mutagenesis of the Columbia (Co) and Landsberg *erecta* (Ler) ecotypes. Both are examples of early summer-annual races, i.e. they grow and flower in early spring as the days are getting longer, and have little vernalisation requirement. They behave as facultative LDP; flowering is induced by LD but although flowering is delayed it is not prevented in SD. The critical

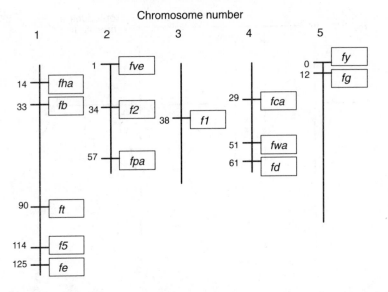

FIG. 9.1. Location of induced late flowering loci in *Arabidopsis thaliana*. After Law *et al.* 1993.

daylength for induction is about 8–10 h and plants flower after about three weeks with 4–7 leaves under LD and 7–10 weeks with about 20 leaves under SD. The precise behaviour of the WT lines varies between laboratories, probably reflecting differences in growth conditions, especially light conditions and temperature.

Arabidopsis flowering time mutants fall into two classes. Firstly, late-flowering mutants in which flowering is delayed in LD and maybe also in SD, and secondly, early-flowering mutants where flowering occurs earlier in SD and perhaps in LD. Most of the initial attempts to isolate flowering time mutants were carried out with Co or Ler under greenhouse conditions using natural photoperiods which, in general, are in excess of the critical daylength of 8–10 h. The WT lines flower quite rapidly under these conditions and the most obvious mutants were those which flowered later than the non-mutated lines. From a series of flowering screens, induced mutations for late flowering were isolated by Redei (1962), Hussein and Van der Veen (1965, 1968) and Koorneef *et al.* (1991). The latter authors carried out the most comprehensive analysis, placing 42 independently isolated mutants in Ler into 11 complementation groups and placing them on the *Arabidopsis* linkage map (Fig. 9.1). The mutants were designated *fca, fd, fha, fe, fb, fg, ft, fy, fpa, fwa* and *fve*. The two mutants, *co* and *gi*, isolated by Redei (1962) in the Co background, were found to be allelic to *fg* and *fb* respectively. The mutants could be assigned to one of three phenotypic groups based on their response to daylength and vernalisation.

- **Group I:** Late flowering – vernalisation sensitive. This group includes *fy, fpa, fve, fca, fe* and is characterised by a much greater requirement for vernalisation than the WT in order to show early flowering under LD. A further gene, *ld*, originally identified by Redei (1962) and subsequently also by Lee *et al.* (1994) also falls into this group.
- **Group II:** Late flowering – daylength sensitive. This group, which includes *fd, ft*

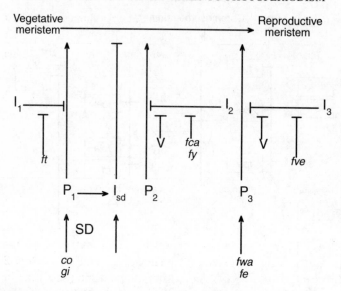

FIG. 9.2. Possible causal model for the interaction of genetic and environmental factors controlling the initiation of flowering in *Arabidopsis thaliana*. SD = short day treatment, V = vernalization treatment, I = inhibition of flowering, Isd = inhibitor produced by SD, P = flowering promoter. Gene symbols indicate the action of the wild-type allele with → = promotive effect and ⊣ = inhibitory effect. After Koornneef *et al.*, 1991.

and *fwa* flowers later than the WT lines in both SD and LD. They are not very sensitive to vernalisation but still retain a significant response to daylength.

- **Group III:** Late flowering – daylength insensitive. This group includes *fha, co* (*fg*) and *gi* (*fb*). They flower at about the same time as WT controls in SD but although flowering is accelerated by LD it is much less so than the WT and thus the mutants are late flowering under normal greenhouse, i.e. LD, conditions. They show little response to vernalisation.

Of these mutants, only those in group III can be assumed to have genetic lesions directly in their daylength-response mechanism, i.e. they are potential photoperiod mutants. Two group III mutants, *co* and *gi*, have been characterised in some detail. The *CO* gene has been studied by Coupland's group (Coupland *et al.* 1993). Plants which have the mutant *co* allele flower with an increased leaf number under LD but with a similar leaf number to the WT under SD. This indicates that the *CO* gene encodes a product required in the promotion of flowering by LD. The earliest difference between *co* and WT can be seen in microscopical sections of the apex of plants grown in LD when the first 2–3 vegetative leaves are longer than 1mm, indicating that *CO* acts before this time in development. The *CO* gene has been isolated by Coupland's group and has the characteristic of a transcription factor. What is not known at the time of writing this book is the location in the plant where *CO* brings about its effects. This is a key question. If *CO* acts in the leaves, it would be a clear indication that *CO* is required for the induction process, whereas if it acts at the apex it must be required for the response to an inductive signal.

A more detailed physiological evaluation of the *GI* locus was performed by Araki

TABLE 9.1 Comparison of photoperiodic responses of GI mutant alleles in Co or Ler backgrounds.

Allele	Short days (8 h)	Long days (16 h)	Continuous light (24 h)
GI (Co)	47	10	12
gi-1	49	25.1	25.6
gi-2	47.8	54.8	48.3
GI (Ler)	15.4	5.7	7.2
gi-3	25.8	22.3	24.1

Flowering response is expressed as the number of rosette leaves at flowering.
From data of Araki and Komeda (1993).

and Komeda (1993). They compared three alleles of the gi mutation, *gi-1* and *gi-2* in the Co background and *gi-3* in Ler (Table 9.1). The Co mutants *gi-1* and *gi-2* behaved similarly to *co*, flowering later than WT in LD but at about the same time in SD. The *gi-3* allele was delayed in LD but also flowered later than the WT in SD. The flowering time was similar in LD and SD for *gi-2* and *gi-3*, but *gi-1* still flowered much earlier in LD than SD. This type of result indicates that the three phenotypic groupings used for the late-flowering mutants need to be interpreted with some care. Some reassignments, especially between groups I and II, may be needed when more alleles of the mutations have been compared. An interesting feature of the *gi-2* allele, which showed the strongest phenotype, was that the response diminished at higher temperatures, being much less severe at 28°C than at 22°C. By transferring *gi-2* from 28 to 22°C during development in LD a sharp break point was noted in the time taken to flower. Transferring before this time point caused a much greater delay in flowering, indicating that GI normally operates in the WT before that particular developmental stage. The critical period was the time at which the plant had produced between 2 and 4 leaves longer than 0.5 mm, i.e. very similar to the time at which CO is believed to act. This could mean that either the CO and GI genes are acting to make the plants competent to respond to induction by LD, which takes place at the 2–4 leaf stage under continuous LD, or it could mean that one or other of the genes is involved directly in the induction process itself.

Based on the flowering responses of the 11 mutant late-flowering alleles, either alone or in mutant × mutant crosses, Koornneef *et al.* (1991) proposed a model in which late or delayed flowering in LD was caused by interference with the synthesis or action of promoters or enhancement of the inhibition of promoters (Fig. 9.2). They further proposed that the mutants in which promotive factors are inhibited fail to produce suppressors of the inhibitors, thus explaining the gain of function, such as vernalisation requirement, through a loss of function in the mutants. While this might be important for the remaining loci, which act independently of the daylength response, *co*, *gi* and *fha* can be considered as simple loss of function mutants (loss of the ability to respond to LD) and could equally well be explained by an inability to make or respond to a flowering promoter.

The second class of flowering time mutants are those which flower earlier than WT

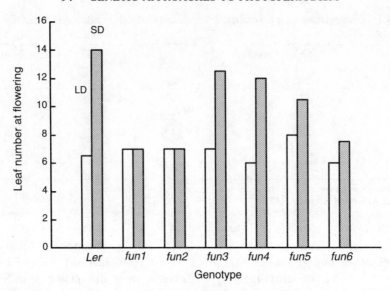

FIG. 9.3. Leaf number at flowering of *fun* early flowering mutants of *Arabidopsis thaliana*. Data of Thomas and Mozley (1994).

under SD. These so-called early-flowering mutants have been isolated by at least four laboratories and are described by different nomenclatures. They include *tfl* (*terminal flower*) and *elf* (*early flowering*; Zagotta *et al.*, 1992), *fun* (*flowering uncoupled*; Thomas and Mozley, 1994) and *esd* (*early short day*) (Coupland *et al.*, 1993, Koorn-neef, unpublished) mutants. The *tfl* and *elf-1* and *elf-2* mutations were isolated as plants which flower early in both LD and SD but retain some photoperiodic sensitivity, whereas *elf-3* flowers early in both LD and SD, with the same leaf number in both conditions. The *elf* mutants all flower earlier and with a lower leaf number than WT in LD and therefore do not appear to represent photoperiodic mutants as such. Although the *lf-3* mutant does not show a response to daylength this appears to be because it becomes autonomously induced at a stage before LD induction takes place. The *esd* series describe four genes and appear to be similar to the *elf* series, except that they are irradiation induced as compared to the *elf* mutations which are chemically induced. The *fun* series of chemically induced mutants were isolated as plants which showed enhanced responsivity to LD inductive treatments or reduced delay by a SD treatment. All of the mutants flowered at about the same time as WT in LD, but earlier in SD (see Fig. 9.3) and in that respect may be distinct from the *esd* and *elf/tfl* mutants. The *fun* mutants fall into two distinct groups. The first, containing *fun-1* and *fun-2*, flower at about the same time as LD controls in both LD and SD. The plants are smaller than controls, with leaves, in particular, being reduced in size. These mutants also flower early when grown in darkness on agar supplemented with sucrose, indicating that flowering is more or less uncoupled from light treatments. The second group, containing *fun-3 – fun-6*, flower at the same time as WT in LD but are not delayed as much as WT in SD. For *fun-3* and *fun-4*, early flowering in SD occurred but with about the same leaf number as WT. In contrast *fun-6* had about half as many leaves as the WT in SD and *fun-5* was intermediate both for flowering time and leaf number. Unlike most early flowering mutants there was little difference in visible appearance between *fun-6* and WT. The time at which full photoperiodic sensitivity became established was earlier in this mutant than in the WT but the ability to respond to daylength and the critical daylength were not altered. It is suggested that this second group of mutations are related to sensitivity to the photoperiodic stimulus and may represent transduction mutants (Thomas and Mozley, 1994). Most early flowering mutants are recessive, which suggests that they represent components required for the production or action of an inhibitor of flowering under SD.

Triticum aestivum (wheat) is a classic quantitative LDP in which the flowering response is a function of daylength, light quantity and light quality (Carr-Smith *et al.*, 1989). In genetic terms the wheat species cultivated for bread flour is an allohexaploid species in which there are 21 pairs of homologous chromosomes. It can be considered as three separate genomes, A, B and D, each having been derived from a different ancestral diploid species. Because genes are present in triplicate, wheat will tolerate the loss or acquisition of entire chromosomes to produce aneuploid lines (Sears, 1954). These lines provide useful tools for identifying chromosomes which carry genes important for particular processes. Two dominant genes, *Ppd1* and *Ppd2*, responsible for a level of insensitivity to daylength or early flowering under SD were first identified by using conventional crosses (Keim *et al.*, 1973; Klaimi and Qualset, 1973). These genes were also detected in subsequent studies in which substituting chromosomes from the homeologous group 2 (chromosomes 2A, 2B and 2C) in the wheat variety Chinese Spring (CS) with those from the wild goat grasses *Aegilops umbellulata* and *Ae. comosa* or perennial rye, *Secale montanum*, produced significant effects on the interaction between the lines and daylength (Law *et al.*, 1978). The *Ppd1* and *Ppd2* genes are located on the short arms of chromosomes 2D and 2B respectively and are probably duplicate genes arising from the polyploid nature of wheat (Scarth and Law, 1983; Worland and Law, 1986). The *Ppd* genes are dominant for insensitivity to daylength. Increasing the gene dosage leads to early flowering under SD conditions but plants lacking chromosomes carrying *Ppd* genes flower late in SD. The major effect of the gene appears to be on the rate of development of the ear and spike rather than on the time of floral initiation (Scarth *et al.*, 1985) and this makes it unlikely that *Ppd* is affecting the induction process. The ability to substitute 'alien' chromosomes, which contain similar loci, from different species such as *Aegilops* and *Secale* along with observations on additional lines of barley suggests that the homology of *Ppd* genes extends to a range of cultivated species.

Pisum sativum: A considerable amount is known about the genetics of flowering in peas based largely on the work of Murfet and co-workers. Pea is a quantitative LDP in which some genotypes also respond to vernalisation. A wide range of genotypes is available and the ability to combine genotypes by grafting enables the site of action of particular genes to be deduced. Also, because grafting experiments give information about the transmission of signals from sensitive to insensitive tissues and *vice versa* it is possible to interpret the function of certain genes in producing promoters or inhibitors of flowering. A number of genes have been found which influence the onset of flowering, some of which operate through an interaction with daylength

TABLE 9.2   Genes responsible for controlling flowering time in *Pisum sativum*.

Gene	Description	Function
Sn	Sterile nodes	Interacts with *Dne* and *Ppd* to confer daylength sensitivity
Dne	Day neutral	Interacts with *Sn* and *Ppd* to confer daylength sensitivity
Ppd	Photoperiod	Interacts with *Sn* and *Dne* to confer daylength sensitivity
E	Early	Reduces expression of the *Sn Dne Ppd* system during early development
Hr	High response	Acts later in development to prolong *Sn Dne Ppd* activity
Gi (fsd)	Gigas (flowering short days)	Recessive alleles greatly delay flowering
Lf	Late flowering	Confers sensitivity to flowering stimuli, four naturally occurring alleles
Veg	Vegetative	Required for flowering. Recessive homozygotes remain vegetative
Dm	Diminutive	Recessive alleles delay flowering and affect plant size and fertility

Based on Murfet (1990) and Arumingyas and Murfet (1994).

(Table 9.2). The most important for daylength responses seem to be *Sn* for sterile nodes (Murfet, 1971a) and *Dne* for daylength neutral (King and Murfet, 1985). These genes act in combination to modify plant response to photoperiod. The recessive alleles at these two loci give rise to daylength insensitivity whereas the combination of *Sn* and *Dne* gives delayed flowering under SD but response to LD and low temperatures is retained. Grafting experiments indicate that the combination of *Sn* and *Dne* is needed in the same part of the plant (e.g. leaf or cotyledon) for the production of an inhibitor which is required to suppress flowering under SD (King and Murfet, 1985). Recently, *Ppd*, a third gene required for the production of the inhibitor, has been identified (Arumingyas and Murfet, 1994). The *Sn, Dne* and *Ppd* system for inhibitor production probably acts in leaf tissues and may be switched off during photoperiodic induction. The inhibitor produced by the *Sn, Dne* and *Ppd* system may modify the pattern of assimilate distribution in the pea (see Chapter 6) leading to a pleiotrophic phenotype, modified in branching, flower number and life span, rate of flower and fruit development, maturity date and yield (Murfet, 1990). Two additional modifying genes, *E* (early) and *Hr* (high response) interact with the *Sn Dne* homozygotes to promote earliness or delay flowering further, respectively, under SD conditions. As suggested by their names, *E* operates in the cotyledons to reduce *Sn Dne* activity in the early stages of seedling growth but HR acts later in the life cycle to prolong the duration of *Sn* and *Dne* activity.

Although early grafting experiments suggested the presence of a transmissible flowering promoter in peas (Murfet, 1971b) it is only recently that genetic evidence for a gene acting in the floral stimulus pathway has been found. The recessive allele *gi* (gigas) was originally obtained from the cultivar Virtus, where it causes plants to flower much later than the initial line (Murfet, 1990). Another recessive mutant, *fsd*

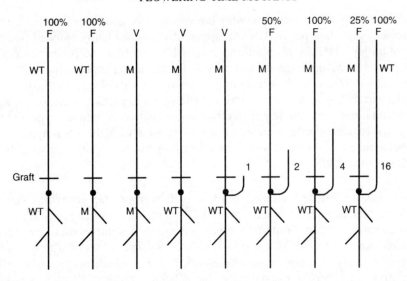

FIG. 9.4. The effect of number of WT (cv Porta) donor leaves on flowering in mutant (M) gi^{fsd} scions in an 18 h photoperiod. Grafts were made epicotyl to epicotyl of 7-day-old seedlings. The WT side shoots arose from the cotyledonary node; the number of leaves present is indicated next to the lateral shoot. Lateral shoots with 1, 2 or 4 leaves were obtained by decapitating the shoot above the specified leaf. The lateral shoots in the treatment on the far right were not decapitated and they produced a mean of 16 leaves. F = flowering, V = vegetative; the percentage of plants flowering is given above. After Taylor and Murfet 1994.

(*flowering short days*) shows obligate SD flowering behaviour and remains vegetative under LD (Taylor and Murfet, 1994). Allelism tests suggest that *fsd* is a more severe allele at the *Gi* locus and a new symbol gi^{fsd} has been proposed for the *fsd* locus. The delay in flowering in *gi* is prevented when the mutant is grafted to the WT stock. Similarly, grafting of gi^{fsd} scions on to WT stocks also overcomes the inability to flower in LD. The WT stock is effective providing it carries at least two foliage leaves (Fig. 9.4). This result is consistent with the WT shoots providing a transmissible substance necessary for flowering, i.e. a floral promoter, which is missing in the mutant shoots. The *gi* or gi^{fsd} phenotypes have many similarities with the *veg* (*vegetative*) mutants which do not flower under any environmental or genetic circumstances. Grafting evidence in this case indicates that *veg* acts at the apex and blocks some aspect of flower initiation (Reid and Murfet 1984). The late flowering phenotype of *Gi* mutants and its alleles is similar to the late flowering phenotype of *co* or *gi* in *Arabidopsis*, although the site of action of the latter two genes is not as yet known. A further gene, *Lf* for late flowering has been identified and controls the response to both vernalisation and LD. A number of alleles at this locus have been found and these are characterised by different minimum node number at which flowers are formed. In this case it is thought that the gene acts at the apex and establishes the sensitivity to flowering signals (Murfet, 1971b).

Hordeum vulgare L. (Barley): At least four recessive *ea* (*early maturity*) loci have been found to influence flowering time in barley. Of the recessive homozygotes, ea_{sp} confers the earliest flowering phenotype while ea_k suppresses the other *ea* loci. A dominant enhancing allele interacts with the ea_k homozygous recessive to give the earliest flowering phenotype ea_k/EN. An apparently independently isolated flowering genotype of barley (*BMDR-1*) is allelic to the ea_k/EN genotype and is characterised by complete photoperiod insensitivity but retains FR mediated promotion of flowering (Principe *et al.*, 1992). A biochemical comparison of the BMDR-1 and the corresponding WT showed differences in two peptides when protein extracts were separated by two-dimensional polyacrylamide gel electrophoresis. However, neither appeared to be pysiologically regulated in a manner which correlated with the phenotypic photoperiodic behaviour of the mutant alleles and they may not be important in establishing photoperiod sensitivity.

Soyabean: Almost all of the detailed analysis of flowering time genes in relation to photoperiodic mechanisms is based on the LDP described above. For SDP, genetic analysis has largely been limited to descriptive genetics in relation to breeding requirements. An example of this sort is soyabean (*Glycine max*) of which there has been extensive analysis. Four genes (E_1 E_2, E_3 and E_4) which, as recessive alleles confer relative photoperiod insensitivity, have been identified (Palmer and Kilen 1987). Of these, the E_3 locus appears to be very important in establishing photoperiod differentials. The dominant E_3 gene causes a strong delay by long photoperiods and the recessive e_3 makes plants insensitive to photoperiod and causes early flowering. The e_4 locus also causes insensitivity to long photoperiods (Buzzell and Voldeng, 1980).

PHOTORECEPTOR MUTANTS

Light-sensing is an integral part of photoperiodism and it might be expected that photoreceptor mutants will show altered photoperiodic responses.

Arabidopsis: As described in Chapter 3, a series of photomorphogenetic mutants (*hy-1* to *hy-8*) have been isolated from *Arabidopsis*. Of these, *hy-3* lacks phytochrome B (Somers *et al.*, 1991) and the *hy-2* mutant is a phytochrome chromophore mutant and probably affects all phytochrome species. Both of these mutants show earlier flowering and reduced leaf number under both LD and SD conditions (Goto *et al.*, 1991). This pattern of responses resembles those of the *elf*, *esd* and *fun-1* and *fun-2* early flowering mutants. Further similarity can be seen between *elf-3* and the *hy-2* and *hy-3* mutants inasmuch as *elf-3* has a long-hypocotyl and pale green phenotype, especially pronounced under SD conditions. *elf-3* complemented alleles of *hy-1* to *hy-6* (Zagotta *et al.*, 1992) demonstrating that *elf-3* is not a phytochrome photoreceptor mutant but it could well be a phytochrome signal transduction mutant with its effect mediated through the same process as *hy-3*. Mozley and Thomas (1995) found that the main reason for the lower leaf number for *hy-3* compared to WT was a slower rate of leaf production and that the ability to be florally induced in LD was affected to only a small degree in the mutant. They proposed that phytochrome-B had a positive role in leaf production and that effects on flowering, especially as measured by the number of leaves, was partly an indirect consequence of a photomorphogenetic function of the photoreceptor. This directly contradicts the use of leaf number as a specific criterion for flowering as given in Chapter 1. The leaf number criterion derives mainly from physiological studies in which imposed environmental or chemical treatments might be expected to have a minimal effect on leaf production, other than through its arrest at the conversion of the apex from a vegetative to floral structure. Under these conditions leaf number is the most specific indicator of when the initiation of flowering has occured. However, mutations which alter the rate of leaf development or production in the vegetative growth phase could indirectly affect flowering time as indicated by leaf number. That impaired leaf production and development can cause early flowering is consistent with the properties of the *fun-1* and *fun-2* mutants, which are characterised by small, poorly developed leaves.

In contrast, phytochrome A-deficient mutants of *Arabidopsis* (*hy-8*) (Parks and Quail, 1993; Nagatani *et al.*, 1993; Whitelam *et al.*, 1993) show little change in photomorphogenetic responses but are relatively insensitive to floral promotion when grown from germination to flowering under day extensions with incandescent light (Johnson *et al.*, 1994). The inductive response of these mutants to extensions with FR and R/FR mixtures during a single photoperiod is also lost (see Chapter 5). Thus, phytochrome A is needed to detect long days and the *hy-8* class of mutants is probably the first in which a biochemical function of the gene product can be specifically defined as being part of the photoperiodic mechanism.

There is considerable evidence for a blue-absorbing photoreceptor-mediated component of daylength-sensing in *Cruciferae*, including *Arabidopsis* (see Chapter 5). However the evidence from mutants is, as yet, contradictory. When Goto *et al.* (1991) studied the *hy-4* mutant, which is deficient in a putative blue photoreceptor (Ahmad and Cashmore, 1993), they found little difference from the WT. Mozley and Thomas (1995), on the other hand found that *hy-4* flowered later than WT and with a greater number of leaves in SD and was less sensitive than WT to induction by a single LD. These observations are consistent with the blue photoreceptor being coded for by *hy-4* being involved in sensing daylength. Differences between the studies might reflect the use of different lamp types and irradiation protocols and more work will be needed in future to resolve this point.

Sorghum bicolor: *Sorghum* is a monocotyledonous SD annual plant of tropical origin and exhibiting C4 photosynthetic structure and metabolism. Early work by Quinby and colleagues showed that photoperiodism in *sorghum* was drastically altered by four genes, three of which involving seven alleles were collected into milo cultivar background. When Pao and Morgan (1986) grew the lines in 12 h photoperiods, which is longer than the critical daylength for most daylength-sensitive cultivars, one group, containing the recessive allele ma_3^R were not delayed in their flowering response. The ma_3^R lines also caused plants to be taller, have narrower leaves with reduced surface area and to show enhanced apical dominance with reduced tiller production. Many of these pleiotropic effects of ma_3^R were mimicked by GA3 and so it was proposed that ma_3^R was responsible for over-production of gibberellins. Childs *et al.* (1991) found that the ma_3^R allele also resulted in reduced anthocyanin content and failure of R-mediated de-etiolation, which are symptoms of a phytochrome mutation. Analysis of the phytochrome content of lines with the ma_3^R allele with antibodies which distinguish between phytochrome A and light-stable phytochromes indicated that it was characterised by the absence of a light-stable phytochrome B-like phytochrome. Because added gibberellins failed to restore anthocyanin levels, it was concluded that gibberellin overproduction was a result of the phytochrome deficiency rather than the reverse. The $ma_3{}^R$ mutation has enabled the role of the missing phytochrome to be assessed in situations involving biological timekeeping. Absence of the stable phytochrome did not prevent the expression or alter the phase of a circadian rhythm in *cab* mRNA but did reduce the abundance of the message (Morgan, 1994; Childs *et al.*, 1995). It was also found that, even in the mutant, daylengths longer than 12 h progressively delayed flowering. The $ma_3{}^R$ is best described as relatively photoperiod insensitive in comparison to the ma_3 or Ma_3 genotypes which flower more than 80 d later than $ma_3{}^R$ plants under 14 h d (Childs *et al.* 1995). The absence of the light-stable phytochrome does not remove the requirement for a 10–12 h dark period to show the earliest flowering behaviour; rather the $ma_3{}^R$ plants initiate flowers earlier under all daylengths. This probably indicates that some other phytochrome is involved in night-length measurement, but if there is redundancy caused by functional overlap between different phytochromes the evidence from *Sorghum* is insufficient to rule this out as a potential function for the missing phytochrome. The daylength-insensitivity of *Sorghum* $ma_3{}^R$ mutants is manifested over a narrow range of daylengths and suggests, along with the *Arabidopsis* *hy-3* mutants, that the particular light-stable phytochromes missing either have only a minor role in daylength-sensing or that their function is duplicated by another photoreceptor in each case (phytochrome or blue photoreceptor) to provide a fail-safe mechanism for the plant. Earlier flowering under non-inductive daylengths represents another similarity between $ma_3{}^R$ *Sorghum* and *hy-3* *Arabidopsis* and may indicate a general property of certain light-stable phytochromes in delaying flowering.

RHYTHM MUTANTS

There are, to the authors' knowledge, no mutations which have been shown to affect the rhythm in light sensitivity which acts as the timekeeping component of the photoperiodic perception mechanism. It might be expected that some components of endogenous rhythms, such as the components of underlying oscillators, might be common to all circadian rhythms operating within a single cell or organism. If this is the case, the type of approach recently developed by Millar *et al.* (1995a,b) may prove to be a way of isolating genes involved in photoperiodic timekeeping. Their method has been to fuse a fragment of the *Arabidopsis CAB2* promoter to the firefly luciferase (*Luc*) gene and transform the construct into *Arabidopsis*. The resulting luciferase enzyme activity can be visualised remotely with a low-light video system. The luciferase activity correlated closely with the *cab2*::*Luc* mRNA abundance and accurately reported the temporal and spatial regulation of *CAB2* transcription. The system thus provides a non-invasive method for remotely monitoring a circadian rhythm in gene expression. The next step was then to mutate the transgenic reporter plants and monitor the progeny to identify new mutants with aberrant cycling patterns. Using this method the authors identified 26 'timing of *CAB* expression' (*toc*) lines, representing at least 21 independent mutations. Both long- and short-period and amplitude mutants were identified in this study. One of these, a semidominant short-period mutant, *toc1*, was mapped to chromosome 5. It was shown to shorten the period of the period of circadian rhythms regulating cab expression and the movements of primary leaves. The *toc1* mutants do not show pleiotrophy for other phenotypes such as photomorphogenesis or flowering time (Millar *et al.*, 1995b) but it is not yet known whether *toc1* mutants have altered critical daylengths or altered rhythms in light sensitivity for photoperiodically inductive treatments.

GROWTH SUBSTANCE MUTANTS

Genetic studies, particularly with *Pisum*, in combination with grafting experiments have identified genes which appear to affect sensitivity to daylength through the synthesis of inhibitors (Murfet, 1985) and possibly also promoters (Murfet, 1987). If the major growth substances were involved in the balance of promoters and inhibitors controlling flowering, it might be expected that mutations affecting production or sensitivity to them would affect daylength-dependent flowering. Once again, *Arabidopsis* is the photoperiodic species in which a range of relevant (i.e. growth substance) mutations has been identified. The fact that the major growth substances are involved in multiple aspects of development makes it difficult to separate out any phenotype which shows modified growth from specific alterations in the daylength response system. Gibberellins have a significant influence on flowering in *Arabidopsis*. Exogenous GA partially overcomes the delaying effect of SD in WT (Langridge, 1957). A series of gibberellin biosynthesis mutants, such as *ga1*, and sensitivity mutants, such as *gai* (Koornneef *et al.*, 1985), both result in a similar, dwarfed phenotype in which flowering is slightly delayed in LD but considerably delayed or prevented in SD (Wilson *et al.*, 1992; Coupland *et al.*, 1993). The *ga1-3* allele exhibits normal flowering in continuous light but does not flower in SD unless it is treated with exogenous gibberellin. This suggests the operation of two independent pathways to flowering in *Arabidopsis*. A gibberellin-dependent pathway which operates in SD, i.e. the autonomous flowering pathway, and a LD-dependent pathway which does not depend on gibberellins (Wilson *et al.*, 1992). Double mutants of *co* and *gai* flower much later than either parental line and in some cases not at all. This suggests that *CO* and *GAI* can partially compensate for each other, perhaps because they operate in the two pathways leading to flowering; *CO* in the LD-dependent pathway and *GAI* in the gibberellin-dependent autonomous pathway. This is not certain, however, because the double mutant also showed a more extreme phenotype than *gai* with respect to dwarfing and aberrant flower development, which could suggest that they operate within a common pathway.

The *amp-1* (*altered meristem programme*) mutant is pleiotrophic, showing a phenotype altered in three different aspects of plant development; spatial pattern of development, constitutive photomorphogenetic development in darkness and an altered flowering phase. The *amp-1* mutants have about six times as much cytokinin as WT plants. Chaudhury *et al.* (1993) describe the mutant as being precocious for flowering, consistent with cytokinin being a floral stimulus in *Arabidopsis*. This is based on the observation that it took about 16 d before flower buds were visible in WT plants in LD, but they were visible after about 12 d in the mutant. On the other hand, the number of rosette leaves in the *amp-1* mutant at flowering was about 20 as compared with about 7 for the WT plants. Based on the criterion of leaf number, *amp-1* is actually a late-flowering rather than an early-flowering mutant. The effect of cytokinin therefore appears to be largely on leaf production and it is difficult to reconcile this with a specific role for cytokinin in floral promotion in *Arabidopsis*. Mutants affecting the ethylene system in *Arabidopsis* have been reported in which flowering time in LD is delayed (Bleeker *et al*., 1988; Guzman and Ecker, 1990) but there is no evidence that photoperiodic responses as such are altered.

HOMEOTIC MUTANTS

The transition from vegetative to reproductive growth is the best characterised consequence of photoperiodic regulation and genes involved in this change of developmental pattern can be regarded as early targets of induction-related changes. The transition from vegetative to floral development can be separated into two distinct stages. In the first the pattern of apical organisation is altered including an increase in apical volume, modification of phyllotaxis and alterations in primordium spacing. The result is an altered three-dimensional arrangements of growing points to form an inflorescence structure. The second phase is floral differentiation and involves reduced internode elongation and the formation of modified leaf structures as whorls of sepals, petals, stamen and carpels. Genes required during the early stages of floral initiation have been identified through the isolation of mutants in which meristems intermediate between inflorescences and floral meristems occur in positions normally occupied by floral meristems. These mutants include *floricaula* (*flo*) and *squamosa* (*squa*) in *Antirrhinum* and the equivalent genes *leafy* (*lfy*) and *apetala1* (*ap1*) in *Arabidopsis* (Coen, 1991). The functions of *FLO* and *LFY* are similar in that they are required for the transition from the early inflorescence meristem to the initiation of flower development although they are not, on their own, sufficient for the complete transition from vegetative to floral development. For *Antirrhinum*, both *FLO* and *SQUA* are required for the correct sequence of changes during floral initiation and early development and in *Arabidopsis* differentiation of lateral meristems as flowers requires *LFY*, *AP1* and *AP2* (Huala and Sussex, 1992). In *Arabidopsis*, phenotypes of *lfy*, *ap1* and *ap2* are more severe under SD than LD, suggesting that some factor produced in response to daylength is necessary for the activity or expression of these regulatory genes. Although these observations are consistent with *FLO*, *LFY* and related genes being the targets for the transmitted stimuli from leaves the situation is not entirely clear cut. Unlike in *Arabidopsis* and *Antirrhinum* the *FLO/LFY* homologue is expressed in the vegetative as well as the floral apices of tobacco (Kelly *et al*., 1995) and *Sinapis alba* (Melzer *et al*., 1995), indicating that its expression in relation to its function may be quite complex.

CONCLUSIONS

The application of physiological and molecular genetics to the photoperiodic control of flowering is helping to clarify some of the long-standing questions which have been asked by plant physiologists. In particular *Arabidopsis* mutants offer the prospect of physically isolating genes which code for components of the photoperiodic mechanism. The late flowering mutants *co* and *gi* are good candidates for plants which are modified in their photoperiodic induction processes and some of the early flowering mutants are also probably impaired in photoperiod perception or transduction of the daylength signal.

Photoreceptor mutants have shown a probable role for phytochrome A in sensing long days in the LDP *Arabidopsis* in agreement with physiological experiments which have identified a requirement for FR light at certain times in the photoperiod for the promotion of flowering in light-dominant LDP. As yet there are no SDP, either mutants or transgenic, in which phytochrome A expression is prevented, which would help clarify the role of phytochrome A in SD-sensing species. The role of light-stable phytochromes appears similar in LDP and SDP. Light-stable phytochromes modulate the response to daylength, e.g. phytochrome B in the LDP *Arabidopsis* and a phytochrome B-like light-stable phytochrome in the SDP *Sorghum*. Mutants lacking these light-stable phytochromes flower early under normally non-inductive daylengths, but also earlier under inductive daylengths. There is some evidence from *Arabidopsis* that this earlier flowering might be related to a positive effect of light-stable phytochromes on promoting vegetative growth. The evidence from *Sorghum* indicates that the light-stable phytochrome which is absent in the $ma_3{}^R$ mutant is not required for biological timekeeping because both circadian rhythms of gene expression (for photosynthetic genes) and critical nightlength are relatively unaffected, even though the magnitude of responses are altered in the mutant. However, there have not yet been any rigorous studies of night-break timing in photoreceptor mutants, which is probably a better indicator of the photoperiodic timekeeping function in SDP than critical nightlength. Taken with the physiological evidence from SDP (Chapter 4) that a type II (i.e. a light-stable) phytochrome is probably required for photoperiodic timekeeping it would be premature to dismiss this as a potential function for light-stable phytochromes based on the responses of single gene mutations in only two plant species.

A range of mutants in *Pisum* is giving insights into the importance of transmissible promoters and inhibitors in establishing the response to photoperiod. Growth substance mutants, in *Arabidopsis*, suggest that gibberellins are important for flowering under non-inductive conditions, but that flowering induced by long days may operate through a separate pathway. A specific role for cytokinin in photoperiodic floral initiation is not supported by the properties of *amp-1* mutants, which have high cytokinin levels, but have a greater leaf number at flowering. This is in general agreement with the conclusions of Chapter 7. Methods are also now being developed for identifying mutations and consequently genes which are required for the circadian rhythm components of biological timekeeping.

In general, the most rapid progress in genetic studies has been made with LDP, especially because of the work with *Arabidopsis* and *Pisum*. There is, as yet, no good model SDP system in which it is possible to apply experimental genetic approaches with the prospect of rapid progress. Such systems will be needed, either to establish how far it is possible to extrapolate the findings from the LDP systems, or to identify genes which have specific roles in the photoperiodic response mechanisms of SDP.

PART II
PHOTOPERIODIC CONTROL OF DEVELOPMENT

10 Floral Development

Daylength clearly has a major effect on the initiation of floral development in photoperiodically sensitive species. Permissive daylengths are perceived in the leaves leading to photoperiodic induction in the same organs. The induced leaves generate signals which may be promotive or less inhibitory to flowering as compared to leaves which have not been induced. Evocation of the apex in response to these signals marks the beginning of a series of processes in floral development and differentiation leading to the maturation of the sexually competent floral structures. Although the best characterised aspect of the response to daylength is these early changes, there is good evidence that daylength can continue to play a part in directing the pathway of floral development subsequent to induction and evocation. The two facets of floral development which are most influenced by daylength are reversion to the vegetative state and sex expression.

REVERSION TO VEGETATIVE GROWTH

The ability of a plant to exhibit floral reversion may involve factors unrelated to photoperiodism. The most basic factor concerns the nature of the flower development itself. Heslop-Harrison (1963) postulated that flower development represents an inevitable series of events, which are initiated by the appropriate stimulus, such as daylength. In his relay hypothesis, each stage and its associated gene products triggers the progression to the next until a fully differentiated flower is formed. If this was the case, floral reversion would not be observed under any circumstances. A second factor could be the extent to which evocation proceeds. Between vegetative and floral development there is an intermediate inflorescence state in which phyllotaxis and primordium spacing may be altered prior to the beginning of the differentiation of floral structures and reversions can be classified as flower reversions or inflorescence reversions (Battey and Lyndon, 1990). In inflorescence reversion, vegetative development continues after a temporary phase of inflorescence production. It differs from an incomplete or partial response to daylength in that inflorescence meristems themselves revert and resume active vegetative growth whereas, in partial flowering, subsequent vegetative development occurs from buds which remained vegetative during a non-saturating inductive treatment. Partial flowering is quite common when photoperiodically sensitive plants are returned to non-inductive conditions after suboptimal or subsaturating treatments. In *Nicotiana glutinosa* and *Salvia splendens*, postinductive vegetative development occurs from basal axillary buds (Diomaiuto-Bonnand, 1960; Crawford, 1961). A contrasting common pattern is seen in species such as *Pharbitis nil* or *Glycine max*, where subapical buds are more susceptible to the inductive treatment than the apical meristem. Following a subsaturating inductive treatment, flowers are found at the most susceptible axillary buds but the main shoot apex continues vegetative growth (Borthwick and Parker, 1938; King and Evans, 1969). Inflorescence reversion in *Sinapis alba* has been described in detail by Bagnard *et al.* (1972). They caused reversions by treating plants with inductive LD and then returning them to SD at low irradiances. The consequence was that the main shoot apex produced determinate side shoots with single flowers and then subsequently lateral shoots with several flowers and variable presence of bracts. This represented the reversal of the normal sequence of events associated with flower initiation, which is axillary branch suppression followed by the suppression of bracts. The continuation of normal apical floral development thus requires LD in *Sinapis* and reversion is daylength dependent.

Flower reversion is recognisable as a flower in which some floral organs have been formed but the meristem has subsequently produced vegetative structures terminating with a leaf or vegetative shoot. One example is the SDP *Impatiens balsamina* in which reversion can be brought about at any stage of flower development by returning plants from SD to LD (Battey and Lyndon, 1984). The longer the plants have spent in SD, the more complete the flower development up to the point where stamen formation can be followed by petal or leaf production. Similar observations have been made with *Anagallis arevensis*, except in this case induction and subsequent flower development requires LD (Brulfert and Chouard, 1961). In general, floral reversion is uncommon, although this may reflect the inability of flowers of most species to undergo developmental reversion, rather than reflect daylength insensitivity during later stages of floral development. The situation in *Impatiens* and *Anagallis* may be taken as the most extreme examples of a more common situation in which inductive daylengths often need to be continued subsequent to the first stages in the floral transition in order to ensure a complete floral response.

SEX EXPRESSION

In the majority of angiosperms both male and female parts develop to form perfect (hermaphrodite) flowers but, in others, only one organ develops to maturity to give staminate (male) or carpellate (female) flowers. It has been estimated that unisexual plants represent over 10% of all plant species, distributed among 75% of plant families. Male and female flowers occur on the same plant in monoecious species and on different plants in dioecious species. Unisexuality is usually caused by the reduction or abortion of sex organ primordia (which can occur at any stage of development) and unisexual flowers often pass through a bisexual stage in which all floral organs are initiated (e.g. maize, cucumber). Thus sex determination is traditionally considered to be the selective abortion of the gynoecium or androecium of initially hermaphroditic floral primordia. Only in a few species (e.g., *Mercurialis*, *Cannabis* and *Spinacia*) do the floral primordia lack any vestiges of inappropriate sex organs; however, even in these plants, sexuality can be reversed under certain conditions indicating that the floral primordia are sexually bipotent (Dellaporta and Calderon-Urrea, 1993).

Effects of daylength

Although sex expression is genetically determined, it can often be modified by environmental and chemical factors. Changes in sex expression were among the first observed responses of plants to daylength and several examples are given in Table 10.1. Tournois reported in 1914 that the accelerated flowering of the dioecious species *Humulus japonicus*, which he obtained by sowing in winter SD under glass, was invariably accompanied by anomalies in sexuality (Tournois, 1914). About 50% of the male plants showed some degree of sex inversion during the earlier stages of flowering. Stigmatic apices occurred on otherwise normal anthers and in more extreme cases stamens were transformed into carpel-like structures. As the days lengthened the majority of anomalous male plants reverted to a normal mode of flowering. Another dioecious species, *Cannabis*, responded to SD in the same way, but with even more pronounced symptoms of femaleness in the male plants. Either femaleness or maleness may be enhanced by exposure to short or long photoperiods, depending on the species (see Table 10.1). In the dioecious SDP *Humulus* and *Cannabis*, exposure to SD promoted a shift towards femaleness while, in the LDP *Spinacia*, the same treatment given after floral induction in LD, promoted maleness in genetic females. Similar effects have been observed in monoecious plants. In the SDP *Zea mays*, exposure to SD caused some of the flowers in the terminal, normally entirely staminate inflorescence, to develop into females. In *Xanthium*, the number of staminate and carpellate inflorescences depended on the number of SD given before transfer to non-inductive LD cycles. Following a single SD cycle, flowers developed slowly and produced 10 staminate and 6 carpellate inflorescences, while plants grown continuously in SD produced 4 staminate and about 13 carpellate inflorescences (Naylor, 1941b).

The results of these experiments seem to indicate that stronger photoperiodic induction leads to a greater degree of femaleness. However, this is not invariably the case. In the monoecious *Begonia* × *cheimantha*, for example, the greatest degree of feminisation occurred under nearly critical conditions for flowering, namely LD and high temperature (Table 10.2). Similarly, in the gynomonoecious SDP, chrysanthemum, more female florets are produced in longer photoperiods than in shorter ones (Van Veen, 1969). This effect is of some commercial significance in the production of flowers under glass; when some cultivars are induced to flower late in the year in extremely short natural photoperiods, they develop a large number of hermaphrodite (disc) florets and fewer female (ray) florets, leading to the development of poor-quality blooms, often with a daisy-eyed centre. The florets are initiated on the capitulum in acropetal succession and the outer ring of primordia develop into ray florets in any daylength short enough for anthesis. However, those in the more central positions in some cultivars develop into either hermaphrodite or female florets according to the photoperiod, LD favouring ray and SD favouring disc florets. In several cultivars, the interpolation of several LD after the initiation of florets has taken place in SD, leads to

TABLE 10.1. Effects of daylength and growth regulators on sex expression.

Species	Relative femaleness promoted by [a]	Sex promoted by growth regulator application [a]		
		Auxin and/or ethylene	Gibberellin	Cytokinin
Dioecious				
Cannabis sativa	SD	Female	Male	Female
Humulus japonicus	SD			
Mercurialis annua	–[b]	Male		Female
Morus nigra		Female		
Spinacia oleracea Matador	LD	Female	Female	Female
Vitis spp.				Female
Monoecious				
Ambrosia elatior	SD			
A. trifida	SD			
Begonia × cheimantha	LD	Female	Female	
Coriandrum sativum			Female	
Cucumis anguria (gherkin)	SD			
C. melo (muskmelon)	–[b]	Female	Male	
C. sativus (cucumber)				
most cultivars	SD	Female	Male	Female
Higan fushinari	LD	Female	Male	
Cucurbita maxima		Female		
C. pepo (squash)	SD	Female	Male	
Cupressus arizonica	SD			
Dahlia pinnata	LD			
Dendranthema cvs	LD			
Luffa acutangula			Male	Female
Mercurialis ambigua				Female
Picea		Male		
Pinus banksiana	SD			
P. contorta	SD			
Pseudotsuga		Male		
Thuya plicata	SD		Female	
Xanthium strumarium	SD			
Zea mays	SD	Female	Female	
Hermaphrodite				
Fagopyron esculentum			Male	
Lycopersicon esculentum				
several cvs	SD			
a stamenless mutant	LD		Male	
Hyoscyamus niger		Female	Female	
Silene pendula	LD	Female		

[a] The effects of daylength and growth regulator application are shown only in terms of whether relative maleness or femaleness is promoted. The actual effect may vary considerably. For example: increase in the number of female flowers (*Begonia*, auxin, GA); lower node of first female flower (*Cucurbita pepo, Cucumis sativus*, auxin); female flowers in male inflorescence (*Zea*, auxin, gibberellin); gynoecium reduction (*Fagopyron*, GA) or stamen reduction (*Hyoscyamus*, auxin); male flowers on female plants (*Cannabis*, GA) and vice versa (*Cannabis*, auxin).
[b] Daylength does not modify sex expression.
From Vince-Prue (1975), Durand and Durand (1984), and additional references given in the text.

TABLE 10.2. The effect of temperature and daylength on flowering and sex expression in *Begonia* × *cheimantha*

Daylength (h)	Temperature (°C)	Number of flowers		Percentage females
		Male	Female	
8	15	1114	2	0.2
	21	862	7	0.8
24	15	1182	19	1.6
	21	372	131	26

From Heide (1969).

TABLE 10.3. Effect of long-day interruption on sex expression in the inflorescence of chrysanthemum

Daylength treatment	Number of ray (female) florets	Number of disc (hermaphrodite) florets
Control: short days	474	48
Long day interruption	687	16

An interruption with 10 LD was given after inflorescence initiation had been induced by exposure to 4 SD. From Van Veen (1969).

an increase in the number of ray florets and a decrease in the number of disc florets (Table 10.3) and may increase the commercial value of the flower. In another member of the Compositae, *Dahlia hybrida*, the number of ray florets produced in the inflorescence is also increased by giving longer photoperiods or a night-break (Canham, 1969).

Photoperiodic mechanism

A night-break treatment has been shown to be equivalent to a LD in *Dahlia hybrida* (for an increase in the number of female flowers) and in conifers for the promotion of male cone buds in *Cupressus arizonica* (Pharis *et al.*, 1970). Giving a low-intensity day-extension treatment was also found to be an effective LD in *Zea* for the inhibition of male sex expression (Krüger, 1984). The effect of LD on sex expression in conifers has been obtained in the field by giving a night-break with red light from a laser source, confirming the likely involvement of phytochrome (Durzan *et al.*, 1979). Consequently, it is usually assumed that the photoperiodic mechanism for the control of sex expression in plants is basically the same as for the induction of flowering, although very little direct experimental work has been carried out.

For a number of different species it has been observed that sex expression appears to be linked with the strength of photoperiodic induction, with strong induction promoting relative femaleness in both LDP and SDP (Heslop-Harrison, 1957); this might suggest that a high concentration of the photoperiodic floral stimulus is required for the development of carpellate structures. Some evidence for this comes from grafting experiments between members of the Cucurbitaceae belonging to different photoperiod response groups. In grafts between the strict SDP *Sicyos angulatus* and day-

TABLE 10.4. The effect of photoperiodic induction in donor plants of *Sicyos angulatus* on sex expression in receptor plants of cucumber (cv Matsunomidori).

	Sicyos donors non-induced (in LD)	*Sicyos* donors induced in SD)
Number of nodes with staminate flowers	11.0	6.2
Number of nodes with pistillate flowers	1.5	8.0

Cucumber receptor plants were grafted at the cotyledonary stage to donor plants of *Sicyos* under LD conditions. The donor plants had either been induced to flower by exposure to SD or were vegetative plants which had remained in LD throughout.
Data of Takahashi *et al.* (1982).

neutral cultivars of cucumber (*Cucumis sativus*), induction of the *Sicyos* donors influenced sex expression in the cucumber receptors. Grafting to an induced donor increased the number of pistillate and decreased the number of staminate flowers in the cv Matsunomidori, compared with grafting to a vegetative *Sicyos* donor (Table 10.4). Sex expression in Matsunomidori is influenced by photoperiod, with SD leading to an increase in femaleness. The results suggest that, although this cucumber cultivar is day-neutral for floral initiation, exposure to SD increases the production of floral stimulus and so leads to an increase in femaleness. Moreover, it seems possible to separate effects of photoperiod from those of the growth regulators GA and ethylene (Takahashi *et al.*, 1983), suggesting that some of the effects of daylength on sexuality in cucurbits are dependent on an unidentified substance, perhaps florigen? A role for the floral stimulus in promoting femaleness is also indicated in *Sicyos*, which can be induced to flower by the reciprocal graft with several cultivars of cucumber. Young cucumber seedlings in the cotyledonary stage would be expected to generate relatively small amounts of floral stimulus and, when such plants were used as donors, only a few undeveloped floral buds were formed on the receptors and these were mostly male. However, when *Sicyos* was induced to flower by grafting to older plants of cucumber, the genetically-determined sex of the day-neutral donor rather surprisingly had little influence on sex expression in the *Sicyos* receptor (Fig. 10.1). A possible role for the floral stimulus is also strongly indicated for *Xanthium* by experiments in which the translocation of factors stimulating male or female flower development has been studied by defoliation treatments. Translocation of the stimulus for the initiation of the apical male inflorescences began at the same time as that for the formation of the lateral female inflorescences (Fig. 10.2) indicating that both are probably dependent on the same fast-moving component; however, as noted already, increasing the amount of floral stimulus by increasing the duration of SD induction favours the development of carpellate rather than pistillate inflorescences (Naylor, 1941b; Leonard *et al.*, 1981).

Plant growth regulators

Inevitably, a good deal of attention has been paid to the effects of known plant growth regulators. All of them modify sex expression in certain plants but their action varies with species. For example, GA feminises maize but has the opposite effect in

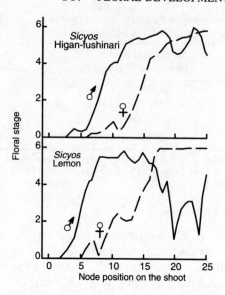

FIG. 10.1. Flower development and sex expression of non-induced *Sicyos* scions grafted onto two different sex types of cucumber under long-day conditions. The donor cucumber stock plants had 8–10 expanded leaves at the time of grafting. ♂, staminate flowers; ♀, pistillate flowers. Higan-fushinari produces mainly pistillate flowers; Lemon is andromonoecious and produces staminate and hermaphrodite flowers, but no pistillate ones. After Takahashi *et al.*, 1982.

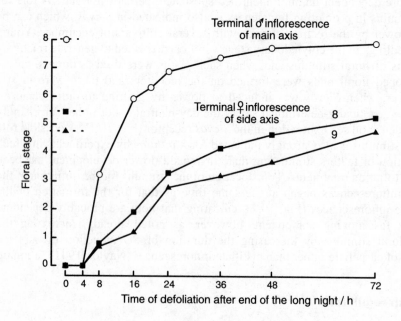

FIG. 10.2. The effect of leaf removal at different times after induction on sex expression in *Xanthium strumarium*. Plants were induced by exposure to a single 16-h dark period and subsequently maintained in 20 h long days. Dotted lines show the floral stage of un-defoliated controls. After Leonard *et al.*, 1981

cucumber, while auxins cause masculinisation in *Mercurialis* but the opposite reaction in other species. In some species, hormone applications have little or no effect on the sexuality of the flowers. Among the most consistent in their effects on sex expression are auxins and ethylene, both of which favour femaleness in several different plants, although they favour maleness in the dioecious species *Mercurialis* (see Table 10.1). The rates of ethylene production and the endogenous auxin content have both been found to be affected by daylength and the greater rate of ethylene evolution and higher auxin level in SD are consistent with the effect of SD to increase femaleness in many cultivars of cucumber (Rudich *et al.*, 1972a). Changes in auxin content with daylength, however, do not always correlate well with the resultant sex changes; for example, LD (which promotes maleness) slightly increased the auxin content of *Zea mays* at the time which was considered to be critical for the determination of the sex of the flower (Heslop-Harrison, 1964).

The application of GA is more variable in its effects (see Table 10.1) and can increase either femaleness (e.g. *Begonia* × *cheimantha*, *Coriandrum*, *Zea*) or maleness (e.g. *Cannabis*, *Luffa*, *Mercurialis*). In hop, GA had no effect on sex expression (Pharis and King, 1985). However, in many cases, the effect of GA is similar to that of LD (see Table 10.1), irrespective of whether the latter increases maleness or femaleness. The well known effect of LD to increase the content and/or turnover of GA (see Chapters 7 and 13) suggests that at least some of the effects of daylength on flower sexuality involve changes in endogenous GAs.

Cytokinins have been shown to transform sexuality in several widely different species and, in general, their effect is to promote femaleness. In *Vitis*, for example, dipping the flower clusters into a cytokinin solution converted male clones of *V. champinii*, *V. berlandieri*, *V. rupestris* and *V. cinerea* to functional hermaphrodites (Moore, 1970). The application of CKs has also been found to cause the production of hermaphrodite and female flowers on otherwise male plants of *Cannabis* and *Mercurialis* and, in the latter, endogenous free zeatin may be associated with the production of female flowers (Dauphin-Guerin *et al.*, 1980). Although these results were not shown to be related to photoperiod, the known variation of CK content with daylength in some plants (see Appendix II and Chapter 7) suggests that CKs should be considered as a possible component of the photoperiodic regulation of sexuality. Indications of a role for cytokinins in the photoperiodic modification of sex expression comes from *Begonia* × *cheimantha*, where marginal induction in LD decreases the proportion of male flowers. In this case, the content of iP plus iPA in leaf extracts was also lower in LD, although there was no consistent effect of daylength on the amount of Z plus ZR (Table 10.5). However, the extracts represent the conditions after 8 weeks of treatment, whereas flower initiation, and presumably also sex determination, occurs within a few days of transfer to SD.

The growth inhibitor ABA has been found to influence sex expression in a few plants. In general, the effect is to promote femaleness as in *Cucurbita pepo* and *Cannabis sativa*, but it has also been found to increase maleness in a monoecious line of cucumber. However, no clear link has been established between the effects of daylength on sex expression and on the modification of endogenous ABA levels, although the latter may vary with photoperiod. A somewhat similar situation is seen with polyamines. Their content and composition is known to be influenced by daylength and there seems to be some relationship between sexuality and polyamine

TABLE 10.5. Effects of daylength on cytokinin activity and sex expression in *Begonia* ✕ *cheimantha*

Photoperiod (h)	Temperature (°C)	Cytokinin (pmol g^{-1} FW)		Flowers per plant	Male:female ratio
		Z + ZR	iP + iPA		
24	24	6.1	4.7	1	5
8	24	13.3	42.2	310	606
24	15	53.1	42.9	41	7
8	15	25.6	84.2	217	19

Cytokin levels were determined in leaf extracts by enzyme immunoassay after plants had been grown for 8 weeks at the indicated photoperiods.
Data of Hansen *et al.* (1988).

content (Kakkar and Rai, 1993); however, possible links between photoperiod-dependent changes in polyamine metabolism and sex determination remain to be investigated. Other substances shown to have some influence on sex expression in particular plants include morphactins, which have been found to increase femaleness in *Cannabis* and *Arachis hypogaea* and maleness in *Ricinus*; it was suggested that these substances may perhaps influence sex expression by altering the GA/auxin ratio (Varkey and Nigam, 1982). The steroidal sex hormones of animals can also modify sex expression in some plants. In *Melandrium rubrum*, a clear positive effect in the female direction was obtained with the female sex hormones, oestrone and oestradiol, and in the male direction with the male hormone, testosterone (Heslop-Harrison, 1957); in contrast, both oestradiol and testosterone increased femaleness in monoecious plants of cucumber. Again, no specific links with photoperiodic effects have been demonstrated, although oestrogen-like substances appeared after photoperiodic induction in both SDP (*Perilla* and *Chenopodium*) and LDP (*Hyoscyamus*); these oestrogens appeared at the time of flower formation (Kopcewicz, 1972a,b).

The widespread influence of PGRs on sex expression in many different plants indicates that the modification of endogenous levels of these substances is a likely explanation of many, although perhaps not all, effects of daylength to alter sexuality. However, specific links between the photoperiodic control of the content of endogenous PGRs and changes in sex expression have been established in only a few species. Some of the best documented of these are discussed in more detail below.

Cucurbitaceae

The control of sex expression has been studied in considerable detail in a number of cucurbits, where the floral primordium can develop into a pistillate, staminate or hermaphrodite flower depending on genetic constitution and environment. At least seven genes are known to be involved in cucumber sex expression (Malepszy and Niemirowicz-Szczytt, 1991). In several members of the Cucurbitaceae, including most cultivars of cucumber, SD and low temperatures enhance femaleness; however, in the cucumber Higan fushinari, the number of pistillate flowers is increased by LD, while some cucumber cultivars and the related species, *Cucumis melo* are insensitive to daylength (Table 10.1; Takahashi *et al.*, 1982).

The application of auxin was found to cause earlier formation of female flowers in cucumber lines that differ genetically in their sex expression. The different forms of sexual expression are

- **monoecious**: staminate and pistillate flowers on the same plant
- **gynoecious**: pistillate only
- **andromonoecious**: staminate and perfect on the same plant
- **hermaphrodite**: perfect flowers only.

The auxin content of the main stem was higher in hermaphrodite than in andromonoecious plants (Galun *et al.*, 1965) and higher in gynoecious than monoecious plants (Rudich *et al.*, 1972a). The endogenous auxin content was found to be higher in SD than in LD, which correlates with the effect of SD to increase femaleness in most cultivars (Rudich *et al.*, 1972a). However, there was no correlation between endogenous auxin levels and femaleness in the related species *Cucurbita pepo*, in which feminisation is also promoted by SD (Chrominski and Kopcewicz, 1972).

Femaleness in cucurbits is increased by ethylene or ethylene-releasing compounds in the same way as by the application of auxins (see Table 10.1). It is unlikely that the effect of ethylene depends on endogenous auxins as it has been shown that auxin levels are decreased in ethylene-treated plants of *Cucurbita pepo* (Chrominski and Kopcewicz, 1972), whereas both auxin and ethylene increased femaleness. Since the application of auxin is known to enhance ethylene production, it is probable that the effects of auxin on sex-expression result from the induction of ethylene formation, rather than from the action of auxin *per se*. For example, the effect of IAA to increase femaleness is nullified by inhibitors of ethylene action or synthesis (Takahashi and Jaffe, 1984).

Several lines of evidence indicate that ethylene is a major endogenous factor for feminisation in the Cucurbitaceae. Exposure to ethylene gas or treatment with the ethylene-releasing agent CEPA causes the production of pistillate flowers in cucumber and many other curcurbits (Takahashi and Jaffe, 1984) and, in both cucumber and muskmelon (*Cucumis melo*), gynoecious lines have been shown to produce more ethylene than monoecious ones (Byers *et al.* 1972). When the ethylene content of tissues was lowered by growing plants at reduced pressure, hermaphrodite flowers were formed on gynoecious plants of muskmelon and the same plants produced pistillate flowers when returned to atmospheric pressure or when the reduced-pressure system was supplemented with ethylene. Enrichment of the atmosphere with carbon dioxide, a competitive inhibitor of ethylene action, also resulted in the development of bisexual flowers. The anti-ethylene agents silver nitrate (which inhibits ethylene action) and AVG (which blocks synthesis) also led to the production of male flowers in gynoecious cucumbers (Takahashi and Jaffe, 1984). The application of ACC, the immediate precursor of ethylene biosynthesis, increased the production of pistillate flowers in cucumber under conditions when none developed on control plants. This effect was nullified by silver nitrate which blocks the action of ethylene but not the conversion of ACC to ethylene; consequently, it was concluded that the action of ACC was due to ethylene and not (as proposed for *Cucurbita pepo*; Hume and Lovell, 1983) to ACC itself. Similar results with different sex types of cucumber cv Beit Alpha were extended to the effect of photoperiod (Rudich *et al.*, 1972b). A female (gynoecious) line produced considerably more ethylene than a monoecious one, and exposure to SD

TABLE 10.6. Ethylene evolution from different sex types of cucumber (*Cucumis sativus*) growing in long days and short days.

Sex type	Daylength		Mean of sex type
	16 h	8 h	
Monoecious (male and female)	1.72	2.04	18.8
Gynoecious (female only)	2.71	5.07	39.9
Mean of daylength	2.22	3.56	

Plants were grown in natural daylight with or without an 8 h extension with low irradiance light. Plants were incubated for 2 h at 25°C in the light before ethylene measurements were made (given as nl ethylene $g^{-1} h^{-1}$)
Data of Rudich *et al.* (1972b).

conditions (which increase femaleness in most cultivars) markedly increased the rate of ethylene evolution. The daylength-induced difference was more marked in the gynoecious than in the monoecious line (Table 10.6). The highest rates of ethylene evolution occurred in excised floral buds, with greater quantities being produced by pistillate than by staminate flowers.

Sexuality in the cucurbits is also influenced by GA which, in contrast to auxin and ethylene, increases maleness (see Table 10.1). GA does not affect ethylene production (Atsmon and Tabbak, 1979) and presumably acts by a basically different mechanism. The formation of staminate flowers on gynoecious lines of cucumber can be induced by treating plants with GA (Peterson and Andher, 1960). GA_7 and GA_4 were reported to be considerably more effective than GA_3 in cucumber (Pike and Peterson, 1969) but in another cucurbit, *Luffa acutangula*, GA_3 was somewhat more effective than $GA_{4/7}$ (Krishnamoorthy, 1972). A positive correlation between maleness and endogenous GA has been demonstrated in both *Cucubita pepo* (Chrominski and Kopcewicz, 1972) and in cucumber, where the phloem exudate of a monoecious line contained many times more GA than a gynoecious one (Friedlander *et al.*, 1977a). The GA content of shoot extracts of cucumber has also been found to be higher in monoecious and andromonoecious lines than in gynoecious ones (Fig. 10.3). Consequently, it seems likely that GA plays a role in the regulation of sexuality in these plants. However, the effects of applying a range of growth retardants in order to depress endogenous GA levels has produced inconsistent results, although in some cases their application has led to an increase in femaleness, as would be expected if a reduction in endogenous GA content is involved (Vince-Prue, 1975; Pharis and King, 1985).

In cucumber, ethylene leads to an increase in femaleness and GA to an increase in maleness. Since ethylene levels have been shown to be higher in SD (see Table 10.6) which promote femaleness, while GA levels are higher in LD (Table 10.7) which promote maleness, it has seemed probable that the photoperiodic control of sexuality in cucumber depends on daylength-induced changes in these two PGRs. This assumption has been specifically examined using two cultivars (Higan-fushinari and Matsu-nomidori) which have different responses to daylength (Takahashi *et al.*, 1983). In both cultivars, the number of pistillate flowers was increased by ethylene and decreased by GA, while the content of GA was highest in SD and ethylene production

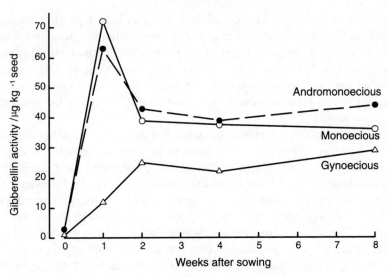

FIG. 10.3. Gibberellin activity of phosphate buffer extracts of three sex types of cucumber (*Cucumis sativa*). Andromonoecious, male and hermaphrodite flowers; monoecious, male and female flowers; gynoecious, female flowers. Gibberellin activity assessed by dwarf d_5 corn bioassay. After Hemphill *et al.*, 1972.

TABLE 10.7. The content of abscisic acid and gibberellin in a gynoecious line of cucumber (*Cucumis sativus*) in long and short photoperiods.

Photoperiod	Abscisic acid (ng g^{-1} DW)	Gibberellin (ng g^{-1} DW)
Short days	160	600
Long days	55	1 100

Extracts were made at the four-leaf stage from a gynoecious line of cucumber originating from the cultivar Beit alpha. The abscisic acid and gibberellin contents were determined by gas–liquid chromatography. Data of Freidlander *et al.* (1977a).

TABLE 10.8. Effect of photoperiod on ethylene production and sex expression in two cultivars of cucumber.

Photoperiod (h)	Ethylene production (nl g^{-1} h^{-1})		Number of nodes with pistillate flowers	
	Higan-fushinari	Matsunomidori	Higan-fushinari	Matsunomidori
24	0.92	3.44	18.0	1.9
8	0.70	1.32	12.6	3.6

Ethylene evolution was determined for plants at the five-leaf stage.
Data of Takahashi *et al.* (1983).

was highest in LD (Table 10.8); nevertheless, the number of pistillate flowers was highest in SD in Matsunomidori and in LD in Higan-fushinari. Although these results differ from the earlier findings on the effect of photoperiod on both ethylene evolution (see Table 10.6) and GA content (see Table 10.7), they indicate that the effect of photoperiod on sex expression in cucumbers may not to be attributable entirely to GA and/or ethylene; it has been suggested (Takahashi *et al.*, 1982) that the floral stimulus may also be involved.

The situation in cucurbits is clearly complex and appears to vary between species. The strict SDP *Sicyos angulatus* can be induced to flower by grafting to day-neutral cucumber plants, but sex expression in the *Sicyos* receptor was not influenced by the genetically-determined sex of the cucumber donor (see Fig. 10.1); even when the donor was the andromonoecious cultivar, Lemon, the *Sicyos* receptors produced both male and female flowers as normal. Neither is sex expression in *Sicyos* modified by GA or ethylene (Takahashi *et al.*, 1982). Thus, while changes in endogenous ethylene and GA may influence sex expression in cucumber, they do not appear to be effective in *Sicyos*. The situation is further complicated by the fact that *Sicyos* donors can influence sex in cucumber receptors, which responded with an increase in the number of pistillate flowers when the *Sicyos* donors were induced to flower by exposure to SD (see Table 10.4), suggesting a role for the floral stimulus. The situation in *Cucumis melo* is also somewhat different. In contrast to cucumber, the gibberellin content was considerably lower in male than in female lines (Hemphill *et al.*, 1972) and, although the application of GA increased male tendency in an andromonoecious line (by decreasing the number of hermaphrodite flowers), it did not increase the number of staminate flowers on a gynoecious line. This species displays very little sex change in response to daylength.

Other components of the endogenous growth regulator complex have also been found to influence sex expression in cucurbits. Cytokinins have been shown to have an effect in some cases; for example, in *Luffa*, where the application of benzyladenine led to an increase in femaleness (Takahashi *et al.*, 1980). In cucumber, the application of ABA via the roots increased femaleness in a gynoecious line grown under conditions which promoted maleness (12 h photoperiod, 18°C) and the ABA content of leaves was twice as high in the gynoecious line as in a monoecious one (Rudich *et al.*, 1972a; Table 10.9A). There was, however, no effect on either gynoecious or monoecious cucumbers when ABA was applied as a foliar spray. In contrast, the application of

TABLE 10.9. Effects of daylength on the content of abscisic acid in monoecious and gynoecious lines of cucumber derived from the cultivar Beit alpha.

Sex type	A	B	
	LD (16 h)	LD (16 h)	SD (8 h)
Monoecious	10.0	25.1	36.8
Gynoecious	22.3	17.1	28.1

A: ABA (ng g^{-1} FW) extracted from leaves of plants with 4–5 expanded leaves (data of Rudich *et al.*, 1972a).
B: ABA (ng g^{-1} FW) extracted from shoot tips of plants with two expanded leaves (data of Friedman *et al.*, 1977a).

ABA to a monoecious cucumber increased maleness (Friedlander *et al.*, 1977b) and the ABA content in the shoot tip and phloem exudate was considerably higher in the monoecious than in a gynoecious line (Friedlander *et al.*, 1977a, Table 10.9B). It is not clear why a high level of ABA should be associated with femaleness in one case and maleness in another since the same genetic lines were employed in both. The differences in ABA contents (Table 10.9) may be related to the tissues used (mature leaves compared with shoot apices) but this does not account for the different responses to ABA application. The trend towards greater femaleness in gynoecious plants and towards greater maleness in monoecious plants might be explained by an optimum curve hypothesis, with the response to exogenous ABA depending on the endogenous content, which varies with sex type and stage of development (Friedlander *et al.*, 1977a). The overall picture is further complicated by the fact that the relationship between sex and ABA content in bisexual lines of cucumber (from Richmond Green Apple) was the opposite of that in the monoecious and gynoecious lines (from Beit Alpha); a higher ABA content was associated with a greater degree of femaleness in the former and maleness in the latter.

Based on changes in their endogenous levels, it has been suggested that sex expression in cucumber depends on interactions between four components of the growth regulating complex, namely auxins, ethylene, GA and ABA. However, although observed correlations between the effects of daylength on sex expression and on the content of GA and ethylene (and possibly also auxin) suggest that the photoperiodic regulation of sex expression in cucumber may be achieved through modifications of the endogenous content and/or metabolism of these growth regulators, specific studies with cultivars which show different responses to daylength suggest that this conclusion may be an oversimplification. Moreover, although there is evidence that ABA is involved in the genetic determination of sex in cucumber, changes in ABA content do not appear to be associated with the photoperiodic control of sex expression. The ABA content was found to be higher in SD (see Table 10.7) and, since this was not consistent with the effects of SD to promote femaleness and ABA to increase maleness (in a monoecious type), it was concluded that the daylength-induced changes in ABA content (which were observed in both gynoecious and monoecious lines) were probably irrelevant to sex determination in these cucumber cultivars (Friedlander *et al.*, 1977a). Modifications of sex expression by PGRs, especially ethylene, auxin and GA, have also been documented in several other cucurbits but correlations with photoperiodic responses have not been investigated in any detail.

Zea mays

In the monoecious plant *Zea mays*, the unisexual flowers are borne on separate inflorescences; the terminal inflorescence (tassel) contains only staminate flowers and the axillary inflorescence (ear) contains only pistillate ones. Sex determination takes place subsequent to a common bisexual stage and normally involves the programmed abortion of the preformed organ primordia of the inappropriate sex (Dellaporta and Calderon-Urrea, 1993). However, primordia can be induced to change their developmental pathway into male or female flowers by the environment or by the application of PGRs. Exposure to LD retards the initiation and development

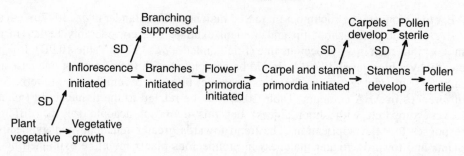

FIG. 10.4. Alternative developmental pathways in the flowering of *Zea mays*. The developmental pathways can be deflected by transfer to short-day conditions as indicated. After Heslop-Harrison 1961

of the terminal staminate inflorescence, while SD promote femaleness (Fig. 10.4, Krüger, 1984). As in cucumber, auxin application promotes femaleness and may lead to the development of female flowers in normally wholly staminate terminal inflorescences (Heslop-Harrison, 1961). Attempts to relate daylength effects to auxin content were, however, unsuccessful in *Zea*. At the time which was considered to be critical for the determination of the sex of the flower, the auxin content was, if anything, higher in LD (which promotes maleness) than in SD, although the difference was small (Heslop-Harrison, 1964).

In contrast to cucumber, the application of GA in *Zea* leads to an increase in femaleness. When plants are treated with GA before the determination of sexuality, male flowers become sterile and female flowers develop in the terminal, male inflorescence (Hansen *et al.*, 1976). Evidence for the involvement of endogenous GAs in sex expression in *Zea* comes from the observation that the GA content drops during the development of the male inflorescence while, at anthesis, the content of GA is much higher in female than male inflorescences (Rood *et al.*, 1980). In dwarf mutants which lack GA, the normal process of stamen abortion in the ear is prevented without affecting the gynoecia. From the available evidence, therefore, it seems likely that the stamen abortion process in maize requires the action of GA (Dellaporta and Calderon-Urrea, 1993). In contrast to the dwarf mutations (which lack GA and fail to suppress stamen development in the ear, thus increasing maleness) are the **tassel seed** mutants in which pistil suppression fails in the tassel, thus producing functional pistillate flowers in the tassel and so increasing femaleness. GA has no effect in such mutants and it has been suggested that the biochemical pathway for this sex phenotype may differ from that of the **dwarf** mutation (Irish and Nelson, 1989).

The feminising effect of GA application occurs during development of the tassel, which takes place early in the life of the plant before microspore meiosis (probably within 3–4 weeks from sowing); later applications of GA after pollen microspore meiosis may lead to an increase in maleness (Pharis and King, 1985). *Zea* is an exception to the general rule that GA mimics the effect of LD on sex expression, since feminisation occurs in response to both GA application and exposure to SD early in the life of the plant. However, the effect of photoperiod on endogenous GA content at this time does not appear to have been reported, although exposure to a low level of incident radiation (which also triggers sex reversal towards femaleness) maintained a high content of GA-like substances (especially non-polar) in the apical tissue while, at high irradiance, endogenous GAs decreased dramatically (Rood *et al.*, 1980). Thus, although both auxin and GA modify sex expression in *Zea*, there is, as yet, no evidence that the effects of daylength result from changes in either of these two growth regulators.

Dioecious plants

Detailed studies of sex expression have been carried out in several dioecious species. In some (e.g. *Silene*, asparagus) sex is determined by X and Y chromosomes with males usually being the heterogamic (XY) sex; in others (e.g. *Rumex, Humulus*), the controlling factor is the X-autosome ratio, rather than the presence or absence of a Y chromosome (Dickinson, 1993; Dellaporta and Calderon-Urrea, 1993).

Daylength has been found to modify sex expression in a number of dioecious plants. In spinach (*Spinacia oleracea*), strong induction by exposure to LD increased feminisation, while returning plants to short photoperiods after floral induction in LD promoted maleness in genetic females. The application of GA_3 also induced feminisation (although an increase in maleness was reported in one case), while growth retardants had the opposite effect (Pharis and King, 1985). The cultivar Matador has a very low percentage of monoecious plants and, when grown in LD, has a stable ratio of male to female plants. In this cultivar, the addition of GA substantially increased the percentage of female flowers when apical buds from young vegetative plants were cultured in LD. The shift in the ratio of male to female may reflect different requirements for the induction of flowering in male and female plants since it appears that, when plants were treated with GA in LD, more female plants were stimulated to flower (Table 10.10), while exposure to a marginally inductive regime in SD at high temperature caused flowering mainly in male plants (Ćulafić and Nešković, 1980). Thus, in this plant as in several others, the most favourable conditions for flowering favour a greater degree of femaleness. The effect of daylength on GA metabolism is well documented for spinach, where exposure to LD increases the metabolism of GA_{19} to GA_{20} (Metzger and Zeevaart, 1982). An increase in the content and turnover of specific GAs in LD is, therefore, correlated with effect of LD to increase femaleness in spinach and a high content of GA is needed for flowering in female plants. However, further information on GA in the flowers, particularly at the time of sex determination, and on the response to the application of GA_{19} and GA_{20} is necessary before it can be concluded that the effects of photoperiod on sex in spinach are dependent on changes in specific GAs.

TABLE 10.10. Effect of gibberellin on sex expression in isolated apical buds of spinach (*Sinacia oleracea*) cv Matador cultivated in 16 h photoperiods.

Treatment	Percentage of plants flowering		Male:female sex ratio
	Female	Male	
Control	37	37	1.00
GA_3	62	30	0.48
GA_{4+7}	45	22	0.49

Gibberellins were applied at 1 mg l^{-1} and plants were scored after 8 weeks in culture.
Data of Ćulafic and Neškovic (1980).

Sex determination has been studied in a number of other dioecious species with no very consistent results. Both photoperiod and treatment with PGRs such as GA, CK and auxin modify sex expression in several cases but there is little evidence for or against the hypothesis that the photoperiodic effects on sex expression are brought about by changes in the content of any particular PGR(s). As in many monoecious plants, auxin and ethylene promote femaleness in hemp (*Cannabis sativa*) and both of these PGRs have been found to cause the formation of female flowers on male plants (Cleland and Ben-Tal, 1983). In contrast, auxin promotes the formation of male flowers in genetic females of *Mercurialis annua*, and the content of endogenous auxin has been found to be higher in male plants, even in callus derived from vegetative apices. In *Cannabis*, both LD and GA_3 lead to masculinisation; the involvement of endogenous GA is indicated by the fact that lowering the GA content with the growth retardant, cycocel, increased femaleness, while leaves of male plants were found to have a higher content of GA-like compounds (Pharis and King, 1985).

From the limited amount of evidence available, the effects of CK appear to be more consistent than those of either GA or auxin. In *Mercurialis*, *Cannabis*, *Vitis*, *Spinacia* and *Asparagus*, the application of CK leads to the development of hermaphrodite or female flowers on genetic male plants (Cleland and Ben-Tal, 1983; Ombrello and Garrison, 1987; Louis *et al.*, 1990). In *Cannabis*, CKs from the roots may be important for the development of female flowers since root removal led to an increase in the number of male flowers, which was reversed by the application of benzyladenine (Chailakhyan and Khryanin, 1978). A similar situation may occur in *Mercurialis* where genetic males are easily feminised by CKs but genetic females can be masculinised only in nodal explants (Irish and Nelson, 1989). The CK content in the apex (as measured by combined gas chromatography–mass spectrometry) was found to vary both qualitatively and quantitatively between genetic male and female plants of *Mercurialis* (Dauphin-Guerin *et al.*, 1980). Free zeatin was detected only in the apices of genetic females and it was suggested that female sex differentiation may be associated with a high concentration of free-base *trans*-zeatin. iPA was detected in both males and females, with a higher content in the former, while zeatin nucleotide was found only in male plants. It seems possible, therefore, that the genetic control of sex expression in these dioecious species is determined, at least in part, by the content of particular cytokinins. Their role, if any, in the photoperiodic modification of sex expression in dioecious plants remains unknown, however.

In dioecious *Mercurialis*, three types of evidence indicate that auxin and cytokinin are both involved in the expression of the sex phenotype. Firstly, the application of these hormones can cause expression of the opposite sex type, a high CK to auxin ratio favouring femaleness. Secondly, measurements of endogenous hormones in sex genotypes are in accordance with their effects on phenotypic expression; male genotypes are rich in IAA and lack *trans*-zeatin, whereas female genotypes contain *trans*-zeatin and are low in IAA. Thirdly, a male-specific cDNA probe disappears as a function of the feminising action of exogenously applied CK (Hamdi *et al.*, 1989). However, daylength has little or no effect on sex expression in *Mercurialis*.

Conifers

The final group of plants in which sex determination has been studied in some detail is the conifers. Most species are monoecious and exhibit a gradient of sexuality, with female cones at the top and male cones at the bottom of the tree; where both sexes occur on the same branch, female cones are found on the outer part and male cones on the inner part. Although generally insensitive to photoperiod for induction, daylength may be a key determinant of sex expression. For example, the proportion of male to female cone buds can be easily manipulated in several Cupressaceae, with LD promoting maleness and SD promoting femaleness. There is also limited evidence that members of the Pinaceae use daylength as a factor in sex determination of the cone bud (Ross *et al.*, 1983). The suggestion has been made that male strobili respond to photoperiod as LDP, while female strobili behave as SDP (Pollard and Portlock, 1984).

Flowering is promoted in the Cupressaceae and Taxodiaceae by a variety of biologically active GAs while, in the Pinaceae, the less polar GA molecules are more effective (see Chapter 7). GAs also modify sex expression in these conifers, although their effect depends on concentration, as well as on environmental conditions at the time of application. In the Cupressaceae and Taxodiaceae, exogenous GA has been shown to increase the proportion of male cone buds in LD, and to increase the proportion of female cone buds in SD (Pharis and King, 1985). In members of the Pinaceae, GA application also increased the proportion of male cone buds produced under long photoperiods, but only in the early stages of shoot ontogeny; in SD and at later stages of development, GA application increase the proportion of female cone buds. The effect of daylength on GA content and metabolism does not appear to have been investigated in conifers.

The manipulation of male/female ratio by means of GA application and photoperiod is of considerable practical importance for breeding. One of the problems in making crosses at a very early age in the Pinaceae is the virtual absence of male flowers on young seedlings. Cultural treatments (such as water stress) together with the application of $GA_{4/7}$ has shown promise in obtaining increased male flowering in seedlings, provided that the GA treatment takes place during long (or lengthening) days. The inhibition of female cones by red laser light in the field, with an interrupted night mimicking a LD, indicates the possibility of using artificial light in seed orchards where $GA_{4/7}$ and cultural treatments have produced little or no male flowering (Ross *et al.*, 1983).

Auxin can also modify sex expression in conifers (Ross *et al.*, 1983) and may promote either femaleness (Cupressaceae) or maleness (*Pseudotsuga* and *Picea*). However, unlike GA, the effect of auxin has not been shown to depend on the photoperiodic conditions. It has been proposed that, for monoecious conifers, internal gradients of auxin and GA control sex expression and that these gradients can be modified by a variety of factors (Pharis and King, 1985). Whether daylength acts in this way remains to be established. Cytokinins also vary between male and female reproductive buds of *Pseudotsuga* (Morris *et al.*, 1990).

CONCLUSIONS

The photoperiodic regulation of sex expression has been studied in far less detail than that of floral initiation and no very coherent picture has so far emerged. All of the main groups of PGRs have some influence on sex determination and environmental factors, including photoperiod, possibly exerting their effect by modifying the metabolism of endogenous PGRs in apical meristems during the course of flower development, preferentially stimulating or suppressing the differentiation of pistils or stamens. It is evident, however, that a variety of distinct strategies may be employed to produce unisexual flowers and that the effects of growth regulators varies with species. At present, there is rather little information on the effect of daylength conditions on the content and turnover of particular PGR molecules in the apex and floral parts, especially in relation to sex expression. Although much circumstantial evidence points to PGRs as the effector molecules, some experiments also indicate a role for the floral stimulus, since the ratio of carpellate to pistillate flowers is modified in some species by the intensity of photoperiodic induction.

11 Bud Dormancy

A period of dormancy during which growth temporarily ceases is a feature of the annual growth cycle of many plants. Such growth inactivity can be imposed directly by the environment (by low winter temperatures, for example) or, for the lateral buds, may be a consequence of apical dominance by the terminal bud. However, many woody plants also show a period of true dormancy when growth is not resumed under favourable environmental conditions, nor when the influence of other buds is removed. Various terms have been used to denote these different types of growth inactivity. **Ecodormancy** or **quiescence** are often used when growth inactivity is a direct consequence of environmental conditions, whereas **ectodormancy** refers to the situation when factors external to the dormant tissue, but within the plants, are the cause of the failure to grow. **Endodormancy**, or **rest**, is caused by factors within the dormant tissue. In many experiments these different types of inactivity are not distinguished; consequently, **dormancy** is used here as a general term for the suspension of visible growth.

In some plants, dormancy involves only a suspension of growth without the development of any specialised structures. In other cases specialised resting organs are formed. The resting buds of woody plants develop when the cessation of normal internode extension leads to a telescoping of the shoot. In some species, the lack of elongation alone leads directly to the formation of a resting bud because the overlapping stipules of the unexpanded leaf primordia form the protective bud scales (e.g. in *Betula*). In other species, the protective outer leaf scales are modified leaves. The modifications may be only slight (*Viburnum*) or quite extensive as in *Ribes*, where the development of the lamina is almost wholly suppressed and the bud scales are formed from the enlarged leaf bases. Even more profound changes occur when resting organs such as bulbs and tubers are developed; these are discussed in Chapter 12. A number of aquatic plants also develop over-wintering structures, called **turions**; in many ways their formation resembles that of resting buds in woody plants and they are included here.

Dormancy is a survival strategy for the organism that displays it and enables the plant to resist unfavourable environments. For species of the temperate zone, low winter temperatures are a major environmental factor and one of the important accompaniments to winter dormancy is a marked increase in the resistance to below freezing conditions. In deciduous species, dormancy is accompanied by the shedding of leaves and this also increases the plant's ability to survive low winter temperatures. Leaf fall is a factor in drought resistance, which may also be associated with the development of a dormant state in species growing in areas with a pronounced dry season.

EFFECTS OF DAYLENGTH

Dormancy in woody plants was among the earliest phenomena shown by Garner and Allard to be under photoperiodic control. It is important to recognise, however, that daylength influences several other aspects of vegetative growth and that these may, in part, be confounded with effects on dormancy. For example, internodes are generally longer in LD in both herbaceous and woody plants and leaves are often larger (see Chapter 13). Thus care must be taken to consider those attributes that are specifically associated with dormancy. In particular, stem length alone is not a good criterion since it is strongly dependent on the length of individual internodes, where differences due to daylength are not necessarily associated with dormancy. A more specific criterion for dormancy is the time at which extension growth ceases; this can be seen in Fig. 11.1, where daylength clearly influences the time at which extension growth ceases (i.e. the onset of dormancy), although the differences in final stem length are partly due to effects on internode length, which are not associated with dormancy. Other criteria include the time at which visible resting buds can be identified or the number of nodes formed during the growing season; the latter mainly reflects the time at which new foliage ceases to expand, although effects on the plastochron during the growing season may also contribute and are not related to the onset of dormancy. A number of

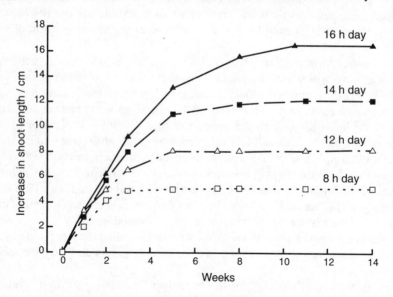

FIG. 11.1. Growth of *Paulownia tomentosa* seedlings in different photoperiods. Plants received 8 h of daylight in a glasshouse. Different photoperiods were obtained by extending the daylength with tungsten-filament lamps at about 1.3 W m^{-2} at plant level. Vince-Prue 1975 (data of Downs and Borthwick, 1956a).

TABLE 11.1   The effect of daylength on growth and dormancy in seedlings of *Liquidambar styraciflua*.

Treatment	Growing season (days)	Internode number	Shoot length (mm)	Mean internode length (mm)
Long days	127.6	16.2	295.9	18.3
Short days	75.9	8.3	20.5	2.5

Plants were grown in natural daylight for 8 h each day (short days); in the long-day treatment this was extended for 6 h with tungsten filament lamps at approximately 5 W m^{-2} to give a total photoperiod of 14 h.
From Lam and Brown (1974).

different criteria have been used in the experiments described in this chapter and their specificity should be taken into account when considering the photoperiodic control of dormancy. For example, in *Liquidambar styraciflua*, both stem length and the final number of nodes formed are influenced by daylength (Table 11.1); the onset of dormancy was clearly accelerated by SD since more than twice as many nodes developed in LD but it is evident that daylength also had a large influence on the final length attained by internodes during the growing season. The wide range of vegetative responses to photoperiod should be born in mind when considering the identity of the endogenous stimuli which control the phenomena specifically associated with dormancy.

In most of the woody species studied so far, particularly those of the temperate zones, photoperiod affects the duration of extension growth and the time at which buds enter dormancy. The rate and duration of elongation growth is usually increased in LD, while the rate of growth is decreased and the onset of dormancy is hastened by SD. In early studies, four major response types were identified (Table 11.2). Plants of some species (e.g. *Robinia pseudacacia*, *Betula pubescens*, *Acer rubrum*) continued to grow more or less indefinitely when maintained in LD, provided that the temperature was favourable for growth. When plants were transferred to SD, growth usually ceased rapidly (Fig. 11.2). In the second group (e.g. *Paulownia tomentosa*, *Acer pseudoplatanus*), the onset of dormancy was delayed but not entirely prevented by LD (see Fig. 11.1). In the third group (e.g. *Pinus sylvestris*), extension growth proceeds in a series of flushes and the duration of the dormant period between successive flushes was shortened by LD, in some cases to such a degree that growth appeared to be continuous. A final group of plants were found to be insensitive to daylength; these included a number of common fruit plants such as apple and pear. In considering the category to which any plant belongs, however, it is important to realise that more recent studies have shown that the critical daylength can vary widely between species and also between ecotypes of the same species. For example, a critical daylength of >20 h has been recorded for several high latitude ecotypes (Table 11.3) and it is possible that growth cessation might have been delayed indefinitely by LD in some of the species that were included in group 2, if a longer photoperiod had been employed. Where the onset of bud dormancy in woody plants of the temperate zone is responsive to photoperiod, it is usually accelerated by SD and prevented or delayed by LD, although dormancy can also be induced by LD, particularly in plants native to regions with hot, dry summers.

Like flowering, the photoperiodic control of dormancy has been shown to be an

TABLE 11.2   Effects of daylength on the induction of dormancy in some common trees and shrubs.

Dormancy induced by short days, prevented by long days

Acer palmatum; rubrum
Albizia julibrissin
Betula pubescens
Catalpa bignonioides
Cercidiphyllum japonicum
Cercis canadensis
Cornus alba; florida; nuttallii
Fagus sylvatica
Larix decidua
Liriodendron tulipifera
Platanus occidentalis
Populus balsamifera (=tacamahaca); robusta
Rhus typhina
Robinia pseudacacia
Tsuga canadensis
Ulmus americana
Weigela florida

Dormancy delayed but not prevented by long days

Acer pseudoplatanus
Aesculus hippocastanum
Liquidambar styraciflua
Paulownia tomentosa

Plants grow in flushes; duration of period between successive flushes is decreased by long days

Camellia japonica
Citrus limon (lemon); *paradisi* (grapefruit)
Ilex opaca
Picea abies
Pinus coulteri; sylvestris; taeda; virginiana
Quercus robur; rubra (=borealis); *suber*
Viburnum opulus

Induction of dormancy not influenced by daylength

Fraxinus excelsior
Fuchsia boliviana
Malus (apple)
Sorbus aucuparia
Pyracantha coccinea

These examples have been taken from Vince-Prue (1975); they are based on the extensive early surveys of Downs and Borthwick (1956a,b), Nitsch (1959, 1962), Nitsch and Somogyi (1958) and Wareing (1959). However, considerable differences may be observed between adults (less sensitive to daylength) and seedlings, and also between ecotypes of the same species.

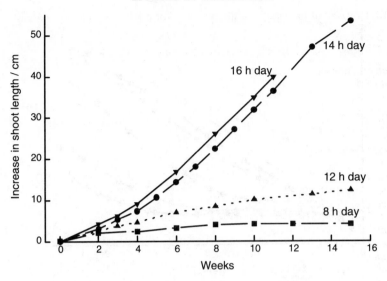

FIG. 11.2. Growth of *Acer rubrum* seedlings in different photoperiods. Lighting treatments were as in Fig. 11.1. Vince-Prue 1975 (data of Downs and Borthwick 1956a).

TABLE 11.3 Examples of critical daylengths for shoot elongation in latitudinal ecotypes.

Species	Latitude at which ecotype was collected (or most northerly population)		
	56°N	63°N	70°N
	Approximate critical daylength (h)		
Acer platanoides	14	16	–
Alnus incana	14	16–18	20–14
Betula nana	–	16–18	20–24
B. pubescens	14–16	16–18	20–24
B. verrucosa	14	16–18	20–24
Corylus avellana	14	16–18	18–20
Hippophae rhamnoides	14	16	20
Picea abies	16	18	20
Salix caprea	14	16–18	20–24
Tilia cordata	14	16	16–18
Ulmus glabra	14	16–18	18–20

From Håbjørg (1978).

inductive phenomenon. Extension growth decreased or ceased entirely in *Picea abies* following exposure to only a few SD (Fig. 11.3). Growth was resumed when plants were returned to continuous light but the duration of growth cessation was related to the number of SD experienced. It was suggested that plants exposed to a limited number of SD entered a state of pre-dormancy since growth was rapidly resumed in LD; the onset of rest was thought to require continuous exposure to SD.

The length of day in any particular latitude may be an extremely important factor in

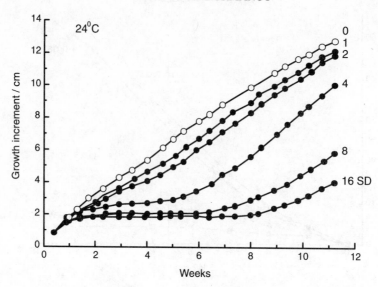

FIG. 11.3.   Apical shoot growth of *Picea abies* seedlings (ecotype F1, 58° 30′) in response to different numbers of short-day cycles. The SD treatments were given from time zero and plants were then returned to continuous light at 24°C. After Heide 1974a.

controlling the time at which dormancy develops under natural conditions. *Robinia pseudacacia* becomes dormant when the photoperiod falls to 12 h, but with additional light will continue to grow until the temperature falls (Kramer, 1936). It is also likely that the onset of dormancy under natural conditions is directly controlled by photoperiod in species such as *Larix, Populus* and *Betula*, where growth continues well into the autumn and which are responsive to SD. The extension growth of many species, however, ceases in mid-summer before the days shorten appreciably and here endogenous conditions appear to override photoperiodic effects, even though many of these species show a response to daylength under experimental conditions. Endogenous conditions seem to be of greater importance in mature trees, which often stop growing earlier than seedlings. In *Picea abies*, for example, the critical daylength for the continuation and resumption of growth in 2-year-old plants was longer than for the maintenance of uninterrupted growth in the first year (Heide, 1974b).

A somewhat surprising after-effect of exposure to shorter photoperiods in *Picea abies* was an increase in the amount of extension growth in the following season (Heide, 1974b). The terminal winter bud in the Pinaceae contains all the primordia for the following year's leading shoot and, after extension growth has ceased, the number of primordia which differentiate depends on the prevailing temperature; if higher temperatures continue for longer, more primordia differentiate and the growth of the following year's shoot is increased. Photoperiod is an important factor here, since shorter days lead to earlier growth cessation when the temperature is higher and, in consequence, more growth in the following year. This after-effect was largely lost after the second year of growth.

Because the length of day varies much less in tropical regions than at higher latitudes, it is often dismissed as being of no importance. Nevertheless, effects of

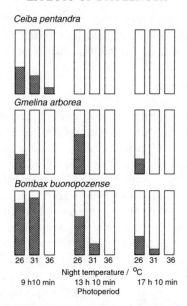

FIG. 11.4. Effects of daylength and night temperatures on dormancy in some tropical trees. The shaded areas represent the number of trees which became recognisably dormant during the experimental period of about 4 months. Plants were grown at a day temperature of 31°C and at the night-temperatures and photoperiods indicated on the figure. Data of Longman, 1969.

daylength on shoot growth have been demonstrated in a number of tropical trees (Longman, 1969, 1978). Although their dormancy is probably not so deep, the responses of tropical trees seem to be broadly similar to those in the temperate zone and a period of growth cessation is often favoured by exposure to SD and cool nights. As with temperate species, the greatly reduced growth made under SD is due in part to the internodes being shorter but, in many cases, plants in SD ceased to produce new leaves and their apices became dormant (Fig. 11.4). As might be expected from the behaviour of other species, internode length may be influenced by daylength (longer in LD), even when there is little or no effect on the production of new leaves (Longman, 1978).

In some tropical species, budbreak and renewed growth occurred after a few weeks even when plants were maintained in SD and it appeared that the trees developed only a temporary form of dormancy (Longman, 1969). However, true dormancy appears to exist in tropical trees, although it has been rather little studied; for example, cut shoots of *Bombax* were more reluctant to grow when taken in the middle part of the leafless period than when taken earlier, or later (Longman, 1978). Photoperiod may, therefore, be a limiting factor for extension growth and leaf production even in the tropics, although care must be taken in attempting to extrapolate from results obtained in growth rooms where the photoperiods are often widely different and may be outside the range experienced under natural conditions. Nevertheless, natural daylengths have been shown to have an influence in Nigeria, where seedlings of *Hildegardia bartei* rapidly became dormant in photoperiods of 11.5 h but continued to grow in 12.5 h (Longman, 1969). The direction of any photoperiod response will depend on the local environment. In high latitude trees, dormancy is induced by autumnal SD and is related to survival in low winter temperatures, whereas in the Mediterranean climate, dormancy is induced by LD and is related to summer drought; in the tropics, the response is probably dependent on the relationship of photoperiod to the dry season.

Cold and Drought Resistance

Dormancy is an adaptation to unfavourable environmental conditions to which the dormant organ is more resistant than the non-dormant one. In species of the temperate zone, an increased resistance to below-freezing temperatures accompanies the winter dormant condition. Although they may be controlled by independent mechanisms, both the formation of winter resting buds and the development of cold resistance depend on exposure to SD conditions. This is sometimes evident in plants growing near street lamps, which may show an increased susceptibility to freezing injury. Even in species such as apple, where the time of onset of dormancy is unaffected by daylength, exposure to SD increased hardiness (Howell and Weiser, 1970). Autumnal short days may thus be important for the survival of plants in which extension growth and the formation of resting buds occur early in the season before the daylength shortens. However, although visually dormant buds appear to be a pre-requisite for the ability to harden in woody plants and both are dependent on SD, the precise relationship between rest and cold acclimation has been shown to vary in different species (Dormling, 1993). In *Pinus sylvestris*, the buds are in a state of quiescence during the early part of SD-induced dormancy and only develop true rest late in the season, in parallel with the development of cold acclimation, which largely depends on exposure to chilling temperatures. In contrast, buds of *Picea abies* enter rest shortly after transfer to SD and release from rest begins 3–4 weeks later simultaneously with the build up of frost tolerance; during most of the winter, therefore, these plants are in a state of quiescence imposed by low temperature. These differences have practical implications. Long night treatments in late summer are used in Sweden to produce *Picea abies* with early hardiness; in *Pinus sylvestris*, the same procedure would not be effective.

Shedding leaves is one way of increasing resistance to winter low temperatures. Leaf fall is directly influenced by SD in many deciduous species of the temperate zone, but temperature is frequently more important. In *Rhus glabra* and *Liriodendron tulipifera* leaves are shed in SD and retained in LD but in *Betula pubescens*, *Robinia pseudacacia*, *Acer pseudoplatanus*, *Liquidambar styraciflua* and *Quercus alba* leaves are retained in SD provided that the temperature does not fall. An interaction with temperature is often important not only for the shedding of leaves but also for the cessation of extension growth. In SD at 10°C *Robinia pseudacacia* stopped growing but did not become dormant, growth being resumed following transfer to LD; at 15°C plants became dormant and the leaves were shed; while at 21°C and 27°C they became dormant but retained their leaves (Nitsch, 1962).

Shedding leaves is also an adaptation to water stress. While this may be a direct response to water shortage, daylength may also be a factor. Many species from Mediterranean climates have evolved mechanisms to reduce the leaf area during the long summer drought. In these regions of hot, dry summers and cool wet winters, the summer deciduous habit is an adaptation allowing plants to survive periods of extreme water stress. In a typical species from California (*Lotus scoparius*), plants lose their leaves in the summer and remain dormant until the rains begin in December (Nilsen

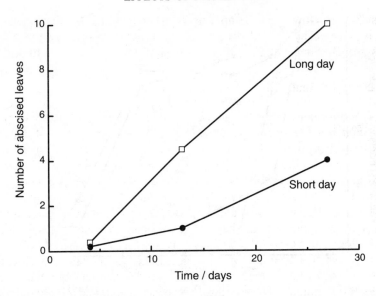

FIG. 11.5. The effect of daylength on the number of abscised leaves on *Lotus scoparius* plants. Plants were grown in a growth chamber at 25°C day; 10°C night temperature and received either a 10 h (SD) or 14 h (LD) photoperiod at 600 μmol m^{-2} s^{-1}. Plants were well-watered for 14 days and then stressed by withholding water at time zero. After Nilson and Muller 1982.

and Muller, 1982). When water stress was imposed under LD, leaf area decreased more rapidly and the leaf abscission rate was higher than in SD (Fig. 11.5); these effects appear to be associated with the induction of dormancy since, in LD, soluble protein synthesis was not re-established following the release from stress. Plants in longer photoperiods also showed a greater ability to adjust osmotically during stress and were able to maintain positive turgor at lower pre-dawn water potentials. This may be an important factor for leaf abscission since turgor is required for several steps in the abscission process. The inhibition of dormancy by SD is also advantageous, since it prevents the development of dormancy during the winter drought conditions that are often experienced in California.

Dormancy may also be induced by LD in geophytes of the Mediterranean region. *Anemone coronaria* grows actively in winter but new leaves cease to emerge during the hot dry summers; growth is then resumed following the first rains in autumn. Leaf emergence rapidly ceased in LD (15 h) or with a night-break of 4 h given in the middle of a 14 h dark period (Fig. 11.6). The critical daylength was between 11 and 12 h (Ben-Hod *et al.*, 1988). There is, however, an interaction with temperature; dormancy is also induced by exposure to high temperature (as in other Mediterranean geophytes) and, in a 27/22°C regime, there was no difference between LD and SD (Ben-Hod *et al.*, 1988). Many herbaceous perennial plants, including grasses, of the arid and semi-arid Mediterranean regions also become dormant in summer. This is frequently induced by LD and, in some cases (e.g. *Poa bulbosa*; Ofir and Kerem, 1982) is associated with the development of bulbs or bulbils. Dormancy is not restricted to flowering plants and photoperiodic control has been recorded in *Lunularia cruciata*, a desert liverwort from Israel (Schwabe and Nachmony-Bascombe,

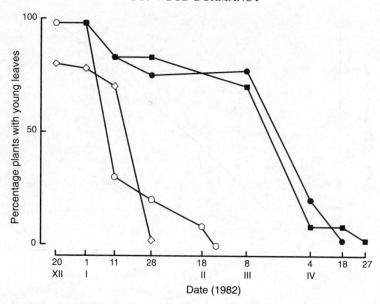

FIG. 11.6. The effect of photoperiod on the onset of dormancy in a wild accession of *Anemone coronaria*. The long-day treatments were: a SD in sunlight followed by (◇), a 4 h night-break with gro-lux lamps (1.5 μmol m^{-2} s^{-1}) or (○) a day-extension for 5 h with quartz-iodide lamps (5 μmol m^{-2} s^{-1}). The SD treatments were: (■) 10 h photoperiods or (l) natural daylengths (minimum 10.25 h). After Kadman-Zahavi *et al.*, 1984.

1963); in this case, dormancy was induced by exposure to LD and is clearly an adaptation to the severe water-stress conditions of the summer desert environment.

Where the induction of cold or drought resistance is influenced by photoperiod, other factors often interact and contribute to the overall level of resistance achieved. For the induction of drought resistance and leaf abscission in *Lotus* it is necessary to expose plants to water-stress in LD (Nilsen and Muller, 1982). Where frost resistance is induced by SD, the relationship between daylength and temperature is often crucial in determining the degree of hardiness achieved. For example, although exposure to SD alone was sufficient to develop a high degree of cold hardiness in a northern ecotype of *Salix pentandra*, the degree of hardiness was significantly enhanced by a short exposure to sub-zero temperatures (Junttila and Kaurin, 1990). In several species it is possible to achieve cold hardiness by exposing plants to either SD or low temperatures (e.g. *Pinus sylvestris*, *Picea abies*; Christersson, 1978; *Cornus stolonifera*; Chen and Li, 1978). The effects of daylength and temperature appeared to be additive and it was suggested that they are probably independent physiological mechanisms for inducing frost hardiness. This is supported by the fact that, in *Pinus* and *Picea*, exposure to low temperature in LD caused hardening without inducing dormancy. In pines, the development of frost hardiness appears to occur in two stages; the first requires exposure to a photoperiod of 11 h, followed by the second stage which develops with exposure to temperatures below a threshold of 5°C (Greer and Warrington, 1982). Thus maximum cold-hardiness was found when SD preceded the exposure to low temperature, as would occur under natural conditions.

In contrast to the situation in most woody plants of the temperate zone, daylength has no effect on cold acclimation in many herbaceous species, where the development of cold hardiness depends on exposure to low temperature. However, acclimation in several clones of white clover (*Trifolium repens*) was significantly greater in plants maintained in 12 h photoperiods compared with those in continuous light (Junttila *et al.*, 1990).

Cambial Activity

In many species, cambial activity is dependent on a supply of growth substances from the actively growing shoot and daylength probably acts indirectly by affecting the activity of the shoot apex. However, in *Pinus sylvestris, Quercus robur, Picea abies* (Heide, 1974a) and *Abies balsamea* (Mellerowicz *et al.*, 1992), cambial activity was found to continue for several weeks after extension growth had ceased while, in *Picea*, the CDL for cambial activity was shorter than for the continuation of apical growth in all ecotypes studied, indicating that photoperiod also has a more direct influence on cambial growth. A direct effect of daylength was also found in *Robinia pseudacacia*; when dormant plants, in which the apices had abscised, were returned to LD, the cambium was found to be active 6–10 weeks later although the shoots had not resumed growth (Wareing, 1956).

Ecotypic Variation

Species with a broad latitudinal distribution frequently show marked ecotypic variation with respect to the effects of daylength on the time of entering dormancy. There is overwhelming evidence for the existence of photoperiodic ecotypes in trees and shrubs of the temperate and arctic zones and critical daylengths for the induction of dormancy can vary widely. A number of studies have been carried out with seeds which have been collected from a variety of different latitudes and grown under different photoperiods in controlled environment conditions. In one such investigation as many species as possible were collected from three different latitudes ranging from 70°N in Norway to 56°N in Denmark; seedlings were raised in a 22 h photoperiod, selected for uniform height and then transferred to a range of daylengths from 12 to 24 h (Håbjørg, 1978). In all species (except two), elongation growth was decreased in SD and apical growth ceased. The critical daylength varied between ecotypes but plants of different species collected from the same geographical area had approximately the same critical daylength for the maintenance of shoot elongation (see Table 11.3). The critical daylengths ranged from a mean of 15 h in ecotypes from 55°N to 23 h in ecotypes from 70°N (Fig. 11.7). Ecotypes collected from different altitudes also showed differences in their critical daylengths but these were smaller, ranging from 14–16 h at 200 m to 18–20 h at approximately 1000 m (Table 11.4). Somewhat different critical daylengths have occasionally been recorded for the same species collected from similar latitudes (e.g. for *Picea abies*, Table 11.3 and Heide, 1974a); this may have been due to variations in the experimental conditions and does not obscure the overall trend. Similar differences in critical daylengths have also been recorded for northern (15 h) and southern (11 h) ecotypes of *Populus trichocarpa* from Canada and California; in addition the northern ecotype had greater photoperiodic sensitivity (defined as change in response per unit change in daylength) than

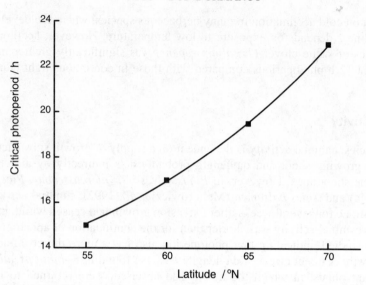

FIG. 11.7. Average critical photoperiods for apical growth cessation in latitudinal ecotypes of Scandinavian trees and shrubs. After Håbjørg 1978.

TABLE 11.4 Effect of altitude and latitude of the seed source on the critical daylength for the induction of dormancy in populations of *Betula pubescens* seedlings.

Latitude	Altitude (m)	Critical daylength for dormancy induction (h)
70°20′ N	50	< 20–24
63°20′ N	50	< 16–18
56°30′ N	50	< 14–16
61°30′ N	1000	< 18–20
61°30′ N	600	< 16–18
61°30′ N	200	< 14–16

Plants were raised in continuous light and the treatments started when they were about 10 weeks old. Plants in all daylength treatments received 12 h of high-intensity light and different photoperiods were obtained by extending the day with tungsten filament light at about 1.05 W m^{-2}. From Håbjørg (1972).

the southern one, indicating that differences in the photoperiod response curves may not be completely described by the photoperiod alone (Howe *et al.*, 1995).

Stunting of growth is often seen when ecotypes adapted to the long photoperiods of northern latitudes are grown in shorter photoperiods further south. The problem has been overcome under nursery conditions by the use of extended photoperiods to raise seedlings; this practice is especially valuable for slow-growing species, such as *Pinus contorta* (Wheeler, 1979). The shorter critical daylengths of southern populations allows them to continue to grow in shorter photoperiods and thus enables them to utilise the longer growing season of their normal habitat. Such lower latitude races often continue growing longer if moved to higher latitudes and may be damaged by

TABLE 11.5 Time of growth cessation and cold acclimation in three latitudinal populations of *Salix pentandra* grown in the open at Tromsø (69°39′ N)

Ecotype	Date for 50% growth cessation	LT$_{50}$ for buds (°C)		
		4 October	11 October	23 October
69°39′ N	7 September	−15.3	<−18.0	<−23.0
63° N	22 September	−7.0	−15.7	−28.0
59°40′ N	14 October	−4.8	−11.7	−18.1

From Junttila and Kaurin (1990).

autumn low temperatures. For example, when grown outdoors in Tromsø at 69° 39′, a southern ecotype (59° 40′ N) of *Salix pentandra* (with a critical daylength of about 14 h) showed delayed cessation of growth and less rapid cold acclimation than a local ecotype with a critical daylength of 22 h (Table 11.5). Such southern ecotypes of *S. pentandra* are severely damaged during the winter when planted in Tromsø. Similarly, delayed dormancy and frost damage have been reported for *Robinia pseudacacia* of southern origin in northern Russia (Moshkov, 1935) and for *Populus tremula* in Sweden (Sylven, 1942). In general, high altitude ecotypes from the same latitude have a longer CDL (see Table 11.4) and so become dormant sooner than their low altitude counterparts; this shortens the growing season in the mountains and is an adaptation to the earlier arrival of autumnal low temperatures.

The relationship of growth and dormancy, on the one hand, and photoperiod and temperature conditions on the other, can be illustrated by comparing graphs for daylength and temperature responses (photherms) of a given ecotype with the photoperiods and temperatures experienced at a particular geographical site (photo-thermographs). The problem of latitudinal transfer is illustrated by comparing two ecotypes of *Picea abies* (Fig. 11.8). At 47° N, the northern ecotype (L1) would be exposed to daylengths well below the critical for growth cessation even in mid-summer and would produce only a brief growth flush in the spring. At 64° N, on the other hand, the southern ecotype (Lankowicz) would continue to grow until September and would be susceptible to frost injury (Heide, 1974a). Even when the actual cessation of growth is largely controlled by endogenous factors, the duration of growth appears to be related to the photoperiod experienced and, if this is lengthened, the time of the onset of dormancy may be delayed (Wareing, 1956). Also, as already noted, exposure to SD may control the induction of frost hardiness even when extension growth has ceased.

Photoperiod is, therefore, a significant and often overriding factor in controlling the latitudinal distribution of tree species. However, temperature is also important and, since this also varies with latitude, it is not always possible to determine which, if either, factor is decisive. Not all species show ecotypic differences in their response to daylength. The broad latitudinal distribution of *Acacia farnesiana* seems to depend on a relative indifference to photoperiod and its distribution is limited by a low tolerance to frost (Peacock and MacMillan, 1968). In contrast the distribution of *Prosopis juliflora* over a similar latitude range has occurred through the natural selection of ecotypes which are adapted to the local photoperiod conditions. Opportunities, therefore, exist to select either for photoperiod indifference or for closely adapted types.

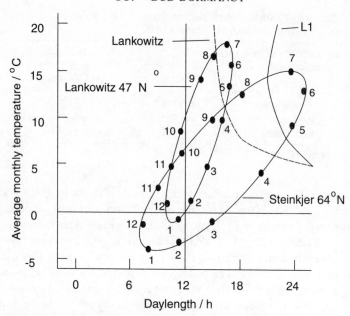

FIG. 11.8. Climate photothermographs for Steinkjer, Norway (64° N) and Lankowitz, Austria (47° N), together with critical phototherms for apical growth cessation of two ecotypes of *Picea abies* (L1 and Lankowitz). The co-ordinates are the average monthly temperatures and photoperiods (including twilight) on the 21st day of each month. Numbers on the photothermographs are months of the year. After Heide 1974a.

The Resumption of Growth

Once they have become dormant many tree species are unable to resume growth until rest has been broken by exposure to a sufficiently long period of low temperature. The effective temperatures (usually between 0 and 10°C) are similar to those for vernalisation (see Chapter 6). Although chilling is the major factor for the breaking of rest, the resumption of bud growth in some species is strongly modified by photoperiod. However, the fully dormant condition develops only gradually and long photoperiods alone may be sufficient to induce the resumption of growth in the pre-dormant state. In *Catalpa bignonioides*, plants resumed growth if they were returned to LD immediately after the terminal bud had abscised; after several more weeks in SD they had become fully dormant and resumed growth only after exposure to 2–3 weeks of temperatures below about 10°C (Downs and Borthwick, 1956a). Similarly, the dormancy between flushes in *Quercus robur* is broken by LD, whereas winter dormancy can only be broken by low temperatures (Leman, 1948). Thus, many trees enter true dormancy through a period of predormancy, in which they may be induced to resume growth by returning them to LD; the longer they remain in SD the less responsive they become until, when they are fully dormant, they fail to resume growth in LD. Continued exposure to SD after extension growth has ceased seems, therefore, to be important for the induction of deep dormancy. Following exposure to low temperatures, plants enter a period of **after rest** when LD may promote bud break. When dormancy is fully broken by low temperature, the resumption of bud growth is usually independent of daylength. For example, in cuttings from several northern tree species taken at monthly intervals from mid-November (including *Betula*, *Populus* and *Alnus*), the thermal time to bud burst was decreased in LD. However, the effect of daylength decreased with increasing duration of chilling and there was no effect when rest had been completely broken by low temperature; during deep rest in winter, LD did not fully substitute for chilling in any of these species (Heide, 1993a).

There is a considerable variation between species in their responsiveness to daylength as a factor inducing the resumption of growth once the dormant condition has developed. Plants such as *Cornus alba* (Whalley and Cockshull, 1976) do not resume growth in LD when they are fully dormant following a long period of exposure to SD, although they may do so in the pre- and post-dormancy states. Other species such as *Populus robusta* and *Liquidambar styraciflua* may slowly resume growth in mid-dormancy when returned to long photoperiods, although often only after a prolonged exposure to continuous light. Some species, such as *Weigela florida*, are very responsive to daylength and the resumption of growth could be brought about by returning the plants to LD at any stage, even in mid-dormancy. In *Cornus florida* and *Betula pubescens*, LD appeared to substitute, at least in part, for the effect of low temperatures on dormancy breaking (Downs and Borthwick, 1956a; Wareing, 1954), while *Fagus sylvatica* (beech) seedlings were reported to have no chilling requirement, so that the resumption of growth is always dependent on exposure to long photoperiods (Wareing, 1953). However, branches cut from an adult beech hedge from October to January did not resume growth in continuous light, whereas they did so in late February (Vince-Prue, 1975). Annual twigs taken from adult trees behaved similarly; in October, a high proportion of shoots was able to resume growth in continuous light but none did so in November and December (Table 11.6). Buds in this state of dormancy were not included in Wareing's (1953) experiments, leading to the erroneous conclusion that *Fagus sylvatica* requires only LD for the breaking of rest. It is evident that, in common with most other temperate woody plants, beech has a chilling requirement for bud-burst; however, the effect of daylength is more pronounced than in many species and even in mid February, bud burst continued to be highly erratic and delayed in SD. The response to increasing photoperiods may be of special importance for the breaking of bud dormancy in woody plants growing in regions of warm winters, where dormancy is only partially broken by low temperature.

TABLE 11.6   The effect of daylength on budburst in cuttings of beech (*Fagus sylvatica*).

Date	% Bud burst		Days to bud burst	
	SD	LD	SD	LD
15 October	0	60	> 50	22
15 November	0	0	> 50	> 50
15 December	0	0	> 50	> 50
15 January	0	46	> 50	31
15 February	13	86	40	22
15 March	40	100	34	13

Twigs were sampled at different times during the winter as indicated. Cuttings were placed in distilled water and incubated at 21°C in continuous light (LD) or in 8 h SD; bud burst was recorded when at least one bud on a cutting was showing green foliage.
From Heide (1993b).

Selection for strong photoperiodic control of bud-burst as in beech could also be an effective strategy for the avoidance of late spring frost injury.

The triggers for the resumption of growth in tropical species which show dormancy are largely unknown but shedding of leaves may be an important factor (Longman, 1969). In *Cedrela odorata*, budbreak occurred in LD but growth did not resume in SD until most, or all of leaves had abscised; similarly, in *Hildegardia bartei*, dormant, leafy plants resumed growth in LD but not in SD until the old leaves had become senescent or abscised.

DAYLENGTH PERCEPTION

Many features of the photoperiodic control of dormancy in woody plants are similar to those recognised in the control of floral initiation. The site of perception is in the leaves and the response occurs at the shoot apical meristems, so that at least one transmissible stimulus must be involved. In some cases, the nightlength rather than the daylength seems to be important; there is a sharply defined critical nightlength and an SD plus night-break is equivalent to an LD. However, in other species, the behaviour is more similar to that observed in many LDP, where an LD response requires a long daily exposure to light and is not very sensitive to a night-break; this light-dominant type of response prevents dormancy. As with other photoperiodic responses in higher plants, phytochrome appears to be the photoreceptor for the perception of daylength in the control of dormancy.

The Site of Perception

The locus of perception for the induction of dormancy by SD seems, as with flowering, to be the youngest fully expanded or partly expanded leaves. In *Cornus florida*, dormancy was induced and growth was inhibited when the leaf which had just attained full size was exposed to SD, even though all the other leaves remained in LD (Waxman, 1957). Similarly, in *Robinia* and *Acer*, active growth was maintained when the leaves alone were exposed to LD (Table 11.7). The importance of the leaves

TABLE 11.7 Site of photoperiodic perception for the onset of bud dormancy in actively growing seedlings.

Photoperiodic treatment given to		Species			
Shoot apex	Mature leaves	*Betula pubescens*	*Acer pseudoplatanus*	*Robinia pseudacacia*	*Quercus robur*
LD	LD	A	A	A	A
SD	SD	D	D	D	D
LD	SD	A	D	D	D
SD	LD	D	A	A	D

LD, plants received continuous illumination; SD, plants received 8–9 h of daylight. A, shoot apex still actively growing or (in *Quercus*) had developed a new flush of growth; D, shoot apex had ceased to expand new leaves or (in *Quercus*) no new flush of growth had developed.
Data of Wareing (1954).

has also been demonstrated in a number of tropical tree species; in *Hildegardia bartei* the presence of mature leaves prevented growth of the terminal bud in SD while, in the absence of leaves, *Cedrela odorata* failed to become dormant in SD, or re-started into growth (Longman, 1978). However, the daylength to which the bud itself is exposed appears to be important in some species. *Quercus* became dormant unless both leaves and the apical bud were in LD, while growth continued in *Betula* when the apex was in LD even though the leaves were exposed to SD (Table 11.7). The development of cold hardiness associated with dormancy has also been shown to depend on exposing the leaves to SD and plants of *Cornus stolonifera* developed cold resistance when only one leaf pair was treated (Hurst *et al.*, 1967).

As with flowering, therefore, evidence based on the site of perception of the photoperiodic signal indicates that one or more transmissible substances are generated by leaves under daylength control. Like the floral stimulus, the hardiness promoter originating in the leaves appears to be translocated through the phloem in *Cornus* (Fuchigami *et al.*, 1971a); the development of cold hardiness in the roots of *Taxus* was also reduced by stem girdling of plants in SD, again indicating transport in the phloem of a SD-dependent factor for the development of frost resistance (Mitya and Lanphear, 1971). A translocatable hardiness-inhibiting factor is also indicated, since exposure of the shoots to LD delayed or depressed the development of cold resistance in the roots of *Potentilla* and *Picea* when they were exposed to low temperature (Johnson and Havis, 1977). In *Weigela florida*, the SD-dependent, dormancy-inducing stimulus has been shown to be transported from a branch in SD to one in LD, especially if the latter was defoliated (Nitsch, 1959). In contrast, the effect of *low temperature* to induce cold hardiness in *Pseudotsuga* was localised to the treated branch and did not appear to involve a translocatable stimulus (Timmis and Worral, 1974). Whether transmissible stimuli are involved in plants such as *Betula*, where the primary site of perception appears to be in the apical bud itself, remains to be established; only the response of the apex was studied (see Table 11.7) and possible transmission of the daylength effect to lateral buds was not excluded.

Fewer studies have been made on the site of perception for the breaking of dormancy by LD. Leafy shoots of *Betula pubescens* resumed growth when only buds were exposed to LD, irrespective of whether the leaves were removed or remained in SD. Similar results were obtained for *Fagus sylvatica* and *Pinus sylvestris* (Wareing, 1954). Thus, in these species, the site of perception for the photoperiodic breaking of dormancy must reside in the bud itself. Leafless seedlings of *Fagus* resumed growth in LD when the outer bud scales were removed (Wareing, 1953), so that the young foliage leaves and possibly the presumptive internodes inside the bud scales appeared to perceive the photoperiodic signal directly. No transmissible substance(s) may, therefore, be involved.

Time Measurement

Although the basis of timekeeping has not been studied in any detail, the photoperiodic control of dormancy exhibits many of the features observed in the control of floral initiation and similar, if not identical mechanisms are assumed. Although temperature interacts with daylength in controlling the onset of dormancy, the photoperiodic timing mechanism appears to be largely insensitive to changes in

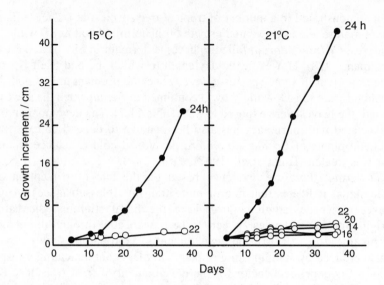

FIG. 11.9. Time course of apical shoot growth in *Salix pentandra* (ecotype Tromsö, 69° 39′ N) at two different temperatures. Seedlings were 2.3 cm in length when they were transferred to constant temperatures at the photoperiods indicated. The main light period (8 h) was given in daylight and the photoperiods were obtained by giving day extensions with tungsten-filament lamps. After Juntilla, 1980.

temperature, at least over a certain range; for example, in *Salix pentandra* (Fig. 11.9) and *Picea abies* (Heide, 1974a) constant temperatures in the range from 9/12 to 21/24°C had little or no effect on the critical daylength.

A fairly sharply defined critical daylength has been observed in those species where continuous growth is maintained in LD. The critical daylength for *Acer rubrum* (see Fig. 11.2), *Weigela florida* and *Catalpa bignoniodes* lies between 12 and 14 h; plants ceased to grow in daylengths of 12 h or less while growth continued indefinitely in daylengths of 14 h or more. A longer critical daylength of between 14 and 16 h was found for the cessation of growth in *Liriodendron tulipifera* and *Betula mandschurica* (Downs and Borthwick, 1956a,b) while in a far northern ecotype of *Salix pentandra* the CDL was >22 h (see Fig. 11.9). As already noted, the critical daylength is often found to vary considerably with ecotype, especially in species with a wide latitudinal or altitudinal dispersion (see Tables 11.3, 11.4). In some plants, the response to daylength appears to be essentially a quantitative one with no clearly defined critical daylength (see Fig. 11.1). Finally, there is often a critical daylength below which growth ceases rapidly, and also a quantitative response to increasing daylengths above this critical value (Junttila and Nilsen, 1993). As with flowering, variations in the critical daylength depend to some extent on the criterion used. In Fig. 11.2, the critical daylength was determined as that below which extension growth ceased; however, the photoperiod for 50% budset is also frequently considered to be the critical daylength.

The first question that can be asked with respect to the underlying timekeeping mechanism is whether plants measure the daily duration of light, or darkness, or both. The approach of varying the duration of darkness and light independently (see Chapter 1) does not appear to have been utilised to study timekeeping in dormancy induction, but a LD effect to induce dormancy-breaking in leafless seedlings of *Fagus sylvatica*

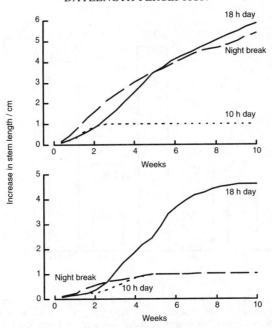

FIG. 11.10. Photoperiodic effects on the growth and dormancy of tree seedlings. Plants of *Tsuga canadensis* (top) and *Fagus sylvatica* (bottom) were grown in 18 h or 10 h photoperiods, or with a night-break treatment in which they received 10 h photoperiods plus 30 min of red light in the middle of the night. Vince-Prue 1975 (data of Nitsch and Somogyi 1958).

was obtained when short (6 h) light periods were associated with short (6 h) nights (Wareing, 1953). In many species, the effectiveness of night-breaks demonstrates that the duration of the daily dark period is a major factor in the photoperiodic regulation of dormancy. In *Tsuga canadensis*, for example, dormancy was prevented in 10 h light/14 h dark cycles when a 30 min night-break with red light was given in the middle of the night (Fig. 11.10) and, in *Weigela florida*, a 30 min night-break in a 16 h dark period was equivalent to continuous light (Fig. 11.11). Similarly, a night-break of 30–60 min at intensities between 0.4 and 1 W m^{-2} from tungsten filament lamps was sufficient to prevent the induction of dormancy by SD in several other species (Table 11.8); a 2 h night-break was as effective as a long photoperiod in maintaining extension growth in the tropical tree species *Terminalia superba* (Longman, 1969). The SD-induction of cold acclimation in *Cornus florida* and *Weigela florida* was also prevented by a 15 min night-break (Table 11.9) and, in leafless seedlings of *Fagus sylvatica*, budbreak occurred when a long dark period was interrupted by a night-break of 60 min (Wareing, 1953). The effect of long nights on cambial activity in *Abies balsamea* could also be negated by briefly interrupting the dark period with low intensity light (Mellerowicz *et al.*, 1992).

The effectiveness of brief night-breaks clearly demonstrates that, in these species, the underlying timekeeping mechanism in the photoperiodic control of dormancy involves measuring the duration of darkness. Night-break treatments have usually been given near the middle of a 14 or 16 h dark period and their timing in relation to light-off and/or light-on signals does not seem to have been investigated. However, evidence both for light-off timing and the involvement of a circadian rhythm has

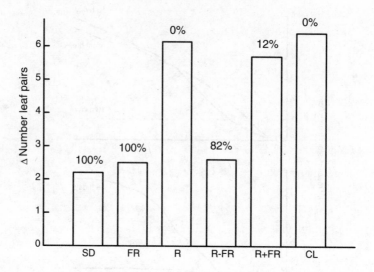

FIG. 11.11. Reversibility by far-red light of the effect of a red night-break to prevent bud dormancy in *Weigela florida*. Leaf counts and visual assessments of dormancy were made after 10 weeks of treatment; figures above bars show percentage of plants visually assessed as being dormant at this time. Plants received 8 h of natural daylight (SD) with a 30 min night-break with far red (FR), red (R) or red plus far red (R+FR), or 30 min R followed by 30 min FR (R-FR). CL, continuous low-intensity light from tungsten-filament lamps throughout the night. Irradiances were 0.45 W m^{-2} in red and in the 700–800 nm band of the far red source. Vince-Prue, 1984.

TABLE 11.8 Examples of species where the onset of dormancy is prevented or delayed by short night-break treatments.

Acer ginnala
Acer palmatum
Betula lutea
Cercidiphyllum japonicum
Diospyros virginia
Platanus occidentalis
Populus tacamahaca
Pseudostuga taxifolia
Rhus typhina
Taxus cuspidata
Tsuga canadensis
Weigela florida
Hibiscus syriacus
Pinus taeda
P. ponderosa

From Vince-Prue (1975).

recently been obtained by Clapham and co-workers at the Uppsala Genetic Center (personal communication) for the control of dormancy in a southern population of *Picea abies*. A 1 h night-break inserted into a 40 h dark period showed a circadian rhythm in the inhibition of budset, with maximum effect occurring at 6 and 28 h after light-off. Thus it seems very likely that the time-measuring clock for the control of

TABLE 11.9   Red/far red reversibility of the induction of freezing resistance in woody plants.

Treatment	Index of injury for stem segments frozen to $-12.5°C$	
	Cornus stolonifera	*Weigela florida*
(a) LD	42	62
(b) SD plus R night-break	37	60
(c) SD plus R/FR night-break	16	49
(d) SD	17	46

Plants were cold acclimated at 20°C (3 weeks), 15°C (1 week), 10°C (1 week) in (a), 16 h daylength; (b), 8 h daylength with a 15 min R night-break in the middle of the night; (c), 8 h daylength with a night-break of 15 min R followed by 15 min FR; (d) 8 h daylength. The index of injury was determined by measuring the conductivity of water in which thawed tissue segments had been held at 5°C for 24 h.
From Williams *et al.* (1972).

dormancy is a circadian rhythm, as in flowering. However, the precise way in which the rhythm is entrained by light remains to be determined.

An interesting situation has been described for *Pinus sylvestris* where long nights of 16 h induced dormancy, but the buds did not enter into rest. However, resting buds developed when the nightlength was progressively increased (by 1 h per week) as under natural conditions (Dormling, 1993). This is one of the few photoperiodic responses where responses to changing, rather than absolute photoperiods have been observed (see Chapter 1); no mechanism has been suggested. A somewhat parallel situation has been recorded for *Picea abies* (Dormling, 1979); when seedlings were raised in continuous light, the subsequent CNL was between 1.7 h (for the most northern population) and 5.8 h (for the most southern); however, when seedlings were initially grown under these photoperiods, a further lengthening of the dark period by 1–3 h was required in order to intiate budset. However, the absolute duration of darkness also influenced the development of rest in both species.

Night-breaks are relatively ineffective, or are much less effective than long photo-periods for the prevention of dormancy in some species (Nitsch and Somogyi, 1958). For example, under similar experimental conditions, a 30 min night-break at 0.8 W m^{-2} was equivalent to an 18 h photoperiod in *Tsuga canadensis* but was almost without effect in *Fagus sylvatica* (see Fig. 11.10). Other species reported to be relatively insensitive to a night-break include *Picea abies* (Fig. 11.12), *Cercis canadensis* and *Juniperus communis*. This lack of response resembles that seen in the control of flowering, especially in LDP where interrupted night treatments are frequently much less effective than long photoperiods (Chapter 5). Thus, as well as counteracting the effect of long dark periods in inducing dormancy, LD may have a specific effect to prevent dormancy in some species.

Photoperception

Despite the fact that a relatively short night-break with low-intensity light is sufficient to prevent dormancy in several species, there has been remarkably little work on the wavelength sensitivity of this response. There is no action spectrum for dormancy and wavelength sensitivity has been studied in rather few species. In part, this is due to the fact that the induction of dormancy may be a rather insensitive process but rapid

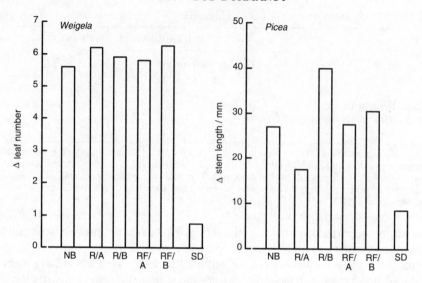

FIG. 11.12. Effects of time of giving day-extension treatments with red or red plus far red light on leaf production in *Weigela florida* and stem extension in *Picea abies*. Plants received day-extension treatment for 8.5 h (*Weigela*) or 9 h (*Picea*) with red (R) or red plus far red (RF), either before (B) or after (A) an 8 h day in sunlight. SD plants received no additional light. NB received a 30 min red night-break in the middle of the dark period. Vince-Prue 1984.

responses (even a single 16 h dark period was sufficient to cause almost 100 % budset in a northern population of *Picea abies* (see Fig. 11.18)) have been recorded and could be studied in more detail. In early work, a night-break with red light was found to be considerably more effective than blue light in preventing dormancy in several species (Phillips, 1941; Nitsch, 1962), indicating the likely involvement of phytochrome. Red/far red reversibility is the usually accepted criterion for a phytochrome-mediated response and has been demonstrated in the control of dormancy by a night-break in *Weigela florida*, as measured both by the number of leaf-pairs produced and by a visual assessment of dormancy (see Fig. 11.11). Red/far red reversibility has also been demonstrated in the photoperiodic induction of cold acclimation; in *Weigela florida* and *Cornus stolonifera*, the effect of a 15 min night-break with R light was fully reversed when 15 min FR was given immediately afterwards (Table 11.9). Axillary bud growth was recorded in the experiment with *Cornus* and the results indicated R/FR reversibility of bud growth as well as of cold acclimation. R/FR reversibility has also been demonstrated for the induction of dormancy by LD in *Lunularia cruciatea*; a 10 min night-break with R in the middle of an 11 h dark period inhibited growth as much as continuous light, and the effect was partly reversed when the red night-break was followed by 1 min FR (Wilson and Schwabe, 1964). Overall, therefore, there is good evidence from a range of species that phytochrome is the photoperceptive pigment for the night-break response in the control of a number of dormancy phenomena, including the prevention of growth cessation, the prevention of cold acclimation, and the induction of dormancy. The reversibility of the R effect by immediate exposure to FR light demonstrates that the effector molecule is Pfr.

It is evident that R is the most effective wavelength for eliciting a night-break response in the species examined so far. However, when the wavelength sensitivity of the photoperiod itself is examined, two different behaviour patterns have been observed (see Fig. 11.12). In *Weigela florida*, a day-extension for 8.25 h with R following a SD in sunlight was fully effective in preventing dormancy and no additional effect was obtained when FR was added. In contrast, a similar day-extension with R was rather ineffective in maintaining growth in *Picea abies* and there was a considerable increase in response when FR was added. There was also a major difference in the response of the two species to the time when the day-extension treatment with R was given. In *Weigela*, all day-extension treatments were equally effective and the response was similar to that obtained with a 30 min night-break. However, in *Picea*, a day-extension with R was most effective when it *preceded* a SD in sunlight; when it *followed* a SD in sunlight, a 9 h extension with R was even less effective than a 30 min night-break. *After* the 8 h day in sunlight, the addition of FR to the R day-extension markedly enhanced the response, whereas *before* the 8 h day in sunlight, the addition of FR actually decreased the response. It is evident that, as with the flowering response in many LDP (see Chapter 5), the sensitivity to added FR for the photoperiodic control of dormancy changes during the course of the daily cycle.

As all day-extension treatments shortened the night to 7 h, the duration of uninterrupted darkness cannot have been the only factor controlling dormancy in *Picea* and, in order to prevent the onset of dormancy, it seems necessary that plants are exposed to a long photoperiod with FR light present during the latter part of the day. This pattern of response contrasts with that observed in *Weigela*, where any day-extension treatment (with R or R plus FR) which shortened the dark period to 7.75 h was fully effective in preventing dormancy. Thus, for *Weigela*, control seemed to depend on the induction of dormancy by exposure to a sufficiently long dark period; when the dark period was shortened, or interrupted with a night-break, dormancy was completely prevented. If these two types of behaviour reflect differences in the underlying photoperiodic mechanism (as has been suggested for the more intensively investigated flowering response, Chapter 5), it appears that similar differences occur in the photoperiodic control of dormancy. Ecotypes of *Salix pentandra* have also been found to differ in their wavelength sensitivity to a day-extension treatment (Table 11.10). All plants of a northern ecotype (CDL = 22 h) ceased growth even in

TABLE 11.10 The effect of 12 h day-extensions with light of different R:R+FR ratios on growth cessation and cold hardiness in seedlings of a northern ecotype of *Salix pentandra*.

Light treatment	Cessation of growth (%)	LT_{50} (°C)	
		Before hardening	After hardening
Tungsten filament	14	−4.0	−9.4
Red fluorescent	100	−8.0	−18.0
Darkness	100	<−12.0	−26.0

The day-extension treatments were given from 20.00 to 08.00 h for 3 weeks at 18°C; after that plants were hardened at 3°C in a 12 h photoperiod for 3 weeks. The data are for a northern ecotype (69° 39′ N) from an experiment carried out in 1986. In a southern ecotype (59° 40′ N), neither the cessation of elongation growth nor the development of frost hardiness was influenced by the type of light used in the day-extension treatments.
From Junttila and Kaurin (1990).

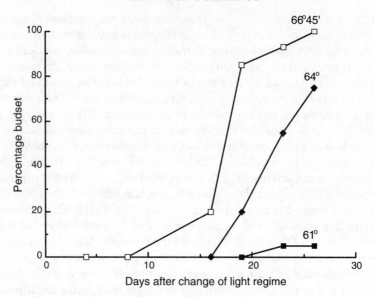

FIG. 11.13.   Response of latitudinal ecotypes of *Picea abies* to day-extensions from lamps with a low far-red content. Seedlings from latitudes ranging from 45° to 66° 45′ N were grown for 4 weeks in continuous light from metal halogen lamps (300 μmol m^{-2} s^{-1}). They were then transferred to a regime consisting of 8 h light from metal halogen lamps (300 μmol m^{-2} s^{-1}) followed by a 16 h day-extension (40 μmol m^{-2} s^{-1}) from Cool White fluorescent lamps (R:FR =10). Seedlings from latitudes 47, 56 and 59° continued to grow without setting buds under this high R:FR day-extension; seedlings from the most northern latitudes, however, rapidly became dormant even in continuous light. Data of D. Clapham, I. Ekberg, M. Qamaruddin and I. Dormling (previously unpublished).

continuous light when the 12 h day-extension was given with fluorescent lamps (R/FR = 10.0) but dormancy was completely prevented when the day-extension was given with tungsten filament light (R/FR = 0.6). A day-extension with R was also considerably less effective in preventing the onset of frost hardening. In contrast, a southern ecotype (CDL 14 h) responded equally to R and R plus FR light. Northern ecotypes are also rather insensitive to a night-break treatment (Junttila and Nilsen, 1993). Similar differences have been shown to occur among latitudinal provenances of *Picea*. When populations from latitudes ranging from 47° to 66° N were grown in continuous light with a 16 h day-extension with cool white fluorescent lamps lacking FR, the response ranged from 100% budset in the most northerly ecotype to 0% budset in ecotypes from 47° to 59° N; ecotypes from latitudes 64° and 61° N were intermediate in their response (Fig. 11.13). It appears, therefore, that adaptation to the northern long day conditions alters the response to light quality (and may change the underlying control mechanism) as well as increasing the CDL. As far as the authors are aware, no work has yet been directed towards identifying which phytochrome(s) is involved in either response type, but the requirement for FR light in the northern ecotypes suggests the participation of phytochrome A.

The control of dormancy can be effected by supplementary light of extremely low intensities and dormancy phenomena have been delayed or prevented by additional light between 0.04 and 1.0 W m^{-2} given as a day-extension (Vince-Prue, 1975). There is some evidence for ecotypic variation in sensitivity; even 0.1 μmol m^{-2} s^{-1}

prevented growth cessation in a southern ecotype of *Salix pentandra*, whereas >1.3 μmol m^{-2} s^{-1} were needed for a LD effect in a northern ecotype (Junttila, 1982a). Similar differences have been reported for *Picea abies*, where populations of southern origin have greater light sensitivity than northern ones (Dormling, 1979). Cold acclimation is also highly sensitive to light; in *Pseudotsuga*, the development of hardiness was prevented by irradiances ≥0.04 W m^{-2} when given during an 8 h day-extension (McCreary *et al.*, 1978). Budbreak in leafless branches of *Fagus* was saturated at 3.9 W m^{-2} which, because of the transmission properties of the outer bud scales, represented only about 0.03 W m^{-2} within the bud; when the outer scales were removed, 0.08 W m^{-2} was sufficient to cause the resumption of growth. There is no evidence of a requirement for high-intensity light in the photoperiodic control of either the onset of dormancy or the resumption of growth.

As with the control of flowering, light given intermittently throughout the night, or during a day-extension has been found to be an effective treatment in the control of dormancy. For example, when given as an all-night lighting treatment, 1–2 min light in every 30 was sufficient to prevent or delay dormancy in several conifers (*Pinus, Picea, Pseudotsuga*, McCreary *et al.* 1978; Arnott, 1974). The effectiveness of intermittent lighting schedules has practical implications for the raising of nursery stock since the cost of lighting is substantially reduced, in terms of both electrical loading and consumption.

Temperature

As with the photoperiodic control of flowering, temperature may be extremely important for dormancy phenomena. High temperatures were found qualitatively to inhibit the SD induction of dormancy in some hybrid seedlings of *Populus* which failed to produce resting buds when grown in a 33°C day/28°C night regime and continued to grow at the same rate as plants in LD (Paton and Willing, 1968). Increased temperature also delayed the onset of dormancy in SD by several weeks in *Acer rubrum* and *Betula* seedlings (Downs and Borthwick, 1956a; Wareing and Black, 1958). The photoperiodic mechanism may also fail to operate if the temperature is too low. At 10°C, plants of *Robinia pseudacacia* stopped growing in SD but failed to become completely dormant and resumed growth when returned to LD conditions. At 15°C, however, complete dormancy was established in SD. This interaction with temperature may be of considerable importance under natural conditions; in northern Russia, for example, SD-induced cold resistance (which is essential for survival) failed to occur if the temperature was too low (8°C). Photoperiodic cold acclimation in *Cornus stolonifera* is also dependent on temperature and plants became more cold resistant at 20°C day/15°C night than at 15°C day/5°C night (Fuchigami *et al.*, 1971a). For any particular locality, therefore, it is necessary that the temperature is within the effective range when the photoperiod shortens sufficiently for the induction of dormancy and cold acclimation.

THE IDENTITY OF DORMANCY-CONTROLLING STIMULI

The cessation of shoot growth and the development of bud dormancy depend on the exposure of the leaves to short photoperiods; consequently, it is assumed that

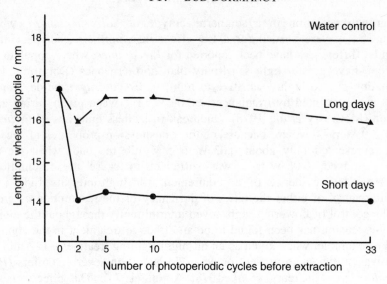

FIG. 11.14.   Inhibitor content of leaves of *Acer pseudoplatanus* at various times after transfer to long or short-days. The inhibitory eluates (R_F 0.55–0.88) from chromatograms of extracts of mature leaves were assayed by the wheat coleoptile test. Vince-Prue 1975 (data of Wareing, 1959).

dormancy-controlliing stimuli are exported from leaves and that these differ in LD and SD.

The association of growth inhibitors with bud dormancy has been recognised for many years (Vince-Prue, 1985; Lavender and Silin, 1987). As plants enter dormancy, an increase in the growth-inhibiting activity of extracts (as measured by a variety of bioassays) has been recorded for a range of species. Evidence that these inhibitors originate in the leaves comes from studies with sycamore (*Acer pseudoplatanus*). Following transfer to SD, an increase in inhibitory activity was detected in leaf extracts after only two SD cycles (Fig. 11.14) and this was followed by an increase in the apex after five cycles, as would be expected of a dormancy-inducing stimulus. Notably, this increase in inhibitory activity in extracts from the apical buds preceded the retardation of growth (Wareing, 1959). Inhibitors have also been detected in the transport systems and there was a pronounced increase in ether-soluble inhibitors in both phloem and xylem of willow (*Salix viminalis*) during entry into dormancy (Bowen and Hoad, 1968).

These results raise two questions. Firstly, does the change in growth-inhibitory activity of the extract result from a change in the content of inhibitor(s) and/or of growth promoters? Secondly, are endogenous growth inhibitors involved in the photoperiodic regulation of bud dormancy?

In sycamore, the inhibitory effect on growth in bioassays was found to be largely due to the presence of abscisic acid (ABA); indeed, this growth inhibitor was originally termed **dormin** because it was thought to be directly involved in the regulation of dormancy. Similarly, the inhibitor in the sap of willow was identified as ABA, which was the main inhibitory component present (Bowen and Hoad, 1968; Alvim *et al.*, 1976). Abscisic acid has also been identified in other dormant tissues.

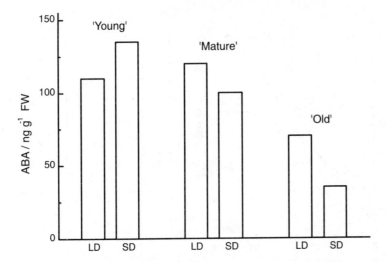

FIG. 11.15. Effect of daylength on levels of free abscisic acid in leaves of different ages from 7 week old seedlings of *Acer pseudoplatanus*. The experimental treatments consisted of 3 cycles of 18 h light/6 h dark (LD) or 8 h light/16 h dark (SD) and the extracts were made at the end of the final dark period. Prior to this, seedlings were grown in long photoperiods for 7 weeks. Leaves were small, not fully expanded (Young), fully expanded with no visible senescence (Mature), or the four lowermost yellowing leaves (Old). After Phillips *et al.*, 1980.

However, the induction of dormancy in willow was not correlated with maximum ABA levels in either buds, leaves or xylem sap (Barros and Neill, 1989). Moreover, when chemical methods rather than bioassays have been used, the ABA content of leaves and buds have not generally been found to increase when plants are transferred to SD (see Appendix II). In *Salix viminalis,* for example, no differences could be detected in the ABA content of young leaves from plants growing in short- or long photoperiods (Alvim *et al.*, 1979). Other workers have also failed to detect increases in endogenous ABA following transfer to SD (in, for example, *Betula pubescens, Acer pseudoplatanus*, Lenton *et al.*, 1972; *Salix pentandra*, Johansen *et al.*, 1986). These results are consistent with those obtained for a number of herbaceous plants. In *Acer pseudoplatanus*, photoperception occurs in mature leaves and so ABA would be expected initially to increase in these if it is associated with the transmissible photoperiodic signal. However, after three cycles, the level of ABA in these leaves was higher in LD than in SD (Fig. 11.15). Nevertheless, a relationship with dormancy and the cessation of extension growth is perhaps indicated by the fact that the ABA content of the young leaves near the apex was somewhat higher after 3 SD. As always, the large increases in ABA content which result from water stress may have caused a problem with some determinations of the effect of daylength on the content of ABA; however, experiments in which the water stress factor was eliminated confirmed that transfer to SD did not increase the content of ABA in strawberry and sycamore (Plancher and Naumann, 1978; Phillips *et al.*, 1980). It seems likely, therefore, that the increase in the growth-inhibitory activity of extracts from shoots exposed to SD resulted from changes in inhibitors other than ABA or from a decrease in growth-promoting substances which co-chromatograph with ABA in some solvent systems.

In contrast to the above results, there are a few recent investigations in which SD-

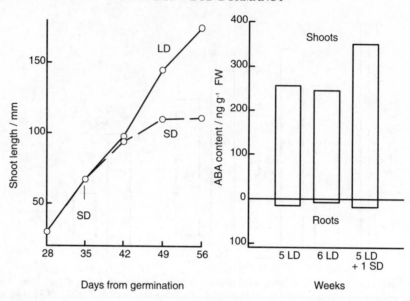

FIG. 11.16. Effects of daylength on shoot growth and the content of free abscisic acid in seedlings of *Pinus sylvestris*. Plants were grown for 35 days in LD (18 h) before they were transferred to SD (8 h) conditions or remained in LD. The irradiance was 470 μmol m^{-2} s^{-1}, and the day and night temperatures were 25°C and 15°C respectively. The abscisic acid extracts were made after plants had received 1 week in SD. After Odén and Dunberg, 1984.

dependent increases in endogenous ABA have been recorded. In birch (*Betula pubescens*), the ABA content in buds increased following transfer to 12 h SD which would cause growth cessation and the onset of dormancy in this ecotype; the increase was observed in both water-stressed and well-watered plants (Rinne *et al.*, 1994). The discrepancy between these and earlier results may be related to differences between ecotypes, since the birch was of a more northern origin than the plants used in the earlier studies. Alternatively, it may have been due to the tissue sampled, since there was no difference in the ABA content of the birch leaves, which would have been the major component in many of the earlier studies where whole shoots were analysed. A large increase in the ABA content has also been recorded in *Pinus sylvestris*, following transfer to SD (Odén and Dunberg, 1984). The increase occurred after one week, before shoot elongation ceased within 2 weeks of SD treatment (Fig. 11.16). There were marked additional increases when plants were subsequently exposed (sequentially) to three weeks of simulated late autumn (low light) and simulated winter (low light and low temperature) conditions in SD (Fig. 11.17). Returning plants to 'summer' conditions reduced the ABA content to a level equal to that found during the first LD period. Thus, in this conifer, a strong correlation was observed between the content of ABA in the (whole) shoot and seasonal changes in the environment; there was also a strong negative correlation between the growth increment of the seedlings and the ABA content of the shoots. However, it seems rather unlikely that the initial increase following transfer to SD was a photoperiodic effect, since plants also received a lower light integral and a lower mean temperature, both of which were found to increase the ABA content without a change in daylength (see Fig. 11.17,

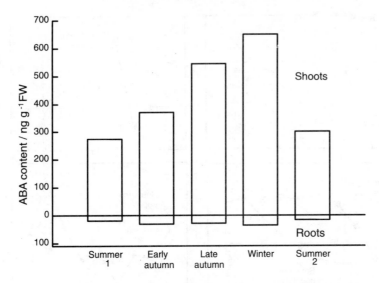

FIG. 11.17. Changes in the content of free abscisic acid in seedlings of *Pinus sylvestris* growing under simulated seasonal climates in controlled environment conditions. The growing sequence from germination was: five weeks in summer, three weeks in early autumn, three weeks in late autumn, three weeks in winter and three weeks in summer. The irradiance was 470 μmol m^{-2} s^{-1}, except in winter (60 μmol m^{-2} s^{-1}) and the photoperiod was 8 h, except in summer (18 h). The day/night temperatures were 25/15°C in summer and early autumn, and 15/10°C in late autumn; winter plants received a constant temperature of 4°C. After Odén and Dunberg, 1984.

early autumn compared with winter and late autumn). Moreover, increasing the duration of the SD treatment without further change in temperature or light integral had no effect. Thus, while the results are suggestive of some kind of causal relationship between the growth of the shoot and its ABA content, they do not provide any evidence to support the hypothesis that the SD induction of dormancy is achieved through changes in ABA. In another conifer, however, a transient increase in the ABA content of needles was observed when seedlings were transferred to SD conditions with no change in temperature (Fig. 11.18). Seedlings of *Picea abies* from two widely dispersed latitudes were grown in continuous light before being transferred to SD (16 h night) conditions at the same temperature, but at a much reduced light integral. A transient peak of free ABA in the needles occurred on the 4th SD in the northern population, but was not observed until the 8th SD in the southern one. The role of this early, transient increase in ABA is not clear, but its timing does appear to be correlated with the photoperiod response of the two populations. The same degree of both budset and budrest required exposure to fewer SD in the northern population than in the southern one, and the peak of ABA occurred earlier (see Fig. 11.18).

Even if ABA is not increased by SD, it could still be a component of the endogenous system which controls dormancy, if it is always exported from leaves to buds thus causing dormancy unless antagonised by compounds whose concentration varies with daylength. This possibility can be explored by considering the response to ABA application. The induction of dormancy in LD by the application of synthetic (±) ABA has been reported for sycamore, *Betula pubescens* and blackcurrant; typical bud scales (modified stipules in *Betula*; modified petioles in sycamore and black-

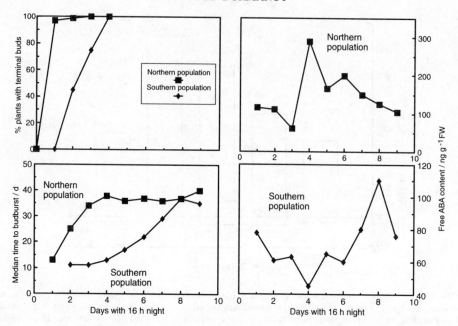

FIG. 11.18.   The effect of transfer to short-days on the cessation of elongation growth and the abscisic acid content of needles in two latitudinal ecotypes of *Picea abies*. Plants were raised in LD conditions at 340 μmol m^{-2} s^{-1}, 20°C, before transfer to SD (16 h dark) for the number of days indicated. Plants were then returned to LD conditions (1 h night for the northern and 5 h night for the southern population). The northern population originated at 66° 45′ N and the southern population at 46° 28′ N. Qamaruddin *et al.* 1995.

currant) were formed in all cases and the buds became fully dormant (El-Antably *et al.*, 1967). The application of ABA in LD increased low-temperature hardiness in *Acer negundo* and a similar effect was obtained with an inhibitor extracted from SD leaves (Irving, 1969). The formation of turions in some aquatic hydrophytes can also be induced by the application of (±) ABA; in *Spirodela polyrrhiza*, for example (where ABA may function *in vivo*, Saks *et al.*, 1980, Smart *et al.*, 1995) but not in *Myriophyllum verticillatum* except under marginally inductive conditions (Weber and Nooden, 1976a). However, (±) ABA applications were found to have no effect in causing dormancy in *Weigela florida* (a species in which dormancy appears to be induced by exposure to a long night and, therefore, to be effected by a dormancy-evoking stimulus) and *Catalpa bignonioides* (Cathey, 1968), nor did it induce cold hardiness in *Cornus stolonifera* (Fuchigami *et al.*, 1971b). In *Salix pentandra*, exogenous ABA did not induced growth cessation in continuous light, although the growth rate was reduced and stomatal resistance increased (Johansen *et al.*, 1986).

One possibility for the lack of effect is failure of sufficient ABA to reach the site of action in the shoot apex. For, example, foliar application of ABA did not induce bud dormancy in either *Betula* or *Alnus glutinosa* in LD, but only a small proportion (<10%) of the applied ABA reached the apical region unaltered (Hocking and Hillman, 1975). Similarly, although a daily foliar spray was ineffective (except at very high concentrations of up to 1000 mg l^{-1}), dormancy could be induced in *Acer palmatum* and *Cornus florida* by immersing the youngest expanded leaf in ABA together with spraying the apical region of the shoot (Cathey, 1968). A foliar spray of ABA was also ineffective in causing dormancy in *Betula* and sycamore (Hocking and Hillman, 1975) whereas dormancy was induced in both species when ABA was applied to both the leaves and apex (El-Antably *et al.*, 1967). However, even where it is effective in preventing elongation, (±) ABA does not always entirely mimic the effect of SD. In *Acer rubrum*, stem extension ceased following the application of ABA but normal winter resting buds failed to develop (Perry and Hellmers, 1973); terminal buds also failed to develop in *Picea abies* seedlings and normal growth was subsequently resumed. Neither did ABA application induce the abscission of leaves (Cathey, 1968) nor of the shoot apex in *Ailanthus glandulosa* (El-Antably *et al.*, 1967), both of which normally accompany the entry to dormancy.

Despite much research effort, the exact role of ABA in the regulation of dormancy induction in tree species remains unclear. Seasonal changes in endogenous ABA have been observed in a number of cases but there is little evidence that these are caused by changes in daylength. However, transient increases in ABA following transfer to SD have been observed in ecotypes of *Picea* and appear to be correlated with their photoperiodic behaviour. The application of exogenous ABA can clearly effect some of the changes associated with dormancy, especially the cessation or reduction of shoot elongation; however, other components of the dormancy complex are not necessarily affected.

Gibberellins

The involvement of GA in dormancy phenomena is fairly well established; dormancy has been shown to be delayed or prevented by the application of GA in several tree species, and the GA content of shoot apices and leaves was found to decrease as plants entered dormancy (Vince-Prue, 1985; Lavender and Silin, 1987). For example, a fall in GA content (determined by bioassay) was recorded in the shoot tips of *B. pubescens* and sycamore, as well as in leaves of peach, following transfer to SD. GA-like substances also greatly decreased in apical buds of *Salix pentandra* before the cessation of apical growth and shoot tip abscission, although changes in the natural photoperiod were small at this time (Junttila, 1982b). The application of GA to plants in SD (especially to the apical bud) prevented the cessation of apical growth in *Salix pentandra* and delayed the onset of dormancy in *Rhus typhina*. Interactions have also been observed between exogenously-applied GA and (±) ABA, with the effect of ABA to induce dormancy being partly, but not entirely overcome by GA_3. Not all species are responsive to GA application, however, and GA is often ineffective when plants have become fully dormant (Lavender and Silin, 1987).

Very similar results have been obtained in cold-hardiness studies. An inhibitor of cold-hardiness appears to be produced in LD leaves, while the application of GA_3 decreased hardiness in SD and the content of GA-like substances was higher in leaves under LD (Irving and Lanphear, 1968; Irving, 1969). Moreover growth retardants such as chlormequat and daminozide increased cold-hardiness in several species, suggesting the involvement of endogenous GAs (Alden and Hermann, 1971). However, hardiness promoters appear also to be produced by leaves; as noted above, (±) ABA can induce cold-hardiness in some plants, but not as effectively as exposure to SD (Irving, 1969). Evidence for translocatable cold-hardiness promoters from the leaves has also been obtained in peach, where application of auxin to half of the tree increased the cold hardiness of the whole tree, although leaf retention was only affected (increased) on the treated side (Seeley *et al.*, 1992).

Although the evidence is limited, a tenable hypothesis is that the photoperiodic induction of bud dormancy is partly achieved through changes in the content and/or composition of GAs in the buds. A decrease in GA-like activity in leaves and/or phloem exudates during the onset of dormancy have been observed in several woody plants (Lavender and Silin, 1987) while, in LD, there may be an increase in the amount of GA, or more likely a change in the composition of the GAs present. Analytic and metabolic studies with *Salix* strongly indicate that early 13-hydroxylation is probably the main biosynthetic pathway for GAs in *Salix* and perhaps in other hardwoods as well (Junttila, 1991). The active GA for shoot elongation in *Salix* appears to be GA_1, and the activity of GA_4 and GA_9 probably results from metabolic conversion to GA_1

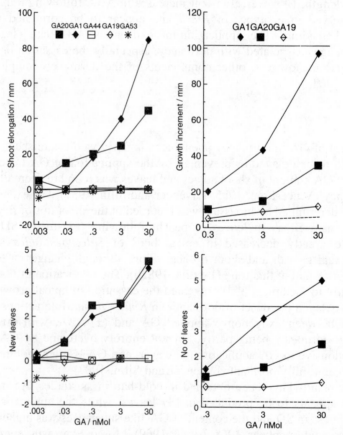

FIG. 11.19.   The effect of different gibberellins on shoot elongation and the development of new leaves in *Salix pentandra* and *Betula pubescens* seedlings under short-day conditions. Gibberellins were applied after 10 SD (12 h) and the response was recorded 14 days later. In *Salix*, the results are expressed as differences from the SD control without GA (shoot length = 19 mm, leaf number = 0.7); in *Betula*, the broken line represents the SD control, and the solid line is the LD control (continuous light). The birch seed originated from 59°N; the *Salix* population originated from 69° 39′ N. Junttila and Jensen, 1988 (*Salix*); Junttila, 1993b (*Betula*).

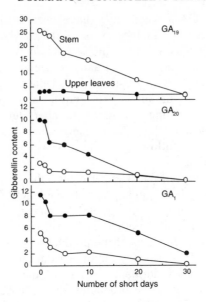

FIG. 11.20. The effect of short-days on the levels of specific gibberellins (ng g^{-1} FW) in young leaves and stems of *Salix pentandra*. Seedlings were grown in continuous light conditions consisting of 12 h high-intensity light (150 µmol m^{-2} s^{-1}) followed by a 12 h day-extension with low-intensity light from tungsten-filament lamps (10 µmol m^{-2} s^{-1}). On transfer to SD cycles, the low-intensity day-extension was discontinued. After Olsen *et al.*, 1995a.

(Junttila, 1993a). Both GA$_1$ and GA$_{20}$ promote shoot elongation and the continued production of new leaves of *Salix* seedlings maintained in SD, while the application of a range of precursors of GA$_1$ indicated that the conversion of GA$_{19}$ to GA$_{20}$ is blocked under these conditions (Fig. 11.19). This conclusion is supported by analytical studies (Olsen *et al.*, 1995a). Following transfer to SD, the levels of GA$_{20}$ and GA$_1$ in young leaves and stem tissue of *Salix* were 30–40% lower than in LD after 2 SD cycles (Fig. 11.20). In *Betula pubescens*, the cessation of stem elongation and leaf expansion in SD was also prevented by GA$_1$, whereas GA$_{19}$ had little or no effect (see Fig. 11.19). However, GA$_{20}$ was significantly less active than GA$_1$ in birch, suggesting that the conversion of GA$_{20}$ to GA$_1$ may also be partly blocked under SD in this case. These results are consistent with those for spinach, where the metabolism of GA$_{19}$ to GA$_{20}$ has been shown to be under photoperiodic control (Metzger and Zeevaart, 1982; Gilmour *et al.*, 1986). A further SD-dependent block between GA$_{53}$ and GA$_{44}$ has been demonstrated for spinach and may also occur in *Salix* since GA$_{19}$ does not accumulate in SD (see Fig. 11.20), although conversion to metabolites other than GA$_{20}$ was not ruled out. Since GA$_{20}$ is the precursor of GA$_1$, a SD-induced block in its biosynthesis (as demonstrated in *Salix*) could be an early process leading to the cessation of shoot elongation and the onset of dormancy in woody plants. However, in *Salix*, no difference was found in the metabolic conversion of applied 16,17-[^{3}H$_2$]GA$_{19}$ in plants grown in LD compared with plants that had been exposed to 14 SD before the application (Olsen *et al.*, 1995b). It was suggested that the discrepancy between these and earlier data indicating photoperiodic control of GA$_{19}$ to GA$_{20}$ in *Salix* might have resulted from the inability to detect relatively small differences in the rate of conversion of GA_{19} which may have been sufficient to result in the observed endogenous differences in the levels of its metabolic products. Alternatively, the use of $16,17-[^3H_2]GA_{19}$ as a mimic of GA_{19} may have been a factor.

In *Salix*, the ability to respond to LD gradually decreased with increase in the number of SD until, finally, LD alone was not able to induce new growth; the response to GA application followed a similar pattern (Junttila and Jensen, 1988). Thus the physiological state of the seedlings appeared to alter during the prolonged SD treatment and it is possible that these later changes are related to the subsequent development of bud dormancy, following the initial GA-dependent cessation of shoot elongation and leaf expansion.

Based on the results of physiological experiments, the existence of a dormancy-evoking stimulus produced by leaves in appropriate photoperiods is strongly indicated for some species. In *Cornus florida*, for example, dormancy was induced by exposing a single leaf to SD, even though all the other leaves were in LD (Waxman, 1957). It is difficult to interpret this except in terms of the production of a dormancy-stimulus by the SD leaves; the absence of a dormancy inhibitor from this single leaf could hardly account for the development of dormancy when all the other leaves, which are maintained in LD, are presumed to be producing this factor. Thus, as with flowering, the control of dormancy does not seem to be associated with changes in a single factor; both a 'dormancy inhibitor' (perhaps GA) and a 'dormancy promoter' appear to be involved. The exact status of ABA as an endogenous regulator of dormancy is still an open question. It seems unlikely that ABA is the SD signal from leaves which induces the onset of dormancy in woody species since most studies so far have failed to detect any increase in the amount of ABA present in leaves when plants are transferred to SD (Lenton *et al.*, 1972; Loveys *et al.*, 1974; Hocking and Hillman, 1975; Alvim *et al.*, 1978; Phillips *et al.*, 1980; Rinne *et al.*, 1994), as would be expected if it were the natural dormancy-inducing stimulus produced by SD-treated leaves. An exception is the transient increase observed in *Picea abies* needles. The extent to which control by a dormancy promoter and/or by a dormancy inhibitor is dominant probably varies with individual species and possibly with different ecotypes.

Growth Regulators and Budbreak

Irrespective of whether it is induced by daylength or develops autonomously, once dormancy has been fully established, the resumption of bud growth will usually not occur until the buds have experienced a period of low temperature. In many species, neither GA nor LD will substitute for a low temperature requirement during deep dormancy, although both may be effective before dormancy is fully established or when it is partly broken. With respect to endogenous GAs, it has been reported that GA-like substances did not increase prior to budbreak in flower buds of peach (Luna *et al.*, 1990), although GAs were abundant in actively growing buds and young shoots of several woody species (Zanewich and Rood, 1994). Thus, although GAs are associated with bud growth, they do not appear to be important for the breaking of rest. There is some evidence, however, that changes in cytokinins are associated with the breaking of dormancy, although they do not appear to be implicated in its induction, at least in woody plants. For example, CK activity was absent from dormant buds of *Betula* and *Populus*, but increased shortly after dormancy was broken, reaching a maximum

TABLE 11.11 The effect of cytokinins on bud growth of excised reproductive buds of *Salix babylonica* cultures *in vitro*.

Date of experiment	Number of buds sprouted	
	on basal medium	on basal medium +1.0 mg l^{-1} kinetin
20 April 1980	2	18
21 May	4	15
15 June	6	16
10 July	7	15
4 August	9	12
29 August	9	15
23 September	8	15
18 October	13	17
12 November	15	16
7 December	19	16
1 January 1981	18	19
16 January	19	20
15 February	20	19

Data of Angrish and Nanda (1982).

shortly before budbreak (Domanski and Kozlowski, 1968; Hewett and Wareing, 1973). A similar increase in CK content prior to budbreak occurred in the xylem sap of willow (Alvim *et al.*, 1976). In some experiments, however, there appeared to be no correlation between budbreak and CK content (Lavender and Silin, 1987). Similarly, the application of synthetic CKs has been found to stimulate the growth of dormant buds in some woody species (e.g. in grape, apple, peach), although it is not effective in others. In *Salix babylonica*, kinetin promoted the growth of excised buds throughout the dormant period; however, it no longer had any effect later in the year when natural budbreak began to increase on the basal medium (Table 11.11), suggesting that the level of endogenous CK was no longer limiting at this time.

In some vascular hydrophytes, CKs have been found both to prevent the development of overwintering, resting structures (turions) and to promote the resumption of growth in these structures. For example, in *Myriophyllum*, the application of CK prevented turion formation in SD and also promoted the resumption of growth (Weber and Nooden, 1976a,b). Since CK activity decreased during the development of turions and the content of CKs was higher in LD, it was suggested that the photoperiodic induction of turion formation in *Myriophyllum* by SD probably involves a decrease in endogenous CKs; however, an increase in acidic inhibitors also accompanies the formation of turions and is likely to be part of the overall control mechanism.

Budbreak in woody species may also be dependent on a decrease in the content of growth inhibitors during the winter. In *Betula*, for example, free ABA was substantially lower in late winter (February/March) than in the autumn (September). Although the content of both free and esterified ABA fluctuated somewhat from month to month, the ratio of free to esterified ABA declined progressively and was well correlated with the time taken for 50% budbreak, which also decreased as winter progressed (Harrison and Saunders, 1975). The ABA content of Japanese pear buds decreased rapidly during rest when maintained at 5°C but only very slowly at 15°C; injection of ABA into excised shoot inhibited bud break following 500 h at 5°C but had no effect after 800 h (Tamura *et al.*, 1993). In contrast, no significant changes were detected during the winter months in either the free or esterified ABA content of almond buds (Lesham *et al.*, 1974). Moreover, although changes in ABA during winter have usually been ascribed to the effect of chilling, this is not necessarily so; in apple buds, the ABA content declined under both high and low temperatures but only the latter alleviated rest (Borowska and Powell, 1982/3). There was no clear effect of chilling on the ABA content of buds in *Betula pubescens* (Rinne *et al.*, 1994). The effect of applying ABA is also somewhat variable and may either inhibit budbreak or have no effect, depending on the experimental conditions and species. In LD, the application of ABA through cut stems depressed budbreak in *Salix alba* and a *S. babylonica/alba* hybrid, but had little effect on *S. fragilis* (McWha and Langer, 1979). There was, however, no response to ABA in the presence of leaves, suggesting that substances from the LD leaves (perhaps GA) were counteracting the inhibitory effect of the applied ABA. The degree of inhibition by ABA also depended on season and, in the absence of leaves, was less in winter than in autumn.

The failure to break dormancy is of considerable importance for fruit trees growing in regions with warm winters, where the duration of winter cold is often insufficient to effect complete dormancy breaking and leads to erratic and/or delayed bud growth. In commercial practice, various treatments are been used to break dormancy, including sprays of mineral oils, often together with dinitroorthocresol; the dormancy breaking effect has been increased by the addition of thiourea and/or potassium nitrate. Although they have been found to promote budbreak in some tree fruits, gibberellins and cytokinins have not been exploited commercially because of high cost and problems of penetration. The dormancy-breaking effect of mineral oils has been associated with the imposition of anaerobic conditions and, in this context, it is interesting that both low temperatures and anaerobic stress have been found to increase the transcriptional activation of alcohol dehydrogenase (Lang, 1994).

MOLECULAR MECHANISMS

Shortly after transfer to SD, return to long photoperiods or the application of GA usually restarts growth but, after more prolonged exposure to SD, a release from dormancy in many plants occurs only after exposure to low temperature. Some progressive change, therefore, seems to take place leading to the induction of deep dormancy in the buds. It has been proposed that the dormant condition is one in which genetic activity is repressed and the synthetic activity of the cells is shut down. However, in *Spirodela*, the induction of dormancy and turion formation by ABA appears to require the production of specific new mRNAs, which result in a sequence of steps leading finally to the shut down of all synthetic processes (Smart and Trewavas, 1984). Following the application of 10^{-4} M ABA, DNA synthesis was inhibited within 3 h, followed by a decrease in protein synthesis after 24 h. However, the protein pattern was also altered and some new, turion-specific proteins were formed; these were found only in tissue capable of developing into turions. Not until 3 days after ABA was applied (by which time the developing primordia were already committed to turion formation) was there an inhibition of total mRNA, leading to the onset of irreversible events leading to the dormant state.

An increase in the synthesis of proteins also occurs at the same time as endodormancy begins in *Populus*; low levels of bark storage proteins are found in LD but transfer to 8 h SD results in an increase in their translatable mRNAs within a week.

The sequence of molecular changes occurring in the needles during the onset of SD-induced dormancy has been studied in two contrasting populations of *Picea abies*, originating from 66° 45′ N and 46° 28′ N (Clapham *et al*, 1994). In both populations, growth cessation and budset were essentially complete 2–3 weeks after transfer to SD. The content of poly(A)$^+$RNA in the needles decreased within the first week of SD in both populations, as did the *in vitro* translatability. The southern population showed a sharp early trough of *in vitro* translability, which coincided with a transient peak in ABA; neither the trough in translatability nor the peak in ABA was seen in the northern population (although a much earlier ABA peak was observed in a subsequent experiment; see Fig. 11.18). In contrast, the total RNA content of the needles (g FWt^{-1}) increased over several weeeks. The relation of these observations to growth cessation and budrest is not clear, since they were followed only in the needles and not in the buds entering dormancy, nor in the subapical regions of the shoot apex where the growth changes occur. In *Picea*, frost tolerance also increases in response to SD; however, there were no obvious correlations between the molecular changes and the degree of hardiness, apart from the upward trend in total RNA over the period of acclimation.

CONCLUSIONS

Bud dormancy is a complex process which, depending on the species, may include the cessation of shoot extension, the development of resting buds with various types of morphological modification, entry into rest (endodormancy), increase in frost hardiness, increase in drought resistance, leaf fall, abscission of the apical bud and, finally, breaking of rest allowing the resumption of growth under suitable environmental conditions. Each of these components of dormancy has been shown to be influenced by photoperiod, although the degree of daylength control differs widely between species. Probably the most consistent photoperiodic effect is the induction of growth cessation and dormancy by SD in woody plants of temperate and cold climates. The mechanisms for the photoperiodic control of dormancy appear to be similar to those for the control of flowering. The site of perception is the leaf, phytochrome is the main photoreceptor and the duration of darkness is the overriding factor. In some cases, however, exposure to a shorter than critical duration of darkness may not alone be sufficient to effect a LD response and there is requirement for far red light during the photoperiod. The underlying timing mechanism(s) has not been studied in any detail.

Since the site of daylength perception is the leaf, it is presumed that dormancy-controlling stimuli produced in long and/or short days are translocated to the shoot apices and buds. The known plant growth regulators (especially ABA, CK and GA) have been found to influence many aspects of dormancy and are generally thought to be the main controlling factors. Specific daylength effects on GA metabolism have been recorded in *Salix* and SD-imposed blocks in GA metabolism may be involved in the cessation of shoot growth, which precedes entry into rest. ABA has also been implicated as a possible dormancy-inducing stimulus, although many experiments have not shown an increase in ABA content following transfer to SD as would be expected if it were the leaf-generated SD signal. The breaking of dormancy normally requires exposure to chilling temperatures, although LD may partially substitute for these; there is, however, no information concerning the biochemical basis for this daylength effect nor whether any transmissible stimuli are involved.

12 Vegetative Storage and Propagation

One of the features of many higher plants is their capability of vegetative (i.e. non-sexual) means for propagating or perpetuating themselves. In many cases this takes place through the formation of resting structures which have both a storage and reproductive function. Vegetative storage organs arise by lateral swellings of a number of different tissues including stems (tubers, corms), roots (tuberous roots or root-tubers) and leaves (bulbs). Physiologically, however, they all have a similar function as perennating organs and are regions into which storage materials are mobilised. Their development is usually accompanied by the cessation of active growth followed by senescence and death of the remainder of the plant. Once formed, storage organs enter into a state of dormancy during which they are more resistant to unfavourable environments such as water stress and high or low temperatures. In some cases plants propagate vegetatively without passing through a dormant storage phase. A good example of this in a commercially important species is strawberry, where daughter plants are produced on runners, which are formed by extreme elongation of internodes in what is otherwise a rosette plant.

Because many vegetatively propagating storage structures are also important food crops, the factors leading to their development have received considerable attention. In several species, the formation of storage organs depends on, or is accelerated by, exposure of the leaves to particular photoperiods. In other cases, endogenous or other environmental factors such as temperature are more important. With the exception of the formation of bulbs in the genus *Allium*, which is favoured by LD, most photoperiodically induced storage organs are favoured by exposure to SD (Table 12.1). Runner formation in strawberries is however favoured by LD and inhibited by SD (Guttridge, 1969).

TUBERS AND CORMS

The formation of storage organs involves a series of morphogenetic events. Stem tubers, such as in the potato, arise by a swelling of a cluster of internodes, often at the

317

TABLE 12.1   Influence of photoperiod on the formation of storage organs.

Species	Storage organ
Favoured by short days	
Apios tuberosa (groundnut)	Root tubers
Begonia evansiana	Aerial stem tubers
B. socotrana	Aerial stem tubers
B. tuberhybrida cv Camelliaflora	Underground stem tubers
B. tuberhybrida cv Multiflora	Underground and aerial stem tubers
Dahlia hybrida	Root tubers
Dioscorea divaricata	Aerial axillary stem tubers
D. alata (yam)	Root tubers
Helianthus tuberosus (artichoke)	Underground stem tubers
Oxalis sp.	Underground stem tubers
Phaseolus multiflorus	Root tubers
S. andigena	Underground stem tubers
S. demissum	Underground stem tubers
S. tuberosum	Underground stem tubers
Trientalis borealis	Underground stem tubers
Ullucus tuberosus	Underground stem tuber
Favoured by long days	
Allium ascalonium (shallot)	Bulbs
A. cepa (onion)	Bulbs
A. sativum (garlic)	Underground and aerial bulbs
Brodiaea laxa	Corms

After Vince-Prue (1975).

tips of underground stems or stolons; the resultant tuber carries leaf scars and buds from which growth is resumed after dormancy is broken. Tubers can also form on above ground parts. In the 'Multiflora' group of the tuberous rooted begonias (*Begonia tuberhybrida*) both aerial and underground tubers developed in short photoperiods of 10–12 h (Lewis, 1953). Aerial tubers were formed by enlargement of the stem behind the growing tip and eventually the stem apex was included in the tuber. Occasionally tubers also formed at a node. When aerial tubers formed, they were fully functional and developed into new plants after a period of dormancy. Some cultivars, however, rarely produced aerial tubers and plants in the related 'Camelliaflora' group of *Begonia tuberhybrida* never did so. *Begonia evansiana* also produces both underground and aerial tubers, but only the latter are under the control of daylength (Esashi, 1960). The first sign of aerial tuber formation is when starch deposition begins in the apical meristem, activity of the distal region is arrested and the vegetative growing points swell and become dome-shaped.

As with many other responses, control of the formation of storage organs by photoperiod may be either quantitative or qualitative. Tuberisation in potato appears always to be hastened by SD (Wassink and Stolwijk, 1953; Pohjakallio *et al.*, 1957), although cultivars differ considerably in the extent to which they are affected. Some have an absolute SD requirement but most of the European and North American cultivars readily form tubers in LD. In SD conditions the stolons are shorter and there is a rapid cessation of vegetative growth with early dieback of the haulms. In LD, vegetative growth is more prolonged. Because of the early senescence of the haulms, tuber yield in potato is often lower in SD despite the fact that the onset of tuberisation is accelerated. Thus, although the time at which tuber formation begins is influenced by daylength, the energy economy of the plant determines its final yield (Pohjakallio, 1953).

Perception of daylength is accomplished by leaves and one or more stimuli are then translocated to the responsive regions. The role of the leaves in the perception of photoperiods for tuber formation was first studied in *Helianthus tuberosus* (artichoke). Tubers developed below ground when only the stem tips were covered to give SD, provided that leaves up to 2 inches long were included in the treated apical region (Hamner and Long, 1939). *Begonia evansiana* showed no tuberising response to SD when leaves were not present. The most sensitive leaves were those which had just fully expanded, or were nearly at the end of their expansion period, and exposure of young folded leaves to SD was quite without effect (Esashi, 1960). Similar findings were made for potato (Chapman, 1958) where the terminal leaf cluster was found to be much more responsive to SD stimuli than expanded leaves near the base of the stem.

The fact that the formation of storage organs can be induced by exposing only a small fraction of the total leaf area to photoinductive cycles may be considered as evidence for the production by these leaves of a transmissible stimulus which evokes the formation of storage organs at specific receptor sites. There is, however, also some evidence for the production of a transmissible inhibitory substance in leaves that are exposed to non-inductive photoperiods. When different areas of leaves in *Begonia evansiana* were exposed to SD or LD, tuber formation decreased as the ratio of the area in SD to that in LD decreased, and fell to zero at a ratio of 1:2 (Esashi, 1961b). Long days were equally inhibitory when given either to the distal or proximal half of the leaf, with the remainder in SD (Fig. 12.1). This compares with the SDP, *Perilla* (Chailakhyan, 1945) and *Kalanchoë* (Harder *et al.*, 1949), where inhibition of flowering occurred only when the *proximal* part of the leaf was in LD (see Table 6.12). If

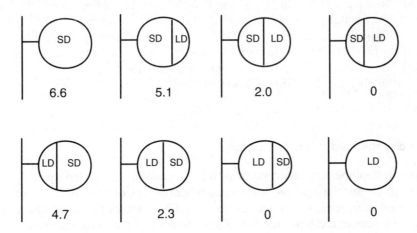

FIG. 12.1. Tuberization in *Begonia evansiana* in response to long- and short-day treatments given to different parts of the leaf cut to a 12 cm² disk in a single node cutting. The areas indicated were exposed to 7 SD and the remainder of the leaf received LD. Tuber formation was scored after 11 further LD and the results are given below the treatment diagrams. After Esashi (1961b).

TABLE 12.2 Effect of length of the photoperiod given prior to SD induction of tuberisation in *Begonia evansiana*

Photoperiod (h)	Tuberising stage
14.25	2.1
15	1.7
16	1.3
18	1.5
20	1.3
24	1.3

Photoperiods of the lengths indicated were given for the 4 days immediately preceding three inductive SD cycles. Tuberisation was recorded after 15 days.
After Esashi (1961b).

cuttings with two leaves were used and different photoperiodic conditions given to each one, the leaf in LD suppressed tuberisation in relation to its area both at its own node and at the node of the leaf in SD. The inhibitory effect occurred irrespective of whether the LD leaf was above or below the SD leaf. Long days given prior to the SD inductive treatment also inhibited tuberisation, and the inhibitory effect increased with increase in length of the daily light period (Table 12.2). These results indicate very strongly that a transmissible tuber inhibitor is produced by *Begonia* leaves exposed to LD. A positive inhibitory effect of non-inductive photoperiods is also indicated for onion where SD intercalated between LD nullified their effect (Kato, 1964) in a manner similar to that observed in flowering, and in potato where removal of the tip and young leaves resulted in the formation of tubers in non-inductive LD cycles (Hammes and Beyers, 1973). In potato, tubers formed when the mature leaves were in SD, even if the tip and young leaves were in LD; it seems, therefore, that a positive stimulus from these leaves in SD is able to overcome the inhibitory effect of the tip. Exposing the tip alone to SD promoted tuberisation but not so effectively as removing it. Tuber formation in response to SD is largely an inductive process. For example only one or two SD cycles were found to be sufficient for the induction of aerial tuber formation in cuttings of *Begonia evansiana* (Esashi, 1961a) and potato plants given 14 SD invariably produced tubers when returned to a continuous light regime (Chapman, 1958).

Short-day induced tuber formation

Studies on photoperiodic perception in tuber formation have been carried out with relatively few species and the most comprehensive set of information refers to the tuberisation of *Begonia evansiana*. In the most direct test for photoperiodic regulation a brief exposure to red light has been shown to prevent the inductive effect of a long night on tuberisation in *Helianthus tuberosus* (Nitsch, 1965), *Solanum tuberosum* (Batutis and Ewing, 1982) and *Begonia evansiana* (Esashi, 1966a). In *Begonia* 1 min of R at 8 W m^{-2} completely prevented tuber formation. The time of maximum sensitivity to a R night-break in *Begonia* was found to be at the 6–8th hour of darkness

TABLE 12.3 Reversibility of the effect of a R night-break on tuberisation in *Begonia evansiana* by FR.

Treatment	Tuberising stage
Control	1.5
R	0
R–FR	0.7
R–FR–R	0
R–FR–R–FR	0.3

R and FR were given for 1 min and 6 min respectively at 8 W m^{-2}.
From Esashi (1966a).

and this was not altered by a pre-illumination with FR at the end of the day (Esashi, 1966a). As with flowering responses, therefore, the time of maximum sensitivity to night-break light does not appear to be related to the time taken for dark reversion of Pfr to Pr but rather to be determined by a timer which is closely coupled to the daily light/dark transition. R/FR reversibility of the night-break has been demonstrated for *Begonia*, but only under some light/dark cycle combinations. When R was given at the 7.5th hour of a 12 h dark period a subsequent exposure to FR resulted in partial reversion which could be repeated several times (Table 12.3, Fig. 12.2a). Under these circumstances giving a dark interval of varying length between the R night-break and subsequent exposure to FR, reversibility by FR was lost after about 90 min at 26°C. The coupling of the Pfr form of the photoreceptor to the transduction chain and propagation of the Pfr signal can therefore be assumed to have occurred within 90 min. When a longer dark period of 16 h was broken by R given after 8 h there was no reversal by FR (Fig. 12.2b) even when the exposure time was kept very short (3 s) and the FR was given immediately. Under these conditions even brief exposures to FR tended to increase the inhibitory effect of R on tuberisation. Thus, in *Begonia*, we find a situation analogous with that in flowering in some SDP where, particularly with shorter days and a longer night, FR reversibility of the R night-break effect cannot be demonstrated. As with flowering, this loss of reversibility by FR light appears to be associated with an increase in the inhibitory effect of FR when it alone is given as a night-break (Esashi, 1963). In *Begonia*, FR given immediately after the end of the main photoperiod may inhibit or promote tuberisation depending on the length of the photoperiod. With cycles of 12 h light/12 h dark, FR at the end of the day either slightly promoted or had no effect on tuberisation. However, with cycles of 8 h light / 16 h dark, FR at the end of the day substantially inhibited tuberisation (see Fig. 12.2). These results indicate that there is a requirement for Pfr in the dark period following an 8 h light period for a strong SD tuberisation response and that the requirement is apparently satisfied during a 12 h photoperiod. This appears to be directly equivalent to the Pfr-requiring reaction for flower initiation seen in several SDP (see Chapter 4).

The light requirement during the main photoperiod has also been studied in some detail for *Begonia* (Esashi, 1961a, 1966b; Esashi and Nagao 1958). In order for a dark period to be inductive it must be preceded by a period of light and no tubers were formed in continuous darkness. In intact plants the minimum duration of light necessary for aerial tuber formation was about 2 h. Cuttings were found to be

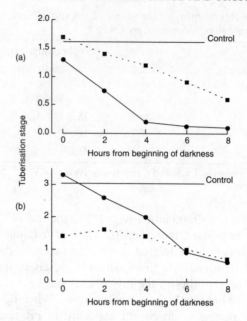

FIG. 12.2. Effects of R or R followed by FR on tuber formation in *Begonia evansiana*. R and FR at 8 W m^{-2} were given for 2 and 5 min respectively. Plants were given R (●) or R/FR (■) as shown during a 12 h (a) or 16 h (b) dark period. Controls were unirradiated during the dark periods. After Esashi (1966a).

TABLE 12.4 Minimum length of photoperiod required for SD-induction of tuberisation in *Begonia evansiana*.

Duration of daily light period	Stage of tuberisation	
	Experiment A	Experiment B
Dark control	0	
1 s	0	
3 s	0.1	
10 s	0.3	
1 min	0.5	
10 min	0.7	
30 min	1.1	
1 h		2
3 h		4.9
5 h		6.5
7 h		7.5

Plants received the photoperiods indicated for 8 d and were then transferred to LD for 12 d before recording the tuber stage.
From Esashi (1961a).

considerably more sensitive to SD induction and some tuber formation occurred with only a few seconds of light each day. Maximum tuber formation, however, required a daily photoperiod of at least 5 h (Table 12.4). Tuberisation in response to SD occurred when the light intensity was below the compensation point (Esashi, 1961a) and the requirement appears to be for light rather than carbon fixation. The effect of the SD photoperiod was not reduced even when the leaf was maintained in an atmosphere without carbon dioxide under nearly anaerobic conditions despite the fact that tuberisation involves the mobilisation of photosynthetic products to the tuber-forming sites. Light quality experiments showed that tuberisation in *Begonia* was induced most strongly by R during the short photoperiod but FR was ineffective.

The photoperiod mechanism for the induction of tuber formation by SD in *Begonia* thus appears to be very similar to that for the induction of flowering in many SDP. A period of light preceding an inductive dark period is essential but only short exposures, or low intensities of light are needed. R is more effective than blue and green, and FR is ineffective. With short photoperiods of 8 h, a terminal exposure to FR reduces the degree of tuberisation. This inhibitory effect of FR given at the end of the day is accompanied by an increased sensitivity to FR given as a night-break, and consequently FR reversibility of a R night-break cannot be demonstrated. With longer photoperiods of 12 h, FR at the end of the day is no longer inhibitory. The effect of the R night-break is then reversed by subsequent exposure to FR showing that phytochrome is the photoreceptor. The induction of tuberisation is primarily determined by the duration of uninterrupted darkness provided that this is preceded by an adequate photoperiod. The inductive effect of the dark period can be nullified by a short night break given 6–8 h after the start. All of the above features are typical of a 'dark-dominant' type of control seen in the induction of flowering in SDP (see Chapter 4). They are consistent with the idea that photoperiodism is a modular process in which daylength perception is a discrete capability which can be coupled to a range of possible end responses as required. If this is the case we would expect that the night break sensitivity is established by a circadian rhythm that is phase-set at the beginning of the dark period. This point has not as yet been addressed in experimental studies on tuberisation.

The availability of genetic transformation in potato potentially opens up new opportunities for studying daylength perception in relation to tuberisation. Potato is much less well characterised than *Begonia* with respect to its photoperiod sensing mechanism but the work so far carried out indicates a dark-dominant type of mechanism. It has been reported that short R treatments given in the middle of an inductive long night inhibit tuberisation (Batutis and Ewing, 1982). These authors also reported reversal of the R effect by FR, although the responses were somewhat variable. Phytochrome genes for phytochromes A and B have been cloned from potato (Heyer and Gatz, 1992a,b) and transgenic potatoes with modified phytochrome levels have been produced (Heyer *et al.*, 1995; Jackson *et al.*, 1996). The suppression of phytochrome B expression by use of antisense methodologies in transgenic *Solanum andigena* has been reported to result in the loss of photoperiodic sensitivity. It may be that this is a similar situation to the ma^3 mutant of *Sorghum* (see Chapters 4 and 9) which is deficient in a stable phytochrome B-like phytochrome and was originally described as a photoperiod mutant (Morgan, 1994). It was later shown that the ma^3 mutant retains the ability to respond to daylength, and is now considered to be partially insensitive to photoperiod. A series of *Solanum tuberosum* plants has been produced in which the potato phytochrome A gene has been introduced in sense and antisense orientation to give a range of phytochrome expression. Although the cultivar used forms tubers in LD as well as in SD, tuber formation was greater in the SD

TABLE 12.5   Effect of daylength on tuberization and stolon formation in transgenic plants of *Solanum tuberosum* over- or under-expressing phytochrome A.

Potato line	Tubers per plant		Stolons per plant	
	Long days	Short days	Long days	Short days
WT	3.0	8.0	0.5	2.0
AP 9	1.0	9.1	0.0	3.8
AP 11	0.0	12.7	1.1	9.1
PS 2	0.0	5.7	0.0	1.7
PS 4	1.7	7.7	0.0	0.0

Plants were grown for 3 weeks in either 16 h (long days) or 8 h (short days) daylengths but with the same light integral. AP 9 and AP 11 are antisense lines and PS 2 and PS 4 are over-expressers. (unpublished data of B. Thomas and D. Mozley).

treatments. The response to SD was enhanced in antisense plants and reduced when phytochrome A was overexpressed (Table 12.5). Increased numbers of tubers in the antisense plants was a result of the increased production, from the axillary buds low down on the stem, of stolons that then produced tubers at their tips. The production of stolons was prevented by LD in the WT and in SD by phytochrome A overexpression. Thus, phytochrome A expression under SD conditions acts to suppress stolon formation in potato. Lack of phytochrome A did not prevent the detection of SD suggesting either that phytochrome A is not involved in SD perception or that there is redundancy between phytochromes in the system. The lack of an obligate role for phytochrome A in the perception of SD for tuberisation is consistent with the lack of evidence for its involvement in SD perception for flowering (see Chapter 4), again suggesting that the perception mechanisms for both responses are likely to be the same.

BULB FORMATION

In bulbs, the storage tissues are modified leaves in which there is no development of the photosynthetic function. Initiation of bulbing in onion is marked by lateral swelling of vacuolated mature cells in the central region at the base of the innermost of the leaves which have already emerged. Cell division and extension slows down or ceases in the basal region of un-emerged leaves, which swell to form the scale leaves of the bulb unit. As a consequence, no new foliage leaf blades are produced (Heath and Holdsworth, 1948). The cells of monocotyledonous leaves such as onion arise from cell divisions at the basal or intercalary meristem of the leaf unit. Cells are produced sequentially and pass initially through the sheath region before reaching the blade, at which time they develop photosynthetic capability. Cells in the sheath are therefore developmentally immature and one way of considering bulb formation is the arrest or redirection of development in the immature, but potentially photosynthetic cells. Swelling of the sheaths and scale leaves is accompanied by the mobilisation of carbohydrate into the bases of the very young leaves, where it is stored as long chain fructans. Eventually the emerged leaves and roots senesce and the bulb remains dormant until the inner leaf initials and new roots emerge.

Bulb initiation and development is dependent on exposure to LD. In lower latitudes

TABLE 12.6 Daylength requirements for bulb formation in different cultivars of onion (*Allium cepa*).

Cultivar	Duration of daily photoperiod (h)				
	10	12	13	14	16
Sweet Spanish	0	29	77	93	100
Yellow Flat Giant	0	13	33	89	100
Yellow Flat Dutch	0	0	15	92	100
Round Madeira	0	0	0	100	100
Yellow Rijnsburg	0	0	9	46	100
Yellow Zittau	0	0	0	40	100

Plants were grown under natural daylight conditions and moved daily into darkened sheds in order to obtain the indicated daylengths. Figures are % plants bulbing.
After Magruder and Allard (1937).

many common cultivars do not form bulbs and the so-called 'short-day bulbing' types must be grown. These also have a LD requirement for bulbing, but the critical daylength is shorter. The daylength responses of a range of onion cultivars is shown in Table 12.6. Both daylength and temperature are important in controlling bulb formation in onion. Higher temperatures favour rapid bulbing, while at lower temperatures (below about 15°C) bulbing is delayed and in some varieties is prevented entirely (Thompson and Smith, 1938; Heath and Holdsworth, 1948). In longer daylengths the onset of bulb growth occurs sooner; only very small bulbs are produced, however, because the emergence of new leaves ceases rapidly and the first formed leaves collapse and senesce with a consequent reduction in photosynthetic capacity (Austin, 1972). As with tuber formation in potato, therefore, a rapid onset of bulbing in onion in response to a strong and early photoperiodic induction treatment leads to a lower bulb yield.

Perception of photoperiod in onions is localised, as one might predict, in the leaves (Heath and Holdsworth, 1948). Photoperiodic sensitivity was lost when leaf blades were removed and only regained when new ones were developed. Later experiments showed that bulbing in onions relies on the perception of LD by young, developing leaves, which were much more sensitive than older leaves even though older plants responded more strongly than young plants (Sobeith and Wright, 1986). In onion, the first detectable signs of swelling of the leaf bases occurs after only 2 days of LD treatment. Austin (1972) found that for onion 14 LD gave an appreciable response but continuing the treatment enhanced the effect and bulbing occurred most rapidly with continuous photoinduction. Similarly, in shallot (*Allium ascalonicum*), small bulbs formed after 7 LD although the size of the bulb increased with an increase in the number of long photoperiods received up to about 28 LD (Jenkins, 1954). The number of LD required before the bulbing process becomes irreversible can be somewhat variable in a particular population but it is clear that the plant does, after sufficient LD cycles, become fully committed to form the resting bulb (Heath and Holdsworth, 1948). There is some evidence that bulb formation in LD is at least partly due to the removal of a bulbing inhibitor produced in SD. Intercalation of SD between LD nullified their effect, indicating a positive inhibitory effect of non-inductive photo-periods similar to that observed in flowering (Kato, 1964).

Early studies with *Allium* found that when LD are obtained by giving low intensity day-extensions with fluorescent light they do not induce bulbing in onion, whereas tungsten filament light of the same order of intensity is highly effective. Even in continuous light, bulbing did not occur under fluorescent lamps (Butt, 1968) except in special colours with a higher content of FR (Austin, 1972). A low-intensity day-extension with light of restricted wavelengths given after a 12 h main light period under white fluorescent lamps only resulted in bulbing with a mixture of R and FR. Neither R nor FR alone was effective (Butt, 1968) and reports of bulbing under day-extensions with FR light (Terabun, 1970) may be attributable to the fact that the source used contained some light of wavelengths shorter than 700 nm. Maximum bulbing was found to occur when the R/FR ratio was about 1:1.65 and a day-extension with blue light did not cause bulbing (Terabun, 1965).

Lercari (1983a) produced an action spectrum for the photoperiodic induction of bulbing in onion. In his experimental system a dark interruption for 3 h in the middle of an 18 h inductive daily light period of mixed fluorescent and incandescent light inhibited the bulbing response. He determined the effect of monochromatic light at different wavelengths in reversing this inhibition (i.e. inducing bulbing). The resultant spectrum had a single peak at 714 nm and no activity below 700 nm (Fig. 12.3). Because the response was fluence-rate dependent and could not be replaced by a FR pulse at the beginning of the dark interruption, Lercari interpreted the response as being a classic FR-HIR, mediated by phytochrome. Additional support for this was that the spectrum of responsivity to monochromatic light at constant irradiance was modified by co-irradiation with light at 660 nm (see Fig. 12.3). This work showed that

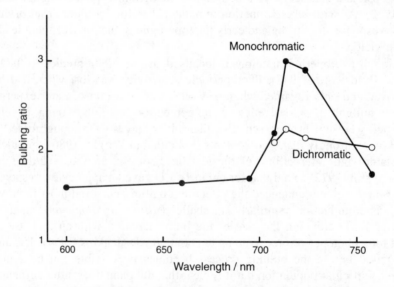

FIG. 12.3. Responsivity spectrum for the stimulation of bulb formation in onion. Plants were given continuous (3 h) monochromatic or dichromatic (simultaneously with 660 nm) irradiation in the middle of an 18 h daily light period. All light treatments were at 2 μmol m^{-2} s^{-1}. After Lercari (1983a).

TABLE 12.7 Effect of lamp type and R/FR ratio on LD-induced bulbing in *Allium cepa* cv Rijnsberger Bola.

Type of fluorescent lamp used	R/FR ratio of lamp	Bulbing ratio (after 76 days)
Colour 29, warm white	22.7	1.92
Colour 34	3.7	3.74
Colour 37	3.6	3.82
De luxe warm white	3.6	4.04

Plants were grown for 14.5 h under the various fluorescent lamps and total photoperiods of 19 h made up by giving low-intensity light from tungsten filament lamps before and after the fluorescent irradiations. Modified from Austin (1972).

not only was phytochrome action through the HIR sufficient for induction, it was also an essential part of the mechanism.

There is evidence that, as with flowering in LDP the time at which FR is given during the photoperiod has a significant influence on its effectiveness. As mentioned above, bulbing was readily obtained with a day-extension of mixed R and FR (or tungsten filament) light following a 12 h main light period under white fluorescent lamps. Bulbing failed, however, when 2.25 h with tungsten filament light was given before and after a 14.5 h main light period with white (colour 29) fluorescent lamps: a total photoperiod of 19 h (Table 12.7). This particular light regime would establish a high Pfr level during the 9–15th hour of the photoperiod, at which time Pfr was found to be apparently inhibitory to floral induction in *Lolium* and several other LDP (see Chapter 5). Using the same light sequence, bulbing was increased when other types of fluorescent lamps with a higher FR/R ratio were employed (see Table 12.7). Lercari and Deitzer (1987) found in a more recent study that the optimum time for the FR treatment was in the middle part of an 18 h photoperiod. The periods from 5 to 9 h and 9 to 13 h were both equally effective and at those times the plants were about five times as sensitive as at the beginning or end of the photoperiod. Even though the sensitivity varied during the photoperiod, the wavelength dependence of the response did not. At all times maximum bulbing response was found at 714 nm (Fig. 12.4).

The photobiology of bulb induction in onion has some clear parallels with the photobiology of floral induction in LDP or light-dominant species. As with flowering, bulb formation in response to long photoperiods apparently involves variable sensitivities to FR treatments at different times in the day. Maximum bulbing response can be obtained by giving a mixture of R and FR throughout the long photoperiod and the provision of some light of wavelengths in the 700–730 nm waveband is an essential requirement for the formation of onion bulbs in controlled environments under artificial light. We would expect from this that phytochrome A, which mediates plant response to continuous FR, is a component of the photoperiodic mechanism for onion bulbing. No studies have been carried out on phytochrome A expression in onion to corroborate this prediction.

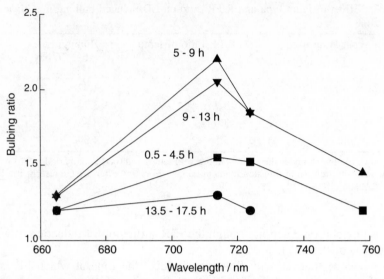

FIG. 12.4. Wavelength dependence for the stimulation of bulbing in onion at different times in the photoperiod. Plants were given continuous (4 h) monochromatic light at 6 μmol m^{-2} s^{-1} at the times indicated during the daily 18 h photoperiod, at other times plants received white fluorescent light. After Lercari and Deitzer (1987).

RUNNER FORMATION

The cultivated strawberry (*Fragaria* $\times$ *ananassa* Duch) has a perennial habit and forms stolons and daughter offsets under LD conditions. Stolons, bearing daughter runner plants, arise in axillary positions and can be regarded as branch crowns in which the first two internodes become extremely elongated. Stolon initiation begins in the spring after winter chilling and is terminated by SD in the autumn. In cultivars where daylength regulates flower initiation, plants require SD for floral initiation. Under a wide range of environmental and cultural treatments vegetative growth, including runner formation is correlated with flower inhibition (Guttridge, 1969). The critical daylength for flower initiation is very similar for that for the inhibition of runner formation (Fig. 12.5) suggesting the possibility of a common regulatory system. Further evidence is that the inhibition of flowering or promotion of runnering by night breaks in a long dark period required one hour with incandescent light, but 20 or 30 min was either less effective or without any effect at all (Borthwick and Parker, 1952).

In general, suppression of flowering by long photoperiods is only effective using light sources which have an appreciable proportion of their energy in the FR part of the spectrum. Flowering in strawberry is therefore inhibited by LD in a typical 'light-dominant' manner rather than promoted by SD as a dark-dominant response. The promotion of runnering also seems to be a light dominant response. A light regime of 10 h fluorescent and incandescent light at a ratio of approximately 30:1 followed by 8 h low intensity incandescent was more effective than 18 h of the combined source on its own (Guttridge, 1969). As with onion bulbing, we would expect that a light-dominant response with sensitivity to FR would be mediated by phytochrome A. Strawberry is another species in which it is now possible to carry out genetic

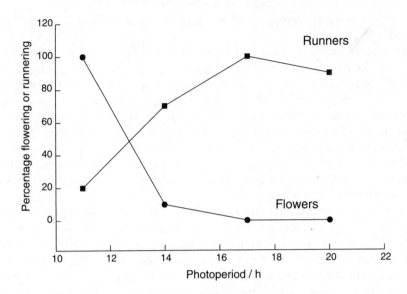

FIG. 12.5. Daylength dependence of runnering (promoted by LD) and floral initiation (promoted by SD) in strawberry. Data of Guttridge (1969).

transformation. An everbearing cultivar, Rapella, has been transformed with rice phytochrome A under the control of a constitutive promoter. The transgenic plants show greatly enhanced runnering, consistent with runnering being promoted by a light-dominant response mediated by phytochrome A (Thomas, unpublished observation).

THE NATURE OF THE PHOTOPERIODIC STIMULUS FOR THE FORMATION OF STORAGE ORGANS

Evidence for a translocated stimulus for the formation of storage organs is mainly based on tuber production in grafting experiments. In potato, a stimulus was shown to move down the stem through the phloem, and girdling resulted in the production of tubers above the ringed portion of the stem (van Schreven, 1949). The tuberising stimulus has been shown to move readily across a graft union in both potato and artichoke. After induction the stimulus continues to be formed or at least persists in the plant for some time (Gregory, 1956) but reversion to the non-induced state has been observed in *Begonia evansiana* (Esashi, 1960) and in potato (Chailakhyan, 1945). The stimulus which induces the development of tubers in artichoke (*Helianthus tuberosus*) is not unique to this plant but is also formed in leaves of sunflower plants (*Helianthus annuus*). When sunflower is grafted to artichoke, the latter forms tubers if the sunflower donor is in SD, although sunflower itself is incapable of forming tubers (Nitsch, 1965). The production of a tuber-forming stimulus by leaves in particular daylengths is apparently not restricted to plants which themselves are capable of forming tubers. As there does not seem to be an unique tuber-forming stimulus it is not surprising that the development, but not necessarily initiation, of storage organs can be influenced by a range of compounds, including photosynthetic products and growth-regulating compounds.

Photosynthates. Because of their storage function, bulbs, corms, tubers and similar organs are sinks for the translocation of assimilates. It has been considered that the increase of assimilates in the storage tissues might act as the signal for stimulating storage organ development. Most evidence, however, indicates that the products of photosynthesis cannot, on their own, replace the daylength signal for initiating the development of tubers or bulbs, as was also the case in floral initiation. Injection of sucrose into the leaf cavities of onion plants growing in SD resulted merely in more vigorous vegetative growth, whereas the same treatment in LD increased bulb thickening (Kato, 1965). Even when daylight was reduced by shading to an intensity so low that several plants died, bulbing occurred in any plants which survived provided that long photoperiods were given (Heath and Hollies, 1965). With onion plants cultured *in vivo*, Kahane *et al.* (1992) found that swelling of the leaf bases could be stimulated at high (80 g l^{-1}) sucrose levels under otherwise non-inductive conditions. However, although the sucrose-treated plants stopped vegetative growth, they did not become dormant and were morphologically different from plants induced to form bulbs under LD treatments. Thus, some additional daylength-dependent signal was needed for the normal bulbing response.

In potato it was suggested that tuber initiation is associated with those conditions which lead to a high concentration of carbohydrate at the stolon tips (Borah and Milthorpe, 1962). However, the observation that tuber formation in potato can be enhanced by defoliation or by darkness suggests that the stimulus which initiates the formation of storage organs is not excess photosynthetic product itself but a substance (or substances) which acts to change the developmental pathway at the target site, leading to the formation of a sink into which photosynthetic products are subsequently mobilised. Even if photosynthetic products are unlikely to be the tuberising stimulus carbohydrate levels have a major effect on the development of potato tubers. For example, high sucrose levels have frequently been found to enhance tuber formation in tissue culture (Palmer and Smith, 1969a; Perl *et al.*, 1991; Simko, 1994). Also, tuber formation is affected in transgenic potato plants in which sucrose availability to sink tissues is modified. Reduction of the level of the phloem-located sucrose transporter by antisense methods or the overexpression of yeast invertase in the apoplast both prevented the export of sucrose from the leaves and caused reduced tuber numbers and yield (Heineke *et al.*, 1992; Riesmeier *et al.*, 1994). Conversely, plants in which sucrose conversion to starch in tubers was blocked by antisense transformation with ADP-glucose pyrophosphorylase accumulated soluble sugars in the tubers. In these plants tuber number and total dry weight was increased although individual tuber dry weight was reduced (Müller-Röber *et al.*, 1992).

Cytokinins. The formation of tubers in tissue cultures of several species has been shown to be promoted or induced by cytokinins. Isolated stolons of potato can readily be induced to form tubers by the addition of kinetin to the medium (Palmer and Smith, 1969b) and starch accumulation in the apical regions accompanies physical swelling. Treatment of cultured stolons of potato with kinetin caused tubers to form on all stolons; the presence of kinetin was only required for 3 days although visible tubers were not seen until 10–12 days after the cytokinin application (Palmer and Smith, 1969b). Tuberisation was also stimulated under non-inductive daylengths by zeatin riboside (ZR) (Mauk and Langille, 1978). It has been suggested that cytokinins may regulate tuber formation in potato by suppressing starch hydrolase activity, or increasing starch synthetase activity (Smith and Palmer, 1970) but cytokinins are also known to be important for cell division and this may be an important factor in their effect on tuber development. In *Dioscorea*, *Ullucus* (Asahira and Nitsch, 1968) and artichoke (Courduroux, 1966) tuberisation *in vitro* is also favoured by kinetin and BA. Mauk and Langille (1978) found a twofold increase in ZR levels in below ground tissues on transfer of *Solanum tuberosum* from LD to SD and cytokinins have been isolated from tuber tissue of both artichoke (Nitsch and Nitsch, 1960) and potato (Tizio, 1966). A spray with BA at 10 mg l^{-1} was effective in increasing the formation of aerial bulblets on inflorescence heads of onions from which the flower buds had been removed (Thomas, 1972) and BA treatment increased root tuber formation in *Ipomoea* (Spence and Humphries, 1972).

The relationship of cytokinin-induced tuber formation *in vitro* to the stimulus which is exported from the leaves is still uncertain. Cytokinin levels are known to increase in leaves of some *Begonia* species in SD (Heide and Skoog, 1967) but the transmissible tuber-inducing stimulus has not been identified as a cytokinin. It could equally well be a substance which affects the production or activity of a cytokinin at the target site, thus establishing a metabolic sink into which photosynthetic products are mobilised. Application of the synthetic cytokinin, BAP, to leaves of intact potato plants did not increase the yield of tubers (Badizadegan *et al.*, 1972), but it is well known that, when synthetic cytokinins are applied to leaves, they do not move readily from their site of application. However, BAP application (by foliar immersion) to leaves of sweet potato, *Ipomoea batatas*, increased the proportion of dry matter in root tubers formed on isolated rooted leaves (Spence and Humphries, 1972).

Cytokinins are present in, and capable of inducing the development of, a range of storage organs (stem tubers in potato and *Ullucus*, bulblets in onion, root tubers in *Ipomoea*) on plants with different photoperiodic requirements (SD in potatoes and *Ullucus*, LD in onion, day-neutral in *Ipomoea*) and may be of general importance in the initiation of storage tissue growth. However, they are unlikely to be the primary daylength-dependent signal. For example in *Solanum andigena*, which has an obligate requirement for SD for tuber formation, cytokinins were not capable of inducing tuberisation in plants which had not been treated with SD (Wareing, 1983). In addition, Lercari and Micheli (1981) found that the initiation of bulbing in onion by transfer to LD was accompanied by a transient increase in cytokinin levels. However, it is known that bulbing requires a persistent supply of the LD stimulus because plants transferred back to SD resume vegetative growth even when they have reached a relatively late stage in the bulbing process. It was therefore not possible to explain the persistent bulbing effect of LD through the observed changes in cytokinin levels.

Inhibitors. Tubers, like dormant buds in woody plants, are resting structures which are often induced to form by SD and which undergo a period of dormancy before growth is resumed. Consequently, the possibility that growth inhibitors such as ABA may regulate their formation needs to be considered. The results obtained when ABA is applied to plants are not altogether easy to interpret. Abscisic acid was found to have no influence on tuber formation in isolation cultures of potato stolons, although the addition of ABA to the medium inhibited stolon growth which is a prerequisite for tuber formation (Palmer and Smith, 1969b). Furthermore, ABA strongly antagonised kinetin-induced tuberisation in potato stolons, and this inhibitory effect of ABA occurred during the early stages of tuberisation. The application of ABA to growing intact plants of potato, however, both inhibited the shoot growth and promoted tuber formation (El-Antably *et al.*, 1967). In a photoperiodically sensitive strain of *Solanum andigena*, a daily spray of ABA at 20 mg l^{-1} markedly increased the number of plants which formed tubers in LD, and tuber yield was also increased in two commercial cultivars of potato. In *S. andigena*, ABA was able to substitute for the presence of the leaf in promoting tuberisation in SD-induced one-node cuttings but had no effect on similar LD-treated (i.e. not induced) material, suggesting that although essential for tuber formation it was not the SD-dependent signal. The application of ABA completely failed to induce the formation of any aerial tubers in *Begonia evansiana* grown in LD conditions (Hashimoto and Tamura, 1969) and, as with *Solanum*, suppressed cytokinin-induced tuber swelling (Esashi and Leopold, 1968). Growth inhibitory substances have been found in stolons and newly formed tubers of potato and have been suggested to be essential for tuber induction (Booth, 1963). However, no differences were found in the ABA content of the leaves or phloem exudates of *S. andigena* in LD or SD (Wareing and Jennings, 1980). In general, therefore, evidence presently available does not support the suggestion that the tuber evoking stimulus from SD leaves is ABA.

Gibberellins. As discussed in Chapter 7, daylength has been found to alter the gibberellin content of a wide range of plant species. In general LD cause an increase in the levels or turnover of gibberellins and in some instances these changes can be related to floral development or bolting in LDP. The formation of tubers is, in most cases, a response to SD, while bulbing in onion is a LD response. If photoperiod regulation of storage organ formation is analogous to regulation of floral initiation we might expect that gibberellins are involved in LD-stimulated processes, e.g. bulbing in onion. However, there is an inherent problem in that while gibberellins usually promote extension growth, bulb scales form as the result of the inhibition of leaf extension. Mita and Shibaoka (1984) found that S-3307, an inhibitor of gibberellin biosynthesis, caused swelling in the bases of young onion seedlings coupled with the reorientation of microtubules from a transverse to a longitudinal direction. Both effects were reversed by gibberellin, which acts to stabilise microtubules in the transverse orientation (i.e. inhibit bulb swelling). Measurements on the endogenous gibberellin content of onions grown in different daylengths showed that bulbing plants in LD in most cases had higher levels than those in SD (Nojiri *et al.*, 1993). There is thus no evidence that gibberellins have a role in the daylength control of bulbing.

As tuber formation is usually SD induced, one might expect that if gibberellins are involved it would be as an inhibitor. This might be considered feasible because tuber formation involves lateral growth at the expense of stem extension. There are several examples of the inhibition of tuber formation by applied gibberellins, not only in potato (Harmey *et al.*, 1966; Hammes, 1971; Tizio, 1971), but also in *Dahlia* (Biran *et al.*, 1972), *Helianthus tuberosus* (Courduroux, 1964) and *Begonia evansiana* (Okagami *et al.*, 1977). Conversely, stimulation of tuber development by the application of inhibitors of gibberellin biosynthesis, including CCC, Amo 1618 or paclobutrazol, to intact plants or tissue cultured material has been reported for potato (Humphries, 1963b; Hammes, 1971; Simko, 1994), artichoke (Courduroux, 1966) and *Begonia evansiana* (Nagao and Okagami, 1966). In addition it was observed that a mutation at the gibberellin-related dwarfing locus ga1 in *Solanum andigena* stimulated tuber yield (Bamberg and Hanneman, 1993), which is consistent with an inhibitory effect of endogenous gibberellins on tuber development.

Gibberellin-like substances have been extracted from developing stolons of potatoes growing in LD; with the onset of tuberisation they disappeared and the level of endogenous inhibitors increased (Okazawa, 1967). In *Begonia evansiana* natural GA-like substances were detected in the buds in LD; on transfer to SD they initially decreased in amount and then increased again (Esashi *et al.*, 1964). Taken together these data indicate that part of the effect of daylength on the formation of tubers may arise from the higher gibberellin content which occurs in LD conditions. However, changes in gibberellin content cannot on their own explain the entire response. In the experiments by Wareing and Jennings (1980) it was found that the requirement for ABA for tuber formation in one-node cuttings from SD-induced plants could be replaced by grafting a LD leaf. However such cuttings will have a higher gibberellin level than the controls and should inhibit tuber formation if gibberellin levels are the decisive factor. It was therefore concluded that some other daylength-dependent signal must have an overriding effect on tuber initiation (Wareing, 1983).

Auxin. Auxin levels may also be higher in LD than in SD but there is little evidence that they play an important part in regulating tuber formation. On transfer to SD, the auxin content in leaves of *Begonia evansiana* fell and, at that stage, the application of auxin as well as gibberellin to the leaves prevented SD-induced tuberisation (Esashi *et al.*, 1964). On the other hand auxin has been used to promote tuber formation *in vitro* in *Solanum tuberosum* (Mader, 1995).

There is some evidence that auxins are part of the regulatory mechanism for bulb formation in onion. Transfer of plants from 8 h days to 16 h days resulted in an increase in extractable auxins from the leaf bases, which rose to a peak at about the fifth LD and then decreased again to the SD value, or below, after 14 LD (Clark and Heath, 1962). The first signs of bulb swelling were visible at about this time. The application of IAA to seedling sections induced a certain amount of bulb development, especially when given along with sucrose and at 10^{-4}M IAA the response occurred rapidly even in the dark. Intact plants were much less responsive but a very slight degree of bulbing was obtained in SD when the roots were immersed in IAA solution. The endogenous auxin which increased in concentration in the leaf bases following transfer to LD was extracted and shown to cause bulb formation when added to seedling sections kept in darkness.

Ethylene. The application of ethylene to certain rapidly elongating tissues results in a retardation of extension and a change in orientation of cell expansion to produce rapid enlargement (Burg and Burg, 1966). Ethylene is, therefore, of particular interest in relation to tuber initiation, which involves cessation of longitudinal growth and radial expansion of cells in the apical and subapical regions of the stolon, and in bulbing, which involves lateral swelling of leaf cells. It has been shown, at least in dark-grown pea stems, that the inhibition of growth and increased lateral swelling, which occurs with high concentrations of exogenous auxin, results from a stimulation of ethylene evolution in the treated tissue. Application of ethrel (2-chloroethyl-phosphonic acid) which causes ethylene evolution, induced a bulbing response under both LD and SD (Levy and Kedar, 1970; Lercari and Ceccarelli, 1975). However, while silver thiosulphate blocked the effect of ethrel on bulb formation in SD, it had no effect on LD-induced bulbing. Additionally, AVG (aminoethoxyvinylglycine), a specific inhibitor of ethylene biosynthesis did not inhibit LD-dependent bulbing (Lercari, 1983b). The promotion of bulbing by exogenous ethylene does not therefore seem to be related to the photoperiodic regulation of bulb formation.

Several studies have been made with potato tubers, but have yielded conflicting results. Treatments of potato plants with ethylene or ethylene-releasing substances was reported both to enhance tuber formation (Catchpole and Hillman, 1969) (although the ethylene-induced tubers were devoid of starch) and to have no effect (Langille, 1972). Direct treatment of cultured stem segments *in vitro* failed to induce any tuber formation (Palmer and Barker, 1973) although some generalised swelling of the entire stolon occurred. It was concluded that ethylene might play a role in tuber formation by acting to inhibit longitudinal stolon growth, which is a prerequisite for tuber initiation. However, since ABA also inhibited stolon growth without causing tuber formation (Palmer and Smith, 1969b), a particular requirement for ethylene is not established.

Jasmonates. Jasmonic acid (JA) and methyl jasmonate are distributed widely among higher plants and exert inhibitory effects on growth and cell division. Both compounds strongly promoted potato tuber formation in a single node segment experimental system (Koda *et al.*, 1991). The same authors also reported that the structure of tuberonic acid, a naturally occurring tuber-inducing compound which was present at higher levels in SD than LD leaves is closely related to that of JA. Jasmonic acid was also found to be strong promoter *in vitro* of tuber formation in *Solanum demissum* (Helder *et al.*, 1993). The levels of JA were not different in LD and SD in this species but a positive correlation was found between the occurrence of 11-OH-JA and 12-OH-JA in leaflets and SD-induced tuber formation. Jackson and Willmitzer (1994) found that JA was unable to replace the requirement for SD in *S. andigena* and *S. demissum* and was itself unlikely to be the transported tuber-inducing signal. They did not, however, measure the levels of hydroxylated jasmonates and their results are not inconsistent with the proposal of Helder *et al.* (1993) that the enzymes responsible for hydroxylation of JA are only active under SD.

Evidence that jasmonates may be involved in bulb formation comes from the report that the development of garlic shoot and bulb formation was promoted by JA *in vitro* (Ravnikar *et al.*, 1993). Levels of JA, as measured by radioimmunoassay, in the sheaths of bulbing onions in LD were about three times higher than non-bulbing onions (Nojiri *et al.*, 1992). Treatments with JA did not, however, promote the bulbing response in SD. The role of jasmonates in LD-dependent bulbing is still not clear. The bulbing stimulus is clearly not JA but the possibility of related compounds such as hydroxylated jasmonates being involved cannot be excluded.

CONCLUSIONS

The formation of tubers and bulbs shows many features of other photoperiodic phenomena, with synthesis in the leaves of a hormonal stimulus (or stimuli) which transmits developmental information to the target sites where morphogenetic events take place. In many species, synthesis of the stimulus is under the control of day-length, but similar substances are formed independently of daylength in some plants and are found even in plants which do not themselves form storage organs. The photoreceptor for the perception of daylength is phytochrome and the actual mechanisms for SD sensing for tuberisation and LD sensing for bulbing show marked similarities to the daylength sensing mechanisms for the photoperiodic control of flowering in SDP and LDP respectively. Changes in the endogenous levels of photosynthetic assimilates, in particular glucose and sucrose, and several plant growth regulators can markedly affect the formation of storage organs in particular species, but as with flowering, no unique tuber or bulb evoking stimulus has yet been identified.

13 Other Effects of Daylength

In this final chapter, we shall consider some other aspects of plant development which are modified by photoperiod. These include

- seed germination
- stem and leaf growth
- assimilate partitioning
- secondary metabolism.

In general, these differ from processes we have considered earlier in the book, such as flowering, the onset of dormancy and vegetative reproduction, in that we would not normally consider that the response to daylength is particularly important in comparison to other environmental factors. There may be some exceptions. One of the most important of these is the bolting response of rosette species to daylength where there are huge changes in stem extension. However, in these cases stem elongation is closely coupled with flowering and the seasonal timing of the bolting response has clear significance for the success of the species. In considering the developmental processes covered in this chapter we will attempt to include only true responses to daylength, although in many cases the design of experiments makes it impossible to distinguish between daylength and other confounding effects, particularly light integral and quality, or temperature interactions.

SEED GERMINATION

Seed germination was the first light response shown to be controlled by phytochrome. Many seeds, such as *Lactuca sativa*, require only a brief exposure to R or white light in order to induce germination; others require repeated short exposures to light or a single prolonged exposure. There are also a few cases where germination has been found to be influenced by the daily duration of light and which are thought to be under photoperiodic control (Attridge, 1990). The photoperiodic requirement for germination is not necessarily the same as that required for flowering; for example, germination is promoted by LD in some cultivars of rice which require SD for flowering (Bhargava, 1975).

The response of birch seeds (*Betula pubescens*) has been studied in some detail (Black and Wareing, 1955). At 15 °C unchilled seeds showed a low percentage of germination in SD (2–8 h), while 90% of the seeds germinated in daylengths of 20 h. The response was studied further by varying the daily duration of light and darkness independently and by giving a 1 h night-break. As with many other photoperiodic effects in plants, it was found that the absolute length of the uninterrupted dark period was the major factor in controlling seed germination in birch. With both 8 h and 20 h photoperiods, the percentage germination was reduced as the dark period was lengthened and, even with a short light period, germination was high provided that the associated dark period was also short. Increasing the duration of the photoperiod to 20 h did not overcome the inhibitory effect of long dark periods. The night-break treatment resulted in an LD type of response. Thus, the control of germination in these birch seeds appears to be a true photoperiodic response since it depends on the timing of the light and dark periods in each 24 h cycle. However, at 20°C, photoperiodic control disappeared and seeds germinated in response to a single exposure to light for 8 h. It appears that LD are required to overcome the inhibitory effect of the pericarp and testa, since isolated embryos were not sensitive to daylength. The fact that leaching also markedly increased germination under SD indicated the presence of water-soluble inhibitors; moreover, a powerful growth inhibitor was present in extracts from birch seeds and, when isolated embryos were placed on filter paper containing the inhibitor, the photoperiodic response was restored (Wareing, 1959).

The control of germination in seeds of *Begonia evansiana* also seems to be a true photoperiodic effect (Nagao *et al.*, 1959). Germination requires exposure to LD and a sharply defined critical daylength of 8 h was found when seeds were given 10 cycles (Fig. 13.1). Night-breaks are also effective and the response was saturated with an exposure of about 10 min (Fig. 13.2); the response is thus unlikely to be related simply

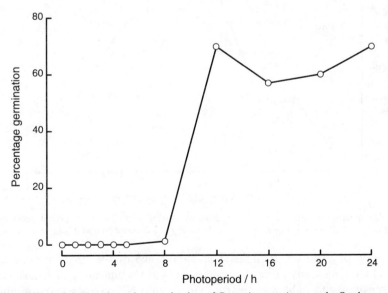

FIG. 13.1. Effect of daylength on the germination of *Begonia evansiana* seeds. Seeds were exposed to various durations of photoperiod for 10 days, under white fluorescent lamps at an intensity of 2.9 W m^{-2}. After Nagao *et al.*, 1959.

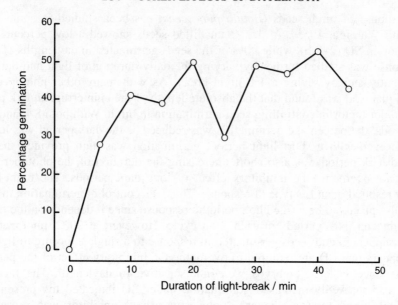

FIG. 13.2.   The effect of night-breaks on the germination of *Begonia evansiana* seeds. Night-breaks were given in the middle of a 17 h dark period with light from white fluorescent lamps, as in Fig. 13.1. After Nagao *et al.*, 1959.

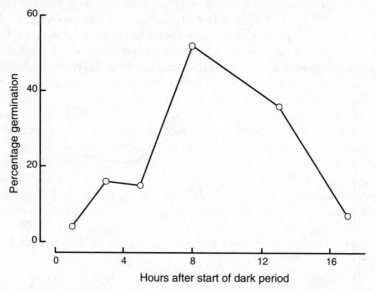

FIG. 13.3.   Germination response of *Begonia evansiana* seeds to a 30 minute night break given at various times during a 17 h dark period. Light conditions as in Fig. 13.1. After Nagao *et al.*, 1959.

to the duration of exposure to light. The importance of the time of exposure was determined by giving seeds a brief night-break at various times in a 17 h dark period; a changing sensitivity to light was clearly evident with NBmax occurring at about the 8th hour of darkness as observed in many flowering responses (Fig. 13.3). However, possible rhythmicity of the night-break effect was not investigated.

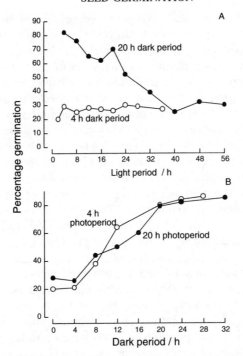

FIG. 13.4. Germination of light-inhibited seeds of *Nemophila insignis* in response to independent variations in the durations of the light and dark periods. Seeds were exposed to light from white fluorescent lamps at an irradiance of 4.4 W m^{-2}. After Black and Wareing, 1960.

Germination may also be *inhibited* by light and such seeds may also be responsive to daylength. For example, seeds of *Nemophila insignis* germinated in SD but not in LD at 21–22 °C, although at a lower temperature they germinated regardless of the photoperiodic regime (Black and Wareing, 1960). This inhibitory effect was also dependent on the absolute duration of the dark period; in short dark periods, germination was inhibited even when the associated photoperiod was also short (Fig. 13.4B). Moreover, long photoperiods of up to about 20 h duration failed to inhibit germination when the accompanying dark period was long, although excessively long photoperiods were inhibitory even when combined with a long dark period (Fig 13.4A). This response pattern, as with the LD-promoted germination in *Betula*, is remarkably similar to that obtained for the photoperiodic control of flowering in the SDP *Glycine* cv Biloxi (see Fig. 1.13). A night-break treatment following a SD resulted in a LD type of response and inhibited germination.

Maximum inhibition of germination in *Nemophila* occurred in the blue (452 nm) and far red (710 nm) regions of the spectrum, while R had almost no inhibitory effect. The photoperiodic effect of blue light was similar to that of white fluorescent lamps and long daily photoperiods were necessary in order to inhibit germination. Far red light was, however, inhibitory even with brief daily exposures of 30 min. A similar response to FR has been found in several dark-germinating seeds which do not exhibit photoperiodic behaviour. In these cases, it has been suggested that a small quantity of Pfr is gradually released from an intermediate form present in the dry seed and is required for germination; photoconversion to the inactive Pr form by FR thus prevents germination, the need for repeated exposures being related to some process of continued release of Pfr within the seed. However, it is pertinent to recall that the promotion of flowering by short-day/long-night cycles also requires the presence of Pfr (and is prevented by brief exposures to FR), even though Pfr is inhibitory at the times when plants are sensitive to a night-break (see Chapter 4); it remains possible, therefore, that the inhibition of germination in *Nemophila* seeds by a brief daily exposure to FR is associated with the photoperiodic control mechanism. Blue light has been shown to be effective in causing a LD response in the control of flowering in several members of the family Cruciferae (see Chapter 5) but is not generally effective in photoperiodic responses; the photoperiodic inhibition of germination by blue light in *Nemophila* (family Hydrophyllaceae) is therefore unusual.

Photoperiodic conditions experienced during the ripening of the seed on the parent plant may carry over to influence germination. Two types of seed dormancy have been found in *Chenopodium album* (Karssen, 1970). The first occurred immediately after harvesting and was only found in seeds from parent plants that were maintained in LD, or in SD with 1 h R night-break; this type of dormancy did not develop when the parent plants were maintained in SD. The second type of dormancy was still seen after 3 months of storage and occurred only in seeds from LD plants. Since it seemed to depend on the total daily duration of light received and was correlated with the development of thick seed coats, it was concluded that this second type of day-length-dependent dormancy was probably a photosynthetic rather than a photoperiodic effect. The germinability of seeds in several other species has also been found to be decreased when the parent plant was exposed to LD during the ripening period and, in many cases, appears to be associated with the development of thicker seed coats (Karssen, 1970; Pourrat and Jacques, 1975). However, in *Chenopodium album*, the seeds from parent plants in LD also had a much higher level of endogenous inhibitors, indicating that the inhibitor levels in the seeds as well as seed coat thickness may be influenced by the daylength treatment during ripening. The effect of parental day-lengths (SD promoted germination compared with LD) was not correlated with seed coat thickness or dry weight in *Amaranthus* (Kigel *et al.*, 1977) and *Chenopodium polyspermum* (Pourrat and Jacques, 1975). Parental SD (8 h) also resulted in faster germination with a higher final percentage than LD (20 h) in *Carrichtera annua;* in this case, the daylength experienced during maturation continued to influence germinability even after 9 years storage (Gutterman, 1989).

In many cases, the effects of daylength given to the parent plant have not been shown to be strictly photoperiodic, since plants received a much greater total light integral as well as being exposed to LD. However, photosynthetic effects have essentially been eliminated in some cases. For example, low irradiances were effective in *Kalmia*, *Chenopodium* (1.0 W m^{-2}; Pourrat and Jacques, 1975; Duncan and Bilderback, 1982), and *Betula* (0.03 W m^{-2}, Black and Wareing, 1955) and night-break treatments have been shown to be effective in a few species, including *Amaranthus retroflexus* (Kigel *et al.*, 1979), *Chenopodium polyspermum* and *C. album* (Pourrat and Jacques, 1975). In *C. polyspermum*, the night-break response was partially R/FR reversible.

The promotion of germination by light is probably not a photoperiodic effect in the majority of cases but rather depends on the fact that Pfr is required for germination and is released or synthesised in the inactive Pr form during imbibition. Dark germination occurs when phytochrome is released as Pfr during imbibition, presumably because Pfr (or an intermediate which is converted to Pfr during imbibition) is the form present when the seed dehydrates (Attridge, 1990). In such cases, repeated exposures to FR prevent germination by converting phytochrome to Pr. However, the inhibition of germination by visible light is less well understood in terms of phytochrome. Direct photoperiodic effects on germination have been demonstrated in only a few species but, where they have been investigated in detail, the characteristics of the germination response to daylength are very similar to those exhibited in other cases and indicate that they may be under the control of a similar photoperiodic mechanism. The site of perception for these photoperiodic effects probably lies in the embryonic shoot tissue, since photoperiodic sensitivity was maintained in isolated embryos of birch, provided that they were tested in the presence of inhibitors leached from the seeds.

STEM ELONGATION

It is possible to distinguish between two types of photoperiod effect on stem growth or extension. The first is the situation where daylength modulates the rate of internode extension or the number of nodes in plants showing a single type of growth habit. In its most common manifestation stem extension is inhibited by SD in woody, perennial plants, as a prelude to the onset of dormancy. Stem elongation in woody plants is discussed in detail in Chapter 11 because it is so closely associated with dormancy. The second case is where photoperiod triggers the switch between a rosette and elongate habit, often in association with flowering.

In the herbaceous species *Lycopersicon esculentum* and *Glycine max* (Downs *et al.*, 1958) and *Cucurbita maxima* (Zack and Loy, 1980) which do not enter dormancy, internode extension was greater when short days were extended with TF light as compared to when they were extended with fluorescent light. However, it is not certain that these are true photoperiodic effects. TF lamps establish a much lower Pfr/Ptot ratio than fluorescent light and for any species which shows a 'shade avoidance' response, stem extension rates will be higher under TF than fluorescent light irrespective of any photoperiod effect. The light treatments will also vary in the blue component, which may contribute to the difference in stem extension. A further confounding effect may be the so-called 'end-of-day response', where a light treatment which lowers the Pfr/Ptot ratio immediately prior to the daily dark period results in greater stem extension and taller plants (see Chapter 3). The most effective treatment is FR which typically establishes a Pfr/Ptot ratio of 0.03, but any treatment which reduces Pfr/Ptot ratio in the range 0 to 0.85 will cause increased stem extension (Vince-Prue, 1975; Fig. 11.18; Gaba and Black, 1985). Extensions with tungsten filament could therefore also result in longer internodes than extensions with fluorescent light because they establish a lower end-of-day Pfr/Ptot ratio in addition to their effect during the photoperiod. Evidence for a photoperiodic effect on stem extension was found in *Fuchsia hybrida* cv Lord Byron where a R night-break resulted in longer internodes than SD controls (Fig. 13.5). The effect of R was maximal at about 6 h into darkness whereas the promotion in response to a short FR treatment decreased linearly as the length of time in darkness before it was applied was increased. The response to R was therefore not

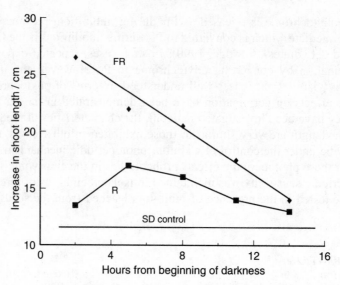

FIG. 13.5.   Response to night-breaks with R or FR given at different times during a dark period of 16 h. Plants of *Fuchsia hybrida* cv Lord Byron were grown in 8 h days and given 1 h white fluorescent (R) or FR at different times in the dark period. The plants treated with R after 4, 8 or 12 h flowered, other plants remained vegetative. After Vince-Prue, 1975.

simply the opposite of the response to FR and could be separated from the 'end-of-day' response.

Daylength control of bolting

In rosette LDP, floral initiation is accompanied by rapid stem extension, often referred to as 'bolting'. Under normal conditions, flowering and stem extension cannot be separated and stem length is sometimes used as a measure of the flowering response. However, close consideration shows that they are, to a considerable extent, independent. A number of species show an apparent obligate relationship between flowering and stem elongation. In *Scrophularia marulandica*, stem elongation always preceded the appearance of flower primordia and in no case were flowers ever observed on rosette plants (Cline and Agatep, 1970). Also, in several other LDP, flowering occurs only with simultaneous elongation of the stems. Examples include *Matricaria parthenoides*, *Centaurea cyanus*, *Coreopsis tinctoria* (Greulach, 1942) and barley, wheat and millet (Downs *et al.*, 1958). Despite the close association between stem elongation and floral initiation, the two processes can be separated in many plants through use of the appropriate treatments. The most common of these is the application of gibberellins, which can cause stem elongation in rosette LDP without inducing flowering (Vince-Prue, 1975; Table 6.1a). Elongation without flowering can be induced by other treatments. For example, in *Hyoscyamus*, by photoperiods of 11 h at 20°C (Lang and Melchers, 1943) and in *Petunia* by FR given throughout the night (Vince-Prue, unpublished data). It is also possible to achieve the opposite condition of flowering without elongation such as in *Rudbeckia bicolor*, grown at high temperature in SD (Murneek, 1940) and in *Silene armeria* grown in LD in the presence of the gibberellin biosynthesis inhibitor AMO 1618 (Cleland and Zeevaart, 1970). Separation of effects of flowering and stem extension had also been demonstrated in *Anagallis arvensis* (Imhoff *et al.*, 1979). In this species a 3 h R night break of a 15 h dark period promoted flowering in SD-grown plants, but was strongly inhibitory to stem elongation. Perception of the daylength signal for promotion of stem elongation takes place in the leaves. In *Silene armeria*, exposure of the mature leaves to LD promoted elongation of the stems and of the immature leaves, regardless of whether the shoot tips were exposed to LD, SD or darkness (Talon and Zeevaart, 1992). The initiation of rapid stem elongation by LD is therefore a classic photoperiodic response, independent of the control of floral initiation, although both may share a common perception mechanism.

Role of gibberellins

Gibberellins and particularly GA_1 appear to play a central role in the daylength-mediated regulation of stem extension in both woody species and rosette LDP. As mentioned earlier, applied gibberellins can replace a LD treatment for stimulating stem extension without causing floral initiation. Also, applied gibberellins (GA_1 and GA_{20}) can replace LD in overcoming the cessation of growth in response to SD in birch (see Fig. 11.19). As discussed in Chapter 7, key steps in the 13-hydroxylation biosynthetic pathway are photoperiodically regulated. Transfer of *Silene armeria* from SD to LD leads to an increase in GA_{53} oxidase, resulting in an increase in GA_{19}, GA_{20} and GA_1 (Talon and Zeevaart, 1992). In spinach, the conversion of GA_{19} to GA_{20} was increased, as well as the further metabolism of GA_{20} to GA_{29}, on transfer from SD to LD (Metzger and Zeevaart, 1982). Stem growth in spinach in response to a LD was inhibited by BX-112 and this was reversible by GA_1 but not GA_{20}. Plants treated with BX-112 had reduced levels of GA_1 and GA_8 but accumulated GA_{53}, GA_{44}, GA_{19} and GA_{20} (Zeevaart *et al.*, 1993). These findings suggest that GA_1 is the primary active gibberellin for LD-induced stem elongation in spinach. In similar experiments, GA_1 was the only gibberellin able to reverse the inhibition of LD-dependent stem elongation in birch, and of gibberellins in the early 13-hydroxylation pathway, GA_1 is the only active gibberellin for stem elongation (Junttila, 1993b). In another woody species, *Salix pentandra*, the levels of GA_1 and its immediate precursor GA_{20} decreased within 2 days of transfer from LD to SD. Plants in SD increasingly lost the ability to respond to GA_{19}, leading the authors to propose that the level of GA_1 is regulated by daylength, with the conversion from GA_{19} to GA_{20} being a likely target for control. In addition to the regulation of specific interconversions between gibberellins, daylength may also have more general effects on gibberellin formation. Biosynthesis of the GA precursor, *ent*-kaurene was also enhanced by LD in the LDP *Agrostemma* and *Spinacia* (Zeevaart and Gage, 1993) and this was concluded to lead to a higher rate of gibberellin biosynthesis, which is essential for stem elongation in rosette plants. Changes in gibberellin levels are greatest in the tissues in which the responses to LD are localised. In *Silene armeria*, the earliest cellular changes following a LD treatment were in the meristematic tissues 1.0–3.0 mm below the shoot apex (Talon *et al.*, 1991a). The number of cells per cell file increased and the length per cell decreased. In the induced tips, cellulose deposition occurred mostly in longitudinal walls, suggesting that many transverse cell divisions had taken place. Similar changes took place in response to treatment with gibberellins which do not promote floral initiation. The response was therefore related to changes in growth and not to the floral transition. The content of GA_1 increased 30-fold in the zone 0.5–3.5 mm below the shoot apex, indicating spatial correlation between the accumulation of GA_1 and the enhanced mitotic activity which occurs in the subapical meristem of elongating *Silene* apices. Levels of GA_1 increased in shoot tips of *Silene* following LD treatment of mature leaves irrespective of whether shoot tips received LD, SD or darkness (Talon and Zeevaart, 1992). When the mature leaves were exposed to SD, maintaining the apices in darkness resulted in an increase in GA_1 at the shoot tips and elongation of the immature leaves. However, this treatment did not lead to stem extension. The response to LD therefore involves a signal which is transmitted from mature leaves to shoot tips where it enhances the effect of GA on stem elongation. Taken together, there is increasing evidence that the photoperiodic control of stem elongation in rosette LDP and woody species is exercised through a similar regulation of gibberellin metabolism, with the level of GA_1 being particularly important. Experiments with the rosette LDP show that, as with flowering, there remains an unknown signal which is transmitted from the leaves to the apex and is essential for the elongation response to daylength. This signal does not seem to be a gibberellin.

LEAF GROWTH

Responses to daylength have been observed in the leaves of many species but only in a relatively few cases have these been demonstrated to be true photoperiodic effects, i.e. to be dependent on the duration and timing of light in each 24 h cycle and not on some other facet such as the total light energy received.

Effects of daylength on leaf growth may be complicated by the transition to flowering. Bracts are usually simpler and smaller than leaves and, consequently, an inductive treatment for flowering may result in a reduced total leaf area per plant, even though the area of individual leaves may be greater in the vegetative phase. In the LDP, *Callistephus chinensis*, for example, the total leaf area after 24 weeks was greater in plants grown in SD than in flowering plants with interrupted nights (Hughes and Cockshull, 1965) whereas, in the vegetative phase, the area of individual leaves were considerably greater on plants grown with a 1 h R night-break (Hughes and Cockshull, 1964). A similar situation was observed in *Sinapis alba* (Humphries, 1963a). The change in leaf form which is linked to the onset of flowering is well illustrated in the SDP *Cannabis sativa*. In the vegetative phase, the trend is towards more highly dissected leaves but is reversed on the transition to flowering when leaves become simple and are reduced to tiny bracts at the tip of the male spike.

Direct effects of daylength on leaf shape have been observed in a number of terrestrial and aquatic plants. Deeply dissected submerged leaves and simple aerial leaves are characteristic of many aquatic species of *Ranunculus*; leaf morphogenesis proceeds along one of two sharply different developmental pathways and no intermediate forms are normally seen. Both types are initiated underwater and photoperiodic treatments given to plants grown under mist in terrestrial environments have indicated that the final leaf form is largely controlled by daylength. In LD, *R. aquatilis* produced simple, land-type leaves whereas typical submerged leaves developed when plants were grown in SD (Davis and Heywood, 1963). In natural environments, deeply dissected submerged leaves develop in *Ranunculus* species under the SD, low temperature conditions of spring while aerial leaves develop in the higher temperatures and LD of summer. Daylength has also been found to influence leaf shape in the semi-aquatic plant *Proserpinaca palustris*, which produced divided, juvenile leaves in 8 h photoperiods and lanceolate-serrate leaves in 12 h photoperiods; in this case, the transition to adult leaves was not correlated with flowering for floral initiation occurred only with daylengths longer than 12 h (Davis, 1967).

The alteration of leaf morphology by daylength can take many forms. The induction of winter resting buds by SD modifies the developmental pathway in such a way that leaf primordia develop into scales instead of into foliage leaves (see Chapter 11). The formation of bulbs in response to LD also involves profound modification of growth to form scales (see Chapter 12). In peach leaves, the growth of mesophyll tissue seems to be favoured by LD (combined with high temperatures) leading to the development of leaves with undulating edges; in SD, low-temperature conditions, vein growth is favoured and long, narrow leaves develop (Nitsch, 1957). A change in the length:breadth ratio is a fairly common response to daylength and in both chrysanthemum (Vince, 1955) and *Chenopodium* (Thomas, 1961) the ratio was greater in SD. Leaf colour changes may accompany the changes in leaf shape; for example, in *Robinia*, leaves in SD were dark in colour and rounded while, in LD, they were yellow-green and elliptical in shape.

In succulent plants, daylength responses have been reported in both the degree of succulence and leaf morphology. In SD, the leaves of *Bryophyllum crenatum* were sessile, rounded and highly succulent while, in LD, the leaves were petiolate, with elliptical blades and reduced succulence (Zeevaart, 1985a). In *Kalanchoë*, leaves in LD were fairly thin and flexible while, in SD, they were thicker and more succulent (Schwabe, 1985). Thus, the influence of LD to decrease the degree of succulence was similar in *Bryophyllum* (LSDP) and *Kalanchoë* (SDP), although they have different photoperiodic requirements for flowering.

The effect of LD in decreasing succulence may be an extreme example of a more general effect of daylength on leaf growth. Whatever the photoperiodic class for flowering, the individual leaves of many plants have been found to be larger in LD than in SD (Table 13.1). In clover, leaf area increased with increase in daylength, reaching a maximum in 21 h photoperiods at 18°C; there was an interaction in temperature, however, and, at 12°C, the maximum leaf area was attained in continuous light (Fig. 13.6). The usual effect of LD is to increase surface expansion and lead to the development of thinner, less succulent leaves with an increased specific leaf area, where specific leaf area = leaf area/total leaf dry weight (Fig. 13.7). Such changes are not, however, invariably associated with LD and, in the woodland species, *Circaea lutetiana*, the specific leaf area was higher when the plants were growing in SD (Frankland and Letendre, 1978). LD stimulation of leaf elongation, including both leaf blade and leaf sheath is well documented for various grass species, including *Dactylis*, *Poa*, *Bromus* and *Triticum* (Junttila, 1989). Needle elongation in *Pinus* has also been shown to be greater in longer daylengths, under conditions when the photon flux density was the same (Oleksyn *et al.*, 1992).

In strawberry, the increased leaf area in LD resulted from an increase in cell number based on counts of epidermal cells (Arney, 1956). Some increase in cell number also

TABLE 13.1   Effects of daylength on leaf expansion.

Species	Long-day treatment (h)	Flowering promoted by
Leaves larger (or longer) in LD		
Amaranthus caudatus		SD
Aphelandra squarrosa		
Arachis	16	
Begonia rex	2 NB	
Beta	16	LD
Bidens pillosa		
Callistephus chinensis	1 NB	LD
Campanula pyramidalis	16	
Ceiba pentandra (26°C)	13 and 17	
Chlorophyta excelsa	13 and 17	
Coleus	14	
Dendranthema grandiflora	17	SD
Dactylis glomerata	16	LD
Fragaria × ananassa	18	SD
Impatiens balsamina		SD
I. parviflora	16	
Ipomoea caerulea		
Kalanchoë blossfeldiana	15	SD
Perilla		SD
Petunia hybrida	2 NB	LD
Phleum pratense	16	LD
Phlox drummondii	2 NB	LD
Pinus sylvestris		
Poa pratensis		LD
Sinapis alba	16	LD
Spinacea oleracea		LD
Taraxacum kok-saghys	18	LD
Trifolium repens	24	LD
Triplochiton	2 NB	
Viola odorata, tricolor		LD
Zea mays		SD
Leaves larger (or longer) in SD		
Chenopodium amaranticolor		SD
Fuchsia hybrida		LD
Oryzopsis mileacea		LD
Plantago lanceolata		LD
No effect of daylength		
Ageratum	2 NB	
Salvia splendens	2 NB	

NB, night-break given in the middle of a long dark period.

occurred in several grass species, but the main effect of LD was to increase the surface area of epidermal cells, especially their length (Heide *et al.*, 1985b).

In *Callistephus chinensis* (Cockshull, 1966), the daylength treatment continued to exert an effect during the final stages of leaf growth and returning plants to SD

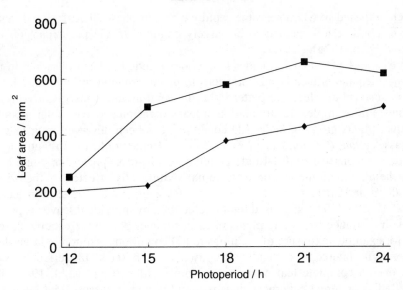

FIG. 13.6. Effects of daylength and temperature on leaf area in white clover (*Trifolium repens*). Plants were exposed to daylight for 12 h. The photoperiodic treatments were established by extending the daylight period with low irradiance tungsten-filament light (150 μmol m^{-2} s^{-1}) and plants were grown at either 18°C (n) or 12°C (u). After Junttila *et al.*, 1990.

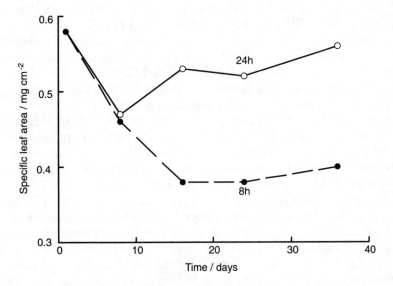

FIG. 13.7. The effect of two different daylengths on the specific leaf area of *Phleum pratense* cv. Engmo (origin 69°N). Plants were grown at 15°C (for light conditions see Heide, 1984). After Heide *et al.*, 1985b.

reduced expansion even when the leaves were already three-quarters grown, suggesting that photoperiod continues to control expansion throughout the development of the leaf. In contrast, the leaves in *Begonia hiemalis* were also larger in LD but the effect decreased as the leaves continued to expand; daylength had only a small effect if plants were exposed to SD during the rapid expansion phase of leaf growth and the response to photoperiod seemed to be largely confined to the pre-expansion phase (Powell and Bunt, 1978).

An increase in leaf expansion under long photoperiods has been observed in many different species and is clearly advantageous to plants growing in high latitudes. The increase in surface area leads to better light interception and, consequently, to faster growth rates. For example, the stimulation of leaf expansion and dry matter production in individual plants growing under LD conditions has been observed in at least four forage grasses (*Poa*, *Phleum*, *Dactylis* and *Bromus*, Heide et al., 1985b) and may well be a general response of high latitude populations of grass species. In white clover, however, the photoperiodic responses were not significantly affected by the origin of the population and all responded to increased daylength by an increase in leaf area (Junttila et al., 1990). The increased light interception by individual leaves is probably of greatest importance for young plants in early summer before the canopy closes. In *Petunia*, for example, exposure of seedlings to LD conditions resulted in larger leaves and better light interception, leading to faster growth rates; the effect was lost, however, once a complete leaf cover was obtained (Merritt and Kohl, 1983). However, in a tall, more erect canopy such as is found in many grasses, light interception may be improved by an increase in specific leaf area even in a closed canopy (Heide et al., 1985b). In this context, it is interesting to note that, in the woodland species *Circaea lutetiana*, the specific leaf area was larger in SD (Frankland and Letendre, 1978); this may be an adaptation allowing for better interception of light during the spring before the canopy closes. The increase in leaf expansion and faster growth rates that can be obtained in many species by exposing young plants to LD (or to a night-break) has considerable potential for the nursery trade; for plants growing under glass during winter and early spring, the more rapid leaf expansion leads to the development of larger plants in the same growing period (Hughes and Cockshull, 1966).

The rate of leaf initiation and/or unfolding may also be influenced by daylength. In spring wheat, the phyllochron (days between appearance of successive leaves) decreased from 124°C d leaf^{-1} in an 8 h day to 97°C d leaf^{-1} in a 16 h day (Mosaad et al., 1995) but, since the irradiance was constant throughout, photosynthetic and photoperiodic effects were confounded. A faster leaf initiation rate in longer photoperiods has been reported for maize whereas, in celeriac and celery, the leaf initiation rate was lower in LD conditions than in SD (Boolj and Meurs, 1994).

Few studies have been made on the underlying mechanism for the photoperiodic control of leaf growth. Many species have been found to be responsive to a relatively brief night-break, which is equivalent to a LD (Table 13.1). In *Callistephus*, the increase in leaf expansion which resulted from a 1 h night-break could not be further increased by intermittent lighting throughout the night or by continuous light, although flowering was promoted by these treatments compared with the response obtained with a night-break (Cockshull and Hughes, 1969). However, a night-break has little effect in some plants; for example in strawberry, where long photoperiods are necessary for both the promotion of leaf growth and the inhibition of flowering (Tafazoli and Vince-Prue, 1978). As has been observed in the photoperiodic control of flowering (see Chapter 5), the effect on leaf growth of the R/FR ratio of light given during a day-extension depends on the species (Soffe et al., 1977). In spinach beet and beet, the increase in mean leaf area in LD was the same irrespective of the R/FR ratio during the 4 h day extension whereas, in celery, the increase in leaf area was much greater with a low ratio of R/FR light (0.7) compared with a higher ratio (4.2), although the latter was at a significantly higher irradiance.

The photoperiodic stimulus affecting leaf growth has been shown to be transmissible in both strawberry (Guttridge, 1969) and *Kalanchoë* (Harder, 1948). For example, when a single leaf of *Kalanchoë* was exposed to SD, the leaf immediately above it in the same orthostichy was small and succulent indicating the production of a translocatable stimulus. Similar experiments indicated that a stimulus for the production of LD-type leaves was translocated from a treated leaf to a branch developing in its axil. The promoter of vegetative growth which is produced in strawberry in response to LD can be translocated along a stolon to a receptor plant maintained in SD. Thus, as with other photoperiodic events, exposing leaves to a particular daylength brings about the production of one or more stimuli which can alter the developmental pattern in other parts of the plant. However, leaves may also respond directly to daylength; in *Kalanchoë*, excised rooted leaves transferred to SD became at least twice as succulent as those remaining in LD (Schwabe, 1985).

The identity of the translocated stimuli for photoperiodic effects on leaf growth has not been investigated in any detail although it has been suggested that LD effects on leaf growth in *Poa pratensis* are mediated by changes in endogenous GAs (Heide *et al.*, 1985a). However, in wheat, LD significantly promoted elongation in a GA-insensitive genotype (cv Siete Cerros), whereas GA_3 application had no effect; in contrast, both GA and LD were effective in a normal genotype, and there was no significant interaction between them. It was concluded from these results that the LD-promotion of leaf length in wheat is not mediated by GA (Junttila, 1989). There is some evidence that photoperiodic effects on leaf shape may depend on gibberellins. In *Proserpinaca*, for example, dipping the shoot twice in GA_3 ($1–10$ mg l^{-1}) resulted in the development of more adult (LD-type) leaves in SD (Davis, 1967). In contrast, GA_3 inhibited flowering at all concentrations used; thus, in this case, it seems unlikely that the induction of flowering by LD depends on an increase in endogenous GA content, although such an increase may be involved in the transition to producing adult-type leaves. However, in strawberry, the application of GA_3 simulated the effect of LD both to inhibit floral induction and increase leaf size (Guttridge, 1969). Possible effects of specific GAs on leaf development remain to be investigated. In *Kalanchoë*, the application of a lipophilic fraction from flowering plants in SD, resulted in the development of succulence in LD leaves but the identity of the active fraction was not determined (Janistyn, 1981).

ASSIMILATE PARTITIONING

Patterns of growth and development need to adjust to the varying demands of different seasons. When alterations in the daylength bring about large developmental changes such as the onset of flowering, dormancy or the formation of storage organs there are certain to be modifications to the distribution of resources within the plant. Where they are a secondary event and a consequence of the developmental change, rather than a direct response to daylength, the changes will take several days to manifest themselves. On the other hand, a rapid change in the partitioning of assimilates between

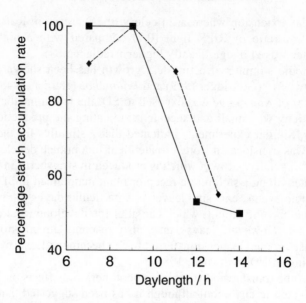

FIG. 13.8. The effect of daylength on starch accumulation rates in *Digitaria decumbens* Stent. Plants were grown in 7 h SD and given daylength extensions at 600 μmol m^{-2} s^{-1} and harvested the following day. Data are from two experiments normalised at 9 h extensions. After Britz *et al.*, 1985a.

different organs or between different storage or structural components would indicate a direct effect of photoperiod on the process of assimilate partitioning. Britz *et al.* (1985a) found that the proportion of photosynthate partitioned into starch in the leaves of *Digitaria decumbens* Stent. adjusts within 24 h of a change in daylength. After a single LD of 14 h, the relative starch accumulation rate was reduced to about half of that under a 7 h SD. The starch accumulation rate was not linearly related to the length of the day but showed a sharp drop when the daylength was extended beyond 11 h (Fig. 13.8). A day extension could be replaced by a 7 h SD displaced so as to start at the time that the SD normally finished. The response therefore established by the timing of the light treatments rather than the quantity of light received. Further evidence for a photoperiodic effect was furnished by the finding that the proportion of photosynthate partitioned into starch depended in a rhythmic fashion on the duration of a dim light treatment given prior to a 7 h SD at high irradiance (Fig. 13.9).

The mechanisms by which daylength modifies patterns of assimilate partitioning have been investigated. Reduced photosynthate partitioning into stored leaf carbohydrates was observed in a gibberellin-deficient dwarf mutant of *Zea mays* (Britz and Saftner, 1988) following transfer from SD to LD. This is good evidence that daylength effects on assimilate partitioning are not mediated by gibberellins, which are often regulated by photoperiod. Huber *et al.* (1984) found a diurnal rhythm in the activity of sucrose phosphate synthase (SPS) in soybean and proposed that the acclimation of sucrose and starch levels to photoperiod was associated with changes in SPS. Rhythmic effects of pretreatments on photosynthesis and accumulation of photosynthates were found in *Sorghum*, although in this case the pretreatments were with darkness (Britz *et al.*, 1987). Net photosynthesis, starch accumulation and sugar storage all depended cyclically on the duration of the dark pre-incubation, but were all out of

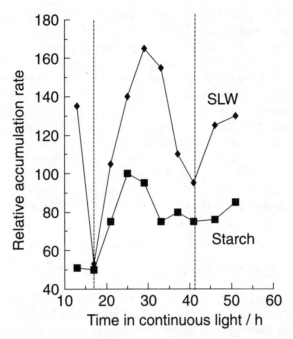

FIG. 13.9. Accumulation of starch and specific leaf weight in leaves of *Digitaria decumbens* Stent. under high irradiance white light after various lengths of pre-incubation in dim white light. Vertical lines are expected time for lights-on for plants maintained in the normal light/dark cycle. After Britz *et al.*, 1985b.

phase. Neither the rhythm in starch nor in sugar accumulation was coupled to the rhythm in photosynthetic rate (Britz, 1990a). In *Sorghum*, the duration of the dark period exerts its effect because the rates of carbohydrate accumulation are determined by the circadian rhythms in participating processes which run in darkness. The effect of light on suppressing the rhythms leads to daylength effects on partitioning. There is some evidence from pulse experiments that R is partially effective and its effect can be reversed by an immediate FR, implicating phytochrome as a photoreceptor for the response (Britz, 1991). Responses to photoperiod appear to be direct effects of daylength on the leaf and there is no evidence for a transmitted signal as in 'classic' photoperiodically regulated processes.

One problem with the experimental system where large shifts in daylength cause rapid changes in patterns of assimilate partitioning is that such changes might arise from transient alterations in growth and sink demand. This raises the question of whether it is possible to demonstrate longer-term changes in partitioning under the influence of daylength. In experiments carried out over a period from May to December, Britz (1990b) showed that the proportion of photosynthate partitioned into starch in soybean leaves was linearly related to daylength at harvest. However, it was not possible to specify the manner in which daylength was perceived or mediated. Although, as discussed in the section on leaf growth, there are several examples of effects of daylength on overall growth and productivity of plants, there is no consistent pattern concerning how this relates to assimilate partitioning. Hay and Heide (1983) found that the stimulation of dry weight, plant height and leaf area by

LD in *Poa pratensis* occurred without any change in the partitioning of assimilates amongst leaves, stems and stolons. On the other hand Gordon *et al.* (1982) found that when barley plants were grown in LD the proportion of assimilates translocated to the roots was greater than when plants were grown in SD at the same irradiance. Eagles (1971) found that with *Dactylis glomerata* patterns of dry matter distribution were influenced by daylength and suggested that these resulted from higher leaf area ratios in LD- than SD- grown plants.

In conclusion, therefore there is good evidence for a specific short term effect of photoperiod on partitioning of assimilates between storage compounds. This arises from interactions between light, darkness and circadian rhythms in component processes of photosynthesis and sucrose biosynthesis. Phasing of the rhythms by light appears to be mediated by phytochrome. Although growth and dry matter accumulation is frequently greater in LD than SD, it is difficult to identify consistent long term effects of photoperiod on partitioning.

CRASSULACEAN ACID METABOLISM

Crassulacean acid metabolism (CAM) is a variant of carbon dioxide assimilation, occurring in succulent plants, and characterised by a diurnal fluctuation of malic acid accumulation during the night and decrease during the day. In some species capable of performing CAM, photoperiodic treatments can enhance or even induce the CAM pathway (Queiroz, 1970). In *Kalanchoë blossfeldiana* cv. Tom Thumb, malate formation is promoted by SD. For plants grown in LD, malate formation was only seen after 7 SD treatment (Brulfert *et al.*, 1975) and maximum malate production was found in continuous SD. When plants were transferred back into non-inductive

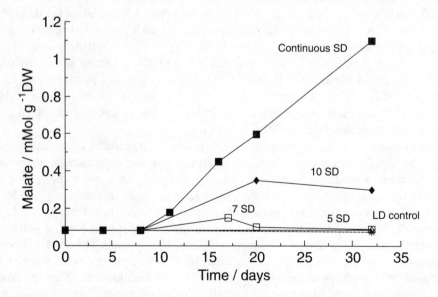

FIG. 13.10.   Variation of content of malate in leaves of *Kalanchoë blossfeldiana* under non-inductive LD conditions (SD with NB), continuous SD or 5, 7 or 10 SD followed by non-inductive conditions. After Brulfert *et al.*, 1975.

conditions after 7 or 10 SD the level of malate remained unchanged, suggesting that CAM metabolism had been 'induced' in these plants (Fig. 13.10). Similar results were obtained for the activity of phosphoenolpyruvate (PEP) carboxylase, the enzyme responsible for the production of malate via the synthesis of oxaloacetate by the dark fixation of carbon dioxide. The non-inductive treatment consisted of a SD + 30 min NB with R light which was FR-reversible. Thus the perception of daylength for modulating the CAM metabolism system appeared to be similar to the photoperiodic floral induction response rather than a direct effect on photosynthesis. The photoperiodic sensitivity of *Kalanchoë blossfeldiana* for CAM is most apparent in young seedlings; leaf ageing under LD eventually leads to CAM and the promotive effect of SD lessens as the plants age (Brulfert *et al.*, 1982). Although CAM is associated with daily rhythms in malate, PEP carboxylase, carbon dioxide fixation, cellular pH and other properties, it is not understood how the induction of CAM by photoperiod is achieved. CAM is also triggered by water stress in *Kalanchoë blossfeldiana* and increases in cellular ABA, which would be expected in drought-stressed tissues, cause an early increase in PEP carboxylase mRNA transcripts (Taybi *et al.*, 1995). It is possible that the photosynthetic behaviour of the leaves depends on the availability of external carbon as governed by stomatal behaviour, although the link with photoperiod for this mode of regulation has not been made.

SECONDARY METABOLISM

Many biochemical constituents of leaves have been shown to vary in plants growing in different photoperiods. Some of these have been discussed in Chapters 7 and 8. Some other compounds are considered here. Anthocyanin pigments are present in the leaves of many plants in amounts which may be influenced by daylength. The response to daylength has been characterized in *Kalanchoë* and *Perilla* and found to be different in the two species. In *Kalanchoë*, control of anthocyanin production appears to be linked to the daylength control of flowering (Neyland *et al.*, 1963). In inductive SD the leaves became red but in LD or with a night-break, the leaves were green with slight reddening at the margins. The leaves maintained in the non-inductive treatments contained large amounts of a colourless *leuco*-anthocyanidin compound, which decreased in SD although the increase in anthocyanin was less than the decrease in the *leuco*-compound. In contrast to *Kalanchoë* the anthocyanin content of the leaves of the SDP, *Perilla* (red), was decreased on transfer to SD but was independent of induction. The content of anthocyanin was always higher in LD than SD irrespective of seedling age and even while the plants were vegetative (Schumacker, 1966). In the LDP *Fuchsia* cv Lord Byron, anthocyanin levels were higher in plants treated with SD than in plants given a 9 h day extension with FR. Levels were not different to SD, however, when a R or R+FD day-extension was given (Vince-Prue, 1975, p. 424). In this situation anthocyanin formation appears to be a typical end-of-day response, in which a stable Pfr fraction is required in darkness for anthocyanin synthesis, rather than a specific response to photoperiod. A further type of interaction between photoperiod and anthocyanin formation has been proposed by Camm *et al.* (1993). They reported that the characteristic purpling of many species of pine seedlings in the autumn can be contingent upon the photoperiod received months previously. For example, a 4 week exposure of *Pinus contorta* seedlings to 10 h days during August and then return to natural photoperiods diminished the frequency of purpling in response to the onset of cold weather in the subsequent November. Phenolic metabolism was hardly affected by the daylength treatment. Purpling thus seemed to be a specific response to two components, namely photoperiod and low temperature, which were temporally separated.

Daylength has been reported to affect a number of other secondary compounds. In *Lycopersicon hirsutum*, levels of the steroidal glycoalkaloid tomatine were higher in LD than SD at the same irradiance and this was correlated with susceptibility to attack by the Colorado potato beetle (Sinden *et al.*, 1978). Higher levels of alkaloids as well as phenolics were also found in tobacco plants grown in LD as compared to SD (Tso *et al.*, 1970). Other examples where lengthening the photoperiod leads to increases in secondary products include fatty acid and hydrocarbon classes of epicuticular lipids of *Cassia obtusifolia* (Wilkinson, 1974), peppermint oils in *Mentha piperita* (Clark and Menary, 1979), oil content of flax (Sairam and Srivastava, 1977) and alkaloid accumulation in *Papaver somniferum* (Bernáth and Tétényi, 1979). In all of these cases, with the exception of the peppermint oils, LD were obtained by extending daylength at the same irradiance and the higher levels of secondary products could be a response to increased daily light integral, rather than to the duration of the daily light or dark period.

Appendix I Photoperiodic Classification of Plants

SHORT-DAY PLANTS (SDP)

1. Qualitative or Absolute SDP = obligate

Amaranthus caudatus albiflorus; *caudatus caudatus*; *tricolor*
Ambrosia elatior H
Andropogon gayanus
Begonia bowerae; × *cheimantha* (>21°C, 9 at 12°C); × *hiemalis* (at >24°C, 2 at LT)
Bidens radiata
Bouvardia (hybrids)
Bothriochloa bladhii (+2)
Brachiaria mutica
Calathea crocata
Caryopteris incana (for completion of floral development, initiation may be promoted by LD especially in low intensity light)
Chamaelaucium uncinatum (at >24/16°C, 2 at lower temperature)
Chenopodium album; *rubrum* (e.g. ecotypes 50°10′N, 34°20′N)
Clerodendrum speciosum
Coffea arabica
Coleus fredericii
Corchorus capsularis; *olitorius*
Cosmos sulphureus (+9, cv Sunset)
Cucumis hardwickii (some lines)
Desmodium intortum (or ?5)
Echeveria carnicolor; *derenbergii* (at 21°C, 9 at 10°C); *elegans* (10–17 at 1°C, 9 at 5°C)
Euphorbia fulgens; *millii* (at >20°C); *pulcherrima*
Fragaria × *ananassa* (>16°C, 2 at <15°C)
Glycine max cv Biloxi_____   soybean
Hibiscus cannabinus; *esculentus*
Humulus japonicus; *lupulus*
Hydrilla verticillata
Hyparrhenia rufa
Hypoestes aristata; *phyllostachya*

355

Impatiens balsamina; *goughi*
Indigofera hirsuta; *spicata*
Ipomaea batatus
Iris ensata (LD promotes flower development)
Kalanchoë blossfeldiana
Keinia articulata (LD induces dormancy)
Lemna aequinoctalis (= *paucicostata*, 6746, T-101; +2,4)
Leptospermum scoparium
Lespedeza cuneata; *stipulacea*
Nenotonia wightii
Nicotiana tabacum Maryland Mammoth (+2,9)
Oryza sativa (+2,9)
Panicum maximum (+2)
Paspalum plicatum (or ?5); *urvillei*
Pennisetum purpureum (or ?2); *typhoides* (late flowering cvs; +2,9)
Perilla (red)
Pharbitis nil
Phaseolus lunatus (+2,9)
Ribes nigrum
Rottboellia cochinchinensis (+9 some populations)
Salvia riparia
Sechium edule
Setaria glauca; *italica*; *moharica*; *verticillata* (+8); *viridis* (+9)
Solidago sempervirens
Stylosanthes humilis
Trifolium humilis
Utricularia inflexa var *stellaris*
Vaccinium angustifolium; *corymbosum*
Viola odorata (for chasmogamous flowers, cleistogamous flowers in LD)   *pansy ?*
Wolffia microscopica
Xanthium strumarium
Zosia japonica; *matrella*; *tenuifolia*

1a. SDP requiring or accelerated by low-temperature vernalisation

Allium fistulosum; *sativa* (at HT, 9a at 9°C)

2. Quantitative SDP

Amaranthus graecizans; *hybridus*; *retroflexus*
Ananas comosus
Andopogon virginicus
Anigozanthus manglesii
Bouteloua eriopoda (for spikelet initiation, LD promotes flower development)
Bougainvillea glabra
Brachiaria ruziziensis
Brassavola nodosa
Cajanus cajan (+9)
Cannabis sativa
Catharanthus roseus
Celosia cristata
Cestrum diurnum (or 7?)
Chenopodium aristatum; *polyspermum*; *rubrum* (ecotypes 60° 46′N, 60° 47′N; +1,8)
Colchicum tunicatum (at 30°C, 4 at 15 °C)
Cosmos bipinnatus
Cucumis sativus (some cvs)

Dahlia (hybrids; +1,9)
Datura stramonium (+9 older plants)
Dendranthema grandiflora (somes cvs are Group 1 or 1a for bud development; +2a)
Desmodium tortuosum
Gardenia jasminoides (LD reduces abortion and promotes flower development)
Gladiolus grandiflorus
Glycine max
Gossypium (most wild spp; +9, modern cvs)
Helianthus annuus (+4,9); *tuberosus* (+9, some clones)
Helicornia stricta; brasiliensis
Holcus sudanensis (9 at LT)
Lagenaria siceraria
Lemna aequinoctialis (151)
Luffa cylindrica
Lycopersicon esculentum (+9?)
Malva verticillata (9 at LT)
Mamillaria longicoma
Nicotiana tabacum (+1,9)   rice
Oryza sativa (+1,9)
Panicum maximum; miliaceum
Paspalum thunbergii
Perilla (green)
Phaseolus vulgaris (LD delay flower development and cause abscission; +8,9)
Portulaca oleracea
Primula malacoides
Psophocarpus tetragonolobus
Punica granatum nana
Rhododendron spp. (Florist's azalea, at > 18°C, 9 at lower temperatures)
Ricinus communis (+9)
Rosa gallica; rugosa
Rubus idaeus
Salsola inermis; volkensii (and several other winter annuals of the Negev desert)
Salvia splendens (+ 4,9)
Sesamum indicum
Setaria italica
Sorghum bicolor; halepense (+9)
Tagetes erecta; patula (+9, Petite Orange); *tenuifolia*
Torenia fournieri
Vigna radiata; unguiculata
Viscaria alba
Zea mays (+9)
Zinnia elegans

2a. Quantitative SDP requiring or accelerated by vernalisation

Allium sativum
Apium graveolens (LD during LT inhibits initiation in some cvs)
Apium graveolens var *rapaceum* (celeriac)
Dendranthema grandiflora (+2)

LONG DAY PLANTS (LDP)

3. Qualitative or absolute LDP   = obligate

Agrostemma githago (also requires LD for development)
Agrostis stolonifera

Allium chinense; *tuberosum*
Anagallis arvensis
Anethum graveolens
Arabidopsis thaliana (annual strains)
Avena sativa (spring strains)
Blitum capitatum; *virgatum*
Bouteloua curtipendula (+7)
Brassica juncea (at 27°C, 4 at 24°C); *carinata*; *pekinensis* (9 at LT)
Calendula officinalis (CDL 6.5 h)
Campanula isophylla
Centaurea cyanus
Chrysantheumum leucanthemum; *superbum*
Cicorium intybus (late cvs)
Coleus lanuginosus
Cynodon dactylon (or ?9)
Delphinium elatum garden hybrids H (at LT, 9 at HT)
Dianthus superbus C; *allwodii* cvs Doris, Joy
Dicentra spectabilis (SD dormancy, require LT to break dormancy)
Echeveria gracilis; *peacockii*
Epilobium adenocaulon; *parviflorum*
Eschscholtzia californica
Foeniculum vulgare
Fuchsia hybrida (+4,9)
Gaillardia grandiflora (4 at 27°C)
Gypsophila paniculata
Helichrysum cassinianum
Helipterum craspedioides
Hibiscus syriacus H
Hieraceum floribundum
Hyoscyamus niger (annual)
Jasminium grandiflorum (at 22 °C, 4 at lower temperatures)
Lemna gibba; *minor*
Lens culinaris (+4)
Lolium temulentum Ceres
Lunaria annua (annual form; non cold-requiring but 9 after vernalisation)
Lychnis × *arkwrightii* (at 8–11°C, 4 at 14–17°C)
Melandrium album
Melilotus alba; *indica*
Mentha citrata; *piperata*; *spicata*
Nicotiana sylvestris
Nigella damascena
Oreganum syriacum var *syriacum*
Papaver somniferum (+4,9)
Phalaris arundinacea H
Phippsia algida (DNP for initiation; LDP for heading)
Phleum nodosum H; *pratense*
Phlox paniculata A
Poa nemoralis
Ranunculus scleratus
Rudbeckia bicolor; *hirta*; *laciniata*; *newmani*; *nitida*; *speciosa*
Salvia splendens (+2,9)
Samolus parviflorus
Scabiosa atropurpurea
Scrophularia arguta (at low light intensity, 4 at high intensity)
Sedum ellacombianum; *spectabile*; *spurium coccineum*; *telephium*
Silene armeria (or 6b): *coeli-rosa*

Stipa barbata
Trifolium pratense (English Montmorency)
Viscaria alba

3a. LDP requiring or accelerated by low-temperature vernalisation

Anagallis tenella
Anthriscus cerefolium
Arctium lappa
Beta vulgaris vulgaris (sugar beet, garden beet)
Campanula pyramidalis
Calamintha acinus
Cheiranthus allionii (with increasing age plants become 4a and finally 9a)
Cichorium intybus (late cvs; SD after sowing accelerates flowering, ?6a)
Cynara scolymus
Dactylis glomerata (requires SD before or during vernalisation)
Deschampsia cespitosa
Dianthus arenarius; gracilis; graniticus C
Hyoscyamus niger (biennial)
Lithospermum arvense
Lysimachia nemorum
Medicago minima
Oenothera biennis; lamarckiana; parviflora
Pisum sativum late cvs (some very late cvs may die without flowering in SD, +4, 4a)   [handwritten: garden pea]
Poa bulbosa (SD after induction → inflorescences with bulblets, LD → sexual inflorescences)
Saxifraga hypnoides C
Silene otites

3b. LDP; vernalisation substitutes (at least partly) for LD (cf. 3a where vernalisation must be followed by LD)

Arabidopsis thaliana (some strains, e.g. La)
Calceolaria hybrids
Lathyrus odoratus (+4a,9a)
Lilium candidum; longiflorum
Lunaria annua (annual strains)
Silene alba
Trifolium pratense; subterraneum

4. Quantitative LDP

Aechmea fasciata
Allium ampeloprasum
Anenome coronaria
Anigozanthos flavidus
Anthemis cotula H (9 at LT)
Antirrhinum majus
Arabidopsis thaliana (summer annual strains)
Avena fastuosa
Baeria chrysostoma
Brassica alba; campestris oleifera (+4a); *napus oleifera; nigra*
Calceolaria hybrids (at HT)
Callistephus chinensis (development accelerated by SD)   [handwritten: aster]
Camellia japonica (+9)
Campanula fragilis

Chamecyparis obtusa (LD more effective in high intensity light)
Chenopodium murale
Chrysanthemum segetum
Cicer arietinum (+4a)
Cleome hassleriana
Coleus blumei (+9)
Coriandrum sativum
Dianthus carthusianorum Napoleon III; *caryophyllus* (glasshouse carnations)
Eustoma grandiflorum
Hemerocallis fulva
Hordeum vulgare (spring strains)
Largerstroemia indica
Liatris spicata (after prolonged cool storage, after shorter cooling flowering is accelerated by SD)
Linum usitatissimum
Lolium temulentum Ba 3081
Lysimachia congestifolia
Manihot esculenta
Matthiola incana (requires cool temperatures for initiation)   stock
Medicago sativa (9 at LT)
Oenothera rosea
Oryzopsis miliacea
Oxalis adenophylla; *deppei*; *pes-caprae*
Pentas lanceolata
Petunia hybrida
Pisum sativum (+3a, 4a)
Salvia splendens (+2, 9)
Schismus arabicus
Schizanthus × *wisetonensis*
Secale cereale (spring cvs)
Setaria sphacelata
Solanum tuberosum (2 for tubers)
Sonchus oleraceus H
Spirodela aequinoctialis (*paucicostata*, S type strains); *punctata*
Stephanotis floribunda
Trachelium caeruleum
Trifolium masaiense
Trigonella arabica; *stellata* (and several other winter annuals of Negev desert)
Trigonella coelesyriaca; *monspeliaca*; *spinosa* (and several other annuals of Mediterranean region of Israel)
- *Triticum aestivum* (spring cvs)   wheat
Vicia faba (+4a, 9)

4a. Quantitative LDP requiring or accelerated by vernalisation

Allium sphaerocephalon
Alstoemeria Regina, Walter Fleming
Aquilegia × *hybrida*
Arenaria serpyllifolia
Aurinia (Alyssum) saxatilis
Campanula alliariifolia; *persicifolia* C
Cynodon dactylon (several US clones; +3a, S. African clones)
Delphinium ajacis
Digitalis purpurea
Hordeum vulgare (some winter strains; +4c)
Lactuca sativa (+9a); *serriola*

Limonium sinuata
Lupinus cosentinii
Myosotis hispida K
Oenothera longiflora; sueveolens
Osteospermum jucundum (Pink Whirls)
Pisum sativum (+3a, 4)
Ranunculus asiaticus
Raphanus sativus (SD during LT delays flowering)
Scabiosa canescens
Scrophularia alata; vernalis
Teucrium scorodonia C
Thlaspi arvensis (LD promotes development, 9a for initiation)
Vicia faba (+4, 9); *sativa*

4b. Quantitative LDP: vernalisation substitutes (at least partly) for LD (cf. 4a where vernalisation must be followed by LD for maximum promotion

Arabidopsis thaliana (winter annual strains)
Cheiranthus cheiri
Lupinus luteus
Sinapis alba (3b at low light intensity)

4c. Quantitative LDP: SD substitutes (at least partly) for a vernalisation requirement. Thus they could also be classified as 6b since, without LT, they respond as SLDP

Avena sativa and other spp.
Chondrilla juncea (fully vernalised plants at 15/10°C are 9)
Festuca arundinacea
Hordeum vulgare (some winter strains, +4a)
Iberis intermedia Durandii C
Poa pratensis
Secale cereale (winter strains)
Triticum aestivum (winter strains)

DUAL DAYLENGTH PLANTS

5. Long-short-day plants (LSDP)

Aloe bulbifera
Aster pilosus (LD required for stem elongation)
Bryophyllum crenatum; daigremontianum; tubiflorum
Cestrum nocturnum (at 26/20°C, SDP at lower temperatures)
Sedum bellum
Stylosanthes guianensis
Wolffia arrhiza (strain Velica Polana)

5a. LSDP requiring or accelerated by vernalisation

Aster novi-belgii Wimbledon

6. Short-long-day plants (SLDP)

Alopecurus pratensis (LD promotes flower development at >12°C, little effect of daylength at 6°C)

Bromis inermis
Dactylis glomerata strain s143
Scabiosa succisa
Symphandra hoffmanni L

6a. SLDP require or accelerated by vernalisation

Apium graveolens (requires SD during vernalisation)

6b. SLDP; vernalisation substitutes at least partly for the SD effect and, after low temperature, plants respond as LDP

Campanula medium
Coreopsis grandiflora
Dactylis glomerata
Echeveria harmsii; setosa (at 15–20°C)
Hordeum bulbosum
Poa annua (some strains, +9) *alpigena; arctica; pratensis; palustris*
Trifolium repens

Iberis intermedia Durandii, winter strains of *Avena, Lolium, Hordeum, Secala* and *Triticum* and
    Festuca could also be placed here instead of 3c or 4c as, without LT, they respond as SLDP

7. Intermediate day plants (flower only in, or are accelerated by intermediate daylengths)

Aristeda contorta
Bouteloua curtipendula (+3)
Capsicum annuum
Cyperus esculentus; rotundus (at 32°C)
Mikania scandens L
Ocimum basilicum L
Paspalum dilatatum (or ?3)
Plectranthus fruticosus L
Saccharum spontaneum
Solidago canadensis (1 for initiation but very SD induce rapid onset of dormancy)
Tephrosia candida H, L (or ?5)
Trifolium usambarense (at >21°C, 1 at 10–21°C)

8. Ambiphotoperiodic plants (inhibited by intermediate daylengths)

Chenopodium rubrum ecotype 62°46 N (at 25°C; 2 at lower temperatures)
Hibiscus sabdariffa
Madia elegans L
Phaseolus vulgaris (some cvs, +2, 9)
Setaria verticillata

9. Day-neutral plants (flower at about the same time in all daylengths)

Achimenes hybrids
Allium cepa
Anigozanthos pulcherrimus; rufus
Aphelandra squarrosa
Arachis hypogaea (? +7)
Asclepias tuberosa
Blepharis linearifolia

Browallia speciosa
Brunfelsia pauciflora
Carica papaya
Cestrum elegans E
Chenopodium foetidum
Coleus blumei Princeton strain
Cornus florida
Cryptomeria japonica (SD increase ♂, LD increase ♀)
Cupressus arizonica (SD increase ♂, LD increase ♀)
Cucumis melo; *sativus* (+2)
Cyclamen persicum
Eucharis grandiflora
Fagopyrum esculentum; *tataricum* A,S
Fragaria × *ananassa* (everbearing); *vesca semperflorens*
Gomphrina globosa S
Gossypium hirsutum (+1, primitive types)
Guzmania monostachys
Hibiscus rosa-sinensis
Hippeastrum hybrids
Ilex aquifolium A
Lemna aequinoctialis (*paucicostata*, K type, strain 351)
Litchi chinensis
Lycopersicon esculentum
Malus sylvestris (apple)
Morus nigra
Olea europaea
Ornithogalum arabicum
Oryza sativa (+1, 2)
Pelargonium × *hortorum*
Phaseolus lunatus (+1, 2); *vulgaris* (+2, 8)
Pinus spp.
Pistia stratioides
Pisum sativum (early flowering strains)
Poa annua (+6b)
Prunus amygdalus (peach, apricot)
Pseudotsuga spp
Pyrus communis (pear)
Ranunculus ficaria; *penicillatus* var calcareus
Rhododendron obtusum P
Rosa (modern cvs, +4)
Saintpaulia ionantha
Schlumbergera truncata (at 10°C, 2 at higher temperatures)
Scrophularia peregrina; *arguta*
Senecio cruentus (4 for development); *vulgaris* C
Simmondsia chinensis (jojoba)
Solanum melongena (?4, small LD response)
Spiraea cantonensis
Spirodela polyrrhiza
Strelitzia reginae
Streptocarpus x hybridus
Thuja plicata (SD increase ♂, LD increase ♀)
Viburnum burkwoodii; *carlesii*
Vicia faba (very early flowering genotypes, +4, 4a)
Vitis vinifera
Vriesea splendens
Zantedeschia aethiopica
Zea mays (+2)

9a. DNP requiring or accelerated by vernalisation

Allium cepa (LD promotes emergence)
Agrimonia eupatoria C
Apium graveolens (see 2a)
Brassica spp. (brussels sprout, cabbage, cauliflower, kale, kohl rabi)
Cardamine amara C
Daucus carota
Dianthus barbatus (very slight acceleration in LD)
Dichondra repens (micrantha)
Digitalis lanata
Draba aizoides; hispanica C
Eryngium variifolium C
Euphorbia lathyris C
Geum bulgaricum; canadense; × *intermedium; macrophyllum* C; *urbanum*
Lunaria annua (biennial strain, or ?4b)
Melandrium rubrum
Nerine flexuosa
Pastinaca sativa
Pyrethrum cinerariifolium C
Rheum rhaponticum
Saxifraga rotundifolia C
Scrophularia umbrosa (syn *alata*); *vernalis* C
Senecio vulgaris C
Setaria viridis
Silene dioica; italica; nutans
Thlaspi perfoliatum K

This appendix was compiled mainly from Halevy (1985). Entries from other sources are indicated as follows:

A, Altman and Dittmer (1964, 1973)
C, Chouard (1960)
E, Evans (1969c)
H, Spector (1956, p. 460)
K, Hajkova and Krekule (1972)
L, Lang (1965)
S, Salisbury (1963b).

The photoperiodic categories used in Vince-Prue 1975 have been much simplified since it is becoming more and more apparent that photoperiodic categories are not fixed but can be considerably modified by temperature and other environmental variables such as irradiance, as well as with age. A few of the interactions with temperature are indicated, especially where low-temperature vernalisation is required or substitutes for daylength. In general, plants have only been included where the photoperiodic response group is based on experiments with a range of daylength; if deduced from field studies, they have been included only when the category is reasonably certain. Thus many plants for which a daylength response group has been proposed on the basis of field observations have been omitted. It is not always possible on the basis of the available evidence to determine if a response is obligate or facultative; allocation to the **absolute** classification in this table means only that flowering is usually delayed for several months in non-inductive conditions.

Crop plants often show a range of behaviour with genotype; where possible, plants have been listed in the most commonly found response groups, with other categories indicated in brackets; e.g. (+9) means that some genotypes are day-neutral. Early cultivars often show a response closer to the wild type, while modern genotypes tend towards less stringent daylength requirements.

Only anthesis has been recorded in many examples and, therefore, effects on initiation may be confounded with effects on development. Where possible, plants have been allocated to response groups on the basis of initiation; specific effects on development are given where these are known.

APPENDIX II Effects of Daylength on the Content of Endogenous Growth Substances

Plant	Response to daylength	Content increased by		Method	Tissue
		LD	SD		
Auxins					
Allium cepa	LD → bulbs	+		B	Leaf base
Begonia × cheimantha	SDP, SD → ♂	+		B	Leaf
Begonia × cheimantha[a]		IAA		C	Leaf
Bryophyllum daigremontianum[b]	LSDP	IAA, T		C	Leaf; shoot-tip
Cucumis sativus Beit Alpha	SD → ♀		+	B	Shoot-tip
Festuca arundinacea		+		B	Shoot (31/17°C)
Glycine max Biloxi	SDP	IAA		C	Shoot-tip
Hyoscyamus niger[c]	LDP	+		B	Leaf
Rhus typhina	SD dormant	+		B	Shoot-tip
Zea mays	SDP, SD → ♀	+		B	Shoot-tip
Cytokinins					
Allium cepa	LD → bulbs	+		B	Leaf blade
Begonia × cheimantha	SDP, SD → ♂		+	B	Leaf
Begonia × cheimantha[d]			+	C	Leaf, 24°C
Bougainvillea San Diego Red[e]	SD → flower development	+			Leaf; shoot-tip
Bryophyllum daigremontianum[b]	LDSP		+	B	Leaf; stem
Chenopodium murale[f]	LDP	=	=	C	Leaf; stem
		+			Shoot-tip

APPENDIX II Continued		LD	SD		
C. rubrum[f]	SDP	=	= + (T)	C	Leaf; stem Shoot-tip
Dactylis glomerata	SD → reduced growth		+	B	Shoot
Myriophyllum verticillatum	SD → turions	+		B	
Hyoscyamus niger[c]	LDP	+		B	Leaf
Nicotiana sylvestris[g]	LDP	+		B/C	Leaf; root
N. tabacum Maryland Mammoth[g]	SDP		+	B/C	Leaf; root
Perilla (red)	SDP		+	B	Xylem sap
Perilla (red)[h]		ZR/iPA		C	Leaf exudate
Phaseolus vulgaris	LD → flowers abscise		+	B	Xylem sap
Sinapis alba[i]	LDP	+		B/C	Phloem sap
Solanum tuberosum	SD → tubers		+ ZR	B/C	Stolon; shoot
Xanthium strumarium	SDP		+	B	Honeydew
Xanthium strumarium		+		B	Bud; leaf; root exudate

Ethylene

Chenopodium rubrum[j]	SDP	+		C	Shoot
Cucumis sativus					
Beit Alpha	SD → ♀		+	C	Shoot-tip
Matsunomidori	SD → ♀	+		C	Leaf
Higan-fushinari	LD → ♀	+		C	Leaf
Dahlia hybrida	SD tubers		+	C	Leaf cutting
Spinacia oleracea[k]	LDP	+		C	Excised leaf

Gibberellins

Acer negundo	SD → cold hardiness	+		B	Leaf
A. pseudoplatanus	SD → dormancy	+		B	Shoot-tip
Agrostemma githago	LDP	+, T		B/C	Shoot
Avena sativa	LDP	+		B	Leaf
Begonia × *cheimantha*	SDP, SD → ♂	+		B	Leaf
Begonia × *cheimantha*[a]				C	Leaf
Betula pubescens	SD → dormancy	+		B	Shoot-tip
Brassica campestris pekinensis[l]	LDP:vernalised	+		B	Seedlings
Bryophyllum daigremontianum	LSDP	+ 20		B/C	Shoot-tip; leaf
Cucumis sativus					
Beit alpha[m]	SD → ♀	+		C	Young leaf
Matsumomidori[n]	SD → ♀		+	B	Shoot
Higan-fushinari[n]	LD → ♀		+	B	Shoot
Cupressus arizonica	LD → ♂, SD → ♀	+		B/C	Shoot
Fagopyrum esculentum	DNP	+		B	Leaf
Fragaria × *ananassa*	SDP	+ ?19		B/C	Crown
Hyoscyamus niger	LDP	+		B	Leaf
Lolium temulentum[o]	LDP	+		C	Shoot apex
Nicotiana sylvestris	LDP	+		B	Leaf

		LD	SD		
N. tabacum Maryland Mammoth	SDP	+		B	Leaf
Perilla (red)	SDP	+		B	Leaf
Pisum sativum G	LDP		+, ?19	B/C	Shoot-tip
Prunus persica	SD → dormancy	+		B	leaf
Rudbeckia bicolor	LDP	+		B	leaf
Salix viminalis	SD → dormancy	+		B	phloem sap
Silene armeria[p]	LDP	19, 20	+ 53	B/C	shoot-tip
Solanum andigena	SD → tubers	+T		B	leaf
Spinacea oleracea	LDP	=	=	B/C	shoot
Spinacea oleracea[q]		+ 20	+19, 53	C	
Trifolium pratense		+, TE	+(D)	B	Shoot; leaf
Vicia faba	DNP	+		B	Shoot

Abscisic acid

		LD	SD		
Acer pseudoplatanus	SD→dormancy	+		C	Shoot-tip
		=	=	C	Leaf
Acer rubrum	SD→dormancy	+		C	Shoot-tip
Betula lutea	SD→dormancy	+		B/C	Leaf
Betula papyrifera	SD→dormancy	+		C	Shoot-tip
B. pubescens	SD→dormancy	+		C	Shoot-tip
B. pubescens[r]		+		C	Leaf
B. pubescens[r]			+	C	Lateral bud
Cucumis sativus[m]	SD→♀		+	C	Young leaf
Dendranthema grandiflora	SDP	+		C	Shoot-tip; leaf
Fragraria × *ananassa*	SDP	+		C	Shoot-tip; leaf
Lolium tementulum	LDP	=	=	C	Apex; leaf
Malus hupehensis	SD→dormancy	=	=	C	Shoot-tip
Perilla (red)	SDP	=	=	C	Leaf
Phaseolus vulgaris	LD → flowers abscise	+		B/C	Leaf; stem; bud
Picea abies[s]	SD→dormancy		+, T	C	Needle
Pinus sylvestris[t]	SD→dormancy		+	C	Shoot
Salix pentandra[u]	SD→dormancy	=	=	C	Shoot
S. viminalis	SD→dormancy				
field-grown		+		C	Shoot-tip
growth room		=	=	B/C	Shoot-tip; xylem sap
Solanum andigena	SD→tubers	=	=	C	Leaf
Spinacia oleracea	LDP	+		C	Shoot
Vitis vinifera	DNP	=	=	C	Leaf
Xanthium strumarium	SDP	=	=	C	Leaf

Xanthoxin

		LD	SD		
Lolium temulentum	LDP	=	=	C	Apex; leaf
Spinacea oleracea	LDP	=	=	C	Shoot

From Table 2 in Vince-Prue 1985, with additional information as indicated.
[a]Oden and Heide (1989); [b]Henson and Wareing (1977c); [c]Kopcewicz *et al.* (1979);
[d]Hansen *et al.* (1988); [e]van Staden (1981); [f]Macháčková *et al.* (1993); [g]Lozhnikova *et al.* (1986);
[h]Grayling and Hanke (1992); [i]Lejeune *et al.* (1988, 1994); [j]Macháčková *et al.* (1988);
[k]Crevecoeur *et al.* (1986); [l]Suge and Takahashi (1982); [m]Friedlander *et al.* (1977a);
[n]Takahashi *et al.* (1983); [o]Pharis *et al.* (1987a); [p]Talon and Zeevaart (1990, 1992); [q]Talon *et al.* (1991b);
[r]Rinne *et al.* (1994); [s]Qamaruddin *et al.* (1995); [t]Oden and Dunsberg (1988); [u]Johansen *et al.* (1986).
Method: B, bioassay; C, chemical or physical method. T, transient increase; E, extractable and D, diffusible growth substance.

References

Abdel-Rahman, M.H. (1982) The involvement of an endogenous circadian rhythm in photoperiodic timing in *Acrochaetium asparagopsis* (Rhodophyta, Acrochaetiales). *Br. Phycol. J.* **17**, 389–400.

Abeles, F.B. (1967) Inhibition of flowering in *Xanthium* by ethylene. *Plant Physiol.* **42**, 608–609.

Aharoni, M., Goldschmidt, E.E. and Halevy, A.H. (1985) Changes in the metabolites of acetate-1-^{14}C following floral induction of *Pharbitis nil* plants. *J. Plant Physiol.* **120**, 145–152.

Ahmad, M. and Cashmore, A. (1993) *HY4* gene of *A. thaliana* encodes a protein with characteristics of a blue-light photoreceptor. *Nature* **366**, 162–166.

Alden, J. and Hermann, R.K. (1971) Aspects of the cold-hardiness mechanism in plants. *Bot. Rev.* **37**, 37–142.

Altamura, M.M. and Tomassi, M. (1994) Transition from absolute to quantitative SD control in *Nicotiana tabacum* cv Maryland. *Physiol. Plant.* **91**, 276–284.

Altman, P.L. and Dittmer, D.S. (1964) *Biology Data Book* 446–448; and (1973) (2nd Edition) vol 2, 893–896. Federation of American Societies for Experimental Biology.

Alvim, R., Hewett, E.W. and Saunders, P.F. (1976) Seasonal variation in the hormone content of willow. I. Changes in abscisic acid content and cytokinin activity in the xylem sap. *Plant Physiol.* **57**, 474–476.

Alvim, R., Thomas, S. and Saunders, P.F. (1978) Seasonal variation in the hormone content of willow. II. Effect of photoperiod on growth and abscisic acid content of trees under field conditions. *Plant Physiol.* **62**, 779–780.

Alvim, R., Saunders, P.F. and Barros, R.S. (1979) Abscisic acid and the photoperiodic induction of dormancy in *Salix viminalis*. *Plant Physiol.* **63**, 774–777.

Andrade, F.H. and La Motte, C.E. (1984) A simple physiological model for flower-induction involving circadian rhythms and phytochrome. *Iowa State J. Res.* **59**, 31–44.

Angrish, R. and Nanda, K.K. (1982) Dormancy and flowering process in reproductive buds of *Salix babylonica* cultured *in vitro*. *Z. Pflanzenphysiol.* **106**, 263–269.

Araki, T. and Komeda, Y. (1990) Electrophoretic analysis of florally-evoked meristems of *Pharbitis nil* Choisy cv. Violet. *Plant Cell Physiol.* **31**, 137–144.

Araki, T. and Komeda, Y. (1993) Analysis of the role of the late-flowering locus, GI, in the flowering of *Arabidopsis thaliana*. *The Plant Journal* **3**, 231–239.

Arney, S.E. (1956) Studies of growth and development in the genus *Fragaria*. VI. The effect of photoperiod and temperature on leaf size. *J. Exp. Bot.* **7**, 65–79.

Arnott, J.T. (1974) Growth response of White Engelmann spruce provenances to extended photoperiod using continuous and intermittent light. *Can. J. For. Res.* **4**, 69–75.

Aronson, B.D., Johnson K.A., Loros J.J. and Dunlap J.C. (1994) Negative feedback defining a circadian clock – autoregulation of the clock gene-frequency. *Science* **263**, 1578–1584.

Arumingyas, E.L. and Murfet, I.C. (1994) Flowering in *Pisum*: a further gene controlling the response to photoperiod. *J. Hered.* **85**, 12–17.

Arumuganathan, K., Dale, P.J. and Cooper, J.P. (1991) Vernalization in *Lolium temulentum* L. Responses of *in vitro* cultures of mature and immature embryos, shoot apices and callus. *Ann. Bot.* **67**, 173–179.

Arzee, T., Gressel, J. and Galun, E. (1970) Flowering in *Pharbitis*: the influence of actinomycin D on growth, incorporation of nucleic acid precursors and autoradiographic growth patterns. In *Cellular and Molecular Aspects of Floral Induction*. (ed. G. Bernier), pp. 93–107, Longman, London.

Asahira, T. and Nitsch, J.P. (1968) Tubérisation *in vitro*: *Ullucus tuberosus* et *Dioscorea*. *Bull. Soc. Bot. Fr.* **115**, 345–352.

Atsmon, D. and Tabbak, C. (1979) Comparative effects of gibberellin, silver nitrate and aminoethoxyvinyl glycine on sexual tendency and ethylene evolution in the cucumber plant (*Cucumis sativus* L.). *Plant Cell Physiol.* **20**, 1547–1555.

Attridge, T.H. (1990) *Light and Plant Responses*. Edward Arnold, London.

Auderset, G., Gahan, P.B., Dawson, A.L., Greppin, H. (1980) Glucose-6-phosphate dehydrogenase as an early marker of floral induction in shoot apices of *Spinacia oleracea* var. Nobel. *Plant Sci. Lett.* **20**, 109–113.

Austin, R.B. (1972) Bulb formation in onions as affected by photoperiod and spectral quality of light. *J. Hort. Sci.* **47**, 493–504.

Badizadegan, M., Tafazoli, E. and Kheradnam, M. (1972) Effect of N^6 benzyladenine on vegetative growth and tuber production in potato. *Am. Potato J.* **49**, 109–116.

Bae, M. and Mercer, E.I. (1970) The effect of long- and short-day photoperiods on the sterol levels in the leaves of *Solanum andigena*. *Phytochemistry* **9**, 63–68.

Bagnard, C., Bernier, G. and Arnal, C. (1972) Étude physiologique et histologique de la reversion chez *Sinapis alba* L. *Physiol. Vég.* **10**, 237–254.

Bailey, L.H. (1893) Greenhouse notes for 1892/91. I. Third report upon electroculture. *Bull. Cornell Univ. Agric. Exp. Stn.* **55**, 147–157.

Baker, C.K., Gallagher, J.N. and Monteith, J.L. (1981) Daylength change and leaf appearance in winter wheat. *Plant Cell Environ.* **4**, 285–287.

Baldev, B. and Lang, A. (1965) Control of flower formation by growth retardants and gibberellin in *Samolus parviflorus*, a long-day plant. *Am. J. Bot.* **52**, 408–417.

Ballard, L.A.T. (1969) *Anagallis arvensis* L. In *The Induction of Flowering* (ed. L.T. Evans), pp. 376–392. Macmillan, Melbourne.

Ballard, L.A.T. and Grant-Lipp, A.E. (1954) Juvenile photoperiodic sensitivity in *Anagallis arvensis* L. subsp. *foemina* (Mill.) Schinz and Thell. *Aust. J. Biol. Sci.* **17**, 323–337.

Bamberg, J.B. and Hanneman, R.E. (1993) Transmission and yield effects of a gibberellin mutant allele in potato. *Potato Res.* **36**, 365–372.

Barros, R.S. and Neill, S.J. (1989) The status of abscisic acid in willow as related to the induction of bud dormancy. *Acta Physiol. Plant.* **11**, 117–123.

Baskin, J.M. and Baskin, C.C. (1976) Effect of photoperiod on germination of *Cyperus inflexus* seeds. *Bot. Gaz.* **137**, 269–273.

Bassett, C.L., Mothershed, C.P. and Galau, G.A. (1991) Polypeptide profiles from cotyledons of developing and photoperiodically induced seedlings of the Japanese morning glory (*Pharbitis* [*Ipomoea*] *nil*). *J. Plant Growth Reg.* **10**, 147–155.

Battey, N.H. and Lyndon, R.F. (1984) Changes in apical growth and phyllotaxis on flowering and reversion in *Impatiens balsamina* L. *Ann. Bot.* **54**, 553–567.

Battey, N.H. and Lyndon, R.F. (1990) Reversion of flowering. *Bot. Rev.* **56**, 162–180.

Batutis, E.J. and Ewing, E.E. (1982) Far-red reversal of red light effect during long-night induction of potato (*Solanum tuberosum* L.) tuberization. *Plant Physiol.* **69**, 672–674.

Bavrina, T.V., Aksenova, N.P. and Konstantinova, T.N. (1969) The participation of photosynthesis in photoperiodism. *Fisiol. Rast.* **16**, 381–391.

Beever, J.E. and Woolhouse, H.W. (1973) Increased cytokinin from root systems of *Perilla frutescens* and flower and fruit development. *Nature* **246**, 31–32.

Beggs, C.J., Holmes, M.G., Jabben, M. and Schäfer, E. (1980) Action spectra for the inhibition of hypocotyl growth by continuous irradiation in light and dark-grown *Sinapis alba* L. seedlings. *Plant Physiol.* **66**, 615–618.

Ben-Hod, G., Kigel, J. and Steinitz, B. (1988) Dormancy and flowering in *Anemone coronaria* L as affected by photoperiod and temperature. *Ann. Bot.* **61**, 623–633.

Bendeck de Cantu, L. and Kandeler, R. (1989) Significance of polyamines for flowering in *Spirodela punctata*. *Plant Cell Physiol.* **30**, 455–458.

Bernáth, J. and Tétényi, P. (1979) The effect of environmental factors on growth, development and alkaloid production of Poppy (*Papaver somniferum* L.). I. Response to day-length and light intensity. *Biochem. Physiol. Pflanzen* **174**, 468–478.

Bernier, G. (1969) *Sinapis alba* L. In *The Induction of Flowering* (ed. L.T. Evans), pp. 304–327. Macmillan, Melbourne.

Bernier, G. (1971) Structural and metabolic changes in the shoot apex in transition to flowering. *Can. J. Bot.* **49**, 803–819.

Bernier, G. (1986) The flowering process as an example of plastic development. In *Plasticity in Plants* (eds D.H. Jennings and A.J. Trewavas), pp. 257–286. The Company of Biologists, Cambridge.

Bernier, G. (1988) The control of floral evocation and morphogenesis. *Ann. Rev. Plant Physiol. Plant Mol. Biol.* **39**, 175–219.

Bernier, G., Kinet, J-M. and Bronchart, R. (1967) Cellular events at the meristem during floral induction in *Sinapis alba* L. *Physiol. Vég.* **5**, 311–324.

Bernier, G., Kinet, J.-M., Sachs, R.M. (1981) *The Physiology of Flowering*, Vol. II. CRC Press, Boca Raton, Florida.

Bernier, G., Lejeune, P., Jacqmard, A. and Kinet, J-M. (1990) Cytokinins and flower initiation. In *Plant Growth Substances 1988* (eds R.P. Pharis and S.B. Rood), pp. 486–491. Springer-Verlag, Berlin.

Bernier, G., Havelange, A., Houssa, C., Petitjean, A. and Lejeune, P. (1993) Physiological signals that induce flowering. *The Plant Cell* **5**, 1147–1155.

Besnard-Wibaut, C., Cochet, T. and Noin, M. (1989) Photoperiod and gibberellic acid-control of the cell cycle in the meristem of *Silene armeria* and its effects on flowering. *Physiol. Plant.* **77**, 352–358.

Bhargava, S.C. (1975) Photoperiodicity and seed germination in rice. *Indian J. Agric. Sci.* **45**, 447–451.

Biran, I., Gur, I. and Halevy, A.H. (1972) The relationship between exogenous inhibitors and endogenous levels of ethylene and tuberisation of dahlias. *Physiol. Plant.* **27**, 226–230.

Bismuth, F. and Miginiac, E. (1984) Influence of zeatin on flowering in root forming cuttings of *Anagallis arvensis* L. *Plant Cell Physiol.* **25**, 1073–1076.

Black, M. and Wareing, P.F. (1955) Growth studies in woody species. VII. Photoperiodic control of germination in *Betula pubescens* Ehsh. *Physiol. Plant.* **8**, 300–316.

Black, M. and Wareing, P.F. (1960) Photoperiodism in the light-inhibited seed of *Nemophila insignis*. *J. Exp. Bot.* **11**, 28–39.

Blakeley, S.D., Thomas, B., Hall, J.L. and Vince-Prue, D. (1983) Regulation of swelling of etiolated wheat-leaf protoplasts by phytochrome and gibberellic acid. *Planta* **158**, 416–421.

Bledsoe, C.S. and Ross, C.W. (1978) Metabolism of mevalonic acid in vegetative and induced plants of *Xanthium strumarium*. *Plant Physiol.* **62**, 683–686.

Bleeker, A.B., Estelle, M.A., Somerville, C.R. and Kende, H. (1988) Insensitivity to ethylene conferred by a dominant mutation in *Arabidopsis thaliana*. *Science* **241**, 1086–1089.

Blondon, F. and Jacques, R. (1970) Action de la lumiére sur l'initiation florale du *Lolium temulentum* L.: spectre d'action et role du phytochrome. *C.R. Acad. Sci., Paris* **270**, 947–950.

Bodson, M. (1975). Variation in the rate of cell division in the apical meristem of *Sinapis alba* during the transition to flowering. *Ann. Bot.* **39**, 547–554.

Bodson, M. (1985a) Changes in adenine-nucleotide content in the apical bud of *Sinapis alba* L. during floral transition. *Planta* **163**, 34–37.

Bodson, M. (1985b) *Sinapis alba*. In *Handbook of Flowering* (ed. A.H. Halevy), vol. IV, pp. 336–354. CRC Press, Boca Raton.

Bodson, M. and Bernier, G. (1985) Is flowering controlled by the assimilate level? *Physiol. Vég.* **23**, 491–501.

Bodson, M. and Outlaw, W.H. Jr. (1985) Elevation in the sucrose content of the shoot apical meristem of *Sinapis alba* at floral evocation. *Plant Physiol.* **79**, 420–424.

Bodson, M. and Remacle, B. (1987) Distribution of assimilates from various source-leaves during the floral transition of *Sinapis alba* L. In *Manipulation of Flowering* (ed J.G. Atherton), pp. 341–350. Butterworths, London.

Bodson, M., King R.W., Evans L.T. and Bernier G. (1977) The role of photosynthesis in flowering of the long-day plant *Sinapis alba*. *Aust. J. Plant Physiol.* **4**, 467–478.

Bollig, I. (1977) Different circadian rhythms regulate photoperiodic flowering response and leaf movement in *Pharbitis nil* (L.) Choisy. *Planta* **135**, 137–142.

Bollig, I., Chandrashekaran, M.K., Engelmann, W. and Johnsson, A. (1976) Photoperiodism in *Chenopodium rubrum*: an explicit version of the Bünning hypothesis. *Int. J. Chronobiol.* **4**, 83–96.

Bonner, J. and Zeevaart, J.A.D. (1962) Ribonucleic acid synthesis in the bud: an essential component of floral induction in *Xanthium*. *Plant Physiol.* **37**, 43–49.

Bonner, J., Heftman, J.E. and Zeevaart, J.A.D. (1963) Suppression of floral induction by inhibition of steroid biosynthesis. *Plant Physiol.* **38**, 81–88.

Bonzon, M. Degli Agosti, R., Wagner, E., Greppin, H. (1985) Enzyme patterns in energy metabolism during flower induction in spinach leaves. *Plant Cell Environ.* **8**, 303–308.

Boolj, R. and Meurs, E.J.J. (1994) Flowering in celeriac (*Apium graveolens* L. var *rapaceum* (Mill.) DC.): effects of photoperiod. *Sci. Hort.* **58**, 271–282.

Booth, A. (1963) The role of growth substances in the development of stolons. In *The Growth of the Potato* (ed. J.D. Ivins and F.L. Milthorpe), pp. 99–113. Butterworth, London.

Booth, A. and Lovell, P.H. (1972) The effect of pre-treatment with gibberellic acid on the distribution of photosynthate in intact and disbudded plants of *Solanum tuberosum* L. *New Phytol.* **71**, 795–804.

Borah, M.H. and Milthorpe, F. (1962) Growth of the potato plant as influenced by temperature. *Indian J. Plant Physiol.* **5**, 53–72.

Borowska, B. and Powell, L.E. (1982/3) Abscisic acid relationships in dormancy of apple buds. *Scientia Hort.* **18**, 111–117.

Borthwick, H.A. (1964) Phytochrome action and its time displays. *Am. Nat.* **95**, 347–355.

Borthwick, H.A. (1972) History of phytochrome. In: *Phytochrome* (eds K. Mitrakos and W. Shropshire Jnr), pp. 4–23. Academic Press, London, New York.

Borthwick, H.A. and Cathey, H.M. (1962) Role of phytochrome in control of flowering of chrysanthemum. *Bot. Gaz.* **123**, 155–162.

Borthwick, H.A. and Downs, R.J. (1964) Roles of active phytochrome in control of flowering of *Xanthium pensylvanicum*. *Bot. Gaz.* **125**, 227–231.

Borthwick, H.A. and Parker, M.W. (1938) Photoperiodic perception in *Biloxi* soybean. *Bot. Gaz.* **100**, 374–387.

Borthwick, H.A. and Parker, M.W. (1952) Light in relation to flowering and vegetative development. *Proc. 13th Int. Hort. Congress.* p 801.

Borthwick, H.A., Hendricks, S.B. and Parker, M.W. (1948) Action spectrum for the photo-periodic control of floral initiation of a long day plant, Wintex barley (*Hordeum vulgare*). *Bot. Gaz.* **110**, 103–118.

Borthwick, H.A., Hendricks, S.B. and Parker, M.W. (1951) Action spectrum for inhibition of stem growth in dark-grown seedlings of albino and non-albino barley (*Hordeum vulgare*). *Bot. Gaz.* **113**, 95–105.

Borthwick, H.A., Hendricks, S.B. and Parker, M.W. (1952a) The reaction controlling floral initiation. *Proc. Nat. Acad. Sci. USA* **38**, 929–934.

Borthwick, H.A., Hendricks, S.B., Parker, M.W., Toole, E.H. and Toole, V.K. (1952b) A reversible photoreaction controlling seed germination. *Proc. Nat. Acad. Sci. USA* **38**, 662–666.

Borthwick, H.A., Hendricks, S.B., Schneider, M.J., Taylorson, R.B. and Toole, V.K. (1969) The high-energy light action controlling plant responses and development. *Proc. Nat. Acad. Sci. USA* **64**, 479–486.

Bowen, M.R. and Hoad, G.V. (1968) Inhibitor content of phloem and xylem sap obtained from willow (*Salix viminalis* L.) entering dormancy. *Planta* **81**, 64–70.

Boylan, M.T. and Quail, P.H. (1989) Oat phytochrome is biologically active in transgenic tomatoes. *The Plant Cell* **1**, 765–773.

Boylan, M.T. and Quail, P.H. (1991) Phytochrome A overexpression inhibits hypocotyl elongation in transgenic *Arabidopsis*. *Proc. Nat. Acad. Sci. USA* **88**, 10806–10810.

Boylan, M.T., Douglas, N. and Quail, P.H. (1994) Dominant negative suppression of *Arabidopsis* photoresponses by mutant phytochrome A sequences identifies spatially discrete regulatory domains in the photoreceptor. *The Plant Cell* **6**, 449–460.

Brest, D.E., Hoshizaki, T. and Hamner, K.C. (1971) Rhythmic leaf movements in Biloxi soybean and their relation to flowering. *Plant Physiol.* **47**, 676–681.

Britz, S. (1990a) Photoperiodic and thermoperiodic regulation of assimilate partitioning into storage carbohydrates (starch and sugar) in leaves of crop plants. In: *Chronobiology: Its Role in Clinical Medicine, General Biology and Agriculture Part B*, pp. 853–866. Wiley-Liss, New York.

Britz, S. (1990b) Regulation of photosynthate partitioning into starch in soybean leaves. *Plant Physiol.* **94**, 350–356.

Britz, S. (1991) Setting the clocks that time photosynthate partitioning into starch and stored sugar in *Sorghum*: Response to spectral quality. In: *Recent Advances in Phloem Transport and Assimilate Compartmentation* (eds J.L. Bonnemain, S. Delrot, W.J. Lucas and J. Dainty), pp. 265–274. Ouest Editions.

Britz, S. and Saftner, R.A. (1988) The effect of daylength on photosynthate partitioning into leaf starch and soluble sugars in a gibberellin-deficient dwarf mutant of *Zea mays*. *Physiol. Plant.* **73**, 245–251.

Britz, S.J., Hungerford, W.E. and Lee, D.R. (1985a) Photoperiodic regulation of photosynthate partitioning in leaves of *Digitaria decumbens* Stent. *Plant Physiol.* **78**, 701–714.

Britz, S.J., Hungerford, W.E. and Lee, D.R. (1985b) Photosynthate partitioning into *Digitaria decumbens* leaf starch varies rhythmically with respect to the duration of prior incubation in continuous dim light. *Photochem Photobiol.* **42**, 741–744.

Britz, S.J., Hungerford, W.E. and Lee, D.R. (1987) Rhythms during extended dark periods determine rates of net photosynthesis and accumulation of starch and soluble sugars in subsequent light periods in leaves of *Sorghum*. *Planta* **171**, 339–345.

Brockman, J., Rieble, S., Kazarinova-Fukshansky, N., Seyfried, M. and Schäfer, E. (1987) Phytochrome behaves as a dimer *in vivo*. *Plant Cell Environ.* **10**, 105–111.

Brulfert, J. and Chouard, P. (1961) Nouvelles observations sur la production experimentale de fleures prolifères chez *Anagallis arvensis* L. *C.R. Acad. Sci. Paris* **253**, 171–181.

Brulfert, J. Guerrier, D. and Queiroz, O. (1975) Photoperiodism and enzyme rhythms. Kinetic characteristics of the photoperiodic induction of Crassulacean acid metabolism. *Planta* **125**, 33–44.

Brulfert, J. Guerrier, D. and Queiroz, O. (1982) Photoperiodism and Crassulacean acid metabolism. II. Relations between leaf ageing and photoperiod in Crassulacean acid metabolism induction. *Planta* **154**, 332–338.

Brulfert, J, Fontaine, D. and Imhoff, C. (1985) *Anagallis arvensis*. In *Handbook of Flowering* (ed. A.H. Halevy), Vol. I, pp. 434–449. CRC Press, Boca Raton.

Bünning, E. (1936) Die endogene Tagesrhythmik als Grundlage der photoperiodischen Reaktion. *Ber. Deut. Bot. Ges.* **54**, 590–607.

Bünning, E. (1960) Circadian rhythms and the time measureent in photoperiodism. *Cold Spr. Harb. Symp. Quant. Biol.* **25**, 249–256.

Bünning, E. and Moser, I. (1966) Unterschiedliche photoperiodische Empfindlichkeit der beiden Blattseiten von *Kalanchoë blossfeldiana*. *Planta* **69**, 296–298.

Bünning, E. and Moser, I. (1969) Interference of moonlight with the photoperiodic measurement of time by plants and their adaptive reaction. *Proc. Natl. Acad. Sci. USA* **62**, 1018–1022.

Burg S.P. and Burg, E.A. (1966) Interaction between auxin and ethylene and its role in plant growth. *Proc. Nat. Acad. Sci. USA* **55**, 262–269.

Burgoyne, R.D. (1978) A model for the molecular basis of circadian rhythms involving monovalent ion-mediated translational control. *FEBS Lett.* **94**, 17–19.

Burn, J.E., Bagnall, D.J., Metzger, J.D., Dennis, E.S. and Peacock, W.J. (1993) DNA methylation, vernalization and the initiation of flowering. *Proc. Nat. Acad. Sci. USA* **90**, 287–291.

Butler, W.L. (1964) Dark transformations of phytochrome *in vivo*. *Q. Rev. Biol.* **39**, 6–10.

Butler, W.L., Norris, K.H., Seigelman, H.W. and Hendricks, S.B. (1959) Detection, assay and preliminary purification of the pigment controlling photoresponsive development of plants. *Proc. Nat. Acad. Sci. USA* **25**, 1703–1708.

Butt, A.M. (1968) Vegetative growth, morphogenesis and carbohydrate content of the onion plant as a function of light and temperature under field and controlled conditions. *Meded. LandbHoogesch. Wageningen* **68**, 1–211.

Buzzell, R.I. and Voldeng, H.D. (1980) Inheritance of insensitivity to long daylength. *Soybean Genet. Newsletter* **7**, 26–29.

Byers, R.E., Baker, L.R., Sell, H.M., Herner, R.C. and Dilley, D.D. (1972) Ethylene: a natural regulator of sex expression of *Cucumis melo* L. *Proc. Nat. Acad. Sci. USA* **69**, 717–720.

Caffaro, S.V. and Vicente, C. (1994) Polyamine implication during soybean flowering induction and early reproductive transition of vegetative buds. *Plant Physiol. Biochem.* **32**, 391–397.

Caffaro, S.V. and Vicente, C. (1995) Early changes in the content of leaf polyamines during the photoperiodic flowering induction in soybean. *J. Plant Physiol.* **145**, 756–758.

Caffaro, S.V., Antognoni, F. Scaramagli, S. and Bagni, N. (1994) Polyamine translocation following photoperiodic flowering induction in soybean. *Physiol. Plant.* **91**, 251–256.

Cameron, J.S. and Dennis, F.G. (1986) The carbohydrate-nitrogen relationship and flowering/fruiting: Kraus and Kraybill revisited. *Hort. Sci.* **21**, 1000–1002.

Camm, E.L., McCallum, J., Leaf, E. and Koupai-Abyazani, M.R. (1993). Cold-induced purpling of *Pinus contorta* seedlings depends on previous daylength treatment. *Plant Cell Environ.* **16**, 761–764.

Canham, A.E. (1969) An effect of daylength on the flowering of dahlias. *Acta Hort.* **14**, 109–115.

Carpenter, B.H. and Hamner, K.C. (1963) Effect of light quality on rhythmic flowering response of Biloxi soybean. *Plant Physiol.* **38**, 608–703.

Carpenter, B.H. and Hamner K.C. (1964) The effect of dual perturbations on the rhythmic flowering response of Biloxi soybean. *Plant Physiol.* **39**, 884–889.

Carr, D.J. (1952) The photoperiodic behaviour of short-day plants. *Physiol. Plant.* **5**, 70–84.

Carr, D.J. (1955) On the nature of photoperiodic induction. III. The summation of the effects of inductive photoperiodic cycles. *Phys. Plant.* **8**, 512–526.

Carr-Smith, H. (1990) The control of flowering in the long-day plant wheat (*Triticum aestivum*) and the role of phytochrome. PhD Thesis, University of Reading.

Carr-Smith, H.D., Thomas, B. and Johnson, C.B. (1989) An action spectrum for the effect of continuous light on flowering in wheat. *Planta* **179**, 428–432.

Carr-Smith, H.D., Thomas, B., Johnson, C.B., Plumpton, C. and Butcher, G. (1994) The kinetics of type 1 phytochrome in green, light-grown wheat (*Triticum aestivum* L.). *Planta* **194**, 136–142.

Catchpole, A.H. and Hillman, J. (1969) Effect of ethylene on tuber initiation in *Solanum tuberosum* L. *Nature* **223**, 1387.

Cathey, H.M. (1968) Response of some ornamental plants to synthetic abscisic acid. *Proc. Am. Soc. Hort. Sci.* **93**, 693–698.

Cathey, H.M. (1969) *Chrysanthemum morifolium* (Ramat.) Hemsl. In *The Induction of Flowering* (ed. L.T. Evans), pp. 268–290. Macmillan, Melbourne.

Cathey, H.M. and Borthwick, H.A. (1957) Photoreversibility of floral initiation in chrysanthemum. *Bot. Gaz.* **119**, 71–76.

Cathey, H.M. and Borthwick, H.A. (1964) Significance of dark reversion of phytochrome in flowering of *Chrysanthemum morifolium*. *Bot. Gaz.* **125**, 232–236.

Chailakhyan, M.Kh. (1936) On the hormonal theory of plant development. *Dokl. Acad. Sci. USSR* **12**, 443–447.

Chailakhyan, M. Kh. (1945) Photoperiodism of individual parts of the leaf, its halves. *Dokl. Akad. Sci. USSR* **48**, 360–364.

Chailakhyan, M.Kh. (1958) Hormonale Faktoren des Pflanzenblühens. *Biol. Zentralbl.* **77**, 641–642.

Chailakhyan, M.Kh. (1975) Substances of plant flowering. *Biol. Plant* **17**, 1–11.

Chailakhyan M.Kh. (1982) Hormonal substances in flowering. In *Plant Growth Substances* 1982 (ed. P.F. Wareing), pp. 645–655. Academic Press, London.

Chailakhyan, M.Kh. and Butenko, R.G. (1957) Movement of assimilates of leaves to shoots under different photoperiodic conditions of leaves. *Fiziol. Rast.* **4**, 450–462.

Chailakhyan, M.Kh. and Khryanin, V.N. (1978) The role of roots in sex expression in hemp plants. *Planta* **138**, 185–187.

Chailakhyan, M.Kh., Aksenova, N.P., Konstantinova, T.N. and Bavrina, T.V. (1975) The callus model of plant flowering. *Proc. R. Soc. Lond. B* **190**, 333–340.

Chailakhyan, M.Kh., Grigorova, N. and Lozhnikova. V. (1977) The effect of leaf extracts from flowering tobacco plants on flowering of seedlings and plants of *Chenopodium rubrum* L (in Russian). *Dokl. Akad. Nauk SSSR* **236**, 773–776.

Chailakhyan, M.Kh., Lozhnikova, V., Seidlova, F., Krekule, J., Dudko, N. and Negretsky, V. (1989) Floral and growth responses in *Chenopodium rubrum* L to an extract from flowering *Nicotiana tabacum* L. *Planta* **178**, 143–146.

Chambers, P.A. and Spence, D.H.N. (1984) Diurnal changes in the ratio of underwater red to far-red light in relation to aquatic plant photoperiodism. *J. Ecol.* **72**, 495–503.

Chapman, H.W. (1958) Tuberization in the potato plant. *Physiol. Plant.* **11**, 215–224.

Chaudhury, A.M., Letham, S., Craig, S. and Dennis, E.S. (1993) *amp1* – a mutant with high cytokinin levels and altered embryonic pattern, faster vegetative growth, constitutive photomorphogenesis and precocious flowering. *The Plant Journal* **4**, 907–916.

Chen, H-H and Li, P.H. (1978) Interactions of low temperature, water stress and short days in the induction of stem frost hardiness in red osier dogwood. *Plant Physiol.* **62**, 833–835.

Cherry J.R. and Vierstra, R.D. (1994) The use of transgenic plants to examine phytochrome structure/function. In *Photomorphogenesis in Plants*, 2nd edn (eds R.E. Kendrick and G.H.M. Kronenberg), pp. 271–297. Kluwer Academic Publishers, Dordrecht.

Cherry, J.R., Hondred, D., Walker, J.M. and Vierstra, R.D. (1992) Phytochrome requires the 6 kDa N-terminal domain for full biological activity. *Proc. Nat. Acad. Sci. USA* **89**, 5039–5043.

Cherry, J.R., Hondred, D., Walker, J.M., Keller, J.M., Hershey, H.P. and Vierstra, R.D. (1993) Carboxy-terminal deletion analysis of oat phytochrome A reveals the presence of separate domains required for structure and biological activity. *The Plant Cell* **5**, 565–575.

✓ Childs, K.L., Pratt, L.H. and Morgan, P.W. (1991) Genetic regulation of development in *Sorghum bicolor*, VI. The ma$_3$^{R} allele results in abnormal phytochrome physiology. *Plant Physiol.* **97**, 714–719.

✓ Childs, K.L., Lu, J-L., Mullet, J.E. and Morgan, P.W. (1995) Genetic regulation of development in *Sorghum bicolor* X. Greatly attenuated photoperiod sensitivity in a phytochrome-deficient sorghum possessing a biological clock but lacking a red light-high irradiance response. *Plant Physiol.* **108**, 345–351.

Chouard, P. (1960) Vernalization and its relation to dormancy. *Ann. Rev. Plant Physiol.* **11**, 191–238.

Christersson, L. (1978) The influence of photoperiod and temperature on the development of frost hardiness in seedlings of *Pinus sylvestris* and *Picea abies*. *Physiol. Plant.* **44**, 288–294.

Chrominski, A. and Kopcewicz, J. (1972) Auxins and gibberellins in 2-chloroethylphosphonic acid-induced femaleness of *Cucurbita pepo* L. *Z. Pflanzenphysiol.* **68**, 184–189.

Clack, T., Mathews, S. and Sharrock, R.A. (1994) The phytochrome apoprotein family in *Arabidopsis* is encoded by 5 genes – the sequences and expression of *phyd* and *phye*. *Plant Mol. Biol.* **25**, 413–427.

Clapham, D.H., von Arnold, S., Dormlin, I., Ekberg, I., Eriksson, G., Larsson, C.T., Norell, L. and Qamaruddin, M. (1994) Variation in total and specific RNA during inwintering of two contrasting populations of *Picea abies*. *Physiol. Plant.* **90**, 504–512.

Clark, J.E. and Heath, O.V.S. (1962) Studies in the physiology of the onion plant. V. An investigation into the growth substance content of bulbing onions. *J. Exp. Bot.* **13**, 227–249.

Clark, R.J. and Menary, R.C. (1979) Effects of photoperiod on the yield and composition of peppermint oil. *J. Amer. Soc. Hort. Sci.* **104**, 699–702.

Cleland, C.F. (1978) The flowering enigma. *Bio. Sci.* **28**, 265–269.

Cleland, C.F. (1984) Biochemistry of induction – the immediate action of light. In *Light and the Flowering Process* (eds D. Vince-Prue, B. Thomas and K.E. Cockshull), pp. 123–136. Academic Press, London.

Cleland, C.F. and Ajami, A. (1974) Identification of flower-inducing factor isolated from aphid honeydew as being salicylic acid. *Plant Physiol.* **54**, 904–906.

Cleland, C.F. and Ben-Tal, Y. (1983) Hormonal regulation of flowering and sex expression. In *Strategies of Plant Reproduction*, BARC Symposium no. 6 (ed. W.J. Meudt), pp. 157–180. Allanheld, Osmun, Totowa.

Cleland, C.F. and Tanaka, O. (1979) Effect of daylength on the ability of salicylic acid to induce flowering in the long-day plant *Lemna gibba* G3 and the short-day plant *Lemna paucicostata* 6746. *Plant Physiol.* **64**, 421–424.

Cleland, C.F. and Zeevaart, J.A.D. (1970) Gibberellins in relation to flowering and stem elongation in the long-day plant *Silene armeria*. *Plant Physiol.* **46**, 392–400.

Clements, H.F. (1968) Lengthening versus shortening dark periods and blossoming in sugar cane as affected by temperature. *Plant Physiol.* **43**, 57–60.

Cline, M.G. and Agatep, A.O. (1970) Control of stem elongation and flowering in *Scrophularia marilandica*. *Physiol. Plant.* **23**, 993–1003.

Cockshull, K.E. (1966) Effects of night-break treatment on leaf area and leaf dry weight in *Callistephus chinensis*. *Ann. Bot.* **30**, 791–806.

Cockshull, K.E. (1979) Effects of irradiance and temperature on flowering of *Chrysanthemum morifolium* Ramat. in continuous light. *Ann. Bot.* **44**, 451–460.

Cockshull, K.E. (1984) The photoperiodic induction of flowering in short-day plants. In *Light and the Flowering Process* (eds D. Vince-Prue, B. Thomas and K.E. Cockshull), pp. 33–49. Academic Press, London.

Cockshull, K.E. and Horridge, J.S. (1978) 2-chloroethanephosphonic and flower initiation by *Chrysanthemum morifolium* Ramat. in short days and in long days. *J. Hort. Sci.* **53**, 85–90.

Cockshull, K.E. and Hughes, A.P. (1969) Growth and dry weight distribution in *Callistephus chinensis* as influenced by lighting treatments. *Ann. Bot.* **33**, 367–379.

Cockshull, K.E. and Hughes, A.P. (1972) Flower formation in *Chrysanthemum morifolium*; the influence of light level. *J. Hort. Sci.* **47**, 113–127.

Coen, E.S. (1991) The role of homeotic genes in flower development and evolution. *Ann. Rev. Plant Physiol. Plant Mol. Biol.* **42**, 241–279.

Colbert, J.T. (1988) Molecular biology of phytochrome. *Plant Cell Environ.* **11**, 305–318.

Colbert, J.T., Hershey, H.P. and Quail, P.H. (1983) Autoregulatory control of translatable phytochrome mRNA levels. *Proc. Nat. Acad. Sci. USA* **80**, 2248–2252.

Colbert, J.T., Hershey, H.P. and Quail, P.H. (1985) Phytochrome regulation of phytochrome mRNA abundance. *Plant Mol. Biol.* **5**, 91–101.

Collins, W.T., Salisbury, F.B. and Ross, C.W. (1963) Growth regulators and flowering. III. Antimetabolites. *Planta* **60**, 131–144.

Collinson, S.T., Summerfield, R.J., Ellis, R.H. and Roberts, E.H. (1993) Durations of the photoperiod sensitive and photoperiod insensitive phases of development to flowering in four cultivars of soya bean, *Glycine max* L. Merrill. *Ann. Bot.* **71**, 389–394.

Coulter M.W. and Hamner, K.C. (1964) Photoperiodic flowering response of Biloxi soybean in 72-hour cycles. *Plant Physiol.* **39**, 846–856.

Coupland, G., Dash, S., Goodrich, J., Lee, K., Long, D., Martin, M., Puangsomlee, P., Puterill, J., Robson, F., Sundberg, E. and Wilson, K. (1993) Molecular and genetic analysis of the control of flowering time in response to daylength in *Arabidopsis thaliana*. *Flowering Newsletter* **16**, 27–32.

Courduroux, J.C. (1964) Inhibition de croissance et tubérisation. *C.R. Acad. Sci. (Paris)* **259**, 4346–4349.

Courduroux, J.C. (1966) Méchanism physiologique de la tubérisation du topinambour. *Bull. Soc. Fr. Physiol. Vég.* **12**, 213–232.

Crawford, R.M.M. (1961) The photoperiodic reaction in relation to development in *Salvia splendens*. *Ann. Bot.* **25**, 78–84.

Crevecoeur, M., Penel, C., Greppin, H. and Gaspar, Th. (1986) Ethylene production in spinach leaves during floral induction. *J. Exp. Bot.* **37**, 1218–1224.

Ćulafić, L. and Nešković, M. (1980) Effect of growth substance on flowering and sex expression in isolated apical buds of *Spinacia oleracea*. *Physiol. Plant.* **48**, 588–591.

Cumming, B.G. (1959) Extreme sensitivity of germination and photoperiodic reaction in the genus *Chenopodium* (Tourn.) L. *Nature* **184**, 1044–1045.

Cumming, B.G. (1963) Evidence of a requirement for phytochrome Pfr in the floral induction of *Chenopodium rubrum*. *Can. J. Bot.* **41**, 901–925.

Cumming, B.G. (1969) Photoperiodism and rhythmic flower induction. Complete substitution of inductive darkness by light. *Can. J. Bot.* **47**, 1241–1250.

Cumming, B.G., Hendricks, S.B. and Borthwick, H.A. (1965) Rhythmic flowering responses and phytochrome changes in a selection of *Chenopodium rubrum*. *Can. J. Bot.* **43**, 825–853.

Cunningham, E.M. and Guiry, M.D. (1989) A circadian rhythm in the long-day photoperiodic induction of erect axis development in the marine red alga *Nemalion helminthoides*. *J. Phycol.* **25**, 705–712.

Dauphin-Guerin, B., Teller, G. and Durand, B. (1980) Different endogenous cytokinins between male and female *Mercurialis annua* L. *Planta* **148**, 124–129.

Davies, P.J., Emshwiller, E., Gianfagna, T.J., Proebsting, W.M., Noma, M. and Pharis, R.P. (1982) The endogenous gibberellins of vegetative and reproductive tissue of G_2 peas. *Planta* **154**, 266–272.

Davis, G.J. (1967) *Proserpinaca*: photoperiodic and chemical differentiation of leaf development and flowering. *Plant Physiol.* **42**, 667–668.

Davis, P.H. and Heywood, V.H. (1963) *Principles of Angiosperm Taxonomy*. Oliver and Boyd.

Deitzer, G.F. (1983) Effect of far-red energy on the photoperiodic control of flowering in Wintex barley *Hordeum vulgare* L. In *Strategies of Plant Reproduction*. BARC Symposium 6 (ed. W. Meudt), pp. 95–115. Allenheld Osmun, Totowa.

Deitzer, G.F. (1984) Photoperiodic induction in long-day plants. In *Light and the Flowering Process* (eds D. Vince-Prue, B. Thomas and K.E. Cockshull), pp. 51–63. Academic Press, London.

Deitzer, G.F., Hayes, R. and Jabben, M. (1979) Kinetics and time dependence of the effect of far red light on the photoperiodic induction of flowering in Wintex barley. *Plant Physiol.* **64**, 1015–1021.

Deitzer, G.F., Hayes, R. and Jabben, M. (1982) Phase shift in the circadian rhythm of floral promotion by far-red light in *Hordeum vulgare* L. *Plant Physiol.* **69**, 597–601.

Dellaporta, S.L. and Calderon-Urrea, A. (1993) Sex determination in flowering plants. *Plant Cell* **5**, 1241–1251.

Deronne, M. and Blondon, F. (1977) Mise en évidence chez le *Perilla ocymoides* L., plante de jours courts typique, d'un autre facteur de l'induction florale: les températures basses. Études de l'état induit acquis par la feuille en jours courts ou au froid. *Physiol. Vég.* **15**, 219–237.

Dickens, C.W.S. and van Staden, J. (1985) *In vitro* flowering and fruiting of soybean explants. *J. Plant Physiol.* **120**, 83–86.

Dickens, C.W.S. and van Staden, J. (1990) The *in vitro* flowering of *Kalanchoë blossfeldiana* Poellniz. II. The effects of growth regulators and gallic acid. *Plant Cell Physiol.* **31**, 757–762.

Dickinson, H.G. (1993) The regulation of sexual development in plants. *Phil. Trans. Roy. Soc. Lond. B* **339**, 147–157.

Diomaiuto-Bonnand, J. (1960) Reversions de l'etat inflorescential vegetatif, chez le *Nicotiana glutinosa* L. sous l'action d'eclairements de faible intensite. *C.R. Acad. Sci. Paris Ser. D* **271**, 45–48.

Diomaiuto-Bonnand, J. and Le Saint, A.-M. (1985) La floraison et les conditions de son renouvellement périodique, chez une espéce vivace polycarpique: la giroflée ravenelle (*Cheiranthus cheiri* L.) II. Analyse des sucres solubles et des acides aminés libres, au niveau des bourgeons. Discussion. *Rev. Cyt. Biol. Vég. Bot.* **8**, 63–87.

Domanski, R. and Koslowski, T.T. (1968) Variations in kinetin-like activity in buds of *Betula* and *Populus* during release from dormancy. *Can. J. Bot.* **46**, 397–403.

Dormling, I (1979) Influence of light intensity and temperature on photoperiodic reponse of Norway spruce provenances. *Proc. IUFRO Norway Spruce Meeting, Bucharest*, pp. 398–408.

Dormling, I. (1993) Bud dormancy, frost hardiness, and frost drought in seedlings of *Pinus sylvestris* and *Picea abies*. In *Advances in Plant Cold Hardiness* (eds P.H. Li and Christersson), pp. 286–298. CRC Press, Boca Raton.

Downs, R.J. (1956) Photoreversibility of flower initiation. *Plant Physiol.* **31**, 279–284.

Downs, R.J. and Borthwick, H.A. (1956a) Effects of photoperiod on the growth of trees. *Bot. Gaz.* **117**, 310–326.

Downs, R.J. and Borthwick, H.A. (1956b) Effect of photoperiod upon the vegetative growth of *Weigela florida* var. variegata. *Proc. Am. Soc. Hort. Sci.* **68**, 518–521.

Downs, R.J. and Thomas, J.F. (1982) Phytochrome regulation of flowering in the long-day plant, *Hyoscyamus niger*. *Plant Physiol.* **70**, 898–900.

Downs, R.J., Hendricks, S.B. and Borthwick, H.A. (1957) Photoreversible control of extension of pinto beans and other plants under normal conditions of growth. *Bot. Gaz.* **118**, 199–208.

Downs, R.J., Borthwick, H.A., Piringer, A.A. (1958) Comparison of incandescent and fluorescent lamps for lengthening photoperiods. *Proc. Am. Soc. Hort. Sci.* **71**, 568–578.

Downs, R.J., Piringer, A.A., Wiebe, G.A. (1959) Effects of photoperiod and kind of supplemental light on growth and reproduction of several varieties of wheat and barley. *Bot. Gaz.* **120**, 170–177.

Dring, M.J. (1988) Photocontrol of development in algae. *Ann. Rev. Plant Physiol. Plant Mol. Biol.* **39**, 157–174.

Dudai, N., Putievsky, E., Palevitsch, D. and Halevy, A.H. (1989) Environmental factors affecting flower initiation and development in *Marjorana syriaca* L (=*Origanum syriacum* var *syriacum*). *Israel J. Bot.* **38**, 229–239.

Duncan, P.J. and Bilderback, T.E. (1982) Effects of irrigation systems, gibberellic acid and photoperiod on seed germination in *Kalmia latifolia* L. and *Rhododendron maximum* L. *Hortscience* **17**, 916–917.

Dunlap, J.C., Taylor, W., Hastings, J.W. (1980) The effects of protein synthesis inhibitors on the *Gonyaulax* clock. I. Phase-shifting effects of cycloheximide. *J. Comp. Physiol.* **138**, 1–8.

Durand, B. and Durand, R. (1984) Sexual differentiation in higher plants. *Physiol. Plant.* **60**, 267–274.

Durzan, D.J., Campbell, R.A. and Wilson, A. (1979) Inhibition of female cone production in white spruce by light treatment during night under field conditions. *Environ. Exp. Bot.* **19**, 133–144.

Eagles, C.F. (1971) Effect of photoperiod on vegetative growth in two natural populations of *Dactylis glomerata* L. *Ann. Bot.* **35**, 75–86.

Eberhard, S., Doubrava, N., Marfa, V., Mohnen, D., Southwick, A., Darvill, A. and Albersheim, P. (1989) Pectic cell wall fragments regulate tobacco thin-cell-layer explant morphogenesis. *The Plant Cell* **1**, 747–755.

Edwards, G.R. (1986) Ammonia, arginine, polyamines and flower induction in apple. *Acta Hort.* **179**, 363–364.

Eichhoff, E. and Rau, W. (1969) Auslosung der Blütenbildung bei der Langtagpflanze *Hyoscyamus niger* in Kurztag durch 2-Thiouracil. *Planta* **87**, 290–303.

El-Antably, H.M.M., Wareing, P.F. and Hillman, J. (1967) Some physiological responses to D,L abscisin (dormin). *Planta* **73**, 74–90.

El Hattab, A.H. (1968) Effects of light quality on flowering and morphogenesis in *Hyoscyamus niger* L. *Meded LandbHoogesch Wageningen.* **68**, 12.

Engelmann, W. (1960) Endogene Rhythmik und photoperiodische Blühinduktion bei *Kalanchoë*. *Planta* **55**, 496–511.

Esashi, Y. (1960) Studies on the formation and sprouting of aerial tubers in *Begonia evansiana* Andr. IV. Cutting method and tuberizing stages. *Sci. Rep. Tohoku Univ.* **26**, 239–246.

Esashi, Y. (1961a) Studies on the formation and sprouting of aerial tubers in *Begonia evansiana*

Andr. VI. Photoperiodic conditions for tuberization and sprouting in the cutting plants. *Sci. Rep. Tohoku Univ.* **27**, 101–112.

Esashi, Y. (1961b) Studies on the formation and sprouting of aerial tubers in *Begonia evansiana* Andr. V. Antagonistic action of long-days to short-day response. *Plant Cell Physiol.* **2**, 117–127.

Esashi, Y. (1963) Studies on the formation and sprouting of aerial tubers in *Begonia evansiana* Andr. VIII. Dual effect on tuberization of near infra-red light given in the dark periods. *Plant Cell Physiol.* **4**, 135–143.

Esashi, Y. (1966a) The relation between red and blue or far-red lights in the night interruption of the photoperiodic tuberization of *Begonia evansiana*. *Plant Cell Physiol.* **7**, 405–414.

Esashi, Y. (1966b) Effects of light quality and gas condition in the main light period on the photoperiodic tuberization of *Begonia evansiana*. *Plant Cell Physiol.* **7**, 465–474.

Esashi, Y. and Leopold, A.C. (1968) Regulation of tuber development in *Begonia evansiana* by cytokinin. In *Biochemistry and Physiology of Plant Growth Substances*. Proc. 6th Int. Congr. Pl. Growth Substances (eds F. Wightman and R. Setterfield), pp. 923–941. Runge Press, Ottawa.

Esashi, Y. and Nagao, M. (1958) Studies on the formation and sprouting of aerial tubers in *Begonia evansiana*. I. Photoperiodic conditions for tuberization. *Sci. Rep. Tohoku Univ.* **24**, 81–88.

Esashi, Y., Eguchi, T. and Nagao, M. (1964) The role of auxin in the photoperiodic tuberization in *Begonia evansiana* Andr. *Plant Cell Physiol.* **4**, 413–427.

Evans, L.T. (1960) Inflorescence initiation in *Lolium temulentum* L. I. Effect of plant age and leaf area on sensitivity to photoperiodic induction. *Aust. J. Biol. Sci.* **13**, 123–131.

Evans, L.T. (1962a) Day-length control of inflorescence initiation in the grass *Rottboellia exaltata* L. *Aust. J. Biol. Sci.* **15**, 291–303.

Evans, L.T. (1962b) Inflorescence initiation in *Lolium temulentum* L. III. The effect of aerobic conditions during photoperiodic induction. *Aust. J. Biol. Sci.* **15**, 281–290.

Evans, L.T. (1964a) Inflorescence initiation in *Lolium temulentum* L. V. The role of auxins and gibberellins. *Aust. J. Biol. Sci.* **17**, 10–23.

Evans, L.T. (1964b) Inflorescence initiation in *Lolium temulentum* L. VI. Effects of some inhibitors of nucleic acid, protein and steroid biosynthesis. *Aust. J. Biol. Sci.* **17**, 24–35.

Evans, L.T. (1969a) *Lolium temulentum* L. In *The Induction of Flowering* (ed. L.T. Evans), pp. 328–349. Macmillan, Melbourne.

Evans, L.T. (1969b) The nature of flower induction. In *The Induction of Flowering* (ed. L.T. Evans), pp. 457–480. Macmillan, Melbourne.

Evans, L.T. (1969c) *The Induction of Flowering*. Macmillan of Australia.

Evans, L.T. (1971) Flower induction and the florigen concept. *Ann. Rev. Plant Physiol.* **22**, 365–394.

Evans, L.T. (1976) Inflorescence initiation in *Lolium temulentum* L. XIV. The role of phytochrome in long day induction. *Aust. J. Biol. Sci.* **3**, 207–217.

Evans, L.T. and King, R.W. (1969) Role of phytochrome in photoperiodic induction of *Pharbitis nil*. *Z. Pflanzenphysiol.* **50**, 277–288.

Evans, L.T. and King, R.W. (1985) *Lolium temulentum*. In *Handbook of Flowering* Vol. III (ed. A.H. Halevy), pp. 306–323. CRC Press, Boca Raton.

Evans, L.T. and Wardlaw, I.F. (1966) Independent translocation of ^{14}C-labelled assimilates and of the floral stimulus in *Lolium temulentum*. *Planta* **68**, 310–326.

Evans, L.T., Borthwick, H.A. and Hendricks, S.B. (1965) Inflorescence initiation in *Lolium temulentum* L. VII. The spectral dependence of induction. *Aust. J. Biol. Sci.* **18**, 745–762.

Evans, L.T., King, R.W., Chu, A., Mander, L.N. and Pharis, R.P. (1990) Gibberellin structure and florigenic activity in *Lolium temulentum*, a long-day plant. *Planta* **182**, 97–106.

Evans, M.R., Wilkins, H.F. and Hackett, W.P. (1992) Meristem ontogenetic age as the controlling factor in long-day floral initiation in poinsettia. *J. Am. Soc. Hort. Sci.* **117**, 961–965.

Evans, L.T., King, R.W., Mander, L.N. and Pharis, R.P. (1994a) The relative significance for stem elongation and flowering in *Lolium temulentum* of 3-hydroxylation of gibberellins. *Planta* **192**, 130–136.

Evans, L.T., King, R.W., Mander, L.N., Pharis, R.P. and Duncan, K.A. (1994b) The differential effects of C-16,17-dihydro gibberellins and related compounds on stem elongation and flowering in *Lolium temulentum. Planta* **193**, 107–114.

Evans, L.T., Blundell, C. and King, R.W. (1995) Developmental responses by tall and dwarf isogenic lines of spring wheat to applied gibberellins. *Aust. J. Plant Physiol.* **22**, 365–371.

Finn, J.C. and Hamner, K.C. (1960) Investigation of *Hyoscyamus niger* L., a long day-plant, for endogenous periodicity in flowering response. *Plant Physiol.* **35**, 982–985.

Flint, L.H. and McAlister, E.D. (1935) Wavelengths of radiation in the visible spectrum inhibiting the germination of light-sensitive lettuce seeds. *Smithson. Misc. Collns.* **94, No. 5**, 1–11.

Francis, D. (1987) Effects of light on cell division in the shoot meristem during floral evocation. In *Manipulation of Flowering* (ed. J.G. Atherton), pp. 269–300. Butterworths, London.

Francis, D. and Lyndon, R.F. (1979) Synchronization of cell division in the shoot apex of *Silene* in relation to flower initiation. *Planta* **145**, 151–157.

Francis, D. and Lyndon, R.F. (1985) The control of cell cycle in relation to floral induction. In: *The Cell Division Cycle in Plants* (eds J.A. Bryant, D. Francis), pp. 199–215, Cambridge University Press.

Frankland, B. and Letendre, R.J. (1978) Phytochrome and effects of shading on growth of woodland plants. *Photochem. Photobiol.* **27**, 223–230.

Fredericq, H. (1962) Le role du gaz carbonique de l'air pendant les jours courts des cycles inductif des *Kalanchoë blossfeldiana* et *Perilla crispa. Bull. Soc. R. Bot. Belg.* **94**, 45–55.

Fredericq, H. (1963) Flower formation in *Kalanchoë blossfeldiana* by very short photoperiods under light of different quality. *Nature* **198**, 101–102.

Fredericq, H. (1964) Conditions determining effects of far-red and red irradiations on flowering response of *Pharbitis nil. Plant Physiol.* **39**, 812–816.

Fredericq, H. (1965) Action of red and far-red light at the end of the short-day and in the middle of the night on flower induction in *Kalanchoë blossfeldiana. Biol. Jaarb.* **33**, 66–91.

Friedlander, M., Atsmon, D. and Galun, E. (1977a) Sexual differentiation in cucumber: abscisic acid and gibberellic acid contents of various sex types. *Plant Cell Physiol.* **18**, 681–691.

Friedlander, M., Atsmon, D. and Galun, E. (1977b) Sexual differentiation in cucumber: the affects of abscisic acid and other growth regulators on various sex genotypes. *Plant Cell Physiol.* **18**, 261–269.

Friedman, H., Goldschmidt, E.E. and Halevy, A.H. (1987) Proteins in phloem exudates from photoperiodically induced *Pharbitis nil* seedlings. *J. Plant Physiol.* **130**, 471–476.

Friedman, H., Goldschmidt, E.E. and Halevy, A.H. (1989) Involvement of calcium in the photoperiodic flower induction process of *Pharbitis nil. Plant Physiol.* **89**, 530–534.

Friedman, H., Goldschmidt, E.E., Spiegelstein, H. and Halevy, A.H. (1992) A rhythm in the flowering response of photoperiodically induced *Pharbitis nil* to agents affecting cytosolic calcium and pH. *Physiol. Plant.* **85**, 57–60.

Friend, D.J.C. (1963) The effect of light intensity and temperature on floral initiation and inflorescence development of Marquis wheat. *Can. J. Bot.* **41**, 1663–1674.

Friend, D.J.C. (1968a) Photoperiodic responses of *Brassica campestris* L. cv Ceres. *Physiol. Plant.* **21**, 990–1002.

Friend, D.J.C. (1968b) Spectral requirements for flower initiation in two long-day plants, rape (*Brassica campestris* cv Ceres), and spring wheat (*Triticum* × *aestivum*). *Physiol. Plant.* **21**, 1185–1195.

Friend, D.J.C. (1969) *Brassica campestris* L. In *The Induction of Flowering* (ed. L.T. Evans), pp. 364–375. Macmillan, Melbourne.

Friend, D.J.C. (1975) Light requirements for photoperiodic sensitivity in cotyledons of dark-grown *Pharbitis nil. Physiol. Plant.* **32**, 286–296.

Friend, D.J.C. (1984) The interaction of photosynthesis and photoperiodism in induction. In *Light and the Flowering Process* (eds D. Vince-Prue, B. Thomas and K.E. Cockshull), pp. 257–275. Academic Press, London.

Friend, D.J.C. (1985) *Brassica*. In *Handbook of Flowering* (ed. A.H. Halevy), vol. II, pp. 48–77. CRC Press, Boca Raton.

Friend, D.J.C., Helson, V.A. and Fisher, J.E. (1961) The influence of the ratio of incandescent to fluorescent light on the flowering response of Marquis wheat grown under controlled conditions. *Can. J. Plant Sci.* **41**, 418–427.

Friend, D.J.C, Bodson, M. and Bernier, G. (1984) Promotion of flowering in *Brassica campestris* L. cv Ceres by sucrose. *Plant Physiol.* **75**, 1085–1089.

Fuchigami, L.H., Weiser, C.J. and Evert, D.R. (1971a) Induction of cold acclimation in *Cornus stolonifera* Michx. *Plant Physiol.* **47**, 98–103.

Fuchigami, L.H., Evert, D.R. and Weiser, C.J. (1971b) A translocatable cold hardiness promoter. *Plant Physiol.* **47**, 164–167.

Fujioka, S. and Sakurai, A. (1992) Effect of L-pipecolic acid on flowering in *Lemna paucicostata* and *Lemna gibba*. *Plant Cell Physiol.* **33**, 419–426.

Fujioka, S., Yamaguchi, I., Murofushi, N., Takahashi, N., Kaihara, S., Takimoto, A. and Cleland, C.F. (1985) The role of benzoic acid and plant hormones in flowering of *Lemna gibba* G3. *Plant Cell Physiol.* **26**, 655–659.

Fujioka, S., Yamaguchi, I., Murofushi, N., Takahashi, N., Kaihara, S., Takimoto, A. and Cleland, C.F. (1986a) The influence of nicotinic acid and plant hormones on flowering in *Lemna*. *Plant Cell Physiol.* **27**, 109–116.

Fujioka, S., Sakurai, A., Yamaguchi, I., Murofushi, N., Takahashi, N., Kaihara, S., Takimoto, A. and Cleland, C.F. (1986b) Flowering and endogenous levels of plant hormones in *Lemna* species. *Plant Cell Physiol.* **27**, 1297–1307.

Fukshansky, L. (1981) A quantitative study of timing in plant photoperiodism. *J. Theor. Biol.* **93**, 63–91.

Funke G.L. (1948) The photoperiodicity of flowering under short days with supplemental light of different wavelengths. *Lotsya* **1**, 79–82.

Furuya, M. (1989) Molecular properties and biogenesis of phytochrome I and II. *Adv. Biophys.* **25**, 133–167.

Furuya, M. and Song, P.-S. (1994) Assembly and properties of holophytochrome. In *Photomorphogenesis in Plants*, 2nd edn (eds R.E. Kendrick and G.H.M. Kronenberg), pp. 105–140. Kluwer Academic Publishers, Dordrecht.

Gaba, V. and Black, M. (1985) Photocontrol of hypocotyl elongation in light-grown *Cucumis sativus* L. *Planta* **164**, 264–271.

Galston, A.W. and Kaur-Sawhney, R. (1990) Polyamines and reproductive activity. *Flowering Newsletter* **9**, 3–8.

Galun, E., Gressel, J. and Keynan, A. (1964) Suppression of floral induction by actinomycin D, an inhibitor of 'messenger' RNA synthesis. *Life Sci.* **3**, 911–915.

Galun, E., Isher, A. and Atsmon, A. (1965) Determination of relative auxin content in hermaphrodite and andromonoecious *Cucumis sativus* L. *Plant Physiol.* **40**, 321–326.

Garg, V.K. and Paleg, L.G. (1986) Changes in the levels and composition of sterols in different tissues of *Lolium temulentum* plants during floral development. *Physiol. Plant.* **68**, 335–341.

Garner, W.W. (1933) Comparative responses of long-day and short-day plants to relative length of day and night. *Plant Physiol.* **8**, 347–356.

Garner, W.W. and Allard, H.A. (1920) Effect of the relative length of day and night and other factors of the environment on growth and reproduction in plants. *J. Agric. Res.* **18**, 553–606.

Garner, W.W. and Allard, H.A (1923) Further studies on photoperiodism, the response of plants to relative length of day and night. *J. Agric. Res.* **23**, 871–920.

Gebhart, J.S. and McDaniel, C.N. (1991) Flowering response of day-neutral and short-day cultivars of *Nicotiana tabacum* L. Interactions among roots, genotype, leaf ontogenetic position and growth conditions. *Planta* **185**, 513–517.

Gibby, D.D. and Salisbury, F.B. (1971) Participation of long-day inhibition in flowering of *Xanthium strumarium* L. *Plant Physiol.* **47**, 784–789.

Gilmour, S., Zeevaart, J.A.D., Schwenen, L. and Graebe. J.E. (1986) Gibberellin metabolism in cell-free extracts from spinach leaves in relation to photoperiod. *Plant Physiol.* **82**, 190–195.

Gollin, D.J., Darvill, A.G. and Albersheim, P. (1984) Plant cell wall fragments inhibit flowering and promote vegetative growth in *Lemna gibba* G3. *Biol. Cell* **51**, 275–280.

Gomez, L.A. and Simon, E. (1995) Circadian rhythm of *Robinia pseudacacia* leaflet movement. Role of calcium and phytochrome. *Photochem. Photobiol.* **61**, 210–215.

Gonthier, R., Jacqmard, A. and Bernier, G. (1987) Changes in cell-cycle duration and growth fraction in the shoot meristem of *Sinapis* during floral transition. *Planta* **170**, 55–59.

Gordon, A.J., Ryle, G.J.A., Mitchell, D.F. and Powell, C.F. (1982) The dynamics of carbon supply from leaves of barley plants grown under long or short days. *J. Expt. Bot.* **33**, 241–250.

Goto, N., Kumagai, T. and Koornneef, M. (1991) Flowering responses to light-breaks in photomorphogenesis mutants of *Arabidopsis thaliana*, a long-day plant. *Physiol. Plant.* **83**, 209–215.

Gottmann, K. and Schäfer, E. (1983) Analysis of phytochrome kinetics in light-grown *Avena sativa* L. seedlings. *Planta* **157**, 392–400.

Grayling, A. (1988) The process of *Perilla* flowering. D. Phil Thesis, University of Cambridge.

Grayling, A. (1990) The action of long days intercalated among short days on flowering in *Perilla frutescens*: a matter of interpretation. *Flowering Newsletter* **9**, 35–37.

Grayling, A. and Hanke, D.E. (1992) Cytokinins in exudates from leaves and roots of red *Perilla*. *Phytochem.* **31**, 1863–1868.

Greene, D.W. (1993) Effects of GA_4 and GA_7 on flower bud formation and russet development on apple. *J Hort. Sci.* **68**, 171–176.

Greer, D.H. and Warrington, I.J. (1982) Effect of photoperiod, night temperature and frost incidence on development of frost hardiness in *Pinus radiata*. *N.Z. J. For.* **13**, 80–86.

Greer, D.H., Stanley, C.J. and Warrington, I.J. (1989) Photoperiod control of the initial phase of frost hardiness development in *Pinus radiata*. *Plant Cell Environ.* **12**, 661–668.

Gregory, F.G. (1948) The control of flowering in plants. *Symp. Soc. Exp. Biol.* **2**, 75–103.

Gregory, L.E (1956) Some factors for tuberization in the potato plant. *Am. J. Bot.* **43**, 281–288.

Greppin, H., Auderset, G., Bonzon, M., Degli Agosti, R., Lenk, R., Penel, C. (1986) Le méchanisme de l'induction florale. *Saussurea* **17**, 71–84.

Gressel, J., Zilberstein A., Porath, D. and Arzee, T. (1980) Demonstration with fiber illumination that *Pharbitis* plumules also perceive flowering photoinduction. In *Photoreceptors and Plant Development* (ed. J. de Greef), pp. 525–530. Antwerpen University Press, Antwerp.

Greulach, V.A. (1942) Photoperiodic after-effects in six Composites. *Bot Gaz.* **103**, 698–709.

Griesel, W.O. (1963) Photoperiodic responses of two *Cestrum* species and non-interchangeability of their flowering hormones. *Plant Physiol.* **38**, 479–482.

Grimm, R., Gast, D. and Rudiger, W. (1989) Characterization of a protein-kinase activity associated with phytochrome from etiolated oat (*Avena sativa* L.) seedlings. *Planta* **174**, 199–206.

Gupta, S. and Maheshwari, S.C. (1969) Induction of flowering by cytokinins in a short-day plant, *Lemna paucicostata*. *Plant Cell Physiol.* **10**, 131–233.

Guron, K., Chandok, M.R. and Sopory, S.K. (1992) Phytochrome-mediated rapid changes in the level of phosphoinositides in etiolated leaves of *Zea mays*. *Photochem. Photobiol.* **56**, 691–695.

Gutterman, Y. (1989) *Carrichtera annua*. In *Handbook of Flowering* vol VI (ed. A.H. Halevy) pp. 157–161. CRC Press, Boca Raton.

Gutterman, Y. and Boeken, B. (1988) Flowering affected by daylength and temperature in the leafless flowering desert geophyte, *Colchicum tunicatum*, its annual life-cycle and vegetative propagation. *Bot. Gaz.* **149**, 382–390.

Gutterman, Y. and Porath, D. (1975) Influences of photoperiodism and light treatments during fruit storage on the phytochrome and on the germination of *Cucumis prophetarum* L. and *Cucumis sativus* L. seeds. *Oecologia* **18**, 37–43.

Guttridge, C.G. (1969) *Fragaria*. In *The Induction of Flowering* (ed. L.T. Evans), pp. 247–267. Macmillan, Melbourne.

Guttridge, C.G. (1985) *Fragaria* × *ananassa*. In *Handbook of Flowering*, Vol. III (ed. A.H. Halevy), pp. 16–33. CRC Press, Boca Raton.

Guttridge, C.G. and Vince-Prue, D. (1973) Floral initiation in strawberry: spectral evidence for the regulation of flowering by long-day inhibition. *Planta* **110**, 165–172.

Guzman, P. and Ecker, J.R. (1990) Exploiting the triple response of *Arabidopsis* to identify ethylene-related mutants. *The Plant Cell* **2**, 513–523.

Håbjørg, A. (1972) Effects of photoperiod and temperature on growth and development of three latitudinal and three altitudinal populations of *Betula pubescens* Ehrh. Meld. *Norges Land-brukhøgesk.,* **51 no. 2,** 1–27.

Håbjørg, A. (1978) Photoperiodic ecotypes in Scandinavian trees and shrubs. *Meld. Norges Landbruksch.* **71,** 1–20.

Hadley, P, Roberts, E.H., Summerfield, R.J. and Minchin, F.R. (1984) Effects of temperature and photoperiod on flowering in soya bean (*Glycine max* (L.) Merrill): a quantitative model. *Ann. Bot.* **53,** 669–681.

Hajkova, L. and Krekule, J. (1972) The developmental pattern in a group of theraphytes. Part 1. Seed dormancy. *Flora, Jena* **161,** 111–120.

Halaban, R. (1968) The flowering response of *Coleus* in relation to photoperiod and the circadian rhythm of leaf movement. *Plant Physiol.* **43,** 1894–1898.

Halaban, R. and Hillman, W.S. (1970) Response of *Lemna perpusilla* to periodic transfer to distilled water. *Plant Physiol.* **46,** 641–644.

Halaban, R. and Hillman, W.S. (1971) Factors affecting the water-sensitive phase of flowering in the short day plant *Lemna perpusilla. Plant Physiol.* **48,** 760–764.

Halevy, A.H. (1984) Light and autonomous induction. In *Light and the Flowering Process* (eds D. Vince-Prue, B. Thomas and K.E. Cockshull), pp. 65–73. Academic Press, London.

Halevy, A.H. (1985) *Handbook of Flowering.* CRC Press, Boca Raton.

Halevy, A.H. (1990) Recent advances in control of flowering in horticultural crops. *Adv. Hort. Sci.* **1,** 39–43.

Halevy, A.H. and Weiss, D. (1991) Flowering control of recently introduced F_1-hybrid cultivars of *Godetia. Scientia Hort.* **46,** 295–299.

Halevy, A.H., Spiegelstein, H. and Goldschmidt, E.E. (1991) Auxin inhibition of flower induction of *Pharbitis* is not mediated by ethylene. *Plant Physiol.* **95,** 652–654.

Hamasaki, N. and Galston, A.W. (1990) The polyamines of *Xanthium strumarium* and their response to photoperiod. *Photochem. Photobiol.* **52,** 181–186.

Hamdi, S., Yu, L.X., Cabre, E. and Delaigue, M. (1989) Gene expression in *Mercurialis annua* flowers; *in vitro* translation and sex genotype specificity. Male specific cDNA cloning and hormonal dependence of a corresponding specific RNA. *Mol. Gen. Genet.* **219,** 168–178.

Hammes, P.S. (1971) Tuber initiation in the potato (*Solanum tuberosum* L.). *Agroplantae* **3,** 73–74.

Hammes, P.S. and Beyers, E.A. (1973) Localization of the photoperiodic perception in potatoes. *Potato Res.* **16,** 68–72.

Hamner, K.C. (1940) Interaction of light and darkness in photoperiodic induction. *Bot. Gaz.* **101,** 658–687.

Hamner, K.C. (1969) *Glycine max* (L) Merrill. In *The Induction of Flowering* (ed. L.T. Evans), pp. 62–89. Macmillan, Melbourne.

Hamner, K.C. and Bonner, J. (1938) Photoperiodism in relation to hormones as factors in floral initiation and development. *Bot. Gaz.* **100,** 388–431.

Hamner, K.C. and Long, E.M. (1939) Localization of photoperiodic perception in *Helianthus tuberosus. Bot. Gaz.* **101,** 81–90.

Hamner, K.C. and Takimoto, A. (1964) Circadian rhythms and plant photoperiodism. *Am. Nat.* **98,** 295–322.

Hanke, J., Hartmann, K.M. and Mohr, H. (1969) Die Wirkung von 'Storlicht' auf die Blüten-bildung von *Sinapis alba* L. *Planta* **86,** 235–249.

Hansen, D.J., Bellman, S.K. and Sacher, R.M. (1976) Gibberellic-controlled sex expression in corn tassels. *Crop Sci.* **16,** 371–374.

Hansen, C.E., Kopperud, C. and Heide, O.M. (1988) Identity of cytokinins in *Begonia* leaves and their variation in relation to photoperiod and temperature. *Physiol. Plant.* **73,** 387–391.

Harder, R. (1948) Vegetative and reproductive development of *Kalanchoë blossfeldiana* as influenced by photoperiodism. *Symp. Soc. Exp. Biol.* **2,** 117–138.

Harder, R., Bode, O. and v. Witsch, H. (1944) Photoperiodische Untersuchungen in köhlen-saurefrier Atmosphäre bei der Kurztagpflanze *Kalanchoë blossfeldiana. Jb. Wiss. Bot.* **91,** 381–394.

Harder, R., Westphal, M. and Behrens, G. (1949) Hemmung der Infloreszenzbildung durch Langtag bei der Kurztagpflanzen *Kalanchoë blossfeldiana. Planta* **36,** 424–438.

Hardin, P.E., Hall J.C. and Rosbash, M. (1992) Circadian oscillations in period gene messenger-RNA levels are transcriptionally regulated. *Proc. Nat. Acad. Sci. USA* **89**, 11711–11715.

Harkness, R.L., Lyons, R.E. and Kushad, M.M. (1992) Floral morphogenesis in *Rudbeckia hirta* in relation to polyamine concentration. *Physiol. Plant.* **86**, 575–582.

Harmey, M.A., Crowley, M.P. and Clinch, P.E.M. (1966) The effect of growth regulators on tuberisation of cultured stem pieces of *Solanum tuberosum*. *Eur. Potato J.* **9**, 146–151.

Harris, G.P. (1968) Photoperiodism in the glasshouse carnation. *Ann. Bot.* **32**, 187–197.

Harris, G.P. (1972) Intermittent illumination and the photoperiodic control of flowering in carnation. *Ann. Bot.* **36**, 345–352.

Harrison, M.A. and Saunders, P.F. (1975) The abscisic acid content of dormant birch buds. *Planta* **123**, 291–298.

Harter, K., Frohnmeyer, H., Kircher, S., Kunkel, T., Muhlbauer, S. and Schäfer, E. (1994a) Light induces rapid changes of the phosphorylation pattern in the cytosol of evacuolated parsley protoplasts. *Proc. Nat. Acad. Sci. USA* **91**, 5038–5042.

Harter, K., Kircher, S., Frohnmeyer, H., Krenz, M., Nagy, F., Schäfer, E. (1994b) Light-regulated modification and nuclear translocation of cytosolic G-box binding factors in parsley. *The Plant Cell* **6**, 545–559.

Hartmann, K.M. (1966) A general hypothesis to interpret 'high energy phenomena' of photomorphogenesis on the basis of phytochrome. *Photochem. Photobiol.* **5**, 349–366.

Hartmann, K.M. (1967) Ein Wirkungspectrum der Photomorphogenese unter Hochenergiebedingungen und seine Interpretation auf der Basis der Phytochroms (Hypokotylwachstumshemmung bei *Lactuca sativa* L.). *Z. Naturforsch.* **22b**, 1172–1175.

Hashimoto, T. and Tamura, S. (1969) Effects of abscisic acid on the sprouting of aerial tubers of *Begonia evansiana* and *Dioscorea batatas*. *Bot. Mag. Tokyo* **82**, 69–75.

Haupt, W. and Reif, G. (1979) 'Ageing' of phytochrome Pfr in *Mesotaenium*. *Z. Pflanzenphysiol.* **92**, 153–161.

Havelange, A. (1980) The quantitative ultrastructure of the meristematic cells of *Xanthium strumarium* during the transition to flowering. *Am. J. Bot.* **67**, 1171–1178.

Havelange, A. and Bernier, G. (1983) Partial evocation by high irradiance in the long-day plant *Sinapis alba*. *Physiol. Plant.* **59**, 545–550.

Havelange, A. and Bernier, G. (1991) Elimination of flowering and most cytological changes after selective long-day exposure of the shoot tip of *Sinapis alba*. *Physiol. Plant.* **81**, 399–402.

Havelange, A., Bodson, M. and Bernier, G. (1986) Partial floral evocation by exogenous cytokinin in the long-day plant *Sinapis alba*. *Physiol. Plant.* **67**, 695–701.

Hay, R.K.M. and Heide, O.M. (1983) Specific photoperiodic stimulation of dry matter production in a high-latitude cultivar of *Poa pratensis*. *Physiol. Plant.* **57**, 135–142.

Hayashi, H., Kaihara, S. and Takimoto, A. (1992) Time-measuring processes involved in the photoperiodic response of *Lemna paucicostata* 441. *Plant Cell Physiol.* **33**, 757–762.

Heath, O.V.S. and Holdsworth, M. (1948) Morphogenic factors as exemplified by the onion plant. *Symp. Soc. Exp. Biol.* **2**, 326–350.

Heath, O.V.S. and Hollies, M.A. (1965) Studies in the physiology of the onion plant. VI. A sensitive morphological test for bulbing and its use in detecting bulb development in sterile culture. *J. Exp. Bot.* **16**, 124–144.

Heide, O.M. (1969) Environmental control of sex expression in *Begonia*. *Z. Pflanzenphysiol.* **61**, 279–285.

Heide, O.M. (1974a) Growth and dormancy in Norway spruce ecotypes (*Picea abies*) I. Interaction of photoperiod and temperature. *Physiol. Plant.* **30**, 1–12.

Heide, O.M. (1974b) Growth and dormancy in Norway spruce ecotypes II. After-effects of photoperiod and temperature on growth and development in subsequent years. *Physiol. Plant.* **31**, 131–139.

Heide, O.M. (1984) Flowering requirements in *Bromus inermis*, a short-long-day plant. *Physiol. Plant.* **62**, 59–64.

Heide, O.M. (1987) Photoperiodic control of flowering in *Dactylis glomerata*, a true short-long-day plant. *Physiol. Plant.* **70**, 523–529.

Heide, O.M. (1993a) Daylength and thermal time responses of budburst during dormancy release in some northern deciduous trees. *Physiol. Plant.* **88**, 541–548.

Heide, O.M. (1993b) Dormancy release in beech buds (*Fagus sylvatica*) requires both chilling and long days. *Physiol. Plant.* **89**, 187–191.

Heide, O.M. (1994) Control of flowering and reproduction in temperate grasses. *New Phytol.* **128**, 347–362.

Heide, O.M. and Skoog, F. (1967) Cytokinin activity in *Begonia* and Bryophyllum. *Physiol. Plant.* **20**, 771–780.

Heide, O.M., Bush, M.G. and Evans, L.T. (1985a) Interaction of photoperiod and gibberellin on growth and photosynthesis of high-latitude *Poa pratensis*. *Physiol. Plant.* **65**, 135–145.

Heide, O.M., Hay, R.K.M. and Baugeröd, H. (1985b) Specific daylength effects on leaf growth and dry-matter production in high-latitude grasses. *Ann. Bot.* **55**, 579–586.

Heide, O.M., Bush. M.G. and Evans, L.T. (1986a) Inhibitory and promotive effects of gibberellic acid on floral initiation and development in *Poa pratensis* and *Bromus inermis*. *Physiol. Plant.* **69**, 342–350

Heide, O.M., King, R.W. and Evans, L.T. (1986b) A semidian rhythm in the flowering response of *Pharbitis nil* to far-red light. I. Phasing in relation to the light-off signal. *Plant Physiol.* **80**, 1020–1024.

Heide, O.M., King, R.W. and Evans, L.T. (1988) The semidian rhythm in flowering response of *Pharbitis nil* in relation to dark period time measurement and to a circadian rhythm. *Physiol. Plant.* **73**, 286–294.

Heineke, D., Wildenberger, K., Sonnewald, U., Bussis, D., Gunter, G., Leidreiter, K., Rashke, K., Willmitzer, L. and Heldt, H.W. (1992) Apoplastic expression of yeast-derived invertase in potato – effects on photosynthesis, leaf solute composition, water relations and tuber composition. *Plant Physiol.* **100**, 301–308.

Heinze, P.H., Parker M.W. and Borthwick H.A. (1942) Floral induction in Biloxi soybean as influenced by grafting. *Bot. Gaz.* **103**, 518–529.

Helder, H., Miersch, O., Vreugdenhil, D. and Sembdner, G. (1993) Occurrence of hydroxylated jasmonic acids under long-day and short-day conditions. *Physiol Plant.* **88**, 647–653.

Hemberg, T. (1970) The action of some cytokinins on the rest period and the content of acid growth-inhibiting substances in potato. *Physiol. Plant.* **23**, 850–858.

Hemphill, D.D., Baker, L.R. and Sell, H.M. (1972) Different sex phenotypes of *Cucumis sativus* L. and *C. melo* L. and their endogenous gibberellin activity. *Euphytica* **21**, 285–291.

Hendricks, S.B. (1960) Rate of change of phytochrome as an essential factor determining photoperiodism in plants. *Cold Spring Harb. Symp. Quant. Biol.* **25**, 245–248.

Hendricks, S.B. and Borthwick, H.A. (1967) The function of phytochrome in regulation of plant growth. *Proc. Nat. Acad. Sci. USA* **58**, 2125–2130.

Hendricks, S.B. and Siegelman, H.W. (1967) Phytochrome and photoperiodism in plants. *Comp. Biochem.* **27**, 211–235.

Henfrey, A. (1852) *The Vegetation of Europe, its Conditions and Causes.* J. van Vooret, London.

Henson, I.E. and Wareing. P.F. (1977a) Cytokinins in *Xanthium strumarium* L.; some aspects of the photoperiodic control of endogenous levels. *New Phytol.* **78**, 35–45.

Henson, I.E. and Wareing, P.F. (1977b) Cytokinins in *Xanthium strumarium* L.; the metabolism of cytokinins in detached leaves and buds in relation to photoperiod. *New Phytol.* **78**, 27–33.

Henson, I.E. and Wareing, P.F. (1977c) Changes in the levels of endogenous cytokinins and indole-3-acetic acid during epiphyllous bud formation in *Bryophyllum daigremontianum*. *New Phytol.* **79**, 225–232.

Hershey, H.P., Barker, R.F., Idler, K.B., Lissemore, J.L. and Quail, P.H. (1985) Analysis of cloned cDNA and genomic sequences for phytochrome: complete amino acid sequences for two gene products in etiolated *Avena*. *Nuc. Acids Res.* **3**, 8543–8559.

Heslop-Harrison, J. (1957) The experimental modification of sex expression in flowering plants. *Biol. Rev.* **32**, 38–90.

Heslop-Harrison, J. (1960) Suppressive effect of 2-thiouracil on differentiation and flowering in *Cannabis sativa*. *Science* **132**, 1943–1944.

Heslop-Harrison, J. (1961) The experimental control of sexuality and inflorescence structure in *Zea mays* L. *Proc. Linn. Soc. Lond.* **172**, 108–123.

Heslop-Harrison, J. (1963) Sex expression in flowering plants. In *Meristems and Differentiation* (Brookhaven Symposium of Biology **16**), pp. 109–125. Brookhaven National Laboratory, Upton, New York.

Heslop-Harrison, J. (1964) The conrol of flower differentiation and sex expression. In *Régulateurs Naturels de la Croissance Végétale*, pp. 597–609. CNRS, Paris.

Hewett, E.W. and Wareing, P.F. (1973) Cytokinins in *Populus* × *robusta*, changes during chilling and bud burst. *Physiol. Plant.* **28**, 393–399.

Heyde, N.M. and Rombach, J. (1988) Flower induction in Norflurazon-treated *Pharbitis nil*: photo-induction of photoperiodic sensitivity in seedlings grown *in vitro* and daylength sensitivity in partly bleached potted plants. *Acta Bot. Neerl.* **37**, 371–377.

Heyer, A. and Gatz, C. (1992a) Isolation and characterization of a cDNA-clone coding for potato type-A phytochrome. *Plant Mol. Biology* **18**, 535–544.

Heyer, A. and Gatz, C. (1992b) Isolation and characterization of a cDNA-clone coding for potato type-B phytochrome. *Plant Mol. Biology* **20**, 589–600.

Heyer, A.G., Mozley, D., Landschutze, V., Thomas, B. and Gatz, C. (1995) Function of phytochrome A in *Solanum tuberosum* as revealed through the study of transgenic plants. *Plant Physiol.* **109**, 53–61.

Hillman, W.S. (1959) Experimental control of flowering in *Lemna* I. General methods. Photoperiodism in *L. perpusilla*. *Am. J. Bot.* **46**, 466–473.

Hillman, W.S. (1961) Photoperiodism, chelating agents and flowering of *Lemna perpusilla* and *L. gibba* in aseptic culture. In *Light and Life* (eds W.D. McElroy and B. Glass), pp. 673–686, Johns Hopkins University Press, Baltimore.

Hillman, W.S. (1967a) Blue light, phytochrome and the flowering of *Lemna perpusilla* 6746. *Plant Cell Physiol.* **8**, 467–473.

Hillman, W.S. (1967b) The physiology of phytochrome. *Ann. Rev. Plant Physiol.* **18**, 301–324.

Hillman, W.S. (1969a) Photoperiodism and vernalization. In *The Physiology of Plant Growth and Development* (ed. M.B. Wilkins), pp. 559–601. McGraw-Hill, New York, London.

Hillman, W.S. (1969b) *Lemna perpusilla* Torr., Strain 6746. In *The Induction of Flowering* (ed. L.T. Evans), pp. 186–204. Macmillan, Melbourne.

Hillman, W.S. (1976) A metabolic indicator of photoperiodic timing. *Proc. Nat. Acad. Sci. USA* **73**, 501–504.

Hilton, J.R. and Thomas, B. (1987) Photoregulation of phytochrome synthesis in germinating embryos of *Avena sativa* L. *J. Exp. Bot.* **38**, 1704–1712.

Hoad, G.V. and Bowen, M.R. (1968) Evidence for gibberellin-like substances in phloem exudate from higher plants. *Planta* **82**, 22–32.

Hocking, T.J. and Hillman, J.R. (1975) Studies in the role of abscisic acid in the initiation of bud dormancy in *Alnus glutinosus* and *Betula pubescens*. *Planta* **125**, 235–242.

Hodson, H.K. and Hamner, K.C. (1970) Floral inducing extracts from *Xanthium*. *Science NY* **167**, 384–385.

Holdsworth, M.L. and Whitelam, G.C. (1987) A monoclonal antibody specific for the red-absorbing form of phytochrome. *Planta* **172**, 539–547.

Holland, R.W.K. (1969) The effects of red and far-red light on floral initiation in *Lolium temulentum* and other long-day plants. Ph.D. Thesis, University of Reading.

Holland, R.W.K. and Vince, D. (1968) Photoperiodic control of flowering and anthocyanin formation in *Fuchsia*. *Nature* **219**, 511–513.

Holland, R.W.K. and Vince, D. (1971) Floral initiation in *Lolium temulentum* L: the role of phytochrome in the responses to red and far-red light. *Planta* **98**, 232–243.

Holmes, M.G. and Smith, H. (1977) The function of phytochrome in the natural environment. IV. Light quality and plant development. *Photochem. Photobiol.* **25**, 551–557.

Horridge, J.S. and Cockshull, K.E. (1979) Size of the *Chrysanthemum* shoot apex in relation to inflorescence initiation and development. *Ann. Bot.* **44**, 547–556.

Horridge, J.S. and Cockshull, K.E. (1989) The effect of the timing of the night-break on flower initiation in *Chrysanthemum morifolium* Ramat. *J. Hort. Sci.* **64**, 183–188.

Hoshizaki, T. and Hamner, K.C. (1969) Interactions between light and circadian rhythms in plant photoperiodism. *Photochem. Photobiol.* **10**, 87–96.

Houssa, P., Bernier, G. and Kinet, J-M. (1991) Qualitative and quantitative analyses of carbohydrates in leaf exudate of the short-day plant, *Xanthium strumarium* L. during floral transition. *J. Plant Physiol.* **138**, 24–28.

Howe, G.T., Hackett, W.P., Furnier, G.R. and Klevorn, R.E. (1995) Photoperiodic responses of a northern and southern ecotype of black cottonwood. *Physiol. Plant.* **93**, 685–708.

Howell, G.S. and Weiser, C.J. (1970) The environmental control of cold acclimation in apple. *Plant Physiol.* **45**, 390–394.

Hsu, J.C.S. and Hamner, K.C. (1967) Studies on the involvement of an endogenous rhythm in the photoperiodic response of *Hyoscyamus niger. Plant Physiol.* **42**, 725–730.

Huala, E. and Sussex, I.M. (1992) Leafy interacts with floral homeotic genes to regulate *Arabidopsis* floral development. *The Plant Cell* **4**, 901–913.

Huber, S.C., Rufty, T.W. and Kerr, P.S. (1984) Effect of photoperiod on photosynthate partitioning and diurnal rhythms in sucrose phosphate synthase activity in leaves of soybean (*Glycine max* L. [Merr.]) and tobacco (*Nicotiana tabacum* L.). *Plant Physiol.* **75**, 1080–1084.

Hughes, A.P. and Cockshull, K.E. (1964) Effects of a night-break of red fluorescent light on leaf growth of *Callistephus chinensis* var. Queen of the Market. *Nature* **201**, 413.

Hughes, A.P. and Cockshull, K.E. (1965) Interactions of flowering and vegetative growth in *Callistephus chinensis* (var. Queen of the Market). *Ann. Bot.* **29**, 131–151.

Hughes, A.P. and Cockshull, K.E. (1966) Effects of night-break lighting on bedding plants. *Exp. Hort.* **16**, 44–52.

Hughes, A.P. and Cockshull, K.E. (1969) Cyclic night-breaks can advance flowering. *Grower* **7**, 601.

Hughes, J.E., Morgan, D.C., Lambton, P.A., Black, C.R. and Smith, H. (1984) Photoperiodic signals during twilight. *Plant Cell Environ.* **7**, 269–277.

Hume, R.J. and Lovell, P.H. (1983) Role of 1-aminocyclopropane-1-carboxylic acid in the control of female flowering in *Cucurbita pepo. Physiol. Plant.* **59**, 324–328.

Humphries, E.C. (1963a) Effects of (2 chloroethyl) trimethylammonium chloride on plant growth. *New Phytol.* **61**, 154–174.

Humphries, E.C. (1963b) Effects of gibberellic acid and CCC on growth of potato. *Rothamsted Exp. Sta. Ann. Rep.* **89**.

Hurst, C., Hall, T.C. and Weiser, C.J. (1967) Reception of the light stimulus for cold acclimation in *Cornus stolonifera* Michx. *Hort. Sci.* **2**, 164–166.

Hussein, H.A.S. and Van der Veen, J.H. (1965) Induced mutants for flowering time. *Arabidopsis Information Service* **2**, 6–8.

Hussein, H.A.S. and Van der Veen, J.H. (1968) Genotypic analysis of induced mutations for flowering time and leaf number in *Arabidopsis thaliana. Arabidopsis Information Service* **5**, 30–32.

Hussey, G. (1954) Experiments with two long-day plants designed to test Bünning's theory of photoperiodism. *Physiol. Plant.* **7**, 253–260.

Ikeda, K. (1974) Photoperiodic control of floral initiation in the rice plant. I. Light requirement during the photoperiod. *Proc. Crop. Sci. Soc. Japan* **43**, 375–381.

Ikeda, K. (1985) Photoperiodic flower induction in rice plants as influenced by light intensity and quality. *JARQ* **18**, 164–170.

Imamura, S., Maramatsu, M., Kitajo, S.I. and Takimoto, A. (1966) Varietal difference in photoperiodic behaviour of *Pharbitis nil* Chois. *Bot. Mag. Tokyo* **79**, 714–721.

Imhoff, C., Brulfert, J. and Jacques, R. (1971) Mise à fleurs en éclairement monochromatique de l'*Anagallis arvensis* L., plante de jour long absolue: spectre d'action et rôle du phytochrome. *C.R. Acad. Sci. Paris* **273**, 737–740.

Imhoff, C., Lecharny, A., Jaques, R. and Brulfert, J. (1979) Two phytochrome-dependent processes in *Anagallis arvensis* L. flowering and stem elongation. *Plant Cell Environ.* **2**, 67–72.

Ireland, C.R. and Schwabe, W.W. (1982) Studies on the role of photosynthesis in the photoperiodic induction of flowering in the short-day plants *Kalanchoë blossfeldiana* Poellniz. and *Xanthium pensylvanicum* Wallr. *J. Exp. Bot.* **33**, 738–760.

Irish, E.E. and Nelson, T. (1989) Sex determination in monoecious and dioecious plants. *The Plant Cell* **1**, 737–744.

Irving, R.M. (1969) Characterization and role of endogenous inhibitor in the induction of cold hardiness in *Acer negundo*. *Plant Physiol.* **44**, 801–805.

Irving, R.M. and Lanphear, F.O. (1968) Regulation of cold hardiness in *Acer negundo*. *Plant Physiol.* **43**, 9–13.

Ishioka, N., Tanimoto, S. and Harada, H. (1991a) Flower-inducing activity of phloem exudates from *Pharbitis* cotyledons exposed to various photoperiods. *Plant Cell Physiol.* **32**, 921–924.

Ishioka, N., Tanimoto, S. and Harada, H. (1991b) Roles of nitrogen and carbohydrate in floral-bud formation in *Pharbitis* apex cultures. *J. Plant Physiol.* **138**, 573–576.

Jabben, M., Heim, B. and Schäfer, E. (1980) The phytochrome system in light- and dark-grown dicotyledonous seedlings. In *Photoreceptors and Plant Development* (ed. J. de Greef), pp. 145–158. Antwerpen University Press, Antwerp.

Jabben, M. and Holmes, M.G. (1983) Phytochrome in light-grown plants. In *Photomorphogenesis. Encyclopaedia of Plant Physiology* (eds W. Shropshire Jr and H. Mohr) New Series Vol. 16, pp. 704–722. Springer-Verlag, Heidelberg.

Jacklett, J.W. (1982) Circadian clock mechanisms. In *Biological Time-Keeping*. Society for Experimental Biology Seminar Series, 14 (ed. J. Brady), pp. 173–188, Cambridge University Press.

Jackson, S.D. and Willmitzer, L. (1994) Jasmonic acid spraying does not induce tuberisation in short-day-requiring potato species kept in non-inducing conditions. *Planta* **194**, 155–159.

Jackson, S.D., Sonnewald, U. and Willmitzer, L. (1993) Characterisation of a gene that is expressed at higher levels upon tuberization in potato and upon flowering in tobacco. *Planta* **189**, 593–596.

Jackson, S.D., Heyer, A., Dietze, J. and Prat, S. (1996) Phytochrome B mediates the photoperiodic control of tuber formation in potato. *Plant Journal* **9**, 159–166.

Jacobs, W.P. (1980) Inhibition of flowering in short-day plants. In *Plant Growth Substances* 1979 (ed. F. Skoog), pp. 301–309. Springer-Verlag, Berlin.

Jacobs, W.P. (1985) The role of auxin in inductive phenomena. *Biol Plant.* **27**, 303–309.

Jacobs, W.P. and Eisinger, D.B. (1990) A cautionary note about incomplete controls in an early experiment on sucrose as a substitute for the floral-inhibiting effect of a long-day leaf on a short-day plant. *Flowering Newsletter* **10**, 30–33.

Jacobs, W.P. and Suthers, H.B. (1974) Effects of leaf excision on flowering of *Xanthium* apical buds in culture under inductive and noninductive photoperiods. *Am. J. Bot.* **61**, 1016–1020.

Jacqmard, A., Houssa, C. and Bernier, G. (1994) Regulation of the cell cycle by cytokinins. In *Cytokinin Metabolism and Activities* (ed. D.W.S. Mok), CRC Press, Boca Raton.

Jacques, M. and Jacques, R. (1989) *Calamintha nepetoides*. In *CRC Handbook of Flowering*, Vol VI (ed. A.H. Halevy), pp. 131–138. CRC Press, Boca Raton, Florida.

Jacques, M. and Leroux, A. (1979) Conditions nécessaires à la transmission du stimulus floral par voie de greffages chez des Chénopodiacées. *C.R. Acad. Sci. Paris Sér. D* **289**, 603–606.

Jaffe, M.J., Bridle, K.A. and Kopcewicz, J. (1987) A new strategy for the identification of native plant photoperiodically regulated flowering substances. In *Manipulation of Flowering* (ed. J.G. Atherton), pp. 267–287. Butterworths, London.

Janistyn, B. (1981) Sukkulenz-Induktion bei *Kalanchoë blossfeldiana* im Langtag durch eine lipophile Fraktion aus blühenden Kalanchoë und MS-Identifikation von Pterosteron. *Z. Naturforsch.* **36**, 455–458.

Jenkins, J.M. (1954) Some effects of different daylengths and temperatures upon bulb formation in shallots. *Proc. Am. Soc. Hort. Sci.* **64**, 311–314.

Johansen, L.G., Odén, P-C. and Junttila, O. (1986) Abscisic acid and cessation of apical growth in *Salix pentandra*. *Physiol. Plant.* **66**, 409–412.

Johnson, C.H. and Hastings, J.W. (1986) The elusive mechanism of the circadian clock. *Am. Sci.* **74**, 29–36.

Johnson, C.H. and Hastings, J.W. (1989) Circadian phototransduction: phase resetting and frequency of the circadian clock of *Gonyaulax* cells in red light. *J. Biol. Rhythms* **4**, 417–437.

Johnson, J.R. and Havis, J.R. (1977) Photoperiod and temperature effects on root cold acclimation. *J. Amer. Soc. Hort. Sci.* **102**, 306–308.

Johnson, E., Bradley, M., Harberd, N.P. and Whitelam, G.C. (1994) Photoresponses of light-grown *phyA* mutants of *Arabidopsis*. *Plant Physiol.* **105**, 141–149.

Jones, T.W.A. (1990) Use of a flowering mutant to investigate changes in carbohydrates during floral transition in red clover. *J. Exp. Bot.* **41**, 1013–1019.

Jones, A.M. and Quail, P.H. (1986) Quaternary structure of 124-kilodalton phytochrome from *Avena sativa* L. *Biochemistry* **25**, 2987–2995.

Jones, J.L. and Roddick, J.G. (1988) Steroidal extrogens and androgens in relation to reproductive development in higher plants. *J. Plant Physiol.* **133**, 156–164.

Jones, M.G. and Zeevaart, J.A.D. (1980) The effect of photoperiod on the levels of seven endogenous gibberellins in the long-day plant *Agrostemma githago* L. *Planta* **149**, 274–279.

Jones, A.M., Vierstra, R.D., Daniels, S.M. and Quail, P.H. (1985) The role of separate molecular domains in the structure of phytochrome from etiolated *Avena sativa*. *Planta* **164**, 501–506.

Jordan, B.R., Partis, M.D. and Thomas, B. (1986) The biology and molecular biology of phytochrome. In *Oxford Surveys of Plant Molecular and Cell Biology* (ed. B.J. Miflin), Vol. 3, pp. 315–362. Oxford University Press.

Joustra, M.K. (1969) Daylength dependence of flower initiation in *Hyoscyamus niger* L. *Meded. LandbHoogesch., Wageningen* **69**(13), 1–10.

Junttila, O. (1980) Effect of photoperiod and temperature on apical growth cessation in two ecotypes of *Salix* and *Betula*. *Physiol. Plant.* **48**, 347–352.

Junttila, O. (1982a) The cessation of apical growth in latitudinal ecotypes and ecotype crosses of *Salix pentandra* L. *J. Exp. Bot.* **33**, 1021–1029.

Junttila, O. (1982b) Gibberellin-like activity in shoots of *Salix pentandra* as related to the elongation growth. *Can. J. Bot.* **60**, 1231–1234.

Junttila, O. (1989) Interaction of gibberellic acid and photoperiod on leaf sheath elongation in normal and gibberellin-insensitive genotype of wheat. *Physiol. Plant.* **77**, 504–506.

Junttila, O (1991) Gibberellins and regulation of shoot elongation in woody plants. In *Gibberellins* (eds N. Takahashi, J. MacMillan and B.O. Phinney), pp. 199–210. Springer-Verlag, Berlin.

Junttila, O. (1993a) Exogenously applied GA_4 is converted to GA_1 in seedlings of *Salix*. *J. Plant Growth Regul.* **12**, 35–39.

Junttila, O. (1993b) Interactions of growth retardants, daylength and gibberellins A_{19}, A_{20} and A_1 on shoot elongation in birch and alder. *J. Plant Growth Regul.* **12**, 123–127.

Junttila, O. and Jensen, E. (1988) Gibberellins and photoperiodic control of shoot elongation in *Salix*. *Physiol. Plant.* **74**, 371–376.

Junttila, O. and Kaurin, A. (1990) Environmental control of cold acclimation in *Salix pentandra*. *Scand. J. For. Res.* **5**, 195–204.

Junttila, O. and Nilsen, J. (1993) Growth and development of northern forest trees as affected by temperature and light. In *Forest Development in Cold Climates* (eds J. Alden, J.L. Mastrantonio and S. Ødum), pp. 43–57. Plenum Press, New York.

Junttila, O., Svenning, M.M. and Solheim, B. (1990) Effects of temperature and photoperiod on frost resistance of white clover (*Trifolium repens*) ecotypes. *Physiol. Plant.* **79**, 435–438.

Kadman-Zahavi, A. and Peiper, D. (1987) Effects of moonlight on flower induction in *Pharbitis nil*, using a single dark period. *Ann. Bot.* **60**, 621–623.

Kadman-Zahavi, A. and Yahel, H. (1971) Phytochrome effects in night-break illuminations on flowering of *Chrysanthemum*. *Physiol. Plant.* **25**, 90–93.

Kadman-Zahavi, A., Horovitz, A. and Ozeri, Y. (1984) Long-day induced dormancy in *Anemone coronaria* L. *Ann. Bot.* **53**, 213–217.

Kahane, R., Delaserve, B.T. and Rancillac, M. (1992) Bulbing in long-day onion (*Allium cepa* L.) cultured *in vitro* comparison between sugar feeding and light induction. *Ann. Bot.* **69**, 551–555.

Kaihara, S. and Takimoto, A. (1985) Flower-inducing activity of Vitamin K in *Lemna paucicostata. Plant Cell Physiol.* **26**, 89–98.

Kaihara, S., Kozaki, A. and Takimoto, A. (1989) Flower-inducing activity of water extracts of various plant species, in particular *Pharbitis nil. Plant Cell Physiol.* **30**, 1023–1028.

Kakkar, R.K. and Rai, V.K. (1993) Plant polyamines in flowering and fruit ripening. *Phytochem.* **33**, 1281–1288.

Kanchanapoom, M.L. and Thomas, J.F. (1987) Quantitative ultrastructural changes in tunica and corpus cells of the shoot apex of *Nicotiana tabacum* during the transition to flowering. *Am. J. Bot.* **74**, 241–249.

Kandeler, R. (1984) Flowering in the *Lemna* system. *Phyton* **24**, 113–124.

Kannangara, T., Durkin, J.P., ApSimon, J. and Wightman, F. (1990) Changes in the polypeptide profiles during photoinduction of *Xanthium strumarium. Physiol. Plant.* **78**, 519–525.

Karssen, C.M. (1970) The light promoted germination of the seeds of *Chenopodium album* L. III. Effect of the photoperiod during growth and development of the plants on the dormancy of the produced seeds. *Acta Bot. Neerl.* **19**, 81–94.

Kasperbauer, M.J., Borthwick, H.A. and Cathey, H.M. (1963a) Cyclic lighting for the promotion of flowering of sweetclover, *Melitotus alba* Desr. *Crop Sci.* **3**, 230–232.

√ Kasperbauer, M.J., Borthwick, H.A. and Hendricks, S.B. (1963b) Inhibition of flowering of *Chenopodium rubrum* by prolonged far-red irradiation. *Bot. Gaz.* **124**, 444–451.

Kasperbauer, M.J., Borthwick, H.A. and Hendricks, S.B. (1964) Reversion of phytochrome 730 (Pfr) to P660 (Pr) assayed by flowering in *Chenopodium rubrum. Bot. Gaz.* **125**, 75–80.

Kato, T. (1964) Physiological studies on the bulbing and dormancy of onion plants. III. Effects of external factors on the bulb formation and development. *J. Jap. Soc. Hort. Sci.* **33**, 53–61.

Kato, T. (1965) Physiological studies on the bulbing and dormancy of onion plants. V. The relations between the metabolism of carbohydrates, nitrogen compounds and auxin and the bulbing phenomenon. *J. Jap. Soc. Hort. Sci.* **34**, 187–195.

Kato, T. (1966) Physiological studies on the bulbing and dormancy of onion plants. IX. Relationship between dormancy of bulb and properties of its juice. *J. Jap. Soc. Hort. Sci.* **35**, 295–303.

Kaufman, L.S., Thompson, W.F. and Briggs, W.R. (1984) Different red light requirements for phytochrome-induced acclimation of cab RNA and rbcs RNA. *Science* **226**, 1447–1449.

Kaur-Sawhney, R., Tiburcio, A.F. and Galston, A.W. (1988) Spermidine and flower-bud differentiation in thin-layer explants of tobacco. *Planta* **173**, 282–284.

Kay, S.A., Keith, B., Shinozaki, K., Chye, M.-L. and Chua, N.-H. (1989a) The rice phytochrome gene: structure, autoregulated expression, and binding of GT-1 to a conserved site in the 5' upstream region. *The Plant Cell* **1**, 351–360.

Kay, S.A., Nagatani, A., Keith, B., Deak, M., Furuya, M. and Chua, N-H. (1989b) Rice phytochrome is biologically active in transgenic tobacco. *The Plant Cell* **1**, 775–782.

Keim, D.L., Welsh, J.R. and McConnell, R.L. (1973) Inheritance of photoperiodic heading response in winter and spring cultivars of bread wheat. *Can. J. Plant Sci.* **53**, 247–250.

Keller, J.M., Shanklin, J., Vierstra, R.D. and Hershey, H.P. (1989) Expression of a functional monocotyledonous phytochrome in transgenic tobacco. *EMBO J.* **8**, 1005–1012.

Kelly, A.J., Zagotta, M.T., White, R.A., Chang, C. and Meeks-Wagner, D.Ry (1990) Identification of genes expressed in the tobacco shoot apex during the floral transition. *The Plant Cell* **2**, 963–972.

Kendrick, R.E. and Hillman, W.S. (1970) Dark reversion of phytochrome in *Sinapis alba* L. *Plant Physiol.* **46**, 596–598.

Kendrick, R.E. and Spruit, C.J.P. (1977) Phototransformations of phytochrome. *Photochem. Photobiol.* **26**, 201–214.

Khudairi, A.K. and Hamner, K.C. (1954) The relative sensitivity of *Xanthium* leaves of different ages to photoperiodic induction. *Plant Physiol.* **29**, 251–257.

Khurana, J.P. and Maheswari, S.C. (1984) Floral induction in short-day *Lemna paucicostata* 6746 by 8-hydroxyquinoline, under long days. *Plant Cell Physiol.* **25**, 77–83.

Kigel, J., Ofir, M and Koller, D. (1977) Control of germination responses of *Amaranthus retroflexus* L. seeds by their parental photothermal environment. *J. Exp. Bot.* **28**, 1125–1136.

Kigel, J., Gibly, A. and Negbi, M. (1979) Seed germination in *Amaranthus retroflexus* L. as affected by the photoperiod and the age during flower induction of the parent plants. *J. Exp. Bot.* **30**, 997–1002.

Kim, I.-S., Bai, U. and Song. P.-S. (1989) A purified 124-kDa oat phytochrome does not possess a protein kinase activity. *Photochem. Photobiol.* **49**, 319–323.

Kimpel, J.A. and Doss, R.P. (1989) Gene expression during floral induction by leaves of *Perilla crispa. Flowering Newsletter* **7**, 20–25.

Kimura, K. (1963) Floral initiation in *Pharbitis nil* subjected to continuous illumination at relatively low temperatures. II. Effect of some factors in culture medium on floral initiation. *Bot. Mag. Tokyo* **76**, 351–358.

Kinet, J.M. (1972) *Sinapis alba*, a plant requiring a single long day or a single short day for flowering. *Nature*, **236**, 406–407.

Kinet, J.M., Bernier, G., Bodson, M. and Jacqmard, A. (1973) Circadian rhythms and the induction of flowering in *Sinapis alba. Plant Physiol.* **51**, 598–600.

King, R.W. (1975) Multiple circadian rhythms regulate photoperiodic flowering responses in *Chenopodium rubrum. Can. J. Bot.* **53**, 2631–2638.

King, R.W. (1976) Implications for plant growth of the transport of regulatory compounds in phloem and xylem. In *Transport and Transfer Processes in Plants* (ed. I.F. Woodlow), pp. 415–431. Academic Press, New York.

King, R.W. (1979) Photoperiodic time measurement and effects of temperature on flowering in *Chenopodium rubrum* L. *Aust. J. Plant. Physiol.* **6**, 417–422.

King, R.W. and Cumming, B.G. (1972a) Rhythms as photoperiodic timers in the control of flowering in *Chenopodium rubrum* L. *Planta* **103**, 281–301.

King, R.W. and Cumming, B.G. (1972b) The role of phytochrome in photoperiodic time-measurement and its relation to rhythmic timekeeping in the control of flowering in *Chenopodium rubrum. Planta* **108**, 39–57.

King, R.W. and Evans, L.T. (1969) Timing of evocation and development of flowers in *Pharbitis nil. Aust. J. Biol. Sci.* **22**, 559–572.

King, R.W. and Evans, L.T. (1977) Inhibition of flowering in *Lolium temulentum* L. by water stress: a role for abscisic acid. *Aust J. Plant Physiol.* **4**, 225–233.

King, R.W. and Evans, L.T. (1991) Shoot apex sugars in relation to long-day induction of flowering in *Lolium temulentum* L. *Aust. J. Plant Physiol.* **18**, 121–135.

King, W.M. and Murfet, I.C. (1985) Flowering in *Pisum*. A sixth locus, Dne. *Ann. Bot.* **56**, 835–846.

King, R.W. and Zeevaart, J.A.D. (1984) Enhancement of phloem exudation from cut petioles by chelating agents. *Plant Physiol.* **53**, 96–103.

King, R.W., Evans, L.T. and Wardlaw, I.F. (1968) Translocation of the floral stimulus in *Pharbitis nil* in relation to that of assimilates. *Z. Pflanzenphysiol.* **59**, 377–385.

King, R.W., Evans, L.T. and Firn, R.D. (1977) Abscisic acid and xanthoxin contents in the long-day plant *Lolium temulentum* L. in relation to daylength. *Aust. J. Plant Physiol.* **4**, 217–223.

King, R.W., Vince-Prue, D. and Quail, P.H. (1978) Light requirement, phytochrome and photoperiodic induction of flowering of *Pharbitis nil* Chois. III. *Planta* **141**, 15–22.

King, R.W., Schäfer, E., Thomas, B. and Vince-Prue, D. (1982) Photoperiodism and rhythmic responses to light. *Plant Cell Environ.* **5**, 395–404.

King, R.W., Pharis, R.P. and Mander, L.N. (1987) Gibberellins in relation to growth and flowering in *Pharbitis nil* Chois. *Plant Physiol.* **84**, 1126–1131.

King, R.W., Blundell, C. and Evans, L.T. (1993) The behaviour of shoot apices of *Lolium temulentum in vitro* as the basis of an assay system for florigenic extracts. *Aust. J. Plant Physiol.* **20**, 337–348.

King, R.W., Shinozaki, M., Takimoto, A. and Swe, K.L. (1994) Genes for dwarfing and photoperiod flowering response in *Pharbitis nil* Choisy. *J. Plant. Res.* **107**, 215–219.

Klaimi, Y.Y. and Qualset, C.O. (1973) Genetics of heading time in wheat (*Triticum aestivum* L.). 1. The inheritance of photoperiodic response. *Genetics* **74**, 139–156.

Klebs, G. (1910) Alterations in the development and forms of plants as a result of environment. *Proc. Roy. Soc. Lond. B* **82**, 547–558.

Klebs, G. (1913) Über das Verhältnis der Aussenwelt zur Entwicklung der Pflanze. *Sber. Akad. Wiss. Heidelberg* **5**, 1–47.

Kloppstech, K. (1985) Diurnal and circadian rhythmicity in the expression of light–induced plant nuclear messenger RNAs. *Planta* **165**, 502–506.

Kluge, M. (1982) Biochemical rhythms in plants. In *Biological Time-Keeping*. Society for Experimental Biology Seminar Series, 14 (ed. J. Brady), pp. 159–172, Cambridge University Press.

Knapp, P.H., Sawhney, S., Grimmett, M.M. and Vince-Prue, D. (1986) Site of perception of the far-red inhibition of flowering in *Pharbitis nil* Choisy. *Plant Cell Physiol.* **27**, 1147–1152.

Knott, J.E. (1934) Effect of a localized photoperiod on spinach. *Proc. Am. Soc. Hort. Sci.* **31**, 152–154.

Koda, Y., Kikuta, Y., Tazaki, H., Tsujino, Y., Sakamura, S. and Yoshihara, T. (1991) Potato tuber-inducing activities of jasmonic acid and related compounds. *Phytochemistry* **30**, 1435–1438.

Kohli, R.K. and Sawhney, S. (1979) Promotory effect of GA_{13} on flowering of *Amaranthus* – a short day plant. *Biol. Plant.* **21**, 206–213.

Koller, D., Kigel, J. and Ovadiah, S. (1977) A kinetic analysis of the facultative photoperiodic response in *Amaranthus retroflexus* L. *Planta* **136**, 13–19.

Kondo, T. and Tsudzuki, T. (1980) Participation of a membrane system in the potassium uptake rhythm in a duckweed, *Lemna gibba* G3. *Plant Cell Physiol.* **21(4)**, 627–635.

Koornneef, M., Elgersma, A., Hanhart, C.J., van Loenen-Martinet, E.P., van Rijn, L. and Zeevaart, J.A.D. (1985) A gibberellin insensitive mutant of *Arabidopsis thaliana*. *Physiol. Plant.* **65**, 33–39.

Koornneef, M., Hanhart, C.J. and Van der Veen, J.H. (1991) A genetic and physiological analysis of late flowering mutants in *Arabidopsis thaliana*. *Mol. Gen. Gen.* **229**, 57–66.

Kopcewicz, J. (1972a) Oestrogens in the long-day plants *Hyoscyamus niger* and *Salvia splendens* grown under inductive and non-inductive light conditions. *New Phytol.* **71**, 129–134.

Kopcewicz, J. (1972b) Estrogens in the short-day plants *Perilla ocimoides* and *Chenopodium rubrum* grown under inductive and non-inductive light conditions. *Z. Pflanzenphysiol.* **67**, 373–376.

Kopcewicz, J., Centrowska, G., Kriesel, K. and Zatorska, Z. (1979) The effect of inductive photoperiod on flower formation and phytohormones level in a long day plant *Hyoscyamus niger* L. *Acta Soc. Bot. Polonae* **XLVIII**, 255–265.

Kozaki, A., Takeba, G. and Tanaka, O. (1991) A polypeptide that induces flowering in *Lemna paucicostata* at a very low concentration. *Plant Physiol.* **93**, 1288–1290.

Kramer, P.J. (1936) The effect of variation in length of day on the growth and dormancy of trees. *Plant Physiol.* **11**, 127–137.

Krebs, O. and Zimmer, K. (1983) Blütenbildumg bei *Begonia boweri* Ziesenh. und einem Abkömmling von *B.* Cleopatra. XII. End-of-day Licht bei sehr kurzen Photoperioden. *Gartenbauwissenschaft.* **48**, 269–271.

Krebs, O. and Zimmer, K. (1984) Blütenbildung bei *Begonia boweri* Ziesenh. und einem Abkömmling von *B.* 'Cleopatra'. XIII. Wirkung von Störlicht auf die Blütenbildung-Zusammenfassung und Diskussion. *Gartenbauwissenschaft.* **49**, 39–45.

Krekule, J. (1979) Stimulation and inhibition of flowering. In *Physiologie de la Floraison* (eds P. Chapagnat and R. Jacques), pp. 19–57. CNRS, Paris.

Krekule, J. and Seidlova, F. (1977) *Brassica campestris* as a model for studying the effects of exogenous growth substances on flowering in long-day plants. *Biol. Plant.* **19**, 462–468.

Krekule, J., Pavlová, L., Soucková, D. and Macháčková, I. (1985) Auxins in flowering of short-day and long-day *Chenopodium* species. *Biol. Plant.* **27**, 310–317.

Kribben, F.J. (1955) Zu den Theorien des Photoperiodismus. *Beitr. Biol. Pfl.* **31**, 97–311.

Krishnamoorthy, H.N. (1972) Effect of GA_3, GA_{4+7}, GA_5 and GA_9 on the sex expression of *Luffa acutangula* var H-2. *Plant Cell Physiol.* **13**, 381–383.

Krüger, G.H.J. (1984) The effect of photoperiod on the initiation of the staminate inflorescence in different hybrids on *Zea mays*. *S. Afr. J. Bot.* **3**, 81–82.

genes for phytochrome and chlorophyll a/b binding protein in *Avena sativa*. *Mol. Cell. Biol.* **8**, 4840–4850.

Lissemore, J.L., Colbert, J.T. and Quail, P.H. (1987) Cloning of cDNA for phytochrome from etiolated *Cucurbita* and co-ordinate regulation of the abundance of two distinct phytochrome transcripts. *Plant Mol. Biol.* **8**, 485–496.

Lona, F. (1949) La fioritura delle brevidiurne a notte continua. *Nuova G. Bot. Ital.* **56**, 479–515.

Longman, K.A. (1969) Dormancy and survival of plants in the humid tropics. *Symp. Soc. Exp. Biol.* **23**, 471–488.

Longman, K.A. (1978) Control of shoot extension and dormancy: external and internal factors. In *Tropical Trees as Living Systems* (eds P.B. Tomlinson and M.H. Zimmerman), pp. 465–493. Cambridge University Press.

Looney, N.E., Pharis, R.P. and Noma, M. (1985) Promotion of flowering in apple trees with gibberellin A_4 and C-3 epi-gibberellin A_4. *Planta* **165**, 292–294.

Louis, J-P., Augur, C. and Teller, G. (1990). Cytokinins and differentiation processes in *Mercurialis annua*. Genetic regulation, relations with auxins, indoleacetic acid oxidases and sexual expression patterns. *Plant Physiol.* **94**, 1535–1541.

Lourtioux, A. (1961) Action des éclairements colorés sur la floraison de *Myosotis palustris*. *Prog. Photobiol. Proc. Int. Congr. 3rd, Copenhagen* **1960**, 406–408.

Loveys, B.R., Leopold, A.C and Kriedemann, P.E. (1974) Abscisic acid metabolism and stomatal physiology in *Betula lutea* following alteration in photoperiod. *Ann. Bot.* **38**, 85–92.

Lozhnikova, V.N., Krekule, J., Dudko, N. and Chailakhyan, M.Kh. (1986) Fluctuation of endogenous cytokinins in leaves and roots of short-day and long-day tobacco associated with photoperiodic induction. *Biol. Plant.* **28**, 43–46.

Lumsden, P.J. (1984) Photoperiodic control of floral induction in *Pharbitis nil*. D. Phil. Thesis, University of Sussex.

Lumsden, P.J. (1991) Circadian rhythms and phytochrome. *Ann. Rev. Plant Physiol. Plant Mol. Biol.* **42**, 351–371.

Lumsden, P.J. and Furuya, M. (1986) Evidence for two actions of light in the photoperiodic induction of flowering in *Pharbitis nil*. *Plant Cell Physiol.* **27**, 1541–1551.

Lumsden, P.J. and Vince-Prue, D. (1984) The perception of dusk signals in photoperiodic time-measurement. *Physiol. Plant.* **60**, 427–432.

Lumsden, P.J., Thomas, B. and Vince-Prue, D. (1982) Photoperiodic control of flowering in dark-grown seedlings of *Pharbitis nil* Choisy. The effect of skeleton and continuous light photoperiods. *Plant Physiol.* **70**, 277–282.

Lumsden, P.J., Yamamoto, K.T., Nagatani, A. and Furuya, M. (1985) Effect of monoclonal antibodies on the *in vitro* Pfr dark-reversion of pea phytochrome. *Plant Cell Physiol.* **26**, 1313–1322.

Lumsden, P.J., Vince-Prue, D. and Furuya, M. (1986) Phase shifting of the photoperiodic response rhythm in *Pharbitis nil* by red-light pulses. *Physiol. Plant.* **67**, 604–607.

Lumsden, P.J., Saji, H. and Furuya, M. (1987) Action spectra confirm two separate actions of phytochrome in the induction of flowering in *Lemna paucicostata* 441. *Plant Cell Physiol.* **28**, 1237–1242.

Lumsden, P.J., Thomas, B. and Vince-Prue, D. (1994) Photoperiodic time-measurement in *Pharbitis nil*. *Flowering Newsletter* **18**, 54–57.

Lumsden, P.J., Youngs, J.A., Thomas, B. and Vince-Prue, D. (1995) Evidence that photoperiodic, dark time-measurement in *Pharbitis nil* involves a circadian rather than a semidian rhythm. *Plant Cell Environ.* **18**, 1403–1410.

Luna, V., Lorenzo, E, Reinoso, H., Tordable, M.C., Abdala, G., Pharis, R.P. and Bottini, R. (1990) Dormancy in peach (*Prunus persica* L) flower buds. I. Floral morphogenesis and endogenous gibberellins at the end of the dormancy period. *Plant Physiol.* **93**, 20–25.

Lyndon, R.F., Jacqmard, A., Bernier, G. (1983) Changes in protein composition of the shoot meristem during floral evocation in *Sinapis alba*. *Physiol. Plant.* **59**, 476–480.

Macháčková, I., Krekule, J., Souckova, D., Zdenek, P. and Ullman, J. (1985) Reversal of IAA-induced inhibition of flowering by aminoethoxyvinyl glycine in *Chenopodium*. *J. Plant Growth Regul.* **4**, 203–209.

Macháčková, I., Ullman, J., Krekule, J. and Stock, M. (1988) Ethylene production and metabolism of 1-aminocyclopropane-1-carboxylic acid in *Chenopodium rubrum* L as influenced by photoperiodic flower induction. *J. Plant Growth Regul.* **7**, 241–247.

Macháčková, I., Krekule, J., Eder, J., Seidlova, F. and Strnad, M. (1993). Cytokinins in photoperiodic induction of flowering in *Chenopodium* species. *Physiol. Plant.* **87**, 160–166.

Mader, J.C. (1995) Polyamines in *Solanum tuberosum in-vitro* – free and conjugated polyamines in hormone-induced tuberization. *J. Plant Phys.* **146**, 115–120.

Magruder, R. and Allard, H.A. (1937) Bulb formation in onions and length of day. *J. Agric. Res.* **54**, 715–752.

Malepszy, S. and Niemirowicz-Szczytt, K. (1991) Sex determination in cucumber (*Cucumis sativus*) as a model system for molecular biology. *Plant Science* **80**, 39–47.

Mancinelli, A.L. (1988) Some thoughts about the use of predicted values of the state of phytochrome in plant photomorphogenesis research. *Plant Cell Environ.* **11**, 429–439.

Mancinelli, A.L. and Rabino, I. (1978) The high irradiance-response of plant photomorphogenesis. *Bot. Rev.* **44**, 129–180.

Mandoli, D.F. and Briggs, W.R. (1981) Phytochrome control of two low-irradiance responses in etiolated oat seedlings. *Plant Physiol.* **67**, 733–739.

Margara, J. (1960) Recherches sur le déterminisme de l'élongation et de la floraison dans la genre *Beta*. *Annls. Inst. Natn. Rech. Agron. Paris B* **10**, 361–471.

Margara, J. and Touraud, G. (1967) Recherches exprimentales sur la néoformation de bourgeons inflorescentiels ou végétatifs *in vitro* a partir d'explantats d'endive (*Cichorium intybus* L) IV. *Ann. Physiol. Vég.* **9**, 339–347.

Marks, M.K. and Prince, S.D. (1979) Induction of flowering in wild lettuce (*Lactuca serriola* L.) II. Devernalization. *New Phytol.* **82**, 357–363.

Martin, C., Vernoy, R. and Paynot, M. (1982) Photopériodisme, tubérisation, floraison et phénolamides. *C.R. Acad. Sci. Paris Ser. III* **295**, 565–568.

Mauk, C.S. and Langille, A.R. (1978) Physiology of tuberization in *Solanum tuberosum* L. Ciszeatin riboside in the potato plant: its identification and changes in endogenous levels as influenced by temperature and photoperiod. *Plant Physiol.* **62**, 438–442.

McCreary, D.D., Tanaka, Y. and Lavender, D.P. (1978) Regulation of Douglas-fir seedling growth and hardiness by controlling photoperiod. *Forest Sci.* **24**, 142–152.

McDaniel, C.N. and Hartnett, L.K. (1993) Floral stimulus activity in tobacco stem pieces. *Planta* **189**, 577–583.

McDaniel, C.N., Sangrey, K.A. and Jegla, D.E. (1989) Cryptic floral determination; stem explants from vegetative tobacco plants have the capacity to regenerate floral shoots. *Dev. Biol.* **134**, 473–478.

McKinney, H.H. and Sando, W.J. (1935) Earliness of sexual reproduction as influenced by temperature and light in relation to growth phases. *J. Agric. Res.* **51**, 621–641.

McMurray, L. and Hastings, J.W. (1972) No desynchronisation among four circadian rhythms in the unicellular alga *Goyaulax polyhedra*. *Science* **175**, 1137–1139.

McWha, J.A. and Langer, H.J. (1979) The effects of exogenously applied abscisic acid on bud burst in *Salix* spp. *Ann. Bot.* **44**, 47–55.

Medford, J.I., Horgan, R., El Sawi, Z. and Klee, H.J. (1989) Alterations of endogenous cytokinins in transgenic tobacco using a chimeric isopentenyl transferase gene. *Plant Cell* **1**, 403–413.

Mellerowicz, E.J., Coleman, W.K., Riding, R.T and Little, C.H.A. (1992) Periodicity of cambial activity in *Abies balsamea* I. Effects of temperature and photoperiod on cambial dormancy and frost hardiness. *Physiol. Plant.* **85**, 515–525.

Melzer, S., Majewski, D.M. and Apel, K. (1990) Early changes in gene expression during the transition from vegetative to generative growth in the long-day plant *Sinapis alba*. *The Plant Cell* **2**, 953–961.

Melzer, S., Menzel, G. and Kania, T. (1995) Molecular analysis of the transition to flowering in *Sinapis alba*. *Flowering Newsletter* **19**, 28–30.

Menhenett, R. and Wareing, P.F. (1977) Effects of photoperiod and temperature on the growth and cytokinin content of two populations of *Dactylis glomerata* (cocksfoot). *New Phytol.* **78**, 17–25.

Merritt, R.H. and Kohl, H.C. (1983) Crop productivity efficiency of petunias in the greenhouse. *J. Amer. Soc. Hort. Sci.* **108**, 544–548.

Metzger, J.D. (1988) Localization of the site of perception of thermoinductive temperatures in *Thlaspi arvense* L. *Physiol. Plant.* **88**, 424–428.

Metzger, J.D. and Zeevaart, J.A.D. (1980) The effect of photoperiod on the levels of endogenous gibberellins in spinach as measured by combined gas chromatography-selected ion current monitoring. *Plant Physiol.* **66**, 844–846.

Metzger, J.D. and Zeevaart, J.A.D. (1982) Photoperiodic control of gibberellin metabolism in spinach. *Plant Physiol.* **69**, 287–291.

Meyer, H., Thienel, U. and Piechulla, B. (1989) Molecular chracterization of the diurnal/circadian expression of the chlorophyll a/b binding proteins in leaves of tomato and other dictotyledonous and monocotyledonous plant species. *Planta* **180**, 5–15.

Michniewicz, M. and Kamienska, A. (1965) Flower formation induced by kinetin and vitamin F treatment in long-day plant (*Arabidopsis thaliana*) grown in short-day. *Naturwissenschaft* **52**, 623.

Milborrow, B.V. (1967) The identification of (+)-abscisin II ((+)-dormin) in plants and the measurement of its concentrations. *Planta* **76**, 93–113.

Millar, A.J., Carré, I.A., Stayer, C.A., Chua, N.-H. and Kay, S.A. (1995a) Circadian clock mutants in *Arabidopsis* identified by luciferase imaging. *Science* **267**, 1161–1163.

Millar, A.J., Straume, M., Chory, J., Chua, Nam-Hai and Kay, S.A. (1995b) The regulation of circadian period by phototransduction pathways in *Arabidopsis*. *Science* **267**, 1163–1166.

Miller, J. and Ross, C.W. (1966) Inhibition of leaf processes by *p*-fluorophenylalanine during induction of flowering in the cocklebur. *Plant Physiol.* **41**, 1185–1192.

Mirolo, C., Bodson, M. and Bernier, G. (1990) Floral induction of *Xanthium strumarium* in long days. *Ann. Bot.* **66**, 475–477.

Mita, T. and Shibaoka, H. (1984) Effects of S-3307, an inhibitor of gibberellin biosynthesis, on swelling of leaf sheath-cells and on the arrangement of cortical microtubules in onion seedlings. *Plant Cell Physiol.* **25**, 1531–1539.

Mitya, H.G. and Lanphear, F.O. (1971) Factors influencing the cold hardiness of *Taxus cuspidata* roots. *J. Amer. Soc. Hort. Sci.*, **96**, 83–86.

Mohnen, D., Eberhard, S., Marfa, V., Darvill, A. and Albersheim, P. (1990) Summary of research at the Complex Carbohydrate Research Center on the effects of oligosaccharins on *de novo* flower formation in tobacco thin cell-layer explants. *Flowering Newsletter* **9**, 16–19.

Montavon, M. and Greppin, H. (1985) Potentiel intracellulaire du mésophylle d'épinard (*Spinacia oleracea* L. cv Nobel) en relation avec la lumière et l'induction florale. *J. Plant Physiol.* **118**, 471–475.

Moore, J.N. (1970) Cytokinin-induced sex conversion in male clones of *Vitis* species. *J. Am. Soc. Hort. Sci.* **95**, 387–393.

Moore, P.H. (1985) *Saccharum*. In *Handbook of Flowering* vol. IV (ed. A.H. Halevy), pp. 243–262. CRC Press, Boca Raton.

Moore, P.H., Reid, H.B. and Hamner, K.C. (1967) Flowering responses of *Xanthium pensylvanicum* to long dark periods. *Plant Physiol.* **42**, 503–509.

Morgan, D.G. and Morgan, C.B. (1984) Photoperiod and the abscission of flower buds in *Phaseolus vulgaris*. In *Light and the Flowering Process* (eds D. Vince-Prue, B. Thomas and K.E. Cockshull), pp. 227–240. Academic Press, London.

Morgan, P.W. (1994) Regulation of flowering in *Sorghum*. *Flowering Newsletter* **17**, 3–11.

Mori, H. (1979) Effect of far-red light pulse on induction and production of flowers in *Lemna paucicostata* 6746 in darkness. *Plant Cell Physiol.* **20**, 639–647.

Morris, J.W., Doumas, P., Morris, R.O. and Zaerr, J.B. (1990) Cytokinins in vegetative and reproductive buds of *Pseudotsuga menziezii*. *Plant Physiol.* **93**, 67–71.

Morse, M.J., Crain, R.C. and Satter, R.L. (1988) Light stimulated inositol phospholipid turnover in *Samanea saman*. *Proc. Nat. Acad. Sci. USA* **86**, 7075–7078.

Mosaad, M.G., Ortiz-Ferrara, G., Mahalakshmi, V. and Fischer, R.A. (1995) Phyllochron response to vernalization and photoperiod in spring wheat. *Crop Sci.* **35**, 168–171.

Moshkov, B.S. (1935) Photoperiodismus und Frosthärte ausdauender Gewächse. *Planta* **23**, 774–803.

Moysset, L. and Simon, E. (1989) Role of calcium in phytochrome-controlled nyctinastic movements of *Albizzia lophantha* leaflets. *Plant Physiol.* **90**, 1108–1114.

Mozley, D. and Thomas, B. (1995) Developmental and photobiological factors affecting photoperiodic induction in *Arabidopsis thaliana* Heynh. Landsberg erecta. *J. Exp. Bot.* **46**, 173–179.

Müller-Röber, B., Sonnewald, U. and Willmitzer, L. (1992) Inhibition of the ADP-glucose pyrophosphorylase in transgenic potatoes leads to sugar-storing tubers and influences tuber formation and expression of tuber storage protein genes. *EMBO J.* **11**, 1229–1238.

Mumford, F.E. and Jenner, E.L. (1966) Purification and characterization of phytochrome from oat seedlings. *Biochem.* **5**, 3657–3662.

Murfet, I.C. (1971a) Flowering in *Pisum*: a three gene system. *Heredity* **27**, 93–110.

Murfet, I.C. (1971b) Flowering in *Pisum*: reciprocal grafts between known genotypes. *Aust. J. Biol. Sci.* **24**, 1089–1101.

Murfet, I.C. (1985) *Pisum sativum.* In *The Handbook of Flowering*, Vol. IV (ed. A.H. Halevy), pp. 97–126. CRC Press, Boca Raton, Florida.

Murfet, I.C. (1987) Linkage data for dne and chromosome 3 markers b and st. *Pisum Newsletter*, **19**, 45.

Murfet, I.C. (1990) Flowering genes in pea and their use in breeding. *Pisum Newsletter* **22**, 78–86.

Murfet, I.C. and Reid, J.B. (1987) Flowering in *Pisum*; gibberellins and the flowering genes. *J. Plant Physiol.* **127**, 23–29.

Murneek, A.E. (1940) Length of the day and temperature effects in *Rudbeckia. Bot. Gaz.* **102**, 269–279.

Nagao M. and Okagami, N. (1966) Effect of (2-chloroethyl) trimethylammonium chloride on the formation and dormancy of aerial tubers of *Begonia evansiana. Bot. Mag. Tokyo.* **79**, 687–692.

Nagao, M., Esashi, Y, Tanaka, T., Kumagai, T. and Fukumoto, S. (1959) Effects of photoperiod and gibberellin on the germination of seeds of *Begonia evansiana* Andr. *Plant Cell Physiol.* **1**, 39–47.

Nagatani, A., Jenkins, G.I. and Furuya, M. (1988) Non-specific association of phytochrome to nuclei during isolation from dark-grown pea (*Pisum sativum* cv Alaska) plumules. *Plant Cell Physiol.* **29**, 1141–1145.

Nagatani, A., Reed, J.W. and Chory, J. (1993) Isolation and initial characterisation of *Arabidopsis* mutants that are deficient in phytochrome A. *Plant Physiol.* **102**, 269–277.

Nakayama, S., Borthwick, H.A. and Hendricks, S.B. (1960) Failure of photoreversible control of flowering in *Pharbitis nil. Bot. Gaz.* **121**, 237–243.

Nanda, K.K. and Jindal, R.K. (1975) Summation of the effect of sub-threshold gibberellin treatment and sub-threshold photo-induction on flowering of *Impatiens balsamina. Indian J. Plant Phys.* **18**, 109–114.

Nanda, K.K., Toky, K.L. and Lata, K. (1969a) Effects of gibberellins A_3, A_{4+7} and A_{13} and of (−)-kaurene on flowering and extension growth of *Impatiens balsamina* under different photoperiods. *Planta* **86**, 134–141.

Nanda, K.K., Toky, K.L., Sawhney, B. and Tewan, N. (1969b) Role of light and dark in the flowering of *Impatiens balsamina. Indian J. Plant. Phys.* **12**, 89–101.

Napp-Zinn, K. (1984) Light and vernalization. In *Light and the Flowering Process* (eds D. Vince-Prue, B. Thomas and K.E. Cockshull), pp. 75–88. Academic Press, London.

Naylor, A.W. (1941a) Effect of nutrition and age upon rate of development of terminal staminate inflorescences of *Xanthium pennsylvanicum. Bot. Gaz.* **103**, 342–353.

Naylor, A.W. (1961) The photoperiodic control of plant behaviour. In *Encyclopedia of Plant Physiology*. Vol. 16 (ed. W. Ruhland), pp. 331–389. Springer-Verlag, Berlin, Heidelberg.

Naylor, F.L. (1941b) Effect of length of induction period on floral development of *Xanthium pennsylvanicum. Bot. Gaz.* **103**, 146–154.

Neuhaus, G., Bowler, C., Kern, R. and Chua, N.H. (1993) Calcium/calmodulin-dependent and -independent phytochrome signal transduction pathways. *Cell* **73**, 937–952.

Newman, I. A. (1981) Rapid electric responses of oats to phytochrome show membrane processes unrelated to pelletability. *Plant Physiol.* **68**, 1494–1499.

Neyland, M, Ng, Y.L. and Thimann, K.V. (1963) Formation of anthocyanin in leaves of *Kalanchoé blossfeldiana* – a photoperiodic response. *Plant Physiol.* **38**, 447–451.

Nilsen, E.T. and Muller, W.H. (1982) The influence of photoperiod on drought induction of dormancy in *Lotus scoparius*. *Oecologia* **53**, 79–83.

Ninneman, H. (1979) Photoreceptors for circadian rhythms. *Photochem. Photobiol. Rev.* **4**, 207–266.

Nitsch, J.P. (1957) Photoperiodism in woody plants. *Proc. Am. Soc. Hort. Sci.* **70**, 526–544

Nitsch, J.P. (1959) Réactions photopériodiques chez les plantes ligneuses. *Bull. Soc. Bot. Fr.* **106**, 259–287.

Nitsch, J.P. (1962) Photoperiodic regulation of growth in woody plants. In *Advances in Horticultural Science and their Applications* (Proc. 16th Int. Hort. Congr.) III, pp. 14–22. Pergamon Press, Oxford.

Nitsch, J.P. (1965) Existence d'un stimulus photopériodique non spécifique capable de provoquer la tubérisation chez *Helianthus tuberosus* L. *Bull. Soc. Bot. Fr.* **112**, 333–340.

Nitsch J.P. and Nitsch, C. (1960) Le problème de l'action des auxines sur la division cellulaire: présence d'un cofacteur de division dans la tubercle de topinambour. *Ann. Physiol. Vég.* **2**, 261–268.

Nitsch, C. and Nitsch, J.P. (1967) The induction of flowering *in vitro* in stem segments of *Plumbago indica* L. II. The production of reproductive buds. *Planta* **72**, 371–384.

Nitsch, C. and Nitsch, J.P. (1969) Floral induction in a short-day plant, *Plumbago indica* L. by 2-chloroethanephosphonic acid. *Plant Physiol.* **44**, 1747–1748.

Nitsch, J.P. and Somogyi, L. (1958) Le photopériodisme des plantes ligneuses. *Ann. Sci. Nat. Hort. France* **16**, 466–490.

Njus, D., Sulzman, F.M. and Hastings, J.W. (1974) Membrane model for the circadian clock. *Nature* **248**, 116–119.

Nojiri, N., Yamane, H., Sewto, H., Yamaguchi, I., Murofushi, N., Yoshihara, T. and Shibaoka, H. (1992) Qualitative and quantitative analysis of endogenous jasmonic acid in bulbing and nonbulbing onion plants. *Plant Cell Physiol.* **33**, 1225–1231.

Nojiri, H., Toyomasu, T., Yamane, H., Shibaoka, H., Murofushi, N. (1993) Qualitative and quantitative analysis of endogenous gibberellins in onion plants and their effects on bulb development. *Biosci. Biotech. Biochem.* **57**, 2031–2035.

Noma, M., Koike, M., Sano, M. and Kawashima, N. (1984) Endogenous indole-3-acetic acid in the stem of tobacco in relation to flower neoformation as measured by mass spectrometry. *Plant Physiol.* **75**, 257–260.

Nongkynrih, P. and Sharma, V.K. (1992) Biological clocks: mechanisms and developments. *J. Photochem. Photobiol. B: Biol.* **13**, 201–217.

Nougarede, A., Rembur, J, Francis, D. and Rondet, P. (1989) Ageing of the *Silene coeli-rosa* L shoot apex under non-inductive conditions: changes in morphology, mitotic index and polypeptide composition. *Protoplasma* **153**, 30–36.

Nwoke, F.I.O. (1986) Effects of intercalated long days on photoperiodic induction and development of flowers and fruits in *Abelmoschus esculentus* (L.) Moench. *J. Plant Physiol.* **125**, 417–425.

Oda, Y. (1969) The action of skeleton photoperiods on flowering in *Lemna perpusilla*. *Plant Cell Physiol.* **10**, 399–409.

Odén, P.-C. and Dunberg, A. (1984) Abscisic acid in shoots and roots of Scots pine (*Pinus sylvestris* L.) grown in controlled long-day and short-day environments. *Planta* **161**, 148–155.

Odén, P.-C. and Heide, O.M. (1989) Quantitation of gibberellins and indoleacetic acid in *Begonia* leaves: relationship with environment, regeneration and flowering. *Physiol. Plant.* **76**, 500–506.

Ofir, M. and Kerem, D. (1982) The effects of temperature and photoperiod on the onset of summer dormancy in *Poa bulbosa* L. *Ann. Bot.* **50**, 259–264.

Ogawa, Y. and King, R.W. (1979a) Establishment of photoperiodic sensitivity by benzyladenine and a brief red irradiation in dark grown seedlings of *Pharbitis nil* Chois. *Plant Cell Physiol* **20**, 115–122.

Ogawa, Y. and King, R.W. (1979b) Indirect action of benzyladenine and other chemicals on flowering on *Pharbitis nil* Chois. *Plant Physiol.* **63**, 643–649.

Ogawa, Y. and King, R.W. (1990) The inhibition of flowering by non-induced cotyledons of *Pharbitis nil*. *Plant Cell Physiol.* **31**, 129–135.

Ohtani, T. and Ishiguri, Y. (1979) Inhibitory action of blue and far-red light in the flowering of *Lemna paucicostata*. *Physiol. Plant.* **47**, 255–259.

Ohtani, T. and Kumagai, T. (1980) Spectral sensitivity of the flowering response in green and etiolated *Lemna paucicostata* T-101. *Plant Cell Physiol.* **21**, 1335–1338.

Okagami, N., Esashi, Y. and Nagao, M. (1977) Gibberellin-induced inhibition and promotion of sprouting in aerial tubers of *Begonia evansiana* and in relation to photoperiodic treatment and tuber storage. *Planta* **136**, 1–6.

Okazawa, Y. (1967) Physiological studies on the tuberization of potato plants. *J. Fac. Agric. Hokkaido Univ.* **55**, 267–336.

Oleksyn, J., Tjoelker, M.G. and Reich, P.B. (1992) Growth and biomass partioning of populations of European *Pinus sylvestris* L. under simulated 50° and 60° N daylengths: evidence for photoperiodic ecotypes. *New Phytol.* **120**, 561–574.

Olsen, J.E., Jensen, E., Junttila, O. and Moritz, T. (1995a) Photoperiodic control of endogenous gibberellins in seedlings of *Salix pentandra*. *Physiol. Plant.* **93**, 639–644.

Olsen, J.E., Moritz, T.k., Jensen, E. and Junttila, O. (1995b) Lack of effect of photoperiod on metabolism of exogenous GA_{19} and GA_1 in *Salix pentandra* seedlings. *Physiol. Plant.* **94**, 522–528.

Ombrello, T.M. and Garrison, S.A. (1987) Endogenous gibberellins and cytokinins in spear tips of *Asparagus officinalis* in relation to sex expression. *J. Am. Soc. Hort. Sci.* **112**, 539–544.

O'Neill, S.D. (1989) Molecular analysis of floral induction in *Pharbitis nil*. In *Plant Reproduction: From Floral Induction to Pollination. Proceedings of the 12th Annual Riverside Symposium in Plant Physiology* (eds E. Lord and G. Bernier), pp. 19–28. The American Society of Plant Physiologists.

O'Neill, S.D. (1993) Changes in gene expression associated with floral induction and evocation. In *The Molecular Biology of Flowering* (ed. B.R. Jordan), pp. 69–92. C.A.B. International, Wallingford, UK.

O'Neill, S.D., Zhang, X.S. and Zheng, C.C. (1994) Dark and circadian regulation of mRNA accumulation in the short-day plant *Pharbitis nil*. *Plant. Physiol.* **104**, 569–580.

Ono, M., Okazaki, M., Harada, H. and Hirofumi, U. (1988) *In vitro* translated polypeptides of different organs of *Pharbitis nil* Chois strain Violet under flower-inductive and noninductive conditions. *Plant Sci.* **58**, 1–7.

Ono, M., Ono, K.S., Yasui, M., Okasaki, M. and Harada, H. (1993) Changes in polypeptides in *Pharbitis* cotyledons during the first flower-inductive photoperiod. *Plant Sci.* **89**, 135–145

Oota, Y. (1983a) Floral inhibition in *Lemna paucicostata* 6746 due to night interruption. *Plant Cell Physiol.* **24**, 327–332.

Oota, Y. (1983b) Physiological structure of the critical photoperiod of *Lemna paucicostata*. *Plant Cell Physiol.* **24**, 1503–1510.

Oota, Y. (1984) Physiological function of night-interruption in *Lemna paucicostata* 6746: action of light as a phaser on the photoperiodic clock. *Plant Cell Physiol.* **25**, 323–331.

Oota, Y. (1985) Measurement of the critical nyctoperiod by *Lemna paucicostata* 6746 grown in continuous light. *Plant Cell Physiol.* **26**, 923–929.

Oota, Y. and Kondo, J. (1974) Removal by cyclic AMP of the inhibition of duckweed flowering due to ammonium and water treatment. *Plant Cell Physiol.* **15**, 403–411.

Oota, Y. and Nakashima, H. (1978) Photoperiodic flowering in *Lemna gibba* G3 – Time measurement. *Bot. Mag. Tokyo.* Special Issue No. **1**, 177–198.

Ormrod, J.C. and Bernier, G. (1987). Cell cycle changes in the shoot apical meristem of *Lolium temulentum* cv. Ceres during the transition to flowering. *Arch. Int. Physiol. Biochim.* **95**, 17.

Osborne, D.J. (1965) Interactions of hormonal substances in the growth and development of plants. *J. Sci. Food Agric.* **16**, 1–13.

Otto, V., Schafer, E., Nagatani, A., Yamamoto, K.T. and Furuya, M. 1984. Phytochrome control of its own synthesis in *Pisum sativum*. *Plant Cell Physiol.* **25**, 1579–1584.

Palmer, C.E. and Barker, W.G. (1973) Influence of ethylene and kinetin on tuberization and enzyme activity in *Solanum tuberosum* L. stolons cultured *in vitro*. *Ann. Bot.* **37**, 85–93.

Palmer, R.G. and Kilen, T.C. (1987) Qualitative genetics and cytogenetics. In *Soybeans: Improvement, Production and Uses* (ed. J.R. Wilcox), pp. 135–209. American Society of Agronomy, Madison, Wisconsin.

Palmer, C.E. and Smith, O.E. (1969a) Cytokinins and tuber initiation in the potato *Solanum tuberosum* L. *Nature* **221**, 279–280.

Palmer, C.E. and Smith, O.E. (1969b) Effect of abscisic acid on elongation and kinetin induced tuberization of isolated stolons of *Solanum tuberosum* L. *Plant Cell Physiol.* **10**, 657–664.

Pao C.I. and Morgan, P.W. (1986) Genetic-regulation of development in *Sorghum bicolor*. 1. Role of the maturity genes. *Plant Physiol.* **82**, 575–580.

Papafotiou, M. and Schwabe, W.W. (1990) Studies on the long-day inhibition of flowering in *Xanthium* and *Kalanchoë*. *Physiol. Plant.* **80**, 177–184.

Papenfuss, H.D. and Salisbury, F.B. (1967) Aspects of clock resetting in flowering of *Xanthium*. *Plant Physiol.* **42**, 1562–1568.

Paré, J.R., ApSimon, J.W. and Wightman, F. (1989) Kinetic studies on the photoinduction of *Xanthium strumarium* L and isolation of a flower-primordia growth-regulating fraction. *Plant Cell Physiol.* **30**, 1145–1152.

Parker, M.W. and Borthwick, H.A. (1940) Floral initiation in Biloxi soy-beans as influenced by photosynthetic activity during the induction period. *Bot. Gaz.* **102**, 256–268.

Parker, M.W., Hendricks, S.B., Borthwick, H.A. and Scully, N.J. (1946) Action spectrum for the photoperiodic control of floral initiation of short-day plants. *Bot. Gaz.* **108**, 1–26.

Parker, M.W., Hendricks, S.B., Borthwick, H.A. and Went, F.W. (1949) Spectral sensitivities of leaf and stem growth in etiolated pea seedlings and their similarity to action spectra for photoperiodism. *Am. J. Bot.* **36**, 194–204.

Parker, M.W., Hendricks, S.B. and Borthwick, H.A. (1950) Action spectrum for the photoperiodic control of floral initiation of the long day plant *Hyoscyamus niger*. *Bot. Gaz.* **111**, 242–252.

Parks, B.M. and Quail, P.H. (1993) *hy8*, a new class of *Arabidopsis* long hypocotyl mutants deficient in functional phytochrome A. *The Plant Cell* **5**, 39–48.

Paton, D.M. (1971) Photoperiodic induction of flowering in the late pea cultivar Greenfeast: the role of exposed cotyledons and leaves. *Aust. J. Biol. Sci.* **24**, 609–618.

Paton, D.M. and Willing, R.R. (1968) Bud dormancy in *Populus*. *Aust. J. Biol. Sci.* **21**, 157–159

Peacock, J.T. and MacMillan, C. (1968) The photoperiodic response of American *Prosopis* and *Acacia* from a broad latitudinal distribution. *Amer. J. Bot.* **55**, 153–159.

Penel, C., Auderset, G., Bernardini, N., Castillo, F.J., Greppin, H. and Morre, D.J. (1988) Compositional changes associated with plasma membrane thickening during floral induction of spinach. *Physiol. Plant.* **731**, 134–146.

Périlleux, G., Bernier, G. and Kinet, J.-M. (1994) Circadian rhythms and the induction of flowering in the long-day grass *Lolium temulentum* L. *Plant Cell Environ.* **17**, 755–761.

Perl, A., Aviv, D., Willmitzer, L. and Galun, E. (1991) *In vitro* tuberization in transgenic potatoes harboring beta-glucuronidase linked to a patatin promoter – effects of sucrose levels and photoperiods. *Plant Sci.* **73**, 87–95.

Perry, T.O. and Hellmers, H. (1973) Effects of ABA on growth and dormancy of two races of red maple (*Acer rubrum*) *Bot. Gaz.* **134**, 283–288.

Peterson, C.E. and Andher, L.D. (1960) Induction of staminate flowers in gynoecious cucumbers with GA_3. *Science NY* **131**, 1673–1674.

Peterson, E.L. and Saunders, D.S. (1980) The circadian eclosion rhythm in *Sarcophaga agyrostoma*: a limit cycle representation of the pacemaker. *J. Theor. Biol.* **86**, 265–277.

Peterson, R.L. and Yeung, E.C. (1972) Effect of two gibberellins on species of the rosette plant *Hieracium*. *Bot. Gaz.* **133**, 190–198.

Pharis, R.P. (1972) Flowering of chrysanthemum under non-inductive long-days by gibberellins and by N^6-benzyladenine. *Planta* **105**, 205–212.

Pharis, R.P. and King, R.W. (1985) Gibberellins and reproductive development in seed plants. *Ann. Rev. Plant Physiol.* **36**, 517–568.

Pharis, R.P., Ruddat, M., Glenn, J.L. and Morf, W. (1970) A quantitative requirement for long day in the induction of staminate strobili by gibberellin in the conifer *Cupressus arizonica*. *Can. J. Bot.* **48**, 653–658.

Pharis, R.P., Evans, L.T., King, R.W. and Mander, L.N. (1987a) Gibberellins, endogenous and applied, in relation to flower induction in the long-day plant *Lolium temulentum*. *Plant Physiol.* **84**, 1132–1138.

Pharis, R.P., Webber, J.E. and Ross, D. (1987b) The promotion of flowering in forest trees by gibberellin $A_{4/7}$ and cultural treatments: a review of the possible mechanisms. *For. Ecol. Manage.* **19**, 65–84.

Pharis, R.P., Evans, L.T., King, R.W. and Mander, L.M. (1989) Gibberellins and flowering in higher plants; differing structures yield highly specific effects. In *From Floral Induction to Pollination* (eds E. Lord and G. Bernier), pp. 29–41. American Society of Plant Physiologists Symposium Series, Vol. 1.

Phillips, D.A. and Cleland, C.F. (1972) Cytokinin activity from the phloem sap of *Xanthium strumarium* L. *Planta* **102**, 173–178.

Phillips, I.D.J., Miners, J. and Roddick, J.G. (1980) Effect of light and photoperiodic conditions on abscisic acid in leaves and roots of *Acer pseudoplatanus* L. *Planta* **149**, 118–122.

Phillips, J.F. (1941) Effect of daylength on dormancy in tree seedlings. *J. For.* **39**, 55–59.

Pike, L.M. and Peterson, C.E. (1969) Gibberellin A_4/A_7 for the induction of staminate flowers on the gynoecious cucumber (*Cucumis sativus*). *Euphytica* **18**, 106–109.

Piringer, A.A. and Cathey, H.M. (1960) Effects of photoperiod, kind of supplemental light, and temperature on growth and flowering of petunia plants. *Proc. Am. Soc. Hort. Sci.* **76**, 649–660.

Pittendrigh, C. (1966) The circadian oscillation in *Drosophila pseudo-obscura* pupae: a model for the photoperiodic clock. *Z. Pflanzenphysiol.* **54**, 275–307.

Pittendrigh, C. (1979) Some functional aspects of circadian pacemakers. In *Biological Rhythms and their Central Mechanisms* (eds M. Suda, O. Hayaishi and H. Nakagawa), pp. 3–12. Elsevier/North-Holland, Amsterdam.

Plancher, B. and Naumann, W.D. (1978) Influence of water supply and day length on abscisic acid content of *Fragaria ananassa*. *Gartenbauwissenschaft.* **43**, 126–136.

Pohjakallio, O. (1953) On the effect of daylength on the yield of potato. *Physiol. Plant.* **6**, 140–149.

Pohjakallio O., Salonen, A. and Antila, S. (1957) Analysis of earliness in the potato. *Acta Agric. Scand.* **7**, 361–388.

Pollard, D.F.W. and Portlock, F.T. (1984) The effects of photoperiod and temperature on gibberellin $A_{4/7}$ induced strobolus production of western hemlock. *Can. J. For. Res.* **14**, 291–294.

Pourrat, Y. and Jacques, R. (1975) The influence of photoperiodic conditions received by the mother plant on morphological and physiological characteristics of *Chenopodium polyspermum* L. seeds. *Plant Sci. Lett.* **4**, 274–279.

Powell, M.C. and Bunt, A.C. (1978) Leaf production and growth in *Begonia* × *hiemalis* under long and short days. *Scientia Hort.* **8**, 289–296.

Pratt, L.H. (1979) Phytochrome: Function and Properties. In *Photochemical and Photobiological Reviews* (ed. K.C. Smith) Vol. 4, pp. 59–124. Plenum Press, New York.

Pratt, L. H. (1994) Distribution and localization of phytochrome within the plant. In *Photomorphogenesis in Plants* 2nd edn (eds R.E. Kendrick and G.H.M. Kronenberg), pp. 163–185. Kluwer Academic Publishers, Dordrecht.

Pratt, L.H. (1995) Phytochromes – differential properties, expression patterns and molecular evolution. *Photochem. Photobiol.* **61**, 10–21.

Pressman, E. and Negbi, M. (1980) The effect of daylength on the response of celery to vernalization. *J. Exp. Bot.* **31**, 1291–1296.

Principe, J.M., Hruschka, W.R., Thomas, B. and Deitzer, G.F. (1992) Protein differences between two isogenic cultivars of barley (*Hordeum vulgare* L.) that differ in sensitivity to photoperiod and far-red light. *Plant Physiol.* **98**, 1444–1450.

Proebsting, W.M. and Heftmann, E. (1980) The relationship of (^{3}H)GA$_9$ metabolism to photoperiod-induced flowering in *Pisum sativum* L. *Z. Pflanzenphysiol.* **98**, 305–309.

Purse, J.G. (1984) Phloem exudate of *Perilla crispa* and its effects on flowering of *P. crispa* shoot explants. *J. Exp. Bot.* **35**, 227–238.

Purves, W.K. (1961) Dark reactions in the flowering of *Lemna perpusilla* 6746. *Planta* **56**, 684–690.

Purvis, O.N. (1940) Vernalization of fragments of embryo tissue. *Nature* **195**, 462.

Purvis, O.N. (1961) The physiological analysis of vernalization. In *Encyclopedia of Plant Physiology* Vol. 16 (ed. W. Ruhland), pp. 76–122. Springer-Verlag, Berlin.

Qamaruddin, M., Dormling, I., Ekberg, I., Eriksson, G. and Tillberg, E. (1993) Abscisic acid content at defined levels of bud dormancy and frost tolerance in two contrasting populations of *Picea abies* grown in a phytotron. *Physiol. Plant.* **87**, 203–210.

Qamaruddin, M., Ekberg, I., Dormling, I., Norell, L., Clapham, D. and Eriksson, G. (1995) Early effects of long nights on budset, bud dormancy and abscisic acid content in two populations of *Picea abies*. *For. Gen.* **2**, 207–216.

Quail, P.H. (1994) Phytochrome genes and their expression. In *Photomorphogenesis in Plants*, 2nd edition (eds R.E. Kendrick and G.H.M. Kronenberg) pp. 71–104. Kluwer Academic Publishers, Dordrecht.

Queiroz, O. (1970) Sur le métabolisme acide des Crassulacées IV. Réflexions sur les phénomènes oscillatoires au niveau enzymatique et sur la compartimentation métabolique, sous l'action du photopériodisme. *Physiol. Vég.* **8**, 75–110.

Rajeevan, M.S. and Lang, A. (1993) Flower-bud formation in explants of photoperiodic and day-neutral *Nicotiana* biotypes and its bearing on the regulation of flower formation. *Proc. Nat. Acad. Sci. USA* **90**, 4636–4640.

Ramin, A.A. and Atherton, J.G. (1994) Manipulation of bolting and flowering in celery (*Apium graveolens* L. var. dulce). III Effects of photoperiod and irradiance. *J. Hort. Sci.* **69**, 861–868.

Rane, F.W. (1894) Electroculture with the incandescent lamp. *W. Virginia Agric. Exp. Stn. Bull.* **37**, 1–27.

Raskin, I. (1992) Salicylate, a new plant hormone. *Plant Physiol.* **99**, 799–803.

Rasumov, V.I. (1933) The significance of the qualitative composition of light in photoperiodical response. *Bull. Appl. Bot. Genetics and Plant Breeding III. Biochem. and Anat. Plants* **3**, 217–251.

Ravnikar M., Zel J., Plaper I. and Spacapan A. (1993) Jasmonic acid stimulates shoot and bulb formation of garlic *in-vitro*. *J. Plant Growth Regul.* **12**, 73–77.

Redei, G. (1962) Supervital mutants of *Arabidopsis*. *Genetics* **47**, 443–460.

Reid, H.B., Moore, P.H. and Hamner, K.C. (1967) Control of flowering of *Xanthium pensylvanicum* by red and far-red light. *Plant Physiol.* **42**, 532–540.

Reid, J.B. and Murfet, I.C. (1977) Flowering in *Pisum*; the effect of genotype, plant age, photoperiod and number of inductive cycles. *J. Exp. Bot.*, **28**, 811–819.

Reid, J.B. and Murfet, I.C. (1984) Flowering in *Pisum* a fifth locus, Veg. *Ann. Bot.* **53**, 369–382.

Ren, X-c., Zhang, J-y., Luo, W-h. and Jin, S-P. (1982) Flower inhibitory effects of long-days preceding the inductive short-days on SDP *Hibiscus cannabinus* cv South-Selected. *Acta Phytophysiol. Sinica* **8**, 215–221.

Riesmeier, J.W., Willmitzer, L. and Frommer, W.B. (1994) Evidence for an essential role of the sucrose transporter in phloem loading and assimilate partitioning. *EMBO J.* **13**, 1–7.

Rinne, P. Saarelainan, A. and Junttila, O. (1994) Growth cessation and bud dormancy in relation to ABA levels in seedlings and coppice shoots of *Betula pubescens* as affected by a short photoperiod, water stress and chilling. *Physiol. Plant.* **90**, 451–458.

Roberts, E.H. and Summerfield, R.J. (1987) Measurement and prediction of flowering in annual crops. In *Manipulation of Flowering* (ed. J.G. Atherton), pp. 17–50. Butterworths, London.

Roberts, E.H., Hadley, P. and Summerfield, R.J. (1985) Effects of temperature and photoperiod on flowering in chickpea (*Cicer arietinum* L.). *Ann. Bot.* **55**, 881–892.

Robertson, C.R. (1995) The use of quantitative RT-PCR techniques to examine the expression of PHY genes: The role of phytochrome A in the photoperiodic induction of flowering in the long-day plant *Sinapis alba* and the short-day plant *Pharbitis nil*. PhD Thesis, University of Reading.

Robinson, L.W. and Wareing, P.F. (1969) Experiments on the juvenile-adult phase change in some woody species. *New Phytol.* **68**, 67–78.

Rombach, J. (1986) Phytochrome in Norflurazon-treated seedlings of *Pharbitis nil*. *Physiol. Plant.* **68**, 231–237.

Rombach, J., Bensink, J. and Katsura, N. (1982) Phytochrome in *Pharbitis nil* during and after de-etiolation. *Physiol. Plant.* **56**, 251–258.

Rood, S.B., Pharis, R.P. and Major, D.J. (1980) Changes of endogenous gibberellin-like substances with sex reversal of the apical inflorescence of corn. *Plant Physiol.* **66**, 793–796.

Ross, C. (1970) Antimetabolite studies and the possible importance of leaf protein synthesis during induction of flowering in the cocklebur. In *Cellular and Molecular Aspects of Floral Induction* (ed. G. Bernier), pp. 139–151. Longman, London.

Ross, S.D., Pharis, R.P. and Binder, W.D. (1983) Growth and regulators and conifers: their physiology and potential use in forestry. In *Plant Growth Regulating Chemicals* vol. 2 (ed. L.G. Nickell), pp. 35–78. CRC Press, Boca Raton.

Rudich, J., Halevy, A.H. and Kedar, N. (1972a) The level of phytohormones in monoecious and gynoecious cucumbers as affected by photoperiod and ethephon. *Plant Physiol.* **50**, 585–590.

Rudich, J., Halevy, A.H. and Kedar, N. (1972b) Ethylene evolution from cucumber plants as related to sex expression. *Plant Physiol.* **49**, 998–999.

Rünger, W. (1957) Untersuchungen über den Einfluss verscheiden langer Kurztags-Perioden nach dem Schnitt der Blattstecklinge auf die Entwicklung der adventiven Triebe von *Begonia* 'Konkurrent' und 'Marina'. *Gartenbauwiss.* **4**, 352–357.

Rünger, W. and Patzer, H. (1986) Interaction between duration of main light period and moment of night-break on flower formation of chrysanthemum. *Gartenbauwissen.* **51**, 37–39.

Ryan, C.A. (1987) Oligosaccharide signalling in plants. *Ann. Rev. Cell Biol.* **3**, 295–317.

Sachs, R.M. (1977) Nutrient diversion; an hypothesis to explain the chemical control of flowering. *Hort. Sci.* **12**, 220–222.

Sachs, R.M. (1985) *Cestrum*. In *Handbook of Flowering* vol II (ed. A.H. Halvery) pp. 176–184. CRC Press, Boca Raton.

Sachs, R.M. (1987) Roles of photosynthesis and assimilate partitioning in flower initiation. In *Manipulation of Flowering* (ed. J.G. Atherton), pp. 317–340. Butterworths, London.

Sachs, R.M., Kofranek, A.M. and Shyr, S.Y. (1967) Gibberellin-induced inhibition of floral initiation in *Fuchsia*. *Am. J. Bot.* **54**, 921–929.

Sage, L.C. (1992) *Pigment of the Imagination*. Academic Press, San Diego.

Sairam, R.K. and Srivastava, G.C. (1977) Changes in oil and fatty acid composition of linseed (*Linum usitatisimum* L.) under varying photoperiods. *Curr. Sci.* **46**, 115.

Saji, H., Takimoto, A. and Furuya, M. (1982) Spectral dependence of night-break effect on photoperiodic floral induction in *Lemna paucicostata* 441. *Plant Cell Physiol.* **26**, 623–629.

Saji, H., Vince-Prue, D. and Furuya, M. (1983) Studies of the photoreceptors for the promotion and inhibition of flowering in dark-grown seedlings of *Pharbitis nil* Choisy. *Plant Cell Physiol.* **24**, 1183–1189.

Saji, H., Furuya, M. and Takimoto, A. (1984) Role of the photoperiod preceding a flower-inductive dark period in dark-grown seedlings of *Pharbitis nil* Choisy. *Plant Cell Physiol.* **25**, 715–720.

Saks, Y., Negbi, M. and Ilan, I. (1980) Involvement of native abscisic acid in the regulation of onset of dormancy in *Spirodela polyrrhiza*. *Aust. J. Plant. Physiol.* **7**, 73–79.

Salisbury, F.B. (1955) The dual role of auxin in flowering. *Plant Physiol.* **30**, 327–334.

Salisbury, F.B. (1963a) Biological timing and hormone synthesis in flowering of *Xanthium*. *Planta* **59**, 518–534.

Salisbury, F.B. (1963b) *The Flowering Process*. Pergamon Press, Oxford.

Salisbury, F.B. (1965) Time measurement and the light period in flowering. *Planta* **66**, 1–26.

Salisbury, F.B. (1969) *Xanthium strumarium* L. In *The Induction of Flowering* (ed. L.T. Evans), pp. 14–21. Macmillan, Melbourne.

Salisbury, F.B. (1981) Twilight effect; initiating dark time-measurement in photoperiodism of *Xanthium*. *Plant Physiol.* **67**, 1230–1238.

Salisbury, F.B. (1985) *Xanthium strumarium*. In *Handbook of Flowering* (ed. A. Halevy) vol. IV, pp. 473–522. CRC Press, Boca Raton.

Salisbury, F.B. (1990) The use of *Xanthium* in flowering research. In *Analysis of Growth and Development in Xanthium* (ed. R. Maksymowych), pp. 153–194. Cambridge University Press.

Salisbury, F.B. and Ross, C.W. (1969) *Plant Physiology* (1st Edition). Wadsworth, Belmont, California (4th edition 1991).

Sampson, D.R. and Burrows, V.D. (1972) Influence of photoperiod, short-day vernalization and cold vernalization on days to heading in *Avena* species and cultivars. *Can. J. Plant Sci.* **52**, 471–482.

Sato, N. (1988) Nucleotide sequence and expression of the phytochrome gene in *Pisum sativum*: Differential regulation by light of multiple transcripts. *Plant Mol. Biol.* **11**, 697–710.

Sato, N. and Furuya, M. (1985) Synthesis of translatable mRNA for phytochrome during imbibition in embryonic axes of *Pisum sativum* L. *Plant Cell Physiol.* **26**, 1511–1517.

Satter, R.L. and Galston, A.W. (1981) Mechanisms of control of leaf movements. *Ann. Rev. Plant Physiol.* **32**, 83–110.

Satter, R.L., Guggino, S.E., Lonergan, T.A. and Galston, A.W. (1981) The effect of blue and far-red light on rhythmic leaf movements in *Samanea* and *Albizia*. *Plant Physiol.* **67**, 965–968.

Saunders, D.S. (1981) Insect photoperiodism: entrainment within the circadian system as a basis for time measurement. In *Biological Clocks in Seasonal Reproductive Cycles* (eds B.K. Follett and D.E. Follett), pp. 67–82. Scientechnica, Bristol.

Sawhney, N. and Sawhney, S. (1985) Role of gibberellin in flowering of short-day plants. *Indian J. Plant Physiol.* **28**, 24–34.

Sawnhey, S., Sawhney, N. and Nanda, K.K. (1976) Gel electrophoresis studies of proteins of photo-induced and vegetative plants of *Impatiens balsamina*. *Plant Cell Physiol.* **17**, 751–755.

Sawhney, S., Sawhney, N. and Nanda, K.K. (1978) Studies on the transmission of floral effects of photoperiod and gibberellin from one branch to another in *Impatiens balsamina*. *Biol. Plant.* **20**, 344–350.

Scarth, R. and Law, C.N. (1983) The location of the photoperiodic gene, Ppd2, and an additional genetic factor for ear-emergence on chromosome 2B of wheat. *Heredity* **51**, 607–619.

Scarth, R., Kirby, E.J.M. and Law, C.N. (1985) Effects of the photoperiod genes, Ppd1 and Ppd2, on growth and development of the shoot apex in wheat. *Ann. Bot.* **55**, 357–359.

Schäfer, E. (1975) A new approach to explain the 'high irradiance responses' of photo-morphogenesis on the basis of phytochrome. *J. Math. Biol.* **2**, 41–56.

Schäfer, E., Schmidt, W. and Mohr, H. (1973) Comparative measurements of phytochrome in cotyledons and hypocotyl hooks of mustard (*Sinapis alba* L.). *Photochem. Photobiol.* **18**, 331–334.

Schmidt, W. and Schäfer, E. (1974) Dependence of phytochrome dark reactions on the initial photostationary state. *Planta* **116**, 267–272.

Schneider, M.J. and Stimson, W.R. (1971) Further evidence for photosynthetic involvement in a high energy reaction (HER) response. *Plant Physiol.* **47** (Suppl), p 2, Abstr. No. 12 (A).

Schneider, M.J. and Stimson, W.R. (1972) Phytochrome and photosystem I: A model for their interaction in high energy responses. *Plant Physiol.* **49** (Suppl), p 53, Abstr. No. 297 (A).

Schneider, M.J., Borthwick, H.A. and Hendricks, S.B. (1967) Effects of radiation on flowering of *Hyoscyamus niger*. *Am. J. Bot.* **54**, 1241–1249.

Schopfer, R. (1977) Phytochrome control of enzymes. *Ann. Rev. Plant Physiol.* **28**, 223–252.

Schumacker, R. (1966) Le métabolisme des anthocyanes en relation avec la lumière et la floraison chez *Perilla nankinensis* (Lour.) Decne. *Photochem. Photobiol.* **5**, 413–422.

Schwabe, W.W. (1956) Evidence for a flowering inhibitor produced in long days in *Kalanchoë blossfeldiana*. *Ann. Bot.* **20**, 1–14.

Schwabe, W.W. (1959) Studies of long-day inhibition in short-day plants. *J. Exp. Bot.* **10**, 317–329.

Schwabe, W.W. (1968) Studies on the role of the leaf epiderm in photoperiodic perception in *Kalanchoë blossfeldiana*. *J. Exp. Bot.* **19**, 108–113.

Schwabe, W.W. (1969) *Kalanchoë blossfeldiana* Poellniz. In *The Induction of Flowering* (ed. L.T. Evans), pp. 227–241. Macmillan, Melbourne.

Schwabe, W.W. (1972) Flower inhibition in *Kalanchoë blossfeldiana*. Bioassay of an endogenous long-day inhibitor and inhibition by ([+/−]) abscisic acid and xanthoxin. *Planta* **103**, 18–23.

Schwabe, W.W. (1984) Photoperiodic induction-flower inhibiting substances. In *Light and the Flowering Process* (eds D. Vince-Prue, B. Thomas and K.E. Cockshull), pp. 143–153. Academic Press, London.

Schwabe, W.W. (1985) *Kalanchoë blossfeldiana*. In *Handbook of Flowering* III (ed. A.H. Halevy), pp. 217–235. CRC Press, Boca Raton.

Schwabe, W.W. and Naschmony-Bascombe, S. (1963) Growth and dormancy in *Lunularia cruciata* (L.) Dum. II. the response to daylength and temperature. *J. Exp. Bot.* **14**, 353–378.

Searle, N.E. (1961) Persistence and transport of flowering stimulus in *Xanthium*. *Plant Physiol.* **36**, 656–662.

Sears, E.R. (1954) The aneuploids of common wheat. *Res. Bull. Mo. Agric. Exp. Stn.* **572**, 1–59.

Seeley, S.D., Damavandy, H., Anderson, J.L., Renquist, R. and Callan, N.W. (1992) Autumn-applied growth regulators influence leaf retention, bud hardiness, bud and flower size and endodormancy in peach and cherry. *J. Am. Soc. Hort. Sci.* **117**, 203–208.

Seidlova, F. (1970) The effect of 2-thiouracil and actinomycin D on the shoot apex and on flower induction in *Chenopodium rubrum*. In: *Cellular and Molecular Aspects of Floral Induction* (ed. G. Bernier), pp. 108–116. Longman, London.

Seidlova, F. (1980) Sequential steps of transition to flowering of *Chenopodium rubrum* L. *Physiol. Vég.* **18**, 477–487.

Seidlova, F. and Krekule, J. (1973) The negative response of photoperiodic floral induction in *Chenopodium rubrum* L to preceding growth. *Ann. Bot.* **37**, 605–614.

Seidlova, F. and Opatrna, J. (1978) Change of growth correlation in the shoot meristem as the cause of age dependence of flowering in *Chenopodium rubrum*. *Z. Pflanzenphysiol.* **89**, 377–392.

Shacklock, P.S., Read, N.D, Trewavas, A.J. (1992) Cytosolic free calcium mediates red light induced photomorphogenesis. *Nature* **358**, 735–755.

Sharrock, R.A. and Quail, P.H. (1989) Novel phytochrome sequences in *Arabidopsis thaliana*: structure, evolution and differential expression of a plant regulatory photoreceptor family. *Genes Develop.* **3**, 1745–1757.

Sharrock, R.A., Parks, B.M., Koornneef, M. and Quail, P.H. (1988) Molecular analysis of the phytochrome deficiency in an aurea mutant of tomato. *Mol. Gen. Genet.* **213**, 9–14.

Shimazaki, Y. and Pratt, L.H. (1985) Immunochemical detection with rabbit polyclonal and mouse monoclonal antibodies of different pools of phytochrome from etiolated and green *Avena* shoots. *Planta* **164**, 333–344.

Shimazaki, Y., Cordonnier, M. and Pratt, L.H. (1983) Phytochrome quantitation in crude extracts of *Avena* by enzyme-linked immunosorbent assay with monoclonal antibodies. *Planta* **159**, 534–544.

Shinozaki, M. and Takimoto, A. (1982) The role of cotyledons in flower induction of *Pharbitis nil* at low temperatures. *Plant Cell Physiol.* **23**, 403–408.

Shinozaki, M., Hikichi, M., Yoshida, K., Watanabe, K. and Takimoto, A. (1982) Effect of high-intensity light given prior to low-temperature treatment on the long-day flowering of *Pharbitis nil*. *Plant Cell Physiol.* **23**, 473–477.

Simko, I. (1994) Sucrose application causes hormonal changes associated with potato-tuber induction. *J. Plant Growth Reg.* **13**, 73–77.

Simmonds, J. (1982) In vitro flowering on leaf explants of *Streptocarpus nobilis*. The influence of culture medium components on vegetative growth and reproductive development. *Can. J. Bot.* **60**, 1461–1468.

Simon, E., Satter, R. and Galston, A.W. (1976) Circadian rhythmicity in excised *Samanea* pulvini. II. Resetting the clock by phytochrome conversion. *Plant Physiol.* **58**, 421–425.

Sinclair T.R. and Hinton, K. (1992) Soybean flowering in response to the long-juvenile trait. *Crop Sci.* **32**, 1242–1248.

Sinden, S.L., Schalk, J.M. and Stoner, A.K. (1978) Effects of daylength and maturity of tomato plants on tomatine content and resistance to the Colorado Potato Beetle. *J. Amer. Soc. Hort. Sci.* **103**, 596–600.

Singer, S.R., Hannon, C.H. and Huber, S.C. (1992) Acquisition of competence for floral development in *Nicotiana* buds. *Planta* **188**, 546–550.

Slafer, G.A., Halloran, G.M. and Connor, D.J. (1994a) Development rate in wheat as affected by duration and rate of change of photoperiod. *Ann. Bot.* **73**, 671–677.

Slafer, G.A., Connor, D.J. and Halloran, G.M. (1994b) Rate of leaf appearence and final number of leaves in wheat; effects of duration and rate of change of photoperiod. *Ann. Bot.* **74**, 427–436.

Smart, C.C. and Trewavas, A.J. (1984) Abscisic-acid-induced turion formation in *Spirodela polyrrhiza* L. III. Specific changes in protein synthesis and translatable RNA during turion development. *Plant Cell and Environ.* **7**, 121–132.

Smart, C.C., Fleming, A.J., Chaloupková, K. and Hanke, D.E. (1995) The physiological role of abscisic acid in eliciting turion morphogenesis. *Plant Physiol.* **108**, 623–632.

Smith, H. (1975) *Phytochrome and Photomorphogenesis*. McGraw-Hill, London.

Smith, H. (1982) Light quality, photoperception and plant strategy. *Ann. Rev. Plant Physiol.* **33**, 481–518.

Smith, H. (1992) The ecological functions of the phytochrome family. Clues to a transgenic programme of crop improvement. *Photochem. Photobiol.* **56**, 815–822.

Smith, H., Jackson, G.M. and Whitelam, G.C. (1988) Photoprotection of phytochrome. *Planta* **175**, 471–477

Smith, O.E. and Palmer, C.E. (1970) Cytokinin induced tuber formation on stolons of *Solanum tuberosum*. *Physiol. Plant.* **23**, 599–606.

Sobeith, W.Y. and Wright, C.J. (1986) The photoperiodic regulation of bulbing in onions (*Allium cepa* L.). Effects of plant age and size. *J. Hort. Sci.* **61**, 337–341

Soffe, R.W., Lenton, J.R. and Milford, G.F.J. (1977) Effects of photoperiod on some vegetable species. *Ann. Appl. Biol.* **85**, 411–415.

Somers, D.E., Sharrock, R.A., Tepperman, J.M. and Quail, P.H. (1991) The *hy3* long hypocotyl mutant of *Arabidopsis* is deficient in phytochrome B. *The Plant Cell* **3**, 1263–1274.

Sotta, B. (1978) Interaction du photopériodisme et des effets de la zéatine, du saccharaose et du l'eau dans la floraison du *Chenopodium polyspermum*. *Physiol Plant.* **43**, 337–342.

Sotta, B., Lejeune, P., Maldiney, R., Kinet, J.-M., Miginiac, E. and Bernier, G. (1992) Cytokinin and auxin levels in apical buds of *Sinapis alba* following floral induction. In *Physiology and Biochemistry of Cytokinins in Plants* (eds M. Kaminek, D.W.S. Mok and E. Zazimalová), pp. 377–379. SPB Academic Publishing, The Hague.

Spector, W.S. (1956) *Handbook of Biological Data*, p. 460. W.B. Saunders.

Spector, C. and Paraska, J.R. (1973) Rhymicity of flowering in *Pharbitis nil*. *Physiol. Plant.* **29**, 402–405.

Spence, J.A. and Humphries, E.C. (1972) Effect of moisture supply, root temperature and growth regulators on photosynthesis of isolated rooted leaves of sweet potato (*Ipomoea batatas*). *Ann. Bot.* **36**, 115–121.

Spruit, C.J.P. (1970) Spectrophotometers for the study of phytochrome *in vivo*. *Meded. LandbHoogesch. Wageningen*, 70–14.

Stanley, C.J. and Cockshull, K.E. (1989) The site of ethephon application and its effect on flower initiation and growth of chrysanthemum. *J. Hort. Sci.* **64**, 341–350.

Steffen, J.D., Cockshull, K.E., Navissano, G. and Sachs, R.M. (1988) Inflorescence development in *Chrysanthemum morifolium* as a function of light on the inflorescence. *Ann. Bot.* **61**, 409–413.

Stiles, J.T. Jr. and Davies, P.J. (1976) Qualitative analysis by isoelectric focusing of the protein content of *Pharbitis nil* apices and cotyledons during floral induction. *Plant Cell Physiol.* **17**, 855–857.

Stockhaus, J., Nagatani, A., Halfter, U., Kay, S., Furuya, M. and Chua, N.H. (1992) Serine-to-alanine substitutions at the amino terminal region of phytochrome A result in an increase in biological activity. *Genes Develop.* **6**, 2364–2372.

Stolwijk, J.A.J. (1952a) Photoperiodic and formative effects of various wavelength regions in *Cosmos bipinnatus, Spinacia oleracea, Sinapis alba* and *Pisum sativum.* I. *Proc. K. Ned. Akad. Wet. Ser. C.* **55**, 489–497

Stolwijk, J.A.J. (1952b) Photoperiodic and formative effects of various wavelength regions in *Cosmos bipinnatus, Spinacia oleracea, Sinapis alba* and *Pisum sativum.* II. *Proc. K. Ned. Akad. Wet. Ser. C.* **55**, 498–502.

Suge, H. and Takahashi, H. (1982) The role of gibberellins in the stem elongation and flowering of Chinese cabbage, *Brassica campestris* var. Pekinensis in their relation to vernalization and photoperiod. *Rep. Inst. Agr. Res. Tohoku Univ.* **33**, 15–34.

Sussex, I.M. (1989) Developmental programming of the cell meristem. *Cell* **56**, 225–229.

Sweeney, B.M. (1979) Bright light does not immediately stop the circadian clock of *Gonyaulax. Plant Physiol.* **64**, 341–344.

Sylven, H. (1942) Lång-och kortdagstyper av de svenska skogsträden. *Svensk Pappersmasse-Tidn.* **43**, 317–324 and 332–342.

Tafazoli, E. and Vince-Prue, D. (1978) A comparison of the effects of long days and exogenous growth regulators on growth and flowering in strawberry, *Fragaria* × *ananassa* Duch. *J. Hort. Sci.* **53**, 255–259.

Takahashi, H. and Jaffe, M.J. (1984) Further studies of auxin and ACC induced feminization in the cucumber plant using ethylene inhibitors. *Phyton* **44**, 81–86.

Takahashi, H., Suge, H. and Saito, T. (1980) Sex expression as affected by N^6-benzylamino-purine in staminate inflorescences of *Luffa cylindrica. Plant Cell Physiol.* **21**, 525–536.

Takahashi, H., Saito, T. and Suge, H. (1982) Intergeneric translocation of floral stimulus across a graft in monoecious Cucurbitaceae with special reference to the sex expression of flowers. *Plant Cell Physiol.* **23**, 1–9.

Takahashi, H., Saito, S. and Suge, H. (1983) Separation of the effects of photoperiod and hormones on sex expression in cucumber. *Plant Cell Physiol.* **24**, 147–154.

Takeno, K. (1991) Flowering response of *Ipomaea batatas* scions grafted onto *Pharbitis nil* stocks. *Physiol. Plant.* **63**, 682–686.

Takeno, K. (1994) Perfusion technique, a new method to treat assay plants with test substances. *Flowering Newsletter* **17**, 34–38.

Takeno, K., Takahashi, M. and Watanabe, K. (1995) Flowering response of an intermediate-day plant, *Salsola komarovii* Iljin under different photoperiodic conditions. *Plant Physiol.* **145**, 121–125.

Takimoto, A. (1957) Two processes involved in the light period of inductive photoperiodic cycles in *Silene armeria. Bot. Mag. Tokyo* **70**, 321–326.

Takimoto, A. (1967) Studies on the light affecting the initiation of endogenous rhythms concerned with photoperiodic responses in *Pharbitis nil. Bot. Mag. Tokyo* **80**, 241–247.

Takimoto, A. and Hamner, K.C. (1964) Effect of temperature and pre-conditioning on photoperiodic response of *Pharbitis nil. Plant Physiol.* **39**, 1024–1030.

Takimoto, A. and Hamner, K.C. (1965) Effect of far-red light and its interaction with red light in the photoperiodic response of *Pharbitis nil. Plant Physiol.* **40**, 859–864.

Takimoto, A. and Ikeda, K. (1960) Studies on the light controlling flower initiation of *Pharbitis nil.* VI. Effect of natural twilight. *Bot. Mag. Tokyo* **73**, 175–181.

Takimoto, A. and Ikeda, K. (1961) Effect of twilight on photoperiodic induction in some short-day plants. *Plant Cell Physiol.* **2**, 213–229.

Takimoto, A. and Saji, H. (1984) A role of phytochrome in photoperiodic induction; two phytochrome-pool theory. *Physiol. Plant.* **61**, 675–682.

Takimoto, A., Kaihara, S. and Nishioka, H. (1987) A comparative study of the short-day and the benzoic acid-induced flowering in *Lemna paucicostata. Plant Cell Physiol.* **28**, 503–508.

Takimoto, A., Kaihara, S., Shinozaki, M. and Miura, J. (1991) Involvement of norepinephrine in the production of flower-inducing substance in the water extract of *Lemna. Plant Cell Physiol.* **32**, 283–289.

Talon, M. and Zeevaart, J.A.D. (1990) Gibberellins and stem growth as related to photoperiod in *Silene armeria*. *Plant Physiol.* **92**, 1094–1100.

Talon, M. and Zeevaart, J.A.D. (1992) Stem elongation and changes in the levels of gibberellins in shoot tips induced by differential photoperiodic treatments in the long-day plant *Silene armeria*. *Planta* **188**, 457–461.

Talon, M., Tadeo, F.R. and Zeevaart, J.A.D. (1991a) Cellular changes induced by exogenous and endogenous gibberellins in shoot tips of the long-day plant *Silene armeria*. *Planta* **185**, 487–493.

Talon, M., Zeevaart, J.A.D. and Gage, D.A. (1991b) Identification of gibberellins in spinach and effects of light and darkness on their levels. *Plant Physiol.* **97**, 1521–1526.

Tamot, B.K., Khurana, J.P. and Maheshwari, S.C. (1987) Obligate requirement for salicylic acid for short-day induction of flowering in a new duckweed, *Wolfiella hyalina* 7378. *Plant Cell Physiol.* **28**, 349–353.

Tamura, F., Tanabe, K. and Ikeda, T. (1993) Relationship between intensity of bud dormancy and levels of ABA in Japanese pear Nijisseiki. *J. Japan. Soc. Hort. Sci.* **62**, 75–81.

Tanada, T. (1968) A rapid photoreversible response of barley root tips in the presence of 3-indoleacetic acid. *Proc. Nat. Acad. Sci. USA* **59**, 376–380.

Tanaka, O., Nakayama, Y., Emori, K., Takeba, G., Beppu, T. and Sugino, M. (1994) Flower-inducing activity of exogenous lysine in *Lemna paucicostata* 151 cultured on nitrogen-rich medium. *Plant Cell Physiol.* **35**, 73–78.

Tanimoto, S., Miyazaki, A. and Harada, H. (1985) Regulation by abscisic acid of *in vitro* flower formation in *Torenia* stem segments. *Plant Cell Physiol.* **26**, 675–682.

Taybi, T., Sotta, B., Gehrig, H., Guclu, S., Kluge, M. and Brulfert, J. (1995) Differential effects of abscisic acid on phosphoenolpyruvate carboxylase and CAM operation in *Kalanchoë blossfeldiana*. *Bot. Acta.* **108**, 240–246.

Taylor, S.A. and Murfet, I.C. (1994) A short-day mutant in pea is deficient in the floral stimulus. *Flowering Newsletter* **18**, 39–43.

Terabun, M. (1965) Studies on the bulb formation in onion plants. I. Effects of light quality on the bulb formation and the growth. *J. Jap. Soc. Hort. Sci.* **34**, 196–204.

Terabun, M. (1970) Studies on the bulb formation in onion plants. V. Effects of mixed light of blue, red and far-red light on bulb formation. *J. Jap. Soc. Hort. Sci.* **39**, 35–40.

Terry, M.J., Hall, J.L. and Thomas, B. (1992) The association of type I phytochrome with wheat leaf plasma membranes. *J. Plant Physiol.* **140**, 691–698.

Thigpen, S.P. and Sachs, R.M. (1985) Changes in ATP in relation to floral induction and initiation in *Pharbitis nil*. *Physiol. Plant.* **65**, 156–162.

Thomas, B. (1991) Phytochrome and photoperiodic induction. *Physiol. Plant.* **82**, 571–577.

Thomas, B. and Dickinson, H.G. (1979) Evidence for two photoreceptors controlling growth in de-etiolated seedlings. *Planta* **146**, 545–550.

Thomas, B. and Mozley, D. (1994) Isolation and properties of mutants of *Arabidopsis thaliana* with reduced sensitivity to short days. In *Molecular and Cellular Aspects of Plant Reproduction* (eds R. Scott and A.D. Stead), pp. 31–38. SEB Seminar Series. Cambridge University Press, Cambridge.

Thomas, B. and Vince-Prue, D. (1984) Juvenility, photoperiodism and vernalization. In *Advanced Plant Physiology* (ed. M.B. Wilkins), pp. 408–439. Pitman, London.

Thomas, B. and Vince-Prue, D. (1987) Photoperiodic control of floral initiation in short-day and long-day plants. In *Models in Plant Physiology and Biochemistry* vol II (eds D.W. Newman and K.G. Wilson), pp. 121–125. CRC Press, Boca Raton.

Thomas, B., Crook, N.E. and Penn, S.E. (1984) An enzyme-linked immunosorbent assay for phytochrome. *Physiol. Plant* **60**, 409–415.

Thomas, B., Penn, S.E. and Jordan, B.R. (1989) Factors affecting phytochrome transcripts and apoprotein synthesis in germinating embryos of *Avena sativa* L. *J. Exp. Bot.* **40**, 1299–1304.

Thomas, R.G. (1961) Correlation between growth and flowering in *Chenopodium amaranticolor*. *Ann. Bot.* **25**, 255–269.

Thomas, R.G. (1979) Inflorescence initiation in *Trifolium repens* L.: influence of natural photoperiods and temperatures. *NZ J. Bot.* **17**, 287–299.

Thomas, T.H. (1972) Stimulation of onion bulblet production by N^6 benzyladenine. *Hort. Res.* **12**, 77–79.

Thompson, H.C. and Smith, O. (1938) Seedstalk and bulb development in the onion *Allium cepa* L. *Bull. Cornell Univ. Agric. Exp. Stn.* 708.

▷ Thummler, F., Dufner, M., Kriesl, P. and Dittrich, P. (1992) Molecular cloning of a novel phytochrome gene of the moss *Ceratodon purpureus* which encodes a putative light-regulated protein kinase. *Plant Mol. Biol.* **20**, 1003–1017.

Timmis, R. and Worral, J. (1974) Translocation of dehardening and bud-break promoters in climatically split Douglas-fir. *Can. J. For. Res.* **4**, 229–237.

Tizio, R. (1966) Presence de kinines dans la périderme de tubercles de pomme de terre. *C.R. Hebd. Séanc. Acad. Sci. Paris* **262**, 868–869.

Tizio, R. (1971) Action et rôle probable de certaine gibberellins (A1, A3, A4, A5, A7, A9 et A13) sur la croissance des stolons et la tubérisation de la Pomme de terre (*Solanum tuberosum*). *Potato Res.* **14**, 193–204.

Tokuhisa, J.G., Daniels, S.M. and Quail, P.H. (1985) Phytochrome in green tissue: spectral and immunochemical evidence for two distinct molecular species of phytochrome in light-grown *Avena sativa*. *Planta* **164**, 321–332.

Tokutomi, S., Nakasako, M., Sakai, J., Kataoka, M., Yamamoto, K.T., Wada, M., Tokunaga, F. and Furuya, M. (1989) A model for the dimeric molecular structure of phytochrome based on small-angle X-ray scattering. *FEBS Lett.* **247**, 139–142.

Tournois, J. (1912) Influence de la lumiére sur la floraison du houblon japonais et du chanvre déterminées par des semis haitifs. *C.R. Acad. Sci. Paris* **155**, 297–300.

Tournois, J. (1914) Études sur la sexualité du Houblon. *Ann. Sci. Nat. (Bot.)* **19**, 49–191.

Tran Thanh Van, K., Dien, N.T. and Chlyah, A. (1974) Regulation of organogenesis in small explants of superficial tissue of *Nicotiana tabacum*. *Planta* **119**, 149–159.

Tretyn, A., Cymerski, M., Czaplewska, J., Lukasiewicz, H., Pawlak, A. and Kopcewicz, J. (1990) Calcium and photoperiodic flower induction in *Pharbitis nil*. *Physiol. Plant.* **80**, 388–392.

Tse, A.T.Y., Ramina, A., Hackett, W.P. and Sachs, R.M. (1974) Enhanced inflorescence development in *Bougainvillea* San Diego Red by removal of young leaves and cytokinin treatment. *Plant Physiol.* **54**, 404–407.

Tso, T.C., Kasperbauer, M.J. and Sorokin, T.P. (1970) Effect of photoperiod and end-of-day light quality on alkaloids and phenolic compounds of tobacco. *Plant Physiol.* **45**, 330–333.

Ullman, J., Seidlová, F., Krekule, J. and Pavlová, L. (1985) *Chenopodium rubrum* as a model plant for testing the flowering effects of PGRs. *Biol Plant.* **27**, 367–372.

Van der Woude, W.J. (1985) A dimeric mechanism for the action of phytochrome: evidence from photothermal interactions in lettuce seed germination. *Photochem. Photobiol.* **42**, 665–661.

Van der Woude, W.J. (1987) Application of the dimeric model of phytochrome action to high irradiance responses. In *Phytochrome and Photoregulation in Plants*, (ed. M. Furuya), pp. 249–258. Academic Press, New York.

Van Schreven, D.A. (1949) Premature tuber formation in early potatoes. *Tijdschr. Piziekt.* **55**, 290–308.

Van Staden, J. (1981) Effect of photoperiod and gibberellic acid on flowering and cytokinin levels in *Bougainvillea*. *S. Afr. J. Sci.*, **77**, 327.

Van Veen, J.W.H. (1969) Interrupted bud formation in spray chrysanthemum – shape and quality of the inflorescence. *Acta Hort.* **14**, 39–60.

Varkey, M. and Nigam, R.K. (1982) A gradual reduction of female structures in the pistillate flowers of *Ricunus communis* L. (Castor bean) with chlorflurenol (Morphactin). *Biol. Plant.* **24**, 152–154.

Vaz Nunes, M., Lewis, R.D. and Saunders, D.S. (1991) A coupled oscillator feedback system as a model for the photoperiodic clock in insects and mites. *J. Theor. Biol.* **152**, 287–298 and 299–317.

Vierstra, R.D. and Quail, P.H. (1983) Purification and initial characterization of 124 kilodalton phytochrome from *Avena*. *Biochem.* **22**, 2498–2505.

Vierstra, R.D., Cordonnier, M-M., Pratt, L.H. and Quail, P.H. (1984) Native phytochrome:

immunoblot analysis of relative molecular mass and *in-vitro* proteolytic degradation for several plant species. *Planta* **160**, 521–528.

Vince, D. (1955) Some effects of temperature and daylength on flowering in the chrysanthemum. *J. Hort. Sci.* **30**, 34–42.

Vince, D. (1965) The promoting effect of far-red light on flowering in the long-day plant *Lolium temulentum. Physiol. Plant.* **18**, 474–482.

Vince, D., Blake, J. and Spencer, R. (1964) Some effects of wavelength of the supplementary light on the photoperiodic behaviour of the long-day plants, carnation and lettuce. *Physiol. Plant.* **17**, 119–125.

Vince-Prue, D. (1975) *Photoperiodism in Plants*. McGraw-Hill, Maidenhead.

Vince-Prue, D. (1976) Phytochrome and photoperiodism. In *Light and Plant Development* (ed. H. Smith), pp. 347–369. Butterworths, London.

Vince-Prue, D. (1979) Effect of photoperiod and phytochrome in flowering: time measurement. In *La Physiologie de la Floraison* (eds P. Chapagnat and R. Jaques), pp. 91–127. CNRS, Paris.

Vince-Prue, D. (1981) Daylight and photoperiodism. In *Plants and the Daylight Spectrum* (ed. H. Smith), pp. 223–242. Academic Press, London.

Vince-Prue, D. (1983) Photoperiodic control of plant reproduction. In *Strategies of Plant Reproduction*. BARC Symposium 6 (ed. W.J. Meudt), pp. 73–97. Allanheld, Osmun, Totowa.

Vince-Prue, D. (1983a) Photomorphogenesis and flowering. In *Encyclopedia of Plant Physiology*, New Series, Vol. 16B: *Photomorphogenesis* (eds W. Shropshire and H. Mohr), pp. 457–490. Springer-Verlag, Berlin.

Vince-Prue, D. (1983b) The perception of light-dark transitions. *Phil. Trans. R. Soc. Lond.* **B303**, 523–536.

Vince-Prue, D. (1984) Contrasting types of photoperiodic response in the control of dormancy. *Plant, Cell Environ.* **7**, 507–513.

Vince-Prue, D. (1985) Photoperiod and hormones. In *Hormonal Regulation of Development* III. Encyclopedia of Plant Physiology, New Series, Vol. 11 (eds R.P. Pharis and D.M. Reid), pp. 308–364. Springer-Verlag, Berlin.

Vince-Prue, D. (1991) Phytochrome action under natural conditions. In *Phytochrome Properties and Biological Action* (eds B. Thomas and C. Johnson), pp. 313–319. Springer Verlag, Heidelberg.

Vince-Prue, D. (1994) The duration of light and photoperiodic responses. In *Photomorphogenesis in Plants* (2nd edn, eds R.E. Kendrick and G.H.M. Kronenberg), pp. 447–490. Kluwer, Dordrecht.

Vince-Prue, D. and Gressel, J. (1985) *Pharbitis nil*. In *Handbook of Flowering* vol IV (ed. A.H. Halevy), pp. 47–81. CRC Press, Boca Raton.

Vince-Prue, D. and Lumsden, P.J. (1987) Inductive events in the leaves: time measurement and photoperception in the short-day plant, *Pharbitis nil*. In *Manipulation of Flowering* (ed. J.G. Atherton), pp. 255–268. Butterworths, London.

Vince-Prue, D. and Takimoto, A. (1987) Roles of phytochrome in photoperiodic floral induction. In *Phytochrome and Photoregulation in Plants* (ed. M. Furuya), pp. 259–275. Academic Press, Tokyo.

Vince-Prue, D., King, R.W. and Quail, P.H. (1978) Light requirement, phytochrome and photoperiodic induction of flowering of *Pharbitis nil* Chois. II. A critical examination of spectrophotometric assays of phytochrome transformations. *Planta* **141**, 9–14.

Wada, N., Shinozaki, M. and Iwamura, H. (1994) Flower induction by polyamines and related compounds in seedlings of morning glory (*Pharbitis nil* cv Kidachi). *Plant Cell Physiol.* **35**, 469–472.

Wall, J.K. and Johnson, C.B. (1983) An analysis of phytochrome action in the 'High Irradiance Response'. *Planta* **159**, 387–397.

Wang, Y.C., Stewart, S.J., Cordonnier, M.-M. and Pratt, L.H. (1991) *Avena sativa* contains three phytochromes, only one of which is abundant in etiolated tissue. *Planta* **184**, 96–104.

Wareing, P.F. (1953) Growth studies in woody species V. Photoperiodism in dormant buds of *Fagus sylvatica* L. *Physiol. Plant.* **6**, 692–706.

Wareing, P.F. (1954) Growth studies in woody species VI. The locus of photoperiodic perception in relation to dormancy. *Physiol. Plant.* **7**, 261–277.

Wareing, P.F. (1956) Photoperiodism in woody plants. *Ann. Rev. Plant. Physiol.* **7**, 191–214.

Wareing, P.F. (1959) Photoperiodism in seeds and buds. In *Photoperiodism and Related Phenomena in Plants and Animals* (ed. R.B. Withrow), pp. 73–87. AAAS, Washington.

Wareing, P.F. (1982) Hormonal control of stolon and tuber development, especially in the potato plant. In *Strategies of Plant Reproduction* BARC Symposium 6 (ed. W.J. Meudt), pp. 181–194. Allanheld, Osmun, Totowa.

Wareing, P.F. and Black, M. (1958) In *The Physiology of Forest Trees* (ed. K.V. Thimann), pp. 539–553. Ronald Press, New York.

Wareing, P.F. and Jennings, A.M.V. (1980) The hormonal control of tuberisation in potato. In: *Plant Growth Substances 1979* (ed. F. Skoog), pp. 293–300. Springer-Verlag, Berlin.

Warm, E. (1980) Effect of phytohormones and salicylic acid in the long-day plant *Hyoscyamus niger*. *Z. Pflanzenphysiol.* **99**, 325–330.

Warm, E. (1984) Changes in the composition of *in vitro* translated leaf mRNA caused by photoperiodic flower induction of *Hyoscyamus niger*. *Physiol. Plant.* **61**, 344–350.

Wassink, E.C. and Stolwijk, J.A.J. (1953) Effect of photoperiod on vegetative development and tuber formation in two potato varieties. *Meded. LandbHoogesch. Wageningen* **53**, 99–112.

Wassink, E.C., Sluijsmans, C.M.J. and Stolwijk, J.A.J. (1950) On some photoperiodic and formative effects of coloured light in *Brassica napa, F. oleifera,* subf. *annua. Proc. K. Ned. Akad. Wet. Ser. C* **4**, 421–432.

Wassink, E.C., Stolwijk, J.A.J., Beemster, A.B.R. (1951) Dependence of formative and photoperiodic reactions in *Brassica rapa* var. Cosmos and *Lactuca* on wavelength and time of irradiation. *Proc. K. Ned. Akad. Wet. Ser. C.* **54**, 421–432.

Watanabe, M., Furuya, M., Miyoshi, Y., Inoue, Y., Iwahashi, I. and Matsumoto, K. (1982) Design and performance of the Okazaki Large Spectrograph for photobiological research. *Photochem. Photobiol.* **36**, 491–498.

Watson, J.D. and Mathews, R.E.F. (1966) Effect of actinomycin D and 2-thiouracil on floral induction and nucleic acid synthesis in the bud in *Chenopodium amaranticolor. Aust. J. Biol. Sci.* **19**, 967–980.

Waxman, S (1957) The development of woody plants as affected by photoperiodic treatments. Ph.D. Thesis, Cornell University, Ithaca.

Weber, J.A. and Nooden, L.D. (1976a) Environmental and hormonal control of turion formation in *Myriophyllum verticillatum. Plant Cell Physiol.* **17**, 721–731.

Weber, J.A. and Nooden, L.D. (1976b) Environmental and hormonal control of turion germination in *Myriophyllum verticillatum. Amer. J. Bot.* **63**, 936–94.

Wellensiek, S.J. (1960) Stem elongation and flower initiation. *Proc. K. Ned. Acad. Wet. C* **63**, 159–166.

Wellensiek, S.J. (1964) Dividing cells as a pre-requisite for vernalization. *Plant Physiol.* **39**, 832–835.

Wellensiek, S.J. (1965) Recent developments in vernalization. *Acta Bot. Neerl.* **14**, 308–314

Whalley, D.N. and Cockshull, K.E. (1976) The photoperiodic control of rooting, growth and dormancy in *Cornus alba* L. *Scientia Hort.* **5**, 127–138.

Wheeler, N. (1979) Effect of continuous photoperiod on growth and development of lodgepole pine seedlings and grafts. *Can. J. For. Res.* **9**, 276–283.

Whitelam, G.C. and Harberd, N.P. (1994) Action and function of phytochrome family members revealed through the study of mutant and transgenic plants. *Plant Cell Environ.* **17**, 615–625.

Whitelam, G.C., Johnson, E., Peng, J., Carol, P., Anderson, M.L., Cowl, J.S. and Harberd, N.P. (1993) Phytochrome A null mutants of *Arabidopsis* display a wild-type phenotype in white light. *The Plant Cell* **5**, 757–768.

Wilkins, M.B. (1959) An endogenous rhythm in the rate of carbon dioxide output of *Bryophyllum*. I. Some preliminary experiments. *J. Exp. Bot.* **10**, 377–390.

Wilkins, H.F. and Halevy, A.H. (1985) *Scabiosa*. In *Handbook of Flowering* vol. V (ed. A.H. Halevy), pp. 328–329. CRC Press, Boca Raton.

Wilkinson, R.E. (1974) Sicklepod surface wax response to photoperiod and S-(2,3-dichloro-allyl)diisopropylthiocarbamate (Diallate). *Plant Physiol.* **53**, 269–275.

Williams, B.J., Pellett, N.E. and Klein, R.M. (1972) Phytochrome control of growth cessation and initiation of cold acclimation in selected woody plants. *Plant Physiol.* **50**, 262–265.

Wilson, J.R. and Schwabe, W.W. (1964) Growth and dormancy in *Lunularia cruciata* L. Dum. III. The wavelength of light effective in photoperiodic control. *J. Exp. Bot.* **15**, 368–380.

√ Wilson R.N., Heckman, J.W. and Somerville, C.R. (1992) Gibberellin is required for flowering in *Arabidopsis thaliana* under short days. *Plant Physiol.* **100**, 403–408.

Withrow, R.B. and Benedict, H.M. (1936) Photoperiodic responses of certain greenhouse annuals as influenced by intensity and wavelength of artificial light used to lengthen the daylight period. *Plant Physiol.* **11**, 225–249.

Withrow, R.B. and Biebel, J.P. (1936) Photoperiodic response of certain long and short-day plants to filtered radiation applied as a supplement to daylight. *Plant Physiol.* **11**, 807–819.

√ Withrow, R.B. and Withrow, A.P. (1940) The effect of various wave bands of supplementary radiation on the photoperiodic response of certain plants. *Plant Physiol.* **15**, 609–624.

Worland, A.J. and Law, C.N. (1986) Genetic analysis of chromosome 2D of wheat. 1. The location of genes affecting height, daylength insensitivity, hybrid dwarfism and yellow-rust resistance. *Z. Pflzücht.* **96**, 331–345.

Wurr, D.C.E., Fellows, J.R., Phelps, K. and Reader, A.J. (1994) Testing a vernalization model on field-grown crops of four cauliflower cultivars. *J. Hort. Sci.* **69**, 251–255.

Yang, Y.Y., Yamaguchi, I, Takeno-Wada, K., Suzuki, Y. and Murofishi, N. (1995) Metabolism and translocation of gibberellins in seedlings of *Pharbitis nil*. I. Effect of photoperiod on stem elongation and endogenous gibberellins in cotyledons and their phloem exudates. *Plant Cell Physiol.* **36**, 221–227.

Young, E. and Hanover, J.W. (1977) Effects of quality, intensity, and duration of light-breaks during a long night on dormancy in blue spruce (*Picea pungens* Engelm.) seedlings. *Plant Physiol.* **60**, 271–273.

Zack, C.D. and Loy, J.B (1980) The effect of light quality and photoperiod on vegetative growth of *Cucurbita maxima*. *J. Am. Soc. Hort. Sci.* **105**, 939–943.

Zagotta, M.T., Shannon, S., Jocobs, C. and Meeks-Wagner, D. Ry. (1992) Early-flowering mutants of *Arabidopsis thaliana*. *Aust. J. Plant Physiol.* **19**, 411–418.

Zanewich, K.P. and Rood, S.B. (1994) Endogenous gibberellins in flushing buds of three deciduous trees: alder, aspen and birch. *J. Plant Growth Regul.* **13**, 159–162.

Zeevaart, J.A.D. (1958) Flower formation as studied by grafting. *Meded. LandbHoogesch. Wageningen* **58**(3), 1–88.

Zeevaart, J.A.D. (1962a) DNA multiplication as a requirement for expression of floral stimulus in *Pharbitis nil*. *Plant Physiol.* **37**, 296–304.

Zeevaart, J.A.D. (1962b) The juvenile phase in *Bryophyllum daigremontianum*. *Planta* **58**, 531–542.

Zeevaart, J.A.D. (1962c) Physiology of flowering. *Science* **137**, 723–731.

Zeevaart, J.A.D. (1969a) The leaf as the site of gibberellin action in flower formation in *Bryophyllum daigremontianum*. *Planta* **84**, 339–347.

Zeevaart, J.A.D. (1969b) *Bryophyllum*. In *The Induction of Flowering* (ed. L.T. Evans), pp. 435–456. Macmillan, Melbourne.

Zeevaart, J.A.D. (1969c) *Perilla*. In: *The Induction of Flowering* (ed. L.T. Evans), pp. 116–155, Macmillan, Melbourne

Zeevaart, J.A.D. (1971) Lack of evidence for distinguishing florigen and flower hormone in *Perilla*. *Planta* **98**, 190–194.

Zeevaart, J.A.D. (1974) Levels of (+)-abscisic acid and xanthoxin in spinach under different environmental conditions. *Plant Physiol.* **53**, 644–648.

Zeevaart, J.A.D. (1976) Physiology of flower formation. *Ann. Rev. Plant Physiol.* **27**, 321–348.

Zeevaart, J.A.D. (1978) Flower formation in the short-day plant *Kalanchoë* by grafting with a long-day and a short-long-day *Echeveria*. *Planta* **140**, 289–291.

Zeevaart, J.A.D. (1979) Perception, nature and complexity of transmitted signals. In *La Physiologie de la Floraison* (eds P. Chapagnat and R. Jacques), pp. 59–90. CNRS, Paris 285.

Zeevaart, J.A.D. (1982) Transmission of the floral stimulus from a long-short-day plant,

Bryophyllum daigremontianum, to the short-long-day plant *Echeveria harmsii*. *Ann. Bot.* **49**, 549–552.

Zeevaart, J.A.D. (1984) Photoperiodic induction, the floral stimulus and flower-promoting substances. In *Light and the Flowering Process* (eds D. Vince-Prue, B. Thomas and K.E. Cockshull), pp. 137–141. Academic Press, London.

Zeevaart, J.A.D. (1985a) *Bryophyllum*. In *Handbook of Flowering* II (ed. A.H. Halevy), pp. 89–99. CRC Press, Boca Raton.

Zeevaart, J.A.D. (1985b) *Perilla*. In *Handbook of Flowering* (ed. A.H. Halevy), vol. V, pp. 239–252. CRC Press, Boca Raton.

Zeevaart, J.A.D. and Boyer, G.L. (1987) Photoperiodic induction and the floral stimulus in *Perilla*. In *Manipulation of Flowering* (ed. J.G. Atherton), pp. 269–277. Butterworths, London.

Zeevaart, J.A.D. and Gage, D.A. (1993) *ent*-kaurene biosynthesis is enhanced by long photoperiods in the long-day plants *Spinacia oleracea* L. and *Agrostemma githago* L. *Plant Physiol.* **101**, 25–29.

Zeevaart, J.A.D. and Marushige, K. (1967) Biochemical approaches. In *Physiology of Flowering in Pharbitis nil* (ed. S Imamura), pp. 121–138. Japanese Society of Plant Physiologists.

Zeevaart, J.A.D., Gage, D.A. and Talon, M. (1993) Gibberellin-A$_1$ is required for stem elongation in spinach. *Proc Nat. Acad. Sci. USA* **90**, 7401–7405.

Zheng, C.C., Bui, A.Q., O'Neill, S.D (1993) Abundance of an mRNA encoding a high mobility group DNA-binding protein is regulated by light and an endogenous rhythm. *Plant Mol. Biol.* **23**, 813–823.

Zimmer, K. and Bahnemann, K. (1981) Blütenbildung bei *Begonia boweri* Ziesenh. und einem Abkömmling von *Begonia* Cleopatra. V. Kritische Blattfläche für die induktion. *Gartenbauwissenshaft* **46**, 13–15.

Zimmer, K. and Krebs, O. (1980) Flower formation in *Begonia boweri* and an offspring of *B.* 'Cleopatra'. IV. Influence of the number of leaves and their position on the flower formation. *Gartenbauwissensaft.* **45**, 29–34.

Index

415